T0281469

Lecture Notes in Mathematics   2234

More information about this series at http://www.springer.com/series/304

Toshiyuki Kobayashi • Birgit Speh

Symmetry Breaking for Representations of Rank One Orthogonal Groups II

 Springer

Toshiyuki Kobayashi
Graduate School of Mathematical Sciences
The University of Tokyo
Komaba, Tokyo, Japan
and
Kavli IPMU
Kashiwa, Japan

Birgit Speh
Department of Mathematics
Cornell University
Ithaca, NY, USA

ISSN 0075-8434   ISSN 1617-9692   (electronic)
Lecture Notes in Mathematics
ISBN 978-981-13-2900-5   ISBN 978-981-13-2901-2   (eBook)
https://doi.org/10.1007/978-981-13-2901-2

Library of Congress Control Number: 2015027247

Mathematics Subject Classification (2010): 22E30, 11F70, 53A30, 22E45, 22E46, 58J70

This Springer imprint is published by the registered company Springer Nature Singapore Pte Ltd.
The registered company address is: 152 Beach Road, #21-01/04 Gateway East, Singapore 189721, Singapore

Preface

For a pair $(G, G') = (O(n+1, 1), O(n, 1))$ of reductive groups, we investigate intertwining operators (*symmetry breaking operators*) between principal series representations $I_\delta(V, \lambda)$ of G and $J_\varepsilon(W, \nu)$ of the subgroup G'. The representations are parametrized by finite-dimensional representations V, W of $O(n)$ respectively of $O(n-1)$, characters δ, ε of $O(1)$, and λ, $\nu \in \mathbb{C}$. Denote by $[V : W]$ the multiplicity of W occurring in the restriction $V|_{O(n-1)}$, which is either 0 or 1. If $[V : W] \neq 0$, then we construct a holomorphic family of symmetry breaking operators and prove that $\dim_\mathbb{C} \mathrm{Hom}_{G'}(I_\delta(V, \lambda)|_{G'}, J_\varepsilon(W, \nu))$ is nonzero for all the parameters λ, ν and δ, ε, whereas if $[V : W] = 0$, there may exist *sporadic* differential symmetry breaking operators.

We propose a *classification scheme* to find all matrix-valued symmetry breaking operators explicitly and carry out this program completely in the case $(V, W) = (\bigwedge^i(\mathbb{C}^n), \bigwedge^j(\mathbb{C}^{n-1}))$. In conformal geometry, our results yield the complete classification of conformal covariant operators from differential forms on a Riemannian manifold X to those on a submanifold Y in the model space $(X, Y) = (S^n, S^{n-1})$.

We use this information to determine the space of symmetry breaking operators for any pair of irreducible representations of G and the subgroup G' with trivial infinitesimal character. Furthermore, we prove the multiplicity conjecture by B. Gross and D. Prasad for tempered principal series representations of $(SO(n+1, 1)$ and $SO(n, 1))$ and also for 3 tempered representations Π, π, ϖ of $SO(2m+2, 1)$, $SO(2m+1, 1)$, and $SO(2m, 1)$ with trivial infinitesimal character. In connection with automorphic form theory, we apply our main results to find *periods* of irreducible representations of the Lorentz group having nonzero $(\mathfrak{g}, K)$-cohomologies.

This book is an extension of the recent work in the two research monographs: Kobayashi–Speh [Memoirs Amer. Math. Soc., 2015] for *spherical* principal series representations and Kobayashi–Kubo–Pevzner [Lecture Notes in Math., 2016] for conformally covariant *differential* symmetry breaking operators.

Komaba, Tokyo, Japan   Toshiyuki Kobayashi
Ithaca, NY, USA   Birgit Speh

Contents

Chapter 1
Introduction

A representation Π of a group G defines a representation of a subgroup G' by restriction. In general irreducibility is not preserved by the restriction. If G is compact then the restriction $\Pi|_{G'}$ is isomorphic to a direct sum of irreducible finite-dimensional representations π of G' with multiplicities $m(\Pi, \pi)$. These multiplicities are studied by using combinatorial techniques. We are interested in the case where G and G' are (noncompact) real reductive Lie groups. Then most irreducible representations Π of G are infinite-dimensional, and generically the restriction $\Pi|_{G'}$ is not a direct sum of irreducible representations [28]. So we have to consider another notion of multiplicity.

For a continuous representation Π of G on a complete, locally convex topological vector space $\mathcal{H}$, the space $\mathcal{H}^\infty$ of C^∞-vectors of $\mathcal{H}$ is naturally endowed with a Fréchet topology, and $(\Pi, \mathcal{H})$ induces a continuous representation Π^∞ of G on $\mathcal{H}^\infty$. If Π is an admissible representation of finite length on a Banach space $\mathcal{H}$, then the Fréchet representation $(\Pi^\infty, \mathcal{H}^\infty)$, which we refer to as an *admissible smooth representation*, depends only on the underlying $(\mathfrak{g}, K)$-module $\mathcal{H}_K$. In the context of asymptotic behaviour of matrix coefficients, these representations are also referred to as an admissible representations of moderate growth [62, Chap. 11]. We shall work with these representations and write simply Π for Π^∞. We denote by $\mathrm{Irr}(G)$ the set of equivalence classes of irreducible admissible smooth representations. We also sometimes call these representations "irreducible admissible representations" for simplicity.

Given another admissible smooth representation π of a reductive subgroup G', we consider the space of continuous G'-intertwining operators (*symmetry breaking operators*)

$$\mathrm{Hom}_{G'}(\Pi|_{G'}, \pi).$$

© Springer Nature Singapore Pte Ltd. 2018
T. Kobayashi, B. Speh, *Symmetry Breaking for Representations of Rank One Orthogonal Groups II*, Lecture Notes in Mathematics 2234, https://doi.org/10.1007/978-981-13-2901-2_1

If $G = G'$ then these operators include the Knapp–Stein operators [22] and the differential intertwining operators studied by B. Kostant [44]. If $G \neq G'$ the dimension

$$m(\Pi, \pi) := \dim_{\mathbb{C}} \mathrm{Hom}_{G'}(\Pi|_{G'}, \pi)$$

yields important information of the restriction of Π to G' and is called the *multiplicity* of π occurring in the restriction $\Pi|_{G'}$. In general, $m(\Pi, \pi)$ may be infinite. The finiteness criterion in [39] asserts that the multiplicity $m(\Pi, \pi)$ is finite for all $\Pi \in \mathrm{Irr}(G)$ and for all $\pi \in \mathrm{Irr}(G')$ if and only if a minimal parabolic subgroup P' of G' has an open orbit on the real flag variety G/P, and that the multiplicity is uniformly bounded with respect to Π and π if and only if a Borel subgroup of $G'_{\mathbb{C}}$ has an open orbit on the complex flag variety of $G_{\mathbb{C}}$.

The latter condition depends only on the complexified pairs $(\mathfrak{g}_{\mathbb{C}}, \mathfrak{g}'_{\mathbb{C}})$, of which the classification was already known in 1970s by Krämer [45] and Kostant. In particular, the multiplicity $m(\Pi, \pi)$ is uniformly bounded if the Lie algebras $(\mathfrak{g}, \mathfrak{g}')$ of (G, G') are real forms of $(\mathfrak{sl}(N+1, \mathbb{C}), \mathfrak{gl}(N, \mathbb{C}))$ or $(\mathfrak{o}(N+1, \mathbb{C}), \mathfrak{o}(N, \mathbb{C}))$. On the other hand, the former condition depends on real forms $(\mathfrak{g}, \mathfrak{g}')$, and the classification of such symmetric pairs was recently accomplished in [36]. For instance, let $(G, G') = (O(n+1, 1), O(n+1-k, 1))$. Then the classification theory [36] and the finiteness criterion [39] imply the following upper and lower estimates of the multiplicity $m(\Pi, \pi)$:

(1) For $2 \leq k \leq n+1$,

$$m(\Pi, \pi) < \infty \text{ for every pair } (\Pi, \pi) \in \mathrm{Irr}(G) \times \mathrm{Irr}(G');$$
$$\sup_{\Pi \in \mathrm{Irr}(G)} \sup_{\pi \in \mathrm{Irr}(G')} m(\Pi, \pi) = \infty.$$

(2) For $k = 1$, there exists $C > 0$ such that

$$m(\Pi, \pi) \leq C \text{ for all } \Pi \in \mathrm{Irr}(G) \text{ and for all } \pi \in \mathrm{Irr}(G'). \tag{1.1}$$

B. Sun and C.-B. Zhu [55] showed that one can take C to be one in (1.1), namely, the multiplicity $m(\Pi, \pi) \in \{0, 1\}$ in this case. Thus one of the open problems is to determine when $m(\Pi, \pi) \neq 0$ for irreducible representations Π and π.

In the previous publication [42] we initiated a thorough study of symmetry breaking operators between *spherical* principal series representations of

$$(G, G') = (O(n+1, 1), O(n, 1)). \tag{1.2}$$

In particular, we determined the multiplicities $m(\Pi, \pi)$ when both Π and π are irreducible composition factors of the spherical principal series representations.

In this article we will determine the multiplicities $m(\Pi, \pi)$ for all irreducible representations Π and π with trivial infinitesimal character ρ of $G = O(n+1, 1)$ and $G' = O(n, 1)$, respectively, and also for irreducible principal series representations.

More than just determining the dimension $m(\Pi, \pi)$ of the space of symmetry breaking operators, we investigate these operators for *general* principal series representations of G and the subgroup G', *i.e.*, for representations induced from irreducible finite-dimensional representations of a parabolic subgroup. We construct a holomorphic family of symmetry breaking operators, and present a classification scheme of *all* symmetry breaking operators T in Theorem 3.13 through an analysis of their distribution kernels K_T. In particular, we prove that any symmetry breaking operators in this case is either a sporadic differential symmetry breaking operator (*cf.* [35]) or the analytic continuation of integral symmetry breaking operators and their renormalization in Theorem 3.13.

The proof for the explicit formula of the multiplicity $m(\Pi, \pi)$ is built on the functional equations (Theorems 9.24 and 9.25) satisfied by the regular symmetry breaking operators.

A principal series representation $I_\delta(V, \lambda)$ of $G = O(n+1, 1)$ is an (unnormalized) induced representation from an irreducible finite-dimensional representation $V \otimes \delta \otimes \mathbb{C}_\lambda$ of a minimal parabolic subgroup $P = MAN_+$. In our setting, $M \simeq O(n) \times \mathbb{Z}/2\mathbb{Z}$ and $A \simeq \mathbb{R}_+$. We assume that V is a representation of $O(n)$, $\delta \in \{\pm\}$, and $\lambda \in \mathbb{C}$. In what follows, we identify the representation space of $I_\delta(V, \lambda)$ with the space of C^∞-sections of the G-equivariant bundle $G \times_P \mathcal{V}_{\delta, \lambda} \to G/P$, so that $I_\delta(V, \lambda)^\infty = I_\delta(V, \lambda)$ is the Fréchet globalization having moderate growth in the sense of Casselman–Wallach [62]. The parametrization is chosen so that the representation $I_\delta(V, \frac{n}{2})$ is a unitary tempered representation. The representations $I_\delta(V, \lambda)$ are either irreducible or of composition series of length 2, see Corollary 14.22 in Appendix I.

The group $P' = G' \cap P = M'AN'_+$ is a minimal parabolic subgroup of $G' = O(n, 1)$. For an irreducible representation (τ, W) of $O(n-1)$, a character $\varepsilon \in \{\pm\}$ of $O(1)$, and $v \in \mathbb{C}$ we define the principal series representation $J_\varepsilon(W, v)$ of G'.

We set

$$[V : W] := \dim_\mathbb{C} \mathrm{Hom}_{O(n-1)}(W, V|_{O(n-1)}) = \dim_\mathbb{C} \mathrm{Hom}_{O(n-1)}(V|_{O(n-1)}, W).$$

For principal series representations $I_\delta(V, \lambda)$ of G and $J_\varepsilon(W, v)$ of the subgroup G', we consider the cases $[V : W] \neq 0$ and $[V : W] = 0$ separately. In the first case we obtain a lower bound for the multiplicity.

In what follows, it is convenient to introduce the set of "special parameters":

$$\Psi_{\mathrm{sp}} := \Big\{ (\lambda, v, \delta, \varepsilon) \in \mathbb{C}^2 \times \{\pm\}^2 : v - \lambda \in 2\mathbb{N} \qquad \text{when } \delta\varepsilon = +$$

$$\text{or} \quad v - \lambda \in 2\mathbb{N} + 1 \quad \text{when } \delta\varepsilon = - \Big\}. \tag{1.3}$$

Theorem 1.1 (see Theorems 3.13 (2) and 3.15). *Suppose $(\sigma, V) \in \widehat{O(n)}$ and $(\tau, W) \in \widehat{O(n-1)}$. Assume $[V : W] \neq 0$.*

(1) (existence of symmetry breaking operators) *We have*

$$\dim_\mathbb{C} \mathrm{Hom}_{G'}(I_\delta(V, \lambda)|_{G'}, J_\varepsilon(W, v)) \geq 1 \quad \text{for all } \delta, \varepsilon \in \{\pm\}, \text{ and } \lambda, v \in \mathbb{C}.$$

(2) (generic multiplicity-one)

$$\dim_{\mathbb{C}} \operatorname{Hom}_{G'}(I_\delta(V, \lambda)|_{G'}, J_\varepsilon(W, \nu)) = 1$$

for any $(\lambda, \nu, \delta, \varepsilon) \in (\mathbb{C}^2 \times \{\pm\}^2) - \Psi_{\mathrm{sp}}$.

(3) *Let* $\ell(\sigma)$ *be the "norm" of* σ *defined by using its highest weight (see* (2.21)*). Then we have*

$$\dim_{\mathbb{C}} \operatorname{Hom}_{G'}(I_\delta(V, \lambda)|_{G'}, J_\varepsilon(W, \nu)) > 1$$

for any $(\lambda, \nu, \delta, \varepsilon) \in \Psi_{\mathrm{sp}}$ *such that* $\nu \in \mathbb{Z}$ *with* $\nu \le -\ell(\sigma)$.

We prove Theorem 1.1 by constructing (generically) regular symmetry breaking operators $\widetilde{\mathbb{A}}_{\lambda, \nu, \delta\varepsilon}^{V, W}$: they are nonlocal operators (*e.g.*, integral operators) for generic parameters, whereas for some parameters they are local operators (*i.e.*, differential operators). See Theorem 3.10 for the construction of the normalized operator $\widetilde{\mathbb{A}}_{\lambda, \nu, \pm}^{V, W}$; Theorem 3.9 for "regularity" [42, Def. 3.3] of $\widetilde{\mathbb{A}}_{\lambda, \nu, \pm}^{V, W}$ under a certain generic condition; Theorem 5.45 for a renormalization of $\widetilde{\mathbb{A}}_{\lambda, \nu, \pm}^{V, W}$ when it vanishes; Fact 9.3 for the residue formula of $\widetilde{\mathbb{A}}_{\lambda, \nu, \pm}^{V, W}$ when it reduces to a differential operator.

In the case $[V : W] = 0$, symmetry breaking operators are "rare" but there may exist *sporadic* symmetry breaking operators:

Theorem 1.2. *Assume* $[V : W] = 0$.

(1) (vanishing for generic parameters, Corollary 3.14) *If* $(\lambda, \nu, \delta, \varepsilon) \notin \Psi_{\mathrm{sp}}$, *then*

$$\operatorname{Hom}_{G'}(I_\delta(V, \lambda)|_{G'}, J_\varepsilon(W, \nu)) = \{0\}.$$

(2) (localness theorem, Theorem 3.6) *Any nontrivial symmetry breaking operator*

$$C^\infty(G/P, \mathcal{V}_{\lambda, \delta}) \to C^\infty(G'/P', \mathcal{W}_{\nu, \varepsilon})$$

is a differential operator.

Combining Theorems 1.1 (2) and 1.2 (1) together with the existence condition of differential symmetry breaking operators (see Theorem 5.21), we determine the following multiplicity formulæ for *generic parameters*:

Theorem 1.3. *Suppose that* $(\lambda, \nu, \delta, \varepsilon) \notin \Psi_{\mathrm{sp}}$. *Then there are no differential symmetry breaking operators and*

$$\dim_{\mathbb{C}} \operatorname{Hom}_{G'}(I_\delta(V, \lambda)|_{G'}, J_\varepsilon(W, \nu)) = \begin{cases} 1 & \text{if } [V : W] \neq 0, \\ 0 & \text{if } [V : W] = 0. \end{cases}$$

It deserves to be mentioned that the parameter set $(\mathbb{C}^2 \times \{\pm\}^2) - \Psi_{\text{sp}}$ contains parameters (λ, ν) for which the G-module $I_\delta(V, \lambda)$ or the G'-module $J_\varepsilon(W, \nu)$ is *not* irreducible.

In the major part of this monograph, we focus our attention on the special case

$$(V, W) = (\textstyle\bigwedge^i(\mathbb{C}^n), \bigwedge^j(\mathbb{C}^{n-1})).$$

The principal series representations of G and the subgroup G' are written as $I_\delta(i, \lambda)$ for $I_\delta(\bigwedge^i(\mathbb{C}^n), \lambda)$ and $J_\varepsilon(j, \nu)$ for $J_\varepsilon(\bigwedge^j(\mathbb{C}^{n-1}), \nu)$, respectively. The representations $I_\delta(i, \lambda)$ of G and $J_\varepsilon(j, \nu)$ of G' are of interest in geometry as well as in automorphic forms and in the cohomology of arithmetic groups. In geometry, given an arbitrary Riemannian manifold X, one forms a natural family of representations of the conformal group G on the space $\mathcal{E}^i(X)$ of differential forms, to be denoted by $\mathcal{E}^i(X)_{\lambda', \delta'}$ for $0 \le i \le \dim X$, $\lambda' \in \mathbb{C}$, and $\delta' \in \{\pm\}$. Then the representations $I_\delta(i, \lambda)$ are identified with such conformal representations in the case where $(G, X) = (O(n+1, 1), S^n)$, see e.g., [35, Chap. 2, Sect. 2] for precise statement. In representation theory, all irreducible, unitarizable representations with nonzero $(\mathfrak{g}, K)$-cohomology arise as subquotients of $I_\delta(i, \lambda)$ with $\lambda = i$ for some $0 \le i \le n$ and $\delta = (-1)^i$, see Theorem 2.20 (9).

Our main results of this article include a complete solution to the general problem of constructing and classifying the elements of $\mathrm{Hom}_{G'}(\Pi|_{G'}, \pi)$ (see [33, Prob. 7.3 (3) and (4)]) in the following special setting:

$$(G, G') = (O(n+1, 1), O(n, 1)) \qquad \text{with } n \ge 3,$$

$$(\Pi, \pi) = (I_\delta(i, \lambda), J_\varepsilon(j, \nu)),$$

where $0 \le i \le n$, $0 \le j \le n-1$, $\delta, \varepsilon \in \{\pm\}$, and $\lambda, \nu \in \mathbb{C}$. Thus our main results include a complete solution to the following question in conformal geometry:

Problem 1.4.

(1) *Find a necessary and sufficient condition on 6-tuples $(i, j, \lambda, \nu, \delta, \varepsilon)$ for the existence of conformally covariant, symmetry breaking operators*

$$A \colon \mathcal{E}^i(X)_{\lambda, \delta} \to \mathcal{E}^i(X)_{\nu, \varepsilon}$$

in the model space $(X, Y) = (S^n, S^{n-1})$.

(2) *Construct those operators explicitly in the (flat) coordinates.*

(3) *Classify all such symmetry breaking operators.*

Partial results were known earlier: when the operator A is given by a *differential* operator, Juhl [21] solved Problem 1.4 in the case $(i, j) = (0, 0)$, see also [38], and it was recently extended in Kobayashi–Kubo–Pevzner [35] for the general (i, j). Problem 1.4 was solved for all (possibly, *nonlocal*) operators in our previous paper [42] in the case $(i, j) = (0, 0)$. The complete classification of (continuous) symmetry breaking operators for the general (i, j) is given in Theorem 3.25 (multiplicity) and Theorem 3.26 (construction of explicit generators), and we have thus settled Problem 1.4 in this monograph. For this introduction, we explain only the "multiplicity" (Theorem 3.25). For this, using the same notation as in [42, Chap. 1], we define the following two subsets on $\mathbb{Z}^2$:

$$L_{\mathrm{even}} := \{(-i, -j) : 0 \leq j \leq i \text{ and } i \equiv j \mod 2\},$$

$$L_{\mathrm{odd}} := \{(-i, -j) : 0 \leq j \leq i \text{ and } i \equiv j+1 \mod 2\}.$$

Theorem 1.5 (multiplicity, Theorem 3.25). *Suppose* $\Pi = I_\delta(i, \lambda)$ *and* $\pi = J_\varepsilon(j, \nu)$ *for* $0 \leq i \leq n$, $0 \leq j \leq n-1$, $\delta, \varepsilon \in \{\pm\}$, *and* $\lambda, \nu \in \mathbb{C}$. *Then we have the following.*

(1)

$$
\begin{aligned}
m(\Pi, \pi) &\in \{1, 2\} && \text{if } j = i - 1 \text{ or } i, \\
m(\Pi, \pi) &\in \{0, 1\} && \text{if } j = i - 2 \text{ or } i + 1, \\
m(\Pi, \pi) &= 0 && \text{otherwise.}
\end{aligned}
$$

(2) *Suppose* $j = i - 1$ *or* i. *Then* $m(\Pi, \pi) = 1$ *generically. It is equal to 2 when the parameter belongs to the following exceptional countable set. Without loss of generality, we take* δ *to be* $+$.

(a) *Case* $1 \leq i \leq n - 1$.

$$
\begin{aligned}
m(I_+(i, \lambda), J_+(i, \nu)) &= 2 && \text{if } (\lambda, \nu) \in L_{\mathrm{even}} - \{\nu = 0\} \cup \{(i, i)\}. \\
m(I_+(i, \lambda), J_-(i, \nu)) &= 2 && \text{if } (\lambda, \nu) \in L_{\mathrm{odd}} - \{\nu = 0\}. \\
m(I_+(i, \lambda), J_+(i - 1, \nu)) &= 2 && \text{if } (\lambda, \nu) \in L_{\mathrm{even}} - \{\nu = 0\} \cup \{(n - i, n - i)\}. \\
m(I_+(i, \lambda), J_-(i - 1, \nu)) &= 2 && \text{if } (\lambda, \nu) \in L_{\mathrm{odd}} - \{\nu = 0\}.
\end{aligned}
$$

(b) *Case* $i = 0$.

$$
\begin{aligned}
m(I_+(0, \lambda), J_+(0, \nu)) &= 2 && \text{if } (\lambda, \nu) \in L_{\mathrm{even}}. \\
m(I_+(0, \lambda), J_-(0, \nu)) &= 2 && \text{if } (\lambda, \nu) \in L_{\mathrm{odd}}.
\end{aligned}
$$

(c) *Case* $i = n$.

$$
\begin{aligned}
m(I_+(n, \lambda), J_+(n - 1, \nu)) &= 2 && \text{if } (\lambda, \nu) \in L_{\mathrm{even}}. \\
m(I_+(n, \lambda), J_-(n - 1, \nu)) &= 2 && \text{if } (\lambda, \nu) \in L_{\mathrm{odd}}.
\end{aligned}
$$

(3) *Suppose $j = i - 2$ or $i + 1$. Then $m(\Pi, \pi) = 1$ if one of the following conditions (d)–(g) is satisfied, and $m(\Pi, \pi) = 0$ otherwise.*

    (d) *Case $j = i - 2$, $2 \le i \le n - 1$, $(\lambda, \nu) = (n - i, n - i + 1)$, $\delta\varepsilon = -1$.*
    (e) *Case $(i, j) = (n, n - 2)$, $-\lambda \in \mathbb{N}$, $\nu = 1$, $\delta\varepsilon = (-1)^{\lambda+1}$.*
    (f) *Case $j = i + 1$, $1 \le i \le n - 2$, $(\lambda, \nu) = (i, i + 1)$, $\delta\varepsilon = -1$.*
    (g) *Case $(i, j) = (0, 1)$, $-\lambda \in \mathbb{N}$, $\nu = 1$, $\delta\varepsilon = (-1)^{\lambda+1}$.*

More than just an abstract formula of multiplicities, we also obtain explicit generators of $\mathrm{Hom}_{G'}(I_\delta(i, \lambda)|_{G'}, J_\varepsilon(j, \nu))$ for $j \in \{i - 1, i\}$ in Theorem 3.26. The generators for $j \in \{i - 2, i + 1\}$ are always differential operators (*localness theorem*, see Theorem 1.2 (2)), and they were constructed and classified in [35] (see Fact 3.23).

The principal series representations $I_\delta(i, \lambda)$ and $J_\varepsilon(j, \nu)$ in the above theorem are not necessarily irreducible. For the study of symmetry breaking of the irreducible subquotients, we utilize the concrete generators of $\mathrm{Hom}_{G'}(I_\delta(i, \lambda)|_{G'}, J_\varepsilon(j, \nu))$ and determine explicit formulæ about

- the (K, K')-spectrum of the normalized regular symmetry breaking operators $\widetilde{\mathbb{A}}^{i,j}_{\lambda,\nu,\pm}$ on basic "(K, K')-types" (Theorem 9.8);
- the functional equations among symmetry breaking operators $\widetilde{\mathbb{A}}^{i,j}_{\lambda,\nu,\pm}$ (Theorems 9.24 and 9.25).

Here, the (K, K')-spectrum is defined in Definition 9.7. It resembles eigenvalues of symmetry breaking operators, and serves as a clue to find the functional equations.

We now highlight symmetry breaking of irreducible representations that have the same infinitesimal character ρ as the trivial one-dimensional representation **1**. Denote by $\mathrm{Irr}(G)_\rho$ the (finite) set of equivalence classes of irreducible admissible representations of G with trivial infinitesimal character $\rho \equiv \rho_G$. The principal series representations $I_\delta(i, i)$ of $G = O(n + 1, 1)$ are reducible, and any element in $\mathrm{Irr}(G)_\rho$ is a subquotient of the representations $I_\delta(i, i)$ for some $0 \le i \le n$ and $\delta \in \{\pm\}$. To be more precise, we have the following.

Theorem 1.6 (see Theorem 2.20). *Let $G = O(n + 1, 1)$ $(n \ge 1)$.*

(1) *For $0 \le \ell \le n$ and $\delta \in \{\pm\}$, there are exact sequences of G-modules:*

$$0 \to \Pi_{\ell,\delta} \to I_\delta(\ell, \ell) \to \Pi_{\ell+1,-\delta} \to 0,$$

$$0 \to \Pi_{\ell+1,-\delta} \to I_\delta(\ell, n - \ell) \to \Pi_{\ell,\delta} \to 0.$$

    These exact sequences split if and only if $n = 2\ell$.

(2) *Irreducible admissible representations of G with trivial infinitesimal character can be classified as*

$$\mathrm{Irr}(G)_\rho = \{\Pi_{\ell,\delta} : 0 \le \ell \le n + 1, \delta = \pm\}.$$

(3) *Every $\Pi_{\ell,\delta}$ $(0 \le \ell \le n + 1, \delta = \pm)$ is unitarizable.*

There are four one-dimensional representations of G, and they are given by

$$\{\Pi_{0,+} \simeq \mathbf{1}, \quad \Pi_{0,-} \simeq \chi_{+-}, \quad \Pi_{n+1,+} \simeq \chi_{-+}, \quad \Pi_{n+1,-} \simeq \chi_{--}(=\det)\}.$$

(See (2.13) for the definition of $\chi_{\pm\pm}$.) The other representations $\Pi_{\ell,\delta}$ ($1 \le \ell \le n$, $\delta = \pm$) are infinite-dimensional representations.

For the subgroup $G' = O(n,1)$, we use the letters $\pi_{j,\varepsilon}$ to denote the irreducible representations in $\mathrm{Irr}(G')_\rho$, similar to $\Pi_{i,\delta}$ in $\mathrm{Irr}(G)_\rho$.

With these notations, we determine

$$m(\Pi_{i,\delta}, \pi_{j,\varepsilon}) = \dim_{\mathbb{C}} \mathrm{Hom}_{G'}(\Pi_{i,\delta}|_{G'}, \pi_{j,\varepsilon})$$

for all $\Pi_{i,\delta} \in \mathrm{Irr}(G)_\rho$ and $\pi_{j,\varepsilon} \in \mathrm{Irr}(G')_\rho$ as follows.

Theorem 1.7 (vanishing, see Theorem 4.1). *Suppose* $0 \le i \le n+1$, $0 \le j \le n$, $\delta, \varepsilon \in \{\pm\}$.

(1) *If* $j \ne i, i-1$ *then* $\mathrm{Hom}_{G'}(\Pi_{i,\delta}|_{G'}, \pi_{j,\varepsilon}) = \{0\}$.
(2) *If* $\delta\varepsilon = -$, *then* $\mathrm{Hom}_{G'}(\Pi_{i,\delta}|_{G'}, \pi_{j,\varepsilon}) = \{0\}$.

Theorem 1.8 (multiplicity-one, see Theorem 4.2). *Suppose* $0 \le i \le n+1$, $0 \le j \le n$ *and* $\delta, \varepsilon \in \{\pm\}$. *If* $j = i-1$ *or* i *and if* $\delta\varepsilon = +$, *then*

$$\dim_{\mathbb{C}} \mathrm{Hom}_{G'}(\Pi_{i,\delta}|_{G'}, \pi_{j,\varepsilon}) = 1.$$

We can represent these results graphically as follows. We suppress the subscript, and write Π_i for $\Pi_{i,+}$, and π_j for $\pi_{j,+}$. The first row are representations of G, the second row are representations of G'. The existence of nonzero symmetry breaking operators is represented by arrows.

Theorem 1.9 (see Theorem 4.3). *Symmetry breaking for irreducible representations with infinitesimal character* ρ *is represented graphically in the following form.*
Symmetry breaking for $O(2m+1, 1) \downarrow O(2m, 1)$

$$
\begin{array}{ccccccc}
\Pi_0 & \Pi_1 & \cdots & \Pi_{m-1} & & \Pi_m & \\
\downarrow \swarrow & \downarrow \swarrow & & \swarrow & \downarrow & \swarrow & \downarrow \\
\pi_0 & \pi_1 & \cdots & \pi_{m-1} & & \pi_m &
\end{array}
$$

Symmetry breaking for $O(2m+2, 1) \downarrow O(2m+1, 1)$

$$
\begin{array}{cccccccc}
\Pi_0 & \Pi_1 & \cdots & \Pi_{m-1} & & \Pi_m & & \Pi_{m+1} \\
\downarrow \swarrow & \downarrow \swarrow & & \swarrow & \downarrow & \swarrow & \downarrow & \swarrow \\
\pi_0 & \pi_1 & \cdots & \pi_{m-1} & & \pi_m & &
\end{array}
$$

We believe that we are seeing in Theorem 4.3 only the "tip of the iceberg", and we present a conjecture that a similar statement holds in more generality, see Conjecture 13.15. Suppose that F and F' are irreducible finite-dimensional representations of G and the subgroup G', respectively, and that

$$\text{Hom}_{G'}(F|_{G'}, F') \neq \{0\}.$$

In Chapters 13 and 14 we describe sequences of irreducible representations $\{\Pi_i \equiv \Pi_i(F)\}$ and $\{\pi_j \equiv \pi_j(F')\}$ of G and G' with the same infinitesimal characters with F and F', respectively. We refer to these sequences as *standard sequences* that starting with $\Pi_0(F) = F$ and $\pi_0(F') = F'$, see Definition 13.2. They generalize the standard sequence with trivial infinitesimal character which we used in the formulation of Theorem 1.9. They are an analogue of a diagrammatic description of irreducible representations with regular integral infinitesimal characters for the connected group $G_0 = SO_0(n+1, 1)$ given in Collingwood [11, p. 144, Fig. 6.3]. In this generality, we conjecture that the results of symmetry breaking can be represented graphically exactly as in Theorem 1.9 for the representations with trivial infinitesimal character ρ. Again in the first row are representations of G, and in the second row are representations of G'. Conjecture 13.15 asserts that symmetry breaking operators are represented by arrows.

Symmetry breaking for $O(2m+1, 1) \downarrow O(2m, 1)$

$$
\begin{array}{cccccccc}
\Pi_0(F) & \Pi_1(F) & \cdots & \Pi_{m-1}(F) & \Pi_m(F) \\
\downarrow & \swarrow \; \downarrow & \swarrow & \swarrow & \downarrow & \swarrow & \downarrow \\
\pi_0(F') & \pi_1(F') & \cdots & \pi_{m-1}(F') & \pi_m(F')
\end{array}
$$

Symmetry breaking for $O(2m+2, 1) \downarrow O(2m+1, 1)$

$$
\begin{array}{cccccccc}
\Pi_0(F) & \Pi_1(F) & \cdots & \Pi_{m-1}(F) & \Pi_m(F) & \Pi_{m+1}(F) \\
\downarrow & \swarrow \; \downarrow & \swarrow & \swarrow & \downarrow & \swarrow & \downarrow & \swarrow \\
\pi_0(F') & \pi_1(F') & \cdots & \pi_{m-1}(F') & \pi_m(F')
\end{array}
$$

We present some supporting evidence for this conjecture in Chapter 13.

Applications of our formulæ include some results about periods of representations. Suppose that H is a subgroup of G. Following the terminology used in automorphic forms and the relative trace formula, we say that a smooth representation U of G is H-*distinguished* if there is a nontrivial linear H-invariant linear functional

$$F^H : U \to \mathbb{C}.$$

If the G-module U is H-distinguished, we say that (F^H, H) is a *period* (or an H-period) of U.

Let $(G, H) = (O(n+1, 1), O(m+1, 1))$ with $m \leq n$. For $0 \leq i \leq n+1$ and $0 \leq j \leq m+1$, we denote by Π_i and π_j the irreducible representations $\Pi_{i,+}$ of G and analogous ones of H with trivial infinitesimal character ρ,

Theorem 1.10 (see Theorems 12.4 and 12.6).

(1) *The irreducible representation Π_i is H-distinguished if $i \leq n - m$.*
(2) *The outer tensor product representation*

$$\Pi_i \boxtimes \pi_j$$

has a nontrivial H-period if $0 \leq i - j \leq n - m$.

The period is given by the composition of the normalized regular symmetry breaking operators (see Chapter 5) with respect to the chain of subgroups:

$$G = O(n+1, 1) \supset O(n, 1) \supset O(n-1, 1) \supset \cdots \supset O(m+1, 1) = H.$$

Using the above chain of subgroups we also define a vector v in the minimal K-type of Π_i. We prove

Theorem 1.11 (see Theorem 12.5). *Suppose that $G = O(n+1, 1)$ and Π_i ($0 \leq i \leq n$) is the irreducible representation with trivial infinitesimal character ρ defined as above. Then the value of the $O(n+1-i, 1)$-period on $v \in \Pi_i$ is*

$$\frac{\pi^{\frac{1}{4}i(2n-i-1)}}{((n-i)!)^{i-1}} \times \begin{cases} \frac{1}{(n-2i)!} & \text{if } 2i < n+1, \\ (-1)^{n+1}(2i-n-1)! & \text{if } 2i \geq n+1. \end{cases}$$

We also prove in Chapter 12 a generalization of a theorem of Sun [54].

Theorem 1.12 (see Theorem 12.13). *Let $(G, G') = (O(n+1, 1), O(n, 1))$, $0 \leq i \leq n$, and $\delta \in \{\pm\}$.*

(1) *The symmetry breaking operator $T : \Pi_{i,\delta} \to \pi_{i,\delta}$ in Proposition 10.12 induces bilinear forms*

$$B_T : H^j(\mathfrak{g}, K; \Pi_{i,\delta}) \times H^{n-j}(\mathfrak{g}', K'; \pi_{n-i,(-1)^n\delta}) \to \mathbb{C}$$

for all j.
(2) *The bilinear form B_T is nonzero if and only if $j = i$ and $\delta = (-1)^i$.*

Inspired by automorphic forms and number theory B. Gross and D. Prasad published in 1992 conjectures about the multiplicities of irreducible tempered representations (U, U') of $(SO(p,q), SO(p-1,q))$ [14]. Over time these conjectures have been modified and proved in some cases for automorphic forms and for p-adic orthogonal and unitary groups. See for example Astérisque volumes [12, 51] by W. T. Gan, B. Gross, D. Prasad, C. Mœglin and J.-L. Waldspurger and the references therein as well as the work by R. Beuzart-Plessis [8] for the unitary groups.

We prove the multiplicity conjecture by B. Gross and D. Prasad for tempered principal series representations of $(SO(n+1, 1), SO(n, 1))$ and also for 3 representations Π, π, ϖ of $SO(2m+2, 1)$, $SO(2m+1, 1)$ and $SO(2m, 1)$ with infinitesimal character ρ. More precisely we show:

Theorem 1.13 (Theorem 3.13). *Suppose that* $\Pi = I_\delta(V, \lambda)$, $\pi = J_\varepsilon(W, \nu)$ *are (smooth) tempered principal series representations of* $G = O(n+1, 1)$ *and* $G' = O(n, 1)$. *Then*

$$\dim_{\mathbb{C}} \mathrm{Hom}_{G'}(\Pi|_{G'}, \pi) = 1 \quad \text{if and only if} \quad [V : W] \neq 0.$$

Restricting the principal series representations to special orthogonal groups implies the conjecture of B. Gross and D. Prasad about multiplicities for tempered principal series representations (Theorem 11.5).

In 2000 B. Gross and N. Wallach [15] showed that the restriction of *small* discrete series representations of $G = SO(2p+1, 2q)$ to $G' = SO(2p, 2q)$ satisfies the Gross–Prasad conjectures [14]. In that case, both the groups G and G' admit discrete series representations. On the other hand, for the pair $(G, G') = (SO(n+1, 1), SO(n, 1))$, only one of G or G' admits discrete series representations. Our results confirm the Gross–Prasad conjecture also for **tempered** representations with trivial infinitesimal character ρ (Theorem 11.6).

The article is roughly divided in three parts and an appendix:

In the first part, Chapters 2–4, we give an overview of the notation and the results about symmetry breaking operators. Notations and properties for principal series and irreducible representations of orthogonal groups are introduced in Chapter 2. Important concepts and properties of symmetry breaking operators are discussed in Chapter 3, in particular, a classification scheme of all symmetry breaking operators is presented in Theorem 3.13. This includes a number of theorems about the dimension of the space of symmetry operators for principal series representations which are stated and discussed also in Chapter 3. The classification scheme is carried out in full details for symmetry breaking from principal series representations $I_\delta(i, \lambda)$ of G to $J_\varepsilon(j, \nu)$ of the subgroup G', and is used to obtain results on symmetry breaking of irreducible representations with trivial infinitesimal character ρ in Chapter 4.

The second part, Chapters 5–9, contains the proofs of the results discussed in Part 1. This is the technical heart of this monograph. In Chapter 5 the estimates and results about regular symmetry breaking operators in Theorems 1.1 and 1.2 are proved. Chapter 6 is devoted to differential symmetry breaking operators. In the remaining chapters of this part we concentrate on the symmetry breaking $I_\delta(i, \lambda) \to J_\varepsilon(j, \nu)$. We collect some technical results in Chapters 7 and 8. The analytic continuation of the regular symmetry breaking operators, their (K, K')-spectrum, and the functional equation are discussed in Chapter 9. Many of the results and techniques developed here are of independent interest, and could be applied to other problems.

In the third part, Chapters 11–13, we use the results in Chapters 3 and 4 to prove some of the conjectures of Gross and Prasad about symmetry breaking for tempered representations of orthogonal groups in Chapter 11. We discuss periods of representations and a bilinear form on the $(\mathfrak{g}, K)$-cohomology using symmetry breaking in Chapter 12. It also includes a conjecture about symmetry breaking for a family of irreducible representations with regular integral infinitesimal character in Chapter 13, which we plan to attack in a sequel to this monograph. A major portion of Part 3 can be read immediately after Part 1.

The appendix contains technical results used in the monograph. We provide three characterizations of irreducible representations of the group $G = O(n+1, 1)$: Langlands quotients (or subrepresentations), cohomological parabolic induction, and translation from $\mathrm{Irr}(G)_\rho$. The first two are discussed in Appendix I (Chapter 14) and the third one is in Appendix III (Chapter 16). For the second description, we recall the description of the Harish-Chandra modules of the irreducible representations of $O(n, 1)$ as the cohomological induction from a θ-stable Levi subgroup and introduce θ-stable coordinates for irreducible representations with regular integral infinitesimal character. This notation is used in the formulation of the conjecture in Chapter 13. We discuss the restriction of representations of the orthogonal group $O(n, 1)$ to the special orthogonal group $SO(n, 1)$ in Appendix II (Chapter 15). The results are used in Chapter 11 about the Gross–Prasad conjecture. In Appendix III, we discuss a translation functor of $G = O(n+1, 1)$ which is not in the Harish-Chandra class when n is even.

Acknowledgements Many of the results were obtained while the authors were supported by the Research in Pairs program at the Mathematisches Forschungsinstitut MFO in Oberwolfach, Germany.

Research by the first author was partially supported by Grant-in-Aid for Scientific Research (A) (25247006) and (18H03669), Japan Society for the Promotion of Science.

Research by the second author was partially supported by NSF grant DMS-1500644. Part of this research was conducted during a visit of the second author at the Graduate School of Mathematics of The University of Tokyo, Komaba. She would like to thank it for its support and hospitality during her stay.

Notations

$A - B$	set theoretic complement of B in A
$\mathbb{N}$	$\{\text{integers} \geq 0\}$
$\mathbb{N}_+$	$\{\text{positive integers}\}$
$\mathbb{R}_+$	$\{t \in \mathbb{R} : t > 0\}$
$\mathrm{Image}\,(T)$	image of the operator T
$\mathrm{Ker}\,(T)$	kernel of the operator T
E_{ij}	the matrix unit
$[a]$	the largest integer that does not exceed a
$\mathbf{1}$	the trivial one-dimensional representation
$\pi^\vee$	the contragredient representation of π
$\pi_1 \boxtimes \pi_2$	the outer tensor product representation of a direct product group
$\pi_1 \otimes \pi_2$	the tensor product representation
$\rho (\equiv \rho_G)$	the infinitesimal character of the trivial representation $\mathbf{1}$

Chapter 2
Review of Principal Series Representations

In this chapter we recall results about representations of the indefinite orthogonal group $G = O(n + 1, 1)$.

2.1 Notation

The object of our study is intertwining restriction operators (*symmetry breaking operators*) between representations of $G = O(n + 1, 1)$ and those of its subgroup $G' = O(n, 1)$. Most of main results are stated in a coordinate-free fashion, whereas the concrete description of symmetry breaking operators depends on coordinates. For the latter purpose, we choose subgroups of G and G' in a compatible fashion. The notations here are basically taken from [42].

2.1.1 Subgroups of $G = O(n + 1, 1)$ and $G' = O(n, 1)$

We define G to be the indefinite orthogonal group $O(n + 1, 1)$ that preserves the quadratic form

$$Q_{n+1,1}(x) = x_0^2 + \cdots + x_n^2 - x_{n+1}^2 \tag{2.1}$$

of signature $(n + 1, 1)$. Let G' be the stabilizer of the vector $e_n := {}^t(0, \cdots, 0, 1, 0)$. Then $G' \simeq O(n, 1)$.

© Springer Nature Singapore Pte Ltd. 2018
T. Kobayashi, B. Speh, *Symmetry Breaking for Representations of Rank One Orthogonal Groups II*, Lecture Notes in Mathematics 2234,
https://doi.org/10.1007/978-981-13-2901-2_2

We take maximal compact subgroups of G and G', respectively, as

$$K := O(n+2) \cap G \simeq O(n+1) \times O(1),$$

$$K' := K \cap G' = \left\{ \begin{pmatrix} A & & \\ & 1 & \\ & & \varepsilon \end{pmatrix} : A \in O(n), \ \varepsilon = \pm 1 \right\} \simeq O(n) \times O(1).$$

Let $\mathfrak{g} = \mathfrak{o}(n+1, 1)$ and $\mathfrak{g}' = \mathfrak{o}(n, 1)$ be the Lie algebras of $G = O(n+1, 1)$ and $G' = O(n, 1)$, respectively. We take a hyperbolic element

$$H := E_{0,n+1} + E_{n+1,0} \in \mathfrak{g}', \tag{2.2}$$

and set

$$\mathfrak{a} := \mathbb{R}H.$$

Then $\mathfrak{a}$ is a maximally split abelian subspace of $\mathfrak{g}'$, as well as that of $\mathfrak{g}$. The eigenvalues of $\mathrm{ad}(H) \in \mathrm{End}(\mathfrak{g})$ are ± 1 and 0, and the eigenspaces give rise to the following two maximal nilpotent subalgebras of $\mathfrak{g}$:

$$\mathfrak{n}_+ = \mathrm{Ker}(\mathrm{ad}(H) - 1) = \sum_{j=1}^{n} \mathbb{R}N_j^+, \quad \mathfrak{n}_- = \mathrm{Ker}(\mathrm{ad}(H) + 1) = \sum_{j=1}^{n} \mathbb{R}N_j^-, \tag{2.3}$$

where N_j^+ and N_j^- $(1 \le j \le n)$ are nilpotent elements of $\mathfrak{g}$ defined by

$$N_j^+ = -E_{0,j} + E_{j,0} - E_{j,n+1} - E_{n+1,j},$$

$$N_j^- = -E_{0,j} + E_{j,0} + E_{j,n+1} + E_{n+1,j}.$$

For $b = {}^t(b_1, \cdots, b_n) \in \mathbb{R}^n$, we define unipotent matrices in G by

$$n_+(b) := \exp\left(\sum_{j=1}^{n} b_j N_j^+\right) = I_{n+2} + \begin{pmatrix} -\frac{1}{2}Q(b) & -{}^tb & \frac{1}{2}Q(b) \\ b & 0 & -b \\ -\frac{1}{2}Q(b) & -{}^tb & \frac{1}{2}Q(b) \end{pmatrix}, \tag{2.4}$$

$$n_-(b) := \exp\left(\sum_{j=1}^{n} b_j N_j^-\right) = I_{n+2} + \begin{pmatrix} -\frac{1}{2}Q(b) & -{}^tb & -\frac{1}{2}Q(b) \\ b & 0 & b \\ \frac{1}{2}Q(b) & {}^tb & \frac{1}{2}Q(b) \end{pmatrix}, \tag{2.5}$$

where we set

$$Q(b) \equiv |b|^2 = \sum_{l=1}^{n} b_l^2. \tag{2.6}$$

Then n_+ and n_- give coordinates of the nilpotent groups $N_+ := \exp(\mathfrak{n}_+)$ and $N_- := \exp(\mathfrak{n}_-)$, respectively. Then N_+ stabilizes ${}^t(1, 0, \cdots, 0, 1)$, whereas N_- stabilizes ${}^t(1, 0, \cdots, 0, -1)$.

Since H is contained in the Lie algebra $\mathfrak{g}'$,

$$\mathfrak{n}'_\varepsilon := \mathfrak{n}_\varepsilon \cap \mathfrak{g}' = \sum_{j=1}^{n-1} \mathbb{R}N_j^\varepsilon \quad \text{for } \varepsilon = \pm$$

are maximal nilpotent subalgebras of $\mathfrak{g}'$. We set $N'_+ := N_+ \cap G' = \exp(\mathfrak{n}'_+)$ and $N'_- := N_- \cap G' = \exp(\mathfrak{n}'_-)$.

We define a split abelian subgroup A and its centralizers M and M' in K and K', respectively, as follows:

$$A := \exp(\mathfrak{a}),$$

$$M := \left\{ \begin{pmatrix} \varepsilon & & \\ & B & \\ & & \varepsilon \end{pmatrix} : B \in O(n), \varepsilon = \pm 1 \right\} \simeq O(n) \times \mathbb{Z}/2\mathbb{Z},$$

$$M' := \left\{ \begin{pmatrix} \varepsilon & & \\ & B & \\ & & 1 \\ & & & \varepsilon \end{pmatrix} : B \in O(n-1), \varepsilon = \pm 1 \right\} \simeq O(n-1) \times \mathbb{Z}/2\mathbb{Z}.$$

Then $P = MAN_+$ is a Langlands decomposition of a minimal parabolic subgroup P of G. Likewise, $P' = M'AN'_+$ is that of a minimal parabolic subgroup P' of G'. We note that A is a common maximally split abelian subgroup in P' and P because we have chosen $H \in \mathfrak{g}'$. The Langlands decompositions of the Lie algebras of P and P' are given in a compatible way as

$$\mathfrak{p} = \mathfrak{m} + \mathfrak{a} + \mathfrak{n}_+, \quad \mathfrak{p}' = \mathfrak{m}' + \mathfrak{a} + \mathfrak{n}'_+ = (\mathfrak{m} \cap \mathfrak{g}') + (\mathfrak{a} \cap \mathfrak{g}') + (\mathfrak{n}_+ \cap \mathfrak{g}').$$

We set

$$m_- := \begin{pmatrix} -1 & \\ & I_n & \\ & & -1 \end{pmatrix} \in M'. \tag{2.7}$$

We note that m_- does not belong to the identity component of G'.

2.1.2   Isotropic Cone Ξ

The isotropic cone

$$\Xi \equiv \Xi(\mathbb{R}^{n+1,1}) = \{(x_0, \cdots, x_{n+1}) \in \mathbb{R}^{n+2} : x_0^2 + \cdots + x_n^2 - x_{n+1}^2 = 0\} - \{0\}$$

is a homogeneous G-space with the following fibration:

$$
\begin{array}{ccccccc}
G/O(n)N_+ & \simeq \Xi & & & gO(n)N_+ & \mapsto & gp_+ \\
\mathbb{R}^\times \downarrow & \downarrow \mathbb{R}^\times & & & \downarrow & & \downarrow \\
G/P & \simeq S^n, & & & gP & & \mapsto [gp_+]
\end{array}
$$

where

$$p_+ := {}^t(1, 0, \cdots, 0, 1) \in \Xi. \tag{2.8}$$

The action of the subgroup N_+ on the isotropic cone Ξ is given in the coordinates as

$$n_+(b)\begin{pmatrix} \xi_0 \\ \xi \\ \xi_{n+1} \end{pmatrix} = \begin{pmatrix} \xi_0 - (b, \xi) \\ \xi \\ \xi_{n+1} - (b, \xi) \end{pmatrix} + \frac{\xi_{n+1} - \xi_0}{2}\begin{pmatrix} Q(b) \\ -2b \\ Q(b) \end{pmatrix}, \tag{2.9}$$

where $b \in \mathbb{R}^n$, $\xi \in \mathbb{R}^n$ and $\xi_0, \xi_{n+1} \in \mathbb{R}$.

The intersections of the isotropic cone Ξ with the hyperplanes $\xi_0 + \xi_{n+1} = 2$ or $\xi_{n+1} = 1$ can be identified with $\mathbb{R}^n$ or S^n, respectively. We write down the embeddings $\iota_N : \mathbb{R}^n \hookrightarrow \Xi$ and $\iota_K : S^n \hookrightarrow \Xi$ in the coordinates as follows:

$$\iota_N : \mathbb{R}^n \hookrightarrow \Xi, \; {}^t(x, x_n) \mapsto n_-(x, x_n)p_+ = \begin{pmatrix} 1 - |x|^2 - x_n^2 \\ 2x \\ 2x_n \\ 1 + |x|^2 + x_n^2 \end{pmatrix}, \tag{2.10}$$

$$\iota_K : S^n \to \Xi, \; \eta \mapsto (\eta, 1). \tag{2.11}$$

The composition of ι_N and the projection

$$\Xi \to \Xi/\mathbb{R}^\times \xrightarrow{\sim} S^n, \quad \xi \mapsto \frac{1}{\xi_{n+1}}(\xi_0, \ldots, \xi_n)$$

yields the conformal compactification of $\mathbb{R}^n$:

$$\mathbb{R}^n \hookrightarrow S^n, \quad r\omega \mapsto \eta = (s, \sqrt{1-s^2}\,\omega) = \left(\frac{1-r^2}{1+r^2}, \frac{2r}{1+r^2}\omega\right). \tag{2.12}$$

Here $\omega \in S^{n-1}$ and the inverse map is given by $r = \sqrt{\frac{1-s}{1+s}}$ for $s \neq -1$.

2.1.3 Characters $\chi_{\pm\pm}$ of the Component Group G/G_0

There are four connected components of the group $G = O(n+1, 1)$. Let G_0 denote the identity component of G. Then $G_0 \simeq SO_0(n+1, 1)$ and the quotient group G/G_0 (*component group*) is isomorphic to $\mathbb{Z}/2\mathbb{Z} \times \mathbb{Z}/2\mathbb{Z}$. Accordingly, there are four one-dimensional representations of G,

$$\chi_{ab} \colon G \to \{\pm 1\} \tag{2.13}$$

with $a, b \in \{\pm\} \equiv \{\pm 1\}$ such that

$$\chi_{ab}\,(\mathrm{diag}(-1, 1, \cdots, 1)) = a, \quad \chi_{ab}\,(\mathrm{diag}(1, \cdots, 1, -1)) = b.$$

We note that χ_{--} is given by the determinant, det, of matrices in $O(n+1, 1)$. Then the restriction of χ_{--} to the subgroup $M \simeq O(n) \times O(1)$ is given by the outer tensor product representation:

$$\chi_{--}|_M \simeq \det \boxtimes \mathbf{1}, \tag{2.14}$$

where det in the right-hand side stands for the determinant for n by n matrices.

2.1.4 The Center $\mathfrak{Z}_G(\mathfrak{g})$ and the Harish-Chandra Isomorphism

For a Lie algebra $\mathfrak{g}$ over $\mathbb{R}$, we denote by $U(\mathfrak{g})$ the universal enveloping algebra of the complexified Lie algebra $\mathfrak{g}_\mathbb{C} = \mathfrak{g} \otimes_\mathbb{R} \mathbb{C}$, and by $\mathfrak{Z}(\mathfrak{g})$ its center. For a real reductive Lie group G with Lie algebra $\mathfrak{g}$, we define a subalgebra of $\mathfrak{Z}(\mathfrak{g})$ of finite index by

$$\mathfrak{Z}_G(\mathfrak{g}) := U(\mathfrak{g})^G = \{z \in U(\mathfrak{g}) : \mathrm{Ad}(g)z = z \quad \text{for all } g \in G\}.$$

Schur's lemma implies that the algebra $\mathfrak{Z}_G(\mathfrak{g})$ acts on any irreducible admissible smooth representation of G by scalars, which we refer to as the $\mathfrak{Z}_G(\mathfrak{g})$-*infinitesimal character*. If the reductive group G is of Harish-Chandra class, then the adjoint group $\mathrm{Ad}(G)$ is contained in the inner automorphism group $\mathrm{Int}(\mathfrak{g}_\mathbb{C})$, and consequently, $\mathfrak{z}_G(\mathfrak{g}) = \mathfrak{z}(\mathfrak{g})$. However, special attention is required when G is not of Harish-Chandra class, as we shall see below.

For the disconnected group $G = O(n+1,1)$, $\mathrm{Ad}(G)$ is not contained in $\mathrm{Int}(\mathfrak{g}_\mathbb{C})$ and $\mathfrak{z}_G(\mathfrak{g})$ is of index two in $\mathfrak{z}(\mathfrak{g})$ if n is even, whereas $\mathrm{Ad}(G) \subset \mathrm{Int}(\mathfrak{g}_\mathbb{C})$ and $\mathfrak{z}_G(\mathfrak{g}) = \mathfrak{z}(\mathfrak{g})$ if n is odd. In both cases, via the standard coordinates of a Cartan subalgebra of $\mathfrak{g}_\mathbb{C} \simeq \mathfrak{o}(n+2,\mathbb{C})$, we have the following Harish-Chandra isomorphisms

$$
\begin{array}{ccc}
\mathfrak{z}(\mathfrak{g}) & \simeq & S(\mathbb{C}^{m+1})^{W_\mathfrak{g}} \\
\cup & & \cup \\
\mathfrak{z}_G(\mathfrak{g}) & \simeq & S(\mathbb{C}^{m+1})^{W_G}.
\end{array}
$$

Here we identify a Cartan subalgebra $\mathfrak{h}_\mathbb{C}$ of $\mathfrak{g}_\mathbb{C} \simeq \mathfrak{o}(n+2,\mathbb{C})$ with $\mathbb{C}^{m+1}$ where $m := [\frac{n}{2}]$, and set

$$
W_\mathfrak{g} := W(\Delta(\mathfrak{g}_\mathbb{C}, \mathfrak{h}_\mathbb{C})) \simeq
\begin{cases}
\mathfrak{S}_{m+1} \ltimes (\mathbb{Z}/2\mathbb{Z})^{m+1} & \text{for } n = 2m+1, \\
\mathfrak{S}_{m+1} \ltimes (\mathbb{Z}/2\mathbb{Z})^{m} & \text{for } n = 2m,
\end{cases}
$$

$$
W_G := \mathfrak{S}_{m+1} \ltimes (\mathbb{Z}/2\mathbb{Z})^{m+1}.
$$

We shall describe the $\mathfrak{z}_G(\mathfrak{g})$-infinitesimal character by an element of $\mathbb{C}^N$ modulo W_G via the following isomorphism.

$$
\begin{array}{ccc}
\mathrm{Hom}_{\mathbb{C}\text{-alg}}(\mathfrak{z}(\mathfrak{g}), \mathbb{C}) & \simeq & \mathbb{C}^N/W_\mathfrak{g} \\
\downarrow & & \downarrow \\
\mathrm{Hom}_{\mathbb{C}\text{-alg}}(\mathfrak{z}_G(\mathfrak{g}), \mathbb{C}) & \simeq & \mathbb{C}^N/W_G
\end{array}
\tag{2.15}
$$

To define the notion of "regular" or "singular" about $\mathfrak{z}_G(\mathfrak{g})$-infinitesimal characters, we use the action of the Weyl group $W_\mathfrak{g}$ for the Lie algebra $\mathfrak{g}_\mathbb{C} = \mathfrak{o}(n+2,\mathbb{C})$ rather than the Weyl group W_G for the disconnected group G as below.

Definition 2.1. Let $G = O(n+1,1)$ and $m := [\frac{n}{2}]$. Suppose that $\chi \in \mathrm{Hom}_{\mathbb{C}\text{-alg}}(\mathfrak{z}_G(\mathfrak{g}), \mathbb{C})$ is given by $\mu \in \mathbb{C}^{m+1} \mod W_G$ via the Harish-Chandra isomorphism (2.15). We say χ is *integral* if

$$
\mu - \rho_G \in \mathbb{Z}^{m+1},
$$

see (2.16) below for the definition of ρ_G, or equivalently, if

$$
\mu \in \mathbb{Z}^{m+1} \qquad\qquad \text{for } n = 2m \qquad\qquad \text{(even)},
$$

$$
\mu \in (\mathbb{Z} + \frac{1}{2})^{m+1} \qquad \text{for } n = 2m+1 \qquad \text{(odd)}.
$$

We note that this condition is stronger than the one which is usually referred to as "integral":

$$\langle \mu, \alpha^\vee \rangle \in \mathbb{Z} \quad \text{for all } \alpha \in \Delta(\mathfrak{g}_\mathbb{C}, \mathfrak{h}_\mathbb{C})$$

where $\alpha^\vee$ denotes the coroot of α.

For $\mu \in \mathbb{C}^{m+1}$, we set

$$W_\mu \equiv (W_\mathfrak{g})_\mu := \{w \in W_\mathfrak{g} : w\mu = \mu\},$$
$$(W_G)_\mu := \{w \in W_G : w\mu = \mu\}.$$

We say μ is $W_\mathfrak{g}$-*regular* (or simply, *regular*) if $(W_\mathfrak{g})_\mu = \{e\}$, and W_G-*regular* if $(W_G)_\mu = \{e\}$. These definitions depend only on the W_G-orbit through μ because $\#W_\mu = \#W_{\mu'}$ if $\mu' \in W_G\mu$. We say χ is *regular integral* (respectively, *singular integral*) infinitesimal character if χ is integral and $W_\mu = \{e\}$ (respectively, $W_\mu \neq \{e\}$). In the coordinates of $\mu = (\mu_1, \cdots, \mu_{m+1})$, $W_\mu = \{e\}$ if and only if

$$\mu_i \neq \pm\mu_j \quad (1 \leq \forall i < \forall j \leq m+1) \qquad \qquad \text{for } n \text{ even,}$$
$$\mu_i \neq \pm\mu_j \quad (1 \leq \forall i < \forall j \leq m+1), \; \mu_k \neq 0 \quad (1 \leq \forall k \leq m+1) \quad \text{for } n \text{ odd.}$$

Remark 2.2. Suppose $G = O(n+1, 1)$ with $n \geq 1$. Then the $\mathfrak{Z}_G(\mathfrak{g})$-infinitesimal character of an irreducible finite-dimensional representation of G is regular integral, and conversely, for any regular integral χ, there exists an irreducible finite-dimensional representation F of G such that χ is the $\mathfrak{Z}_G(\mathfrak{g})$-infinitesimal character of F. Here we remind from Definition 2.1 above that by "regular" we mean $W_\mathfrak{g}$-regular, and not W_G-regular.

The $\mathfrak{Z}_G(\mathfrak{g})$-infinitesimal character of the trivial one-dimensional representation **1** of $G = O(n+1, 1)$ is given by

$$\rho \equiv \rho_G = (\frac{n}{2}, \frac{n}{2} - 1, \cdots, \frac{n}{2} - [\frac{n}{2}]) \in \mathbb{C}^{[\frac{n}{2}]+1}/W_G. \tag{2.16}$$

The infinitesimal character ρ_G will be also referred to as the *trivial infinitesimal character*.

Definition 2.3. We denote by $\text{Irr}(G)_\rho$ the set of equivalence classes of irreducible admissible smooth representations of G that have the trivial infinitesimal character ρ.

The finite set $\text{Irr}(G)_\rho$ is classified in Theorem 2.20 (2) for $G = O(n+1, 1)$ and in Proposition 15.11 (3) for the special orthogonal group $SO(n+1, 1)$.

2.2 Representations of the Orthogonal Group $O(N)$

We recall that the orthogonal group $O(N)$ has two connected components. In this section, we review a parametrization of irreducible finite-dimensional representations of the *disconnected* group $O(N)$ following Weyl [64, Chap. V, Sect. 7]. For later reference we include classical branching theorem for the restriction of representations for the pairs $O(N) \supset O(N-1)$ and $O(N) \supset SO(N)$. The results will be applied to the four compact subgroups K, K', M and $M' = M \cap K'$ of G introduced in Section 2.1.1, which satisfy the following obvious inclusive relations:

$$\begin{pmatrix} K \supset K' \\ \cup \quad \cup \\ M \supset M' \end{pmatrix} = \begin{pmatrix} O(n+1) \times O(1) \supset O(n) \times O(1) \\ \cup \qquad\qquad \cup \\ O(n) \times \mathrm{diag}(O(1)) \supset O(n-1) \times \mathrm{diag}(O(1)) \end{pmatrix}.$$

2.2.1 Notation for Irreducible Representations of $O(N)$

For finite-dimensional irreducible representations of orthogonal groups, we use the following notation. We set

$$\Lambda^+(N) := \{\lambda = (\lambda_1, \ldots, \lambda_N) \in \mathbb{Z}^N : \lambda_1 \geq \lambda_2 \geq \cdots \geq \lambda_N \geq 0\}. \tag{2.17}$$

We write $F^{U(N)}(\lambda)$ for the irreducible finite-dimensional representation of $U(N)$ (or equivalently, the irreducible polynomial representation of $GL(N, \mathbb{C})$) with highest weight $\lambda \in \Lambda^+(N)$. If λ is of the form

$$(\underbrace{c_1, \cdots, c_1}_{m_1}, \underbrace{c_2, \cdots, c_2}_{m_2}, \cdots, \underbrace{c_\ell, \cdots, c_\ell}_{m_\ell}, 0, \cdots, 0),$$

then we also write $\lambda = \left(c_1^{m_1}, c_2^{m_2}, \cdots, c_\ell^{m_\ell}\right)$ as usual.

We define a subset of $\Lambda^+(N)$ by

$$\Lambda^+(O(N)) := \{\lambda \in \Lambda^+(N) : \lambda_1' + \lambda_2' \leq N\},$$

where $\lambda_1' := \max\{i : \lambda_i \geq 1\}$ and $\lambda_2' := \max\{i : \lambda_i \geq 2\}$ for $\lambda = (\lambda_1, \ldots, \lambda_N) \in \Lambda^+(N)$. We note that λ_1' equals the maximal column length in the corresponding Young diagram.

It is readily seen that $\Lambda^+(O(N))$ consists of elements of the following two types:

$$\text{Type I: } (\lambda_1, \cdots, \lambda_k, \underbrace{0, \cdots, 0}_{N-k}), \tag{2.18}$$

$$\text{Type II: } (\lambda_1, \cdots, \lambda_k, \underbrace{1, \cdots, 1}_{N-2k}, \underbrace{0, \cdots, 0}_{k}), \tag{2.19}$$

with $\lambda_1 \geq \lambda_2 \geq \cdots \geq \lambda_k > 0$ and $0 \leq k \leq \left[\frac{N}{2}\right]$.

For any $\lambda \in \Lambda^+(O(N))$, there exists a unique $O(N)$-irreducible summand, to be denoted by $F^{O(N)}(\lambda)$, of the $U(N)$-module $F^{U(N)}(\lambda)$ which contains the highest weight vector. Following Weyl [64, Chap. V, Sect. 7], we parametrize the set $\widehat{O(N)}$ of equivalence classes of irreducible representations of $O(N)$ by

$$\Lambda^+(O(N)) \xrightarrow{\sim} \widehat{O(N)}, \quad \lambda \mapsto F^{O(N)}(\lambda). \tag{2.20}$$

By the Weyl unitary trick, we may identify $F^{O(N)}(\lambda)$ with a holomorphic irreducible representation of $O(N, \mathbb{C})$, to be denoted by $F^{O(N, \mathbb{C})}(\lambda)$, on the same representation space.

Definition 2.4. We say $F^{O(N)}(\lambda) \in \widehat{O(N)}$ is of type I (or type II), if $\lambda \in \Lambda^+(O(N))$ is of type I (or type II), respectively.

We shall identify $\widehat{O(N)}$ with $\Lambda^+(O(N))$ via (2.20), and by abuse of notation, we write $\sigma = (\sigma_1, \cdots, \sigma_N) \in \widehat{O(N)}$ when $(\sigma_1, \cdots, \sigma_N) \in \Lambda^+(O(N))$.

Remark 2.5. We shall also use the notation

$$F^{O(N)}(\sigma_1, \cdots, \sigma_k, \underbrace{0, \cdots, 0}_{\left[\frac{N}{2}\right]-k})_+ \text{ instead of } F^{O(N)}(\sigma_1, \cdots, \sigma_k, \underbrace{0, \cdots, 0}_{N-k}),$$

$$F^{O(N)}(\sigma_1, \cdots, \sigma_k, \underbrace{0, \cdots, 0}_{\left[\frac{N}{2}\right]-k})_- \text{ instead of } F^{O(N)}(\sigma_1, \cdots, \sigma_k, \underbrace{1, \cdots, 1}_{N-2k}, \underbrace{0, \cdots, 0}_{k}),$$

by putting the subscript $+$ or $-$ for irreducible representations of type I or of type II, respectively, see Remark 14.1 in Appendix I.

We define a map by summing up the first k-entries ($k \leq \left[\frac{N}{2}\right]$) of σ:

$$\ell : \Lambda^+(O(N)) \to \mathbb{N}, \quad \sigma \mapsto \ell(\sigma) := \sum_{i=1}^{k} \sigma_i, \tag{2.21}$$

which induces a map

$$\ell : \widehat{O(N)} \to \mathbb{N}$$

via the identification (2.20). By (2.23), we have

$$\ell(\sigma) = \ell(\sigma \otimes \det). \tag{2.22}$$

2.2.2 Branching Laws for $O(N) \downarrow SO(N)$

Definition 2.6. We say $\sigma \in \widehat{O(N)}$ is of type X or type Y, if the restriction $\sigma|_{SO(N)}$ to the special orthogonal group $SO(N)$ is irreducible or reducible, respectively.

With the convention as in Definition 2.4, we recall a classical fact about the branching rule for the restriction $O(N) \downarrow SO(N)$.

Lemma 2.7 ($O(N) \downarrow SO(N)$). *Let* $\sigma = (\sigma_1, \cdots, \sigma_N) \in \Lambda^+(O(N))$, *and* $k \ (\leq [\frac{N}{2}])$ *be as in* (2.18) *and* (2.19).

(1) (type X) *The restriction of the irreducible* $O(N)$*-module* $F^{O(N)}(\sigma)$ *to* $SO(N)$ *is irreducible if and only if* $N \neq 2k$. *In this case, the restricted* $SO(N)$*-module has highest weight* $(\sigma_1, \cdots, \sigma_k, 0, \cdots, 0)$.
(2) (type Y) *If* $N = 2k$, *then the restriction* $F^{O(N)}(\lambda)|_{SO(N)}$ *splits into two inequivalent irreducible representations of* $SO(N)$ *with highest weights* $(\sigma_1, \cdots, \sigma_{k-1}, \sigma_k)$ *and* $(\sigma_1, \cdots, \sigma_{k-1}, -\sigma_k)$.

Example 2.8. The orthogonal group $O(N)$ acts irreducibly on the ℓ-th exterior tensor space $\bigwedge^\ell(\mathbb{C}^N)$ and on the space $\mathcal{H}^s(\mathbb{C}^N)$ of spherical harmonics of degree s. Via the parametrization (2.20), these representations are described as follows:

$$\bigwedge^\ell(\mathbb{C}^N) = F^{O(N)}(1^\ell) \qquad (0 \leq \ell \leq N),$$

$$\mathcal{H}^s(\mathbb{C}^N) = F^{O(N)}(s, 0, \cdots, 0) \qquad (s \in \mathbb{N}).$$

The $O(N)$-module $\bigwedge^\ell(\mathbb{C}^N)$ is of type Y if and only if $N = 2\ell$; the $O(N)$-module $\mathcal{H}^s(\mathbb{C}^N)$ is of type Y if and only if $N = 2$ and $s \neq 0$.

Irreducible $O(N)$-modules of types I and II are related by the following $O(N)$-isomorphism:

$$F^{O(N)}(a_1, \cdots, a_k, 1, \cdots, 1, 0, \cdots, 0) = \det \otimes F^{O(N)}(a_1, \cdots, a_k, 0, \cdots, 0). \tag{2.23}$$

Hence we obtain the following:

Lemma 2.9. *Let* $\sigma \in \widehat{O(N)}$. *Then* σ *is of type Y if and only if* $\sigma \otimes \det \simeq \sigma$.

Then the following proposition is clear.

Proposition 2.10. *Suppose $\sigma \in \widehat{O(n)}$.*

(1) *If σ is of type Y, then σ is of type I.*
(2) *If σ is of type II, then σ is of type X.*

2.2.3   Branching Laws $O(N) \downarrow O(N-1)$

Next we recall the classical branching laws for $O(N) \downarrow O(N-1)$. Let $\sigma = (\sigma_1, \cdots, \sigma_N) \in \Lambda^+(O(N))$ and $\tau = (\tau_1, \cdots, \tau_{N-1}) \in \Lambda^+(O(N-1))$.

Definition 2.11. We denote by $\tau \prec \sigma$ if

$$\sigma_1 \geq \tau_1 \geq \sigma_2 \geq \tau_2 \geq \cdots \geq \tau_{N-1} \geq \sigma_N.$$

Then the irreducible decomposition of representations of $O(N)$ with respect to the subgroup $O(N-1)$ is given as follows:

Fact 2.12 (Branching rule for orthogonal groups). *Let $(\sigma_1, \cdots, \sigma_N) \in \Lambda^+(O(N))$. Then the irreducible representation $F^{O(N)}(\sigma_1, \cdots, \sigma_N)$ decomposes into a multiplicity-free sum of irreducible representations of $O(N-1)$ as follows:*

$$F^{O(N)}(\sigma_1, \cdots, \sigma_N)|_{O(N-1)} \simeq \bigoplus_{\tau \prec \sigma} F^{O(N-1)}(\tau_1, \cdots, \tau_{N-1}). \tag{2.24}$$

The commutant $O(1)$ of $O(N-1)$ in $O(N)$ acts on the irreducible summand $F^{O(N-1)}(\tau_1, \cdots, \tau_{N-1})$ by $(\mathrm{sgn})^{\sum_{j=1}^{N} \sigma_j - \sum_{i=1}^{N-1} \tau_i}$.

The following lemma is derived from Lemma 2.9 and Fact 2.13.

Lemma 2.13. *Let $\sigma \in \widehat{O(n)}$ be of type I (see Definition 2.4). Then the following four conditions are equivalent:*

(i) *$\sigma \otimes \det \simeq \sigma$;*
(ii) *$[\sigma|_{O(n-1)} : \tau] = [\sigma|_{O(n-1)} : \tau \otimes \det]$ for all $\tau \in \widehat{O(n-1)}$;*
(iii) *n is even and $\sigma = F^{O(n)}(s_1, \cdots, s_{\frac{n}{2}}, 0, \cdots, 0)$ with $s_{\frac{n}{2}} \neq 0$;*
(iv) *$\sigma|_{SO(n)}$ is reducible, i.e., σ is of type Y (Definition 2.6).*

2.3   Principal Series Representations $I_\delta(V, \lambda)$ of the Orthogonal Group $G = O(n+1, 1)$

We discuss here (nonspherical) principal series representations $I_\delta(V, \lambda)$ of $G = O(n+1, 1)$. We shall use the symbol $J_\varepsilon(W, \nu)$ for the principal series representations of the subgroup $G' = O(n, 1)$.

We recall the structure of principal series representations for rank one orthogonal groups. The main references are Borel and Wallach [9] and Collingwood [11, Chap. 5, Sect. 2] for the representations of the identity component group $G_0 = SO_0(n+1, 1)$. We extend here the results to the disconnected group. For representations of the disconnected group G, see also [42, Chap. 2] for the spherical case (i.e., $V = \mathbf{1}$) and [35, Chap. 2, Sect. 3] for $V = \bigwedge^i(\mathbb{C}^n)$ ($0 \leq i \leq n$).

2.3.1 C^∞-induced Representations $I_\delta(V, \lambda)$

We recall from Section 2.1.1 that the Levi subgroup MA of the minimal parabolic subgroup P of G is a direct product group $(O(n) \times O(1)) \times \mathbb{R}$. Then any irreducible representation of MA is the outer tensor product of irreducible representations of the three groups $O(n)$, $O(1)$, and $\mathbb{R}$.

One-dimensional representations δ of $O(1) = \{1, m_-\}$ are labeled by $+$ or $-$, where we write $\delta = +$ for the trivial representation $\mathbf{1}$, and $\delta = -$ for the nontrivial one given by $\delta(m_-) = -1$. Thus we identify $\widehat{O(1)}$ with the set $\{\pm\}$.

For $\lambda \in \mathbb{C}$, we denote by $\mathbb{C}_\lambda$ the one-dimensional representation of the split group A normalized by the generator $H \in \mathfrak{a}$ (see (2.2)) as

$$A \mapsto \mathbb{C}^\times, \qquad \exp(tH) \mapsto e^{\lambda t}.$$

Let (σ, V) be an irreducible representation of $O(n)$, $\delta \in \{\pm\}$, and $\lambda \in \mathbb{C}$. We extend the outer tensor product representation

$$V_{\lambda, \delta} := V \boxtimes \delta \boxtimes \mathbb{C}_\lambda \tag{2.25}$$

of the direct product group $MA \simeq O(n) \times O(1) \times \mathbb{R}$ to a representation of the parabolic subgroup $P = MAN_+$ by letting the unipotent subgroup N_+ act trivially. The resulting irreducible P-module will be written as $V_{\lambda, \delta} = V \otimes \delta \otimes \mathbb{C}_\lambda$ by a little abuse of notation. We define the induced representation of G by

$$I_\delta(V, \lambda) \equiv I(V \otimes \delta, \lambda) := \operatorname{Ind}_P^G(V_{\lambda, \delta}).$$

We refer to δ as the *signature* of the induced representation. If $\delta = +$ (the trivial character $\mathbf{1}$), we sometimes suppress the subscript.

If $(\sigma, V) \in \widehat{O(n)}$ is given as $V = F^{O(n)}(\sigma_1, \cdots, \sigma_n)$ with $(\sigma_1, \cdots, \sigma_n) \in \Lambda^+(O(n))$ via (2.20), then $I_\delta(V, \lambda)$ has a $\mathfrak{Z}_G(\mathfrak{g})$-infinitesimal character

$$(\sigma_1 + \frac{n}{2} - 1, \sigma_2 + \frac{n}{2} - 2, \cdots, \sigma_k + \frac{n}{2} - k, \frac{n}{2} - k - 1, \cdots, \frac{n}{2} - [\frac{n}{2}], \lambda - \frac{n}{2}) \tag{2.26}$$

in the standard coordinates via the Harish-Chandra isomorphism, see (2.15). We are using in this article unnormalized induction, *i.e.*, the representation $I_\delta(V, \frac{n}{2})$ is a unitarily induced principal series representation. Thus if λ is purely imaginary, the principal series representations $I_\delta(V, \lambda + \frac{n}{2})$ are tempered. If n is even, then every irreducible tempered representation is isomorphic to a tempered principal series representation. If n is odd, then there is one family of discrete series representations parametrized by characters of the compact Cartan subgroup and every irreducible tempered representation is isomorphic to a tempered principal series representation or a discrete series representation.

We denote by

$$\mathcal{V}_{\lambda,\delta} := G \times_P V_{\lambda,\delta} \qquad (2.27)$$

the G-equivariant vector bundle over the real flag manifold G/P associated to the representation $V_{\lambda,\delta}$ of P. We assume from now on that the principal series representations $I_\delta(V, \lambda)$ are realized on the Fréchet space $C^\infty(G/P, \mathcal{V}_{\lambda,\delta})$ of smooth sections for the vector bundle $\mathcal{V}_{\lambda,\delta} \to G/P$. Thus $I_\delta(V, \lambda)$ is the induced representation $C^\infty\text{-Ind}_P^G(V_{\lambda,\delta})$ which is of moderate growth, see [42, Chap. 3, Sect. 4]. As usual, we denote the representation space and the representation by the same letter. We trivialize the vector bundle $\mathcal{V}_{\lambda,\delta}$ over G/P on the open Bruhat cell via the following map

$$\iota_N : \mathbb{R}^n \xrightarrow[n_-]{\sim} N_- \xrightarrow{\sim} N_- \cdot o \subset G/P.$$

Then $I_\delta(V, \lambda)$ is realized in a subspace of $C^\infty(\mathbb{R}^n) \otimes V$ by

$$\iota_N^* : I_\delta(V, \lambda) \hookrightarrow C^\infty(\mathbb{R}^n) \otimes V, \quad F \mapsto f(b) := F(n_-(b)), \qquad (2.28)$$

and this model is referred to as the *noncompact picture*, or the *N-picture*, see Section 8.2.

2.3.2 Tensoring with Characters $\chi_{\pm\pm}$ of G

The character group $(G/G_0)^\wedge$ of the component group $G/G_0 \simeq \mathbb{Z}/2\mathbb{Z} \times \mathbb{Z}/2\mathbb{Z}$ acts on the set of admissible representations Π of G, by taking the tensor product

$$\Pi \mapsto \Pi \otimes \chi \qquad (2.29)$$

for $\chi \in (G/G_0)^\wedge$. This action leaves the subsets $\text{Irr}(G)$ and $\text{Irr}(G)_\rho$ (see Definition 2.3) invariant. We describe the action explicitly on principal series representations in Lemma 2.14 below. The action on $\text{Irr}(G)_\rho$ will be given explicitly in Theorem 2.20 (5), and on the space of symmetry breaking operators in Section 3.7.

Lemma 2.14. *Let $V \in \widehat{O(n)}$, $\delta \in \{\pm\}$, and $\lambda \in \mathbb{C}$. Let $\chi_{\pm\pm}$ be the one-dimensional representations of $G = O(n+1, 1)$ as defined in (2.13). Then we have the following isomorphisms between representations of G:*

$$I_\delta(V, \lambda) \otimes \chi_{+-} \simeq I_{-\delta}(V, \lambda),$$

$$I_\delta(V, \lambda) \otimes \chi_{-+} \simeq I_{-\delta}(V \otimes \det, \lambda),$$

$$I_\delta(V, \lambda) \otimes \chi_{--} \simeq I_\delta(V \otimes \det, \lambda).$$

Proof. For any P-module U and for any finite-dimensional G-module F, there is an isomorphism of G-modules:

$$F \otimes \operatorname{Ind}_P^G(U) \simeq \operatorname{Ind}_P^G(F \otimes U).$$

Then Lemma 2.14 follows from the restriction formula of the character χ of G to the subgroup $M \simeq O(n) \times O(1)$ as below:

$$\chi_{+-}|_M \simeq \mathbf{1} \boxtimes \operatorname{sgn}, \quad \chi_{-+}|_M \simeq \det \boxtimes \operatorname{sgn}, \quad \chi_{--}|_M \simeq \det \boxtimes \mathbf{1}.$$

□

A special case of Lemma 2.14 for the exterior tensor representations $V = \bigwedge^i(\mathbb{C}^n)$ will be stated in Lemma 3.36.

2.3.3 K-structure of the Principal Series Representation $I_\delta(V, \lambda)$

Let $(\sigma, V) \in \widehat{O(n)}$ and $\delta \in \{\pm\}$ as before. By the Frobenius reciprocity law, K-types of the principal series representation $I_\delta(V, \lambda)$ are the irreducible representations of $K = O(n+1) \times O(1)$ whose restriction to $M \simeq O(n) \times O(1)$ contains the representation $V \boxtimes \delta$ of M. The classical branching theorem (Fact 2.12) is used to determine K-types of the G-module $I_\delta(V, \lambda)$. We shall give an explicit K-type formula in the next section when V is the exterior tensor representation $\bigwedge^i(\mathbb{C}^n)$ of $O(n)$. For the general representation $(\sigma, V) \in \widehat{O(n)}$, we do not use an explicit K-type formula of $I_\delta(V, \lambda)$, but just mention an immediate corollary of Fact 2.12:

Proposition 2.15. *The K-types of principal series representations $I_\delta(V, \lambda)$ of $O(n+1, 1)$ have multiplicity one.*

2.4   Principal Series Representations $I_\delta(i,\lambda)$

For $0 \le i \le n$, $\delta \in \{\pm\}$, and $\lambda \in \mathbb{C}$, we denote the principal series representation $I_\delta(\bigwedge^i(\mathbb{C}^n),\lambda) = C^\infty\text{-Ind}_P^G(\bigwedge^i(\mathbb{C}^n) \otimes \delta \otimes \mathbb{C}_\lambda)$ of $G = O(n+1,1)$ simply by $I_\delta(i,\lambda)$. Similarly, we write $J_\varepsilon(j,\nu)$ for the induced representation $C^\infty\text{-Ind}_{P'}^{G'}(\bigwedge^j(\mathbb{C}^{n-1}) \otimes \varepsilon \otimes \mathbb{C}_\nu)$ of $G' = O(n,1)$ for $0 \le j \le n-1$, $\varepsilon \in \{\pm\}$, and $\nu \in \mathbb{C}$. In the major part of this monograph, we focus our attention on special families of principal series representations $I_\delta(i,\lambda)$ of G and $J_\varepsilon(j,\nu)$ of the subgroup G'.

In geometry, $I_\delta(i,\lambda)$ is a family of representations of the conformal group $O(n+1,1)$ of S^n on the space $\mathcal{E}^i(S^n)$ of differential forms (*cf.* [35, Chap. 2, Sect. 2]) on one hand. In representation theory, any irreducible, unitarizable representations with nonzero $(\mathfrak{g}, K)$-cohomologies arise as subquotients in $I_\delta(i,\lambda)$ with $\lambda = i$ for some $0 \le i \le n$ and $\delta = (-1)^i$, see Theorem 2.20 (9), also Proposition 14.45 in Appendix I.

In this section we collect some basic properties of the principal series representations

$$I_\delta(i,\lambda) \qquad \text{for } \delta \in \{\pm\}, 0 \le i \le n, \lambda \in \mathbb{C},$$

which will be used throughout the article.

2.4.1 $\mathfrak{Z}_G(\mathfrak{g})$-infinitesimal Character of $I_\delta(i,\lambda)$

As we have seen in (2.26) in the general case, the $\mathfrak{Z}_G(\mathfrak{g})$-infinitesimal character of the principal series representation $I_\delta(i,\lambda)$ is given by

$$(\underbrace{\frac{n}{2},\frac{n}{2}-1,\cdots,\frac{n}{2}-i+1}_{i};\underbrace{\frac{n}{2}-i-1,\cdots,\frac{n}{2}-[\frac{n}{2}]}_{[\frac{n}{2}]-i};\lambda-\frac{n}{2}) \qquad \text{if } 0 \le i \le \frac{n}{2},$$

$$(\underbrace{\frac{n}{2},\frac{n}{2}-1,\cdots,-\frac{n}{2}+i+1}_{n-i};\underbrace{-\frac{n}{2}+i-1,\cdots,\frac{n}{2}-[\frac{n}{2}]}_{i-[\frac{n+1}{2}]};\lambda-\frac{n}{2}) \quad \text{if } \frac{n}{2} \le i \le n.$$

In particular, the G-module $I_\delta(i,\lambda)$ has the trivial infinitesimal character ρ_G if and only if $\lambda = i$ or $n-i$.

2.4.2 K-type Formula of the Principal Series Representations $I_\delta(i,\lambda)$

By the Frobenius reciprocity, we can compute the K-type formula of $I_\delta(i,\lambda)$ explicitly by using the classical branching law (Fact 2.12) and Example 2.8 as follows:

Lemma 2.16 (*K*-type formula of $I_\delta(i,\lambda)$). *Let $0 \le i \le n$ and $\delta \in \{\pm\}$. With the parametrization (2.20), the K-type formula of the principal series representation $I_\delta(i,\lambda)$ of $G = O(n+1,1)$ is described as below:*

(1) *for $i = 0$,*

$$\bigoplus_{a=0}^{\infty} F^{O(n+1)}(a,0^n) \boxtimes (-1)^a \delta;$$

(2) *for $1 \le i \le n-1$,*

$$\bigoplus_{a=1}^{\infty} F^{O(n+1)}(a,1^i,0^{n-i}) \boxtimes (-1)^a \delta \oplus \bigoplus_{a=1}^{\infty} F^{O(n+1)}(a,1^{i-1},0^{n+1-i}) \boxtimes (-1)^{a+1}\delta;$$

(3) *for $i = n$,*

$$\bigoplus_{a=1}^{\infty} (\det \otimes F^{O(n+1)}(a,0^n)) \boxtimes (-1)^{a+1}\delta.$$

See Proposition 14.29 for a more general K-type formula of the principal series representation $I_\delta(V,\lambda)$.

2.4.3 Basic K-types of $I_\delta(i,\lambda)$

Let $\delta \in \{\pm\}$ and $0 \le i \le n$. Following the notation [35, Chap. 2, Sect. 3], we define two irreducible representations of $K \simeq O(n+1) \times O(1)$ by:

$$\mu^\flat(i,\delta) := \bigwedge\nolimits^i(\mathbb{C}^{n+1}) \boxtimes \delta, \tag{2.30}$$

$$\mu^\sharp(i,\delta) := \bigwedge\nolimits^{i+1}(\mathbb{C}^{n+1}) \boxtimes (-\delta). \tag{2.31}$$

This means:

$$\begin{cases} \mu^\flat(i,+) &= \bigwedge^i(\mathbb{C}^{n+1}) \boxtimes \mathbf{1}, \\ \mu^\flat(i,-) &= \bigwedge^i(\mathbb{C}^{n+1}) \boxtimes \mathrm{sgn}, \end{cases} \qquad \begin{cases} \mu^\sharp(i,+) &= \bigwedge^{i+1}(\mathbb{C}^{n+1}) \boxtimes \mathrm{sgn}, \\ \mu^\sharp(i,-) &= \bigwedge^{i+1}(\mathbb{C}^{n+1}) \boxtimes \mathbf{1}. \end{cases}$$

The superscripts $\sharp$ and $\flat$ indicate that there are the following obvious K-isomorphisms

$$\mu^\sharp(i, \delta) = \mu^\flat(i + 1, -\delta) \qquad (0 \le i \le n), \tag{2.32}$$

which will be useful in describing the standard sequence with trivial infinitesimal character ρ_G (Definition 2.21 below), see also Remark 2.19.

By the K-type formula of the principal series representation $I_\delta(i, \lambda)$ in Lemma 2.16, the K-types $\mu^\flat(i, \delta)$ and $\mu^\sharp(i, \delta)$ occur in $I_\delta(i, \lambda)$ with multiplicity one for any $\lambda \in \mathbb{C}$.

Definition 2.17. We say $\mu^\flat(i, \delta)$ and $\mu^\sharp(i, \delta)$ are *basic K-types* of the principal series representations $I_\delta(i, \lambda)$ of $G = O(n + 1, 1)$.

2.4.4 Reducibility of $I_\delta(i, \lambda)$

The principal series representation $I_\delta(i, \lambda)$ is generically irreducible. More precisely, we have the following.

Proposition 2.18. *Let $G = O(n + 1, 1)$, $0 \le i \le n$, $\delta \in \{\pm\}$, and $\lambda \in \mathbb{C}$.*

(1) *The principal series representation $I_\delta(i, \lambda)$ is reducible if and only if*

$$\lambda \in \{i, n - i\} \cup (-\mathbb{N}_+) \cup (n + \mathbb{N}_+). \tag{2.33}$$

(2) *Suppose $(n, \lambda) \ne (2i, i)$. If λ satisfies (2.33), then the G-module $I_\delta(i, \lambda)$ has a unique irreducible proper submodule (say, A) and has a unique irreducible subquotient (say, B) and there is a nonsplitting exact sequence of G-modules:*

$$0 \to A \to I_\delta(i, \lambda) \to B \to 0.$$

(3) *Suppose $(n, \lambda) = (2i, i)$. Then the $I_\delta(i, \lambda)$ decomposes into the direct sum of two irreducible representations of G which are not isomorphic to each other.*

When $n \ne 2i$, the "only if" part of the first statement and the second one in Proposition 2.18 follow readily from the corresponding results [9, 11, 16] for the connected group $SO_0(n + 1, 1)$ and from Lemma 2.22 below because $\bigwedge^i(\mathbb{C}^n)$ is irreducible as an $SO(n)$-module. We need some argument for $n = 2i$ where $\bigwedge^i(\mathbb{C}^n)$ is reducible as an $SO(n)$-module, see Examples 14.16 and 15.6 in Appendix II for the proof of Proposition 2.18 (1) and (3), respectively. In Section 8.5, we discuss the description of proper submodules of reducible $I_\delta(i, \lambda)$ by using the Knapp–Stein operator (8.14) and its normalized one (8.21). The "if" part of the first statement is proved there, see Lemma 8.16.

The composition series of $I_\delta(i, \lambda)$ with trivial infinitesimal character ρ_G (i.e., for $\lambda = i$ or $n - i$) will be discussed in the next subsection (see Theorem 2.20), which will be extended in Theorem 13.11 to the case of regular integral infinitesimal characters.

2.4.5 Irreducible Subquotients of $I_\delta(i, i)$

Every irreducible representation of $G = O(n + 1, 1)$ with trivial infinitesimal character ρ is equivalent to a subquotient of $I_\delta(i, i)$ for some $0 \le i \le n$ and $\delta \in \{\pm\}$, or equivalently, of $I_+(i, i) \otimes \chi$ with $i \ge n/2$ and $\chi \in (G/G_0)\hat{\ }$. We recall now facts about the principal series representations $I_+(i, i)$, $I_-(i, i)$, $I_+(n - i, i)$ and $I_-(n - i, i)$ of the orthogonal group $O(n + 1, 1)$ and their composition factors.

We denote by $I_\delta(i)^\flat$ and $I_\delta(i)^\sharp$ the unique irreducible subquotients of $I_\delta(i, i)$ containing the basic K-types $\mu^\flat(i, \delta)$ and $\mu^\sharp(i, \delta)$, respectively. Then we have G-isomorphisms:

$$I_\delta(i)^\sharp \simeq I_{-\delta}(i + 1)^\flat \quad \text{for } 0 \le i \le n - 1 \text{ and } \delta \in \{\pm\}, \tag{2.34}$$

see Theorem 2.20 (1) below. For $0 \le \ell \le n + 1$ and $\delta \in \{\pm\}$, we set

$$\Pi_{\ell,\delta} := \begin{cases} I_\delta(\ell)^\flat & (0 \le \ell \le n), \\ I_{-\delta}(\ell - 1)^\sharp & (1 \le \ell \le n + 1). \end{cases} \tag{2.35}$$

In view of (2.34), the irreducible representation $\Pi_{\ell,\delta}$ of G is well-defined.

Remark 2.19. The point here is that each irreducible representation $\Pi_{\ell,\delta}$ ($1 \le \ell \le n$, $\delta = \pm$) can be realized in two different principal series representations:

$$I_\delta(\ell, \ell) = \text{Ind}_P^G(\textstyle\bigwedge^\ell(\mathbb{C}^n) \otimes \delta \otimes \mathbb{C}_\ell),$$

$$I_{-\delta}(\ell - 1, \ell - 1) = \text{Ind}_P^G(\textstyle\bigwedge^{\ell-1}(\mathbb{C}^n) \otimes (-\delta) \otimes \mathbb{C}_{\ell-1}).$$

Theorem 2.20. *Let $G = O(n + 1, 1)$ ($n \ge 1$).*

(1) *For $0 \le \ell \le n$ and $\delta \in \{\pm\}$, we have exact sequences of G-modules:*

$$0 \to \Pi_{\ell,\delta} \to I_\delta(\ell, \ell) \to \Pi_{\ell+1,-\delta} \to 0,$$

$$0 \to \Pi_{\ell+1,-\delta} \to I_\delta(\ell, n - \ell) \to \Pi_{\ell,\delta} \to 0.$$

These exact sequences split if and only if $n = 2\ell$.

(2) *Irreducible admissible smooth representations of G with trivial $\mathfrak{Z}_G(\mathfrak{g})$-infinitesimal character ρ_G can be classified as*

$$\text{Irr}(G)_\rho = \{\Pi_{\ell,\delta} : 0 \le \ell \le n + 1, \delta = \pm\}.$$

(3) *For any $0 \leq \ell \leq n+1$ and $\delta \in \{\pm\}$, the minimal K-type of the irreducible G-module $\Pi_{\ell,\delta}$ is given by $\mu^\flat(\ell, \delta) = \bigwedge^\ell(\mathbb{C}^{n+1}) \boxtimes \delta$.*

(4) *There are four one-dimensional representations of G, and they are given by*

$$\{\Pi_{0,+} \simeq \mathbf{1}, \quad \Pi_{0,-} \simeq \chi_{+-}, \quad \Pi_{n+1,+} \simeq \chi_{-+}, \quad \Pi_{n+1,-} \simeq \chi_{--}(= \det)\}.$$

The other representations $\Pi_{\ell,\delta}$ $(1 \leq \ell \leq n, \delta \in \{\pm\})$ are infinite-dimensional.

(5) *There are isomorphisms as G-modules for any $0 \leq \ell \leq n+1$ and $\delta = \pm$:*

$$\Pi_{\ell,\delta} \otimes \chi_{+-} \simeq \Pi_{\ell,-\delta},$$

$$\Pi_{\ell,\delta} \otimes \chi_{-+} \simeq \Pi_{n+1-\ell,\delta},$$

$$\Pi_{\ell,\delta} \otimes \chi_{--} \simeq \Pi_{n+1-\ell,-\delta}.$$

(6) *Every $\Pi_{\ell,\delta}$ $(0 \leq \ell \leq n+1, \delta = \pm)$ is unitarizable and self-dual.*

(7) *For n odd, there are exactly two inequivalent discrete series representations of $G = O(n+1, 1)$ with infinitesimal character ρ_G. Their smooth representations are given by*

$$\{\Pi_{\frac{n+1}{2},\delta} : \delta = \pm\}.$$

All the other representations in the list (2) are nontempered representations of G.

(8) *For n even, there are exactly four inequivalent irreducible tempered representations of $G = O(n+1, 1)$ with infinitesimal character ρ_G. Their smooth representations are given by*

$$\{\Pi_{\frac{n}{2},\delta}, \Pi_{\frac{n}{2}+1,\delta} : \delta = \pm\}.$$

(9) *Irreducible and unitarizable $(\mathfrak{g}, K)$-modules with nonzero $(\mathfrak{g}, K)$-cohomologies are exactly given as the set of the underlying $(\mathfrak{g}, K)$-modules of $\Pi_{\ell,\delta}$ $(0 \leq \ell \leq n+1, \delta = \pm)$.*

The exact sequences in Theorem 2.20 (1) leads us to a labeling of the finite set $\mathrm{Irr}(G)_\rho$ as follows:

Definition 2.21 (standard sequence). Let $G = O(n+1, 1)$ and $n = 2m$ or $2m - 1$. We refer to the sequence

$$\Pi_{0,+} , \ \Pi_{1,+} , \ \ldots , \ \Pi_{m-1,+} , \ \Pi_{m,+}$$

as the *standard sequence starting with the trivial one-dimensional representation* $\Pi_{0,+} = \mathbf{1}$. Likewise, we refer to the sequence

$$\Pi_{0,-} , \ \Pi_{1,-} , \ \ldots , \ \Pi_{m-1,-} , \ \Pi_{m,-}$$

as the standard sequence starting with the one-dimensional representation $\Pi_{0,-} = \chi_{+-}$. Sometimes we suppress the subscript $+$ and write Π_i for $\Pi_{i,+}$ for simplicity.

More generally, we shall define the standard sequence starting with other irreducible finite-dimensional representations of G in Chapter 13, see Definition 13.2 and Example 13.5. An analogous sequence, which we refer to as the Hasse sequence, will be defined also in Chapter 13, see Definition-Theorem 13.1.

We give some remarks on the proof of Theorem 2.20. Basic references are [9, 11, 35]. Theorem 2.20 (1) generalizes the results proved in Borel and Wallach [9, pp. 128–129 in the new edition; p. 192 in the old edition] for the identity component group $G_0 = SO_0(n + 1, 1)$. (Unfortunately and confusingly the restriction of our representations $I_+(i, i)$ to the connected component G_0 are denoted there by I_i when $n \neq 2i$.) See also Collingwood [11, Chap. 5, Sect. 2] for the identity component group G_0; [35, p. 20] for the disconnected group $G = O(n + 1, 1)$.

For the relationship between principal series representations of G and of its identity component group G_0, we recall from [42, Chap. 5] the following.

Lemma 2.22. *For $G = O(n + 1, 1)$, let $P_0 := P \cap G_0$. Then P_0 is connected, and is a minimal parabolic subgroup of G_0. Then we have a natural bijection:*

$$G_0/P_0 \xrightarrow{\sim} G/P \ (\simeq S^n).$$

Then we can derive results for the disconnected group G from those for the connected group G_0 and *vice versa* by using the action of the Pontrjagin dual $(G/G_0)\widehat{}$ of the component group G/G_0 and the classical branching law $O(N) \downarrow SO(N)$ (Section 2.2.2). In Appendix II (Chapter 15) we discuss restrictions of representations of $O(n + 1, 1)$ with respect to $SO(n + 1, 1)$ in the same spirit.

In Proposition 14.44 of Appendix I, we will give a description of the underlying $(\mathfrak{g}, K)$-modules $(\Pi_{i,\pm})_K$ of the G-irreducible subquotients $\Pi_{i,\pm}$ in terms of the so-called $A_\mathfrak{q}(\lambda)$-modules, *i.e.,* cohomologically induced representations from one-dimensional representations of a θ-stable parabolic subalgebra $\mathfrak{q}$.

By using the description, Theorem 2.20 (9) follows readily from results of Vogan and Zuckerman [61], see Proposition 14.45 in Appendix I. The unitarizability of the irreducible subquotients $\Pi_{i,\pm}$ (Theorem 2.20 (6)) traces back to T. Hirai [16], see also Howe and Tan [17]. Alternatively, the unitarizability in Theorem 2.20 (6) is deduced from the theory on $A_\mathfrak{q}(\lambda)$, see [24, Thm. 0.51].

Remark 2.23. Analogous results for the special orthogonal group $SO(n + 1, 1)$ will be given in Proposition 15.11 in Appendix II, where we denote the group by $\overline{G}$.

Chapter 3
Symmetry Breaking Operators for Principal Series Representations—General Theory

In this chapter we discuss important concepts and properties of symmetry breaking operators from principal series representations $I_\delta(V, \lambda)$ of the orthogonal group $G = O(n + 1, 1)$ to $J_\varepsilon(W, \nu)$ of the subgroup $G' = O(n, 1)$. In particular, we present a classification scheme (Theorem 3.13) of all symmetry breaking operators, which is built on the strategy of the classification in the spherical case [42] and also on a new phenomenon for which we refer to as *sporadic operators* (Section 3.2.3). The classification scheme is carried out in full details for symmetry breaking from principal series representations $I_\delta(V, \lambda)$ of G to $J_\varepsilon(W, \nu)$ of the subgroup G', which will play a crucial role in understanding symmetry breaking of all *irreducible* admissible representations of G having the trivial infinitesimal character (Chapters 4, 11, and 12). Various theorem stated in this chapter will be proved in later chapters, in particular, in Chapter 5.

3.1 Generalities

We refer to nontrivial homomorphisms in

$$\mathrm{Hom}_{G'}(I_\delta(V, \lambda)|_{G'}, J_\varepsilon(W, \nu))$$

as intertwining restriction operators or *symmetry breaking operators*. Here $\delta, \varepsilon \in \mathbb{Z}/2\mathbb{Z}$ in our setting where $(G, G') = (O(n + 1, 1), O(n, 1))$. For a detailed introduction to symmetry breaking operators we refer to [33] and [42, Chaps. 1 and 3].

© Springer Nature Singapore Pte Ltd. 2018
T. Kobayashi, B. Speh, *Symmetry Breaking for Representations of Rank One Orthogonal Groups II*, Lecture Notes in Mathematics 2234,
https://doi.org/10.1007/978-981-13-2901-2_3

3.2 Summary of Results

We keep our setting where $(G, G') = (O(n + 1, 1), O(n, 1))$.

For $(\sigma, V) \in \widehat{O(n)}$, $\delta \in \{\pm\}$, and $\lambda \in \mathbb{C}$, we write $I_\delta(V, \lambda)$ for the principal series representation of G as in Section 2.3. Similarly, let (τ, W) be an irreducible representation of $O(n - 1)$, $\varepsilon \in \{\pm\}$, and $\nu \in \mathbb{C}$. We extend the outer tensor product representation

$$W_{\nu, \varepsilon} := W \boxtimes \varepsilon \boxtimes \mathbb{C}_\nu$$

of the direct product group $M'A \simeq O(n - 1) \times O(1) \times \mathbb{R}$ to $P' = M'AN'_+$ by letting N'_+ act trivially. We also write $W_{\nu, \varepsilon} = W \otimes \varepsilon \otimes \mathbb{C}_\nu$ when we regard it as a P'-module. We form a G'-equivariant vector bundle $\mathcal{W}_{\nu, \varepsilon} := G' \times_{P'} W_{\nu, \varepsilon}$ over the real flag manifold G'/P'. The principal series representation $J_\varepsilon(W, \nu)$ of $G' = O(n, 1)$ is defined to be the induced representation $\operatorname{Ind}_{P'}^{G'}(W_{\nu, \varepsilon})$ on the space $C^\infty(G'/P', \mathcal{W}_{\nu, \varepsilon})$ of smooth sections for the vector bundle.

For $(\sigma, V) \in \widehat{O(n)}$ and $(\tau, W) \in \widehat{O(n - 1)}$, we set

$$[V : W] := \dim_\mathbb{C} \operatorname{Hom}_{O(n-1)}(V|_{O(n-1)}, W). \tag{3.1}$$

If we want to emphasize the subgroup, we also write $[V|_{O(n-1)} : W]$ for $[V : W]$. We recall from Fact 2.12 on the classical branching rule for the restriction $O(N) \downarrow O(N - 1)$ that the multiplicity $[V : W]$ is either 0 or 1.

3.2.1 Symmetry Breaking Operators When $[V : W] \neq 0$

Suppose $[V : W] \neq 0$. In this case we prove the existence of nonzero symmetry breaking operators for all $\lambda, \nu \in \mathbb{C}$ and for all signatures $\delta, \varepsilon \in \{\pm\}$:

Theorem 3.1 (existence of symmetry breaking operators, see Theorem 5.42). *Suppose $(\sigma, V) \in \widehat{O(n)}$ and $(\tau, W) \in \widehat{O(n - 1)}$. Assume $[V : W] \neq 0$. Then we have*

$$\dim_\mathbb{C} \operatorname{Hom}_{G'}(I_\delta(V, \lambda)|_{G'}, J_\varepsilon(W, \nu)) \geq 1 \quad \text{for all } \delta, \varepsilon \in \mathbb{Z}/2\mathbb{Z}, \lambda, \nu \in \mathbb{C}.$$

Theorem 3.1 is proved in Section 5.10 by constructing symmetry breaking operators: generic ones are nonlocal (*e.g.* integral operators) see Theorem 3.9 below, whereas a few are local operators (*i.e.* differential operators, see Theorem 3.5).

Definition 3.2. We say that the quadruple $(\lambda, \nu, \delta, \varepsilon)$ is a *generic parameter* if $(\lambda, \nu) \in \mathbb{C}^2$ and $\delta, \varepsilon \in \{\pm\}$ satisfy

$$\begin{cases} \nu - \lambda \notin 2\mathbb{N} & \text{when} \quad \delta\varepsilon = +; \\ \nu - \lambda \notin 2\mathbb{N} + 1 & \text{when} \quad \delta\varepsilon = -. \end{cases} \tag{3.2}$$

We recall from (1.3) that the set of "special parameters" is given as the complement of "generic parameters", namely,

$$\Psi_{sp} = \Big\{ (\lambda, \nu, \delta, \varepsilon) \in \mathbb{C}^2 \times \{\pm\}^2 : \nu - \lambda \in 2\mathbb{N} \qquad \text{when } \delta\varepsilon = +$$

$$\text{or } \nu - \lambda \in 2\mathbb{N} + 1 \quad \text{when } \delta\varepsilon = - \Big\}. \qquad (3.3)$$

In the case $[V : W] \neq 0$, we also prove the following "generic multiplicity-one theorem", which extends [42, Thm. 1.1] in the scalar case ($V = W = \mathbb{C}$).

Theorem 3.3 (generic multiplicity-one theorem). *Suppose* $(\sigma, V) \in \widehat{O(n)}$, (τ, W) $\in \widehat{O(n-1)}$ *with* $[V : W] \neq 0$. *If* $(\lambda, \nu, \delta, \varepsilon) \in \mathbb{C}^2 \times \{\pm\}^2$ *satisfies the generic parameter condition, namely,* $(\lambda, \nu, \delta, \varepsilon) \notin \Psi_{sp}$, *then*

$$\dim_{\mathbb{C}} \operatorname{Hom}_{G'}(I_\delta(V, \lambda)|_{G'}, J_\varepsilon(W, \nu)) = 1.$$

Theorem 3.3 gives a stronger estimate than what the existing general theory guarantees:

- the dimension ≤ 1 if both $I_\delta(V, \lambda)$ and $J_\varepsilon(W, \nu)$ are irreducible [55],
- the dimension is uniformly bounded with respect to $\sigma, \tau, \delta, \varepsilon, \lambda, \nu$ [39].

We note that $I_\delta(V, \lambda)$ or $J_\varepsilon(W, \nu)$ can be reducible even if $(\lambda, \nu, \delta, \varepsilon) \in \Psi_{sp}$. Theorem 3.3 will be proved in a strengthened form by giving an explicit generator (see Theorem 5.41 in Section 5.10).

3.2.2 Differential Symmetry Breaking Operators When $[V : W] \neq 0$

We realize the principal series representations $I_\delta(V, \lambda)$ and $J_\varepsilon(W, \nu)$ in the Fréchet spaces $C^\infty(G/P, \mathcal{V}_{\lambda,\delta})$ and $C^\infty(G'/P', \mathcal{W}_{\nu,\varepsilon})$.

Definition 3.4 (differential symmetry breaking operator). A linear map

$$D \colon C^\infty(G/P, \mathcal{V}_{\lambda,\delta}) \to C^\infty(G'/P', \mathcal{W}_{\nu,\varepsilon})$$

is called a *differential symmetry breaking operator* if D is a differential operator with respect to the inclusion $G'/P' \hookrightarrow G/P$ and D intertwines the action of the subgroup G'. See Definition 6.3 in Chapter 6 for the notion of differential operators between two different manifolds. We denote by

$$\operatorname{Diff}_{G'}(I_\delta(V, \lambda)|_{G'}, J_\varepsilon(W, \nu))$$

the subspace of $\operatorname{Hom}_{G'}(I_\delta(V, \lambda)|_{G'}, J_\varepsilon(W, \nu))$ consisting of differential symmetry breaking operators.

We retain the assumption that $[V : W] \neq 0$. We give a necessary and sufficient condition for the existence of nonzero differential symmetry breaking operators:

Theorem 3.5 (existence of differential symmetry breaking operators). *Suppose* $(\sigma, V) \in \widehat{O(n)}$ *and* $(\tau, W) \in \widehat{O(n-1)}$ *satisfy* $[V : W] \neq 0$. *Then the following two conditions on the parameters* $\lambda, \nu \in \mathbb{C}$ *and* $\delta, \varepsilon \in \{\pm\}$ *are equivalent:*

(i) *The quadruple* $(\lambda, \nu, \delta, \varepsilon)$ *does not satisfy the generic parameter condition* (3.2), *namely,* $(\lambda, \nu, \delta, \varepsilon) \in \Psi_{\mathrm{sp}}$.
(ii) $\mathrm{Diff}_{G'}(I_\delta(V, \lambda)|_{G'}, J_\varepsilon(W, \nu)) \neq \{0\}$.

We shall prove Theorem 3.5 in Chapter 6, see Theorem 6.1.

3.2.3 Sporadic Symmetry Breaking Operators When $[V : W] = 0$

This section treats the case $[V : W] = 0$. In the holomorphic setting, we found in [40] a phenomenon that all symmetry breaking operators are given by differential operators (*localness theorem*). This phenomenon does not occur in the real setting if both V and W are the trivial one-dimensional representations [42]. However, we shall see that this phenomenon may occur in the real setting for vector bundles. Indeed, the following theorem shows that there may exist *sporadic* symmetry breaking operators which are differential operators in the case $[V : W] = 0$:

Theorem 3.6 (localness theorem). *Assume* $[V : W] = 0$. *Then*

$$\mathrm{Hom}_{G'}(I_\delta(V, \lambda)|_{G'}, J_\varepsilon(W, \nu)) = \mathrm{Diff}_{G'}(I_\delta(V, \lambda)|_{G'}, J_\varepsilon(W, \nu))$$

for all $(\lambda, \nu, \delta, \varepsilon) \in \mathbb{C}^2 \times \{\pm\}^2$, *that is, any symmetry breaking operator (if exists)*

$$C^\infty(G/P, \mathcal{V}_{\lambda, \delta}) \to C^\infty(G'/P', \mathcal{W}_{\nu, \varepsilon})$$

is a differential operator.

Theorem 3.6 is proved in Section 5.5. We call such operators *sporadic* because there is no regular symmetry breaking operator if $[V : W] = 0$, see Theorem 3.9 below. Another localness theorem is formulated in Theorem 3.13 (2-b) (see also Proposition 6.16 in Chapter 6) under the assumption that the parameter $(\lambda, \nu) \in \mathbb{C}^2$ satisfies $\nu - \lambda \in \mathbb{N}$.

Example 3.7. Suppose $(V, W) = (\bigwedge^i(\mathbb{C}^n), \bigwedge^j(\mathbb{C}^{n-1}))$. Then $[V : W] \neq 0$ if and only if $j = i - 1$ or i. Hence Theorem 3.6 tells that there exists a nonlocal symmetry breaking operators $I_\delta(i, \lambda) \to J_\varepsilon(j, \nu)$ only if $j \in \{i - 1, i\}$. (In fact, this is also a sufficient condition, see Theorem 9.1.) On the other hand, there exist nontrivial differential symmetry breaking operators for some $(\lambda, \nu) \in \mathbb{C}^2$ if and only if $j \in \{i-2, i-1, i, i+1\}$, as is seen from the complete classification of differential symmetry breaking operators (Fact 3.22). Thus there exist sporadic (differential) symmetry breaking operators when $j = i - 2$ or $i + 1$.

Remark 3.8. The assumption $[V : W] \neq 0$ in Theorems 3.3 and 3.5 is *not* an intertwining property for $M' = M \cap G' \simeq O(n-1) \times O(1)$ but for the subgroup $O(n-1)$ which is of index two in M'. We note that for $V_\delta := V \boxtimes \delta \in \widehat{M}$ and $W_\varepsilon := W \boxtimes \varepsilon \in \widehat{M'}$,

$$\mathrm{Hom}_{M'}(V_\delta|_{M'}, W_\varepsilon) \neq \{0\} \text{ if and only if } [V : W] \neq 0 \text{ and } \delta = \varepsilon.$$

Indeed the condition $\delta = \varepsilon$ is not included in the assumption of Theorem 3.3 on the construction of regular symmetry breaking operators. The reason is clarified in Theorem 3.9 in the next subsection.

3.2.4 Existence Condition for Regular Symmetry Breaking Operators

A *regular symmetry breaking operator* is an "opposite" notion to a differential symmetry breaking operator in the sense that the support of its distribution kernel contains an interior point in the real flag manifold, see [42, Def. 3.3]. (See also Definition 5.10 in our special setting.) In [42, Cor. 3.6] we give a necessary condition for the existence of regular symmetry breaking operators in the general setting. This condition is also sufficient in our setting:

Theorem 3.9 (existence of regular symmetry breaking operators). *Suppose* $V \in \widehat{O(n)}$ *and* $W \in \widehat{O(n-1)}$. *Then the following three conditions on the pair* (V, W) *are equivalent:*

(i) $[V : W] \neq 0$.

(ii) *There exists a nonzero regular symmetry breaking operator from the G-module* $I_\delta(V, \lambda)$ *to the G'-module* $J_\varepsilon(W, \nu)$ *for some* $(\lambda, \nu, \delta, \varepsilon) \in \mathbb{C}^2 \times \{\pm\}^2$.

(iii) *For any* $(\delta, \varepsilon) \in \{\pm\}^2$, *there is an open dense subset* U *in* $\mathbb{C}^2$ *such that a nonzero regular symmetry breaking operator exists from* $I_\delta(V, \lambda)$ *to* $J_\varepsilon(W, \nu)$ *for all* $(\lambda, \nu) \in U$.

The proof will be given in Section 5.7. The open dense subset U is explicitly given in Proposition 5.39.

3.2.5 Integral Operators, Analytic Continuation, and Normalization Factors

For an explicit construction of *regular* symmetry breaking operators, we use the reflection map ψ_n defined as follows:

$$\psi_n : \mathbb{R}^n - \{0\} \to O(n), \quad x \mapsto I_n - \frac{2x^t x}{|x|^2}. \tag{3.4}$$

Then $\psi_n(x)$ gives the reflection $\psi_n(x)$ with respect to the hyperplane $\{y \in \mathbb{R}^n : (x, y) = 0\}$. Clearly, we have

$$\psi_n(x) = \psi_n(-x), \quad \psi_n(x)^2 = I_n, \quad \text{and} \quad \det \psi_n(x) = -1. \tag{3.5}$$

Suppose $(\sigma, V) \in \widehat{O(n)}$ and $(\tau, W) \in \widehat{O(n-1)}$. For the construction of regular symmetry breaking operators, we need the condition $[V : W] \neq 0$, see Theorem 3.9. So let us assume $[V : W] \neq 0$. We fix a nonzero $O(n-1)$-homomorphism

$$\mathrm{pr}_{V \to W} : V \to W,$$

which is unique up to scalar multiplication by Schur's lemma because $[V : W] = 1$. We introduce a smooth map

$$R^{V,W} : \mathbb{R}^n - \{0\} \to \mathrm{Hom}_{\mathbb{C}}(V, W)$$

by

$$R^{V,W} := \mathrm{pr}_{V \to W} \circ \sigma \circ \psi_n. \tag{3.6}$$

In what follows, we use the coordinates $(x, x_n) \in \mathbb{R}^n = \mathbb{R}^{n-1} \oplus \mathbb{R}$ where $x = (x_1, \cdots, x_{n-1})$, and the n-th coordinate x_n will play a special role.

We set

$$\widetilde{\mathcal{A}}_{\lambda,\nu,+}^{V,W} := \frac{1}{\Gamma(\frac{\lambda+\nu-n+1}{2})\Gamma(\frac{\lambda-\nu}{2})} (|x|^2 + x_n^2)^{-\nu} |x_n|^{\lambda+\nu-n} R^{V,W}(x, x_n), \tag{3.7}$$

$$\widetilde{\mathcal{A}}_{\lambda,\nu,-}^{V,W} := \frac{1}{\Gamma(\frac{\lambda+\nu-n+2}{2})\Gamma(\frac{\lambda-\nu+1}{2})} (|x|^2 + x_n^2)^{-\nu} |x_n|^{\lambda+\nu-n} \mathrm{sgn}\, x_n R^{V,W}(x, x_n). \tag{3.8}$$

Theorem 3.10 (regular symmetry breaking operators). *Suppose $[V : W] \neq 0$ and $\gamma \in \{\pm\}$. Then the distributions $\widetilde{\mathcal{A}}_{\lambda,\nu,\gamma}^{V,W}$, initially defined as $\mathrm{Hom}_{\mathbb{C}}(V, W)$-valued locally integrable functions on $\mathbb{R}^n$ for $\mathrm{Re}\,\lambda \gg |\mathrm{Re}\,\nu|$, extends to P'-invariant*

elements in $\mathcal{D}'(G/P, \mathcal{V}^*_{\lambda,\delta}) \otimes W_{\nu,\varepsilon}$ for all $(\lambda, \nu) \in \mathbb{C}^2$ and $\delta, \varepsilon \in \{\pm\}$ with $\delta\varepsilon = \gamma$.
Then the distributions $\widetilde{\mathbb{A}}^{V,W}_{\lambda,\nu,\gamma}$ induce a family of symmetry breaking operators

$$\widetilde{\mathbb{A}}^{V,W}_{\lambda,\nu,\gamma}: C^\infty(G/P, \mathcal{V}_{\lambda,\delta}) \to C^\infty(G'/P', \mathcal{W}_{\nu,\varepsilon}),$$

which depends holomorphically on (λ, ν) in the entire $\mathbb{C}^2$.

Remark 3.11. The denominator in (3.7) is different from the product of the denominators of the two distributions $\dfrac{(|x|^2+x_n^2)^{-\nu}}{\Gamma(\frac{n-\nu}{2})}$ and $\dfrac{|x_n|^{\lambda+\nu-n}}{\Gamma(\frac{\lambda+\nu-n+1}{2})}$ on $\mathbb{R}^n$ that depend holomorphically on (λ, ν) in the entire $\mathbb{C}^2$. In fact the product

$$\frac{(|x|^2 + x_n^2)^{-\nu}}{\Gamma(\frac{n-\nu}{2})} \times \frac{|x_n|^{\lambda+\nu-n}}{\Gamma(\frac{\lambda+\nu-n+1}{2})} \tag{3.9}$$

does not always make sense as distributions on $\mathbb{R}^n$. For instance, if $(\lambda, \nu) = (-1, n)$, then the multiplication (3.9) means the multiplication (up to nonzero scalar multiplication) of the Dirac delta functions $\delta(x_1, \cdots, x_n)$ by $\delta(x_n)$, which is not well-defined in the usual sense.

Theorem 3.10 will be proved in Section 5.6.

We prove in Theorem 3.19 that the normalization is optimal for $(V, W) = (\bigwedge^i(\mathbb{C}^n), \bigwedge^j(\mathbb{C}^{n-1}))$ in the sense that the zeros of $\widetilde{\mathbb{A}}^{V,W}_{\lambda,\nu,\pm}$ are of codimension > 1 in the parameter space of (λ, ν), namely, discrete in $\mathbb{C}^2$ in our setting. For the general (V, W), we shall give an upper and lower estimate of the null set of the symmetry breaking operators $\widetilde{\mathbb{A}}^{V,W}_{\lambda,\nu,+}$ and $\widetilde{\mathbb{A}}^{V,W}_{\lambda,\nu,-}$ in Theorem 3.15.

3.3 Classification Scheme of Symmetry Breaking Operators: General Case

In this section, we give a general scheme for the classification of all symmetry breaking operators $I_\delta(V, \lambda)|_{G'} \to J_\varepsilon(W, \nu)$ between the two principal series representations of G and the subgroup G' in full generality where $(\sigma, V) \in \widehat{O(n)}$ and $(\tau, W) \in \widehat{O(n-1)}$.

We begin with conditions on the parameter $(\lambda, \nu, \delta, \varepsilon)$ for the existence of differential symmetry breaking operators.

Theorem 3.12 (existence of differential symmetry breaking operators).

(1) (Theorem 5.21) *Suppose* $\lambda, \nu \in \mathbb{C}$ *and* $\delta, \varepsilon \in \{\pm\}$ *satisfy the generic parameter condition* (3.2). *Then,*

$$\mathrm{Diff}_{G'}(I_\delta(V, \lambda)|_{G'}, J_\varepsilon(W, \nu)) = \{0\}$$

for any $(\sigma, V) \in \widehat{O(n)}$ *and* $(\tau, W) \in \widehat{O(n-1)}$.

(2) (Theorem 6.1) *Suppose* $[V : W] \neq 0$. *Then the converse statement holds, namely, if* $(\lambda, \nu, \delta, \varepsilon) \in \Psi_{\mathrm{sp}}$ *(see* (1.3)), *then*

$$\mathrm{Diff}_{G'}(I_\delta(V, \lambda)|_{G'}, J_\varepsilon(W, \nu)) \neq \{0\}.$$

We give a proof for the first statement of Theorem 3.12 in Section 5.4, and the second statement in Section 6.7. Keeping Theorem 3.12 on differential symmetry breaking operators in mind, we state a general scheme for the classification of *all* symmetry breaking operators:

Theorem 3.13 (classification scheme of symmetry breaking operators). *Let* $n \geq 3$, $(\sigma, V) \in \widehat{O(n)}$, $(\tau, W) \in \widehat{O(n-1)}$, $\lambda, \nu \in \mathbb{C}$ *and* $\delta, \varepsilon \in \{\pm\}$.

(1) *Suppose* $[V : W] = 0$. *Then*

$$\mathrm{Hom}_{G'}(I_\delta(V, \lambda)|_{G'}, J_\varepsilon(W, \nu)) = \mathrm{Diff}_{G'}(I_\delta(V, \lambda)|_{G'}, J_\varepsilon(W, \nu)).$$

(2) *Suppose* $[V : W] \neq 0$.

(2-a) (generic case) *Suppose further that* $(\lambda, \nu, \delta, \varepsilon) \notin \Psi_{\mathrm{sp}}$, *namely, it satisfies the generic parameter condition* (3.2). *Then*

$$\mathrm{Hom}_{G'}(I_\delta(V, \lambda)|_{G'}, J_\varepsilon(W, \nu)) = \mathbb{C}\widetilde{\mathbb{A}}^{V,W}_{\lambda,\nu,\delta\varepsilon}.$$

In this case, $\widetilde{\mathbb{A}}^{V,W}_{\lambda,\nu,\delta\varepsilon}$ *is nonzero and is not a differential operator.*

(2-b) (special parameter case I, localness theorem) *Suppose* $\widehat{\mathbb{A}}^{V,W}_{\lambda,\nu,\delta\varepsilon} \neq 0$ *and* $(\lambda, \nu, \delta, \varepsilon) \in \Psi_{\mathrm{sp}}$ *(i.e., does not satisfy the generic parameter condition* (3.2)). *Then any symmetry breaking operator (in particular,* $\widetilde{\mathbb{A}}^{V,W}_{\lambda,\nu,\delta\varepsilon}$) *is a differential operator, hence*

$$\mathrm{Hom}_{G'}(I_\delta(V, \lambda)|_{G'}, J_\varepsilon(W, \nu))$$

$$= \mathrm{Diff}_{G'}(I_\delta(V, \lambda)|_{G'}, J_\varepsilon(W, \nu)) \ni \widetilde{\mathbb{A}}^{V,W}_{\lambda,\nu,\delta\varepsilon}.$$

(2-c) (special parameter case II) *Suppose* $\widetilde{\mathbb{A}}^{V,W}_{\lambda,\nu,\delta\varepsilon} = 0$. *Then* $(\lambda, \nu, \delta, \varepsilon) \in \Psi_{\mathrm{sp}}$, *and the renormalized operator* $\widetilde{\widetilde{\mathbb{A}}}^{V,W}_{\lambda,\nu,\delta\varepsilon}$ *(see Section 5.11.2) gives a nonzero symmetry breaking operator which is not a differential operator.*

We have

$$\mathrm{Hom}_{G'}(I_\delta(V,\lambda)|_{G'}, J_\varepsilon(W,\nu))$$

$$= \mathbb{C}\widetilde{\mathbb{A}}^{V,W}_{\lambda,\nu,\delta\varepsilon} \oplus \mathrm{Diff}_{G'}(I_\delta(V,\lambda)|_{G'}, J_\varepsilon(W,\nu)).$$

In particular,

$$\dim_\mathbb{C} \mathrm{Hom}_{G'}(I_\delta(V,\lambda)|_{G'}, J_\varepsilon(W,\nu)) \geq 2.$$

The first assertion of Theorem 3.13 is a restatement of Theorem 3.6. The case (2-a) is given in Theorem 5.41 and the case (2-b) is in Proposition 6.16. The first statement for the case (2-c) is proved in Theorem 5.45 (1). The direct sum decomposition is given in Corollary 5.46. The last statement follows from the existence of nonzero differential symmetry breaking operators for all special parameters (Theorem 3.12 (2)).

Theorems 3.5 and 3.13 lead us to a vanishing result of symmetry breaking operators as follows:

Corollary 3.14 (vanishing of symmetry breaking operators). *Let $(\sigma, V) \in \widehat{O(n)}$, $(\tau, W) \in \widehat{O(n-1)}$, $\lambda, \nu \in \mathbb{C}$ and $\delta, \varepsilon \in \{\pm\}$. If $[V : W] = 0$ and (λ, ν) satisfies the generic parameter condition* (3.2), *then*

$$\mathrm{Hom}_{G'}(I_\delta(V,\lambda)|_{G'}, J_\varepsilon(W,\nu)) = \{0\}.$$

Proof. By Theorem 3.13 (1), we have

$$\mathrm{Hom}_{G'}(I_\delta(V,\lambda)|_{G'}, J_\varepsilon(W,\nu)) = \mathrm{Diff}_{G'}(I_\delta(V,\lambda)|_{G'}, J_\varepsilon(W,\nu))$$

because $[V : W] = 0$. In turn, the right-hand side reduces to zero by Theorem 3.5 because of the generic parameter condition (3.2).   □

Theorem 3.13 gives a classification of symmetry breaking operators up to the following two problems:

- the location of zeros of the normalized regular symmetry breaking operator $\widetilde{\mathbb{A}}^{V,W}_{\lambda,\nu,\gamma}$;
- the classification of *differential* symmetry breaking operators.

For $(V, W) = (\bigwedge^i(\mathbb{C}^n), \bigwedge^j(\mathbb{C}^{n-1}))$, these two problems are solved explicitly in Theorem 3.19 and Fact 3.22, respectively, and thus we accomplish the complete classification of symmetry breaking operators. This will be stated in Theorem 3.25 (multiplicity formula) and in Theorem 3.26 (explicit generators).

3.4 Summary: Vanishing of Regular Symmetry Breaking Operators $\widetilde{\mathbb{A}}^{V,W}_{\lambda,\nu,\pm}$

As we have seen in the classification scheme (Theorem 3.13) for all symmetry breaking operators, the parameter $(\lambda, \nu, \delta, \varepsilon)$ for which the (generically) regular symmetry breaking operator $\widetilde{\mathbb{A}}^{V,W}_{\lambda,\nu,\pm}$ vanishes plays a crucial role in the classification theory. For $(\lambda, \nu, \delta, \varepsilon) \in \Psi_{\mathrm{sp}}$, we noted:

- when $\widetilde{\mathbb{A}}^{V,W}_{\lambda_0,\nu_0,\pm} = 0$, we can construct a nonzero symmetry breaking operator $\widetilde{\widetilde{\mathbb{A}}}^{V,W}_{\lambda_0,\nu_0,\pm}$ by "renormalization" which is *not* a differential operator (Theorem 5.45);
- when $\widetilde{\mathbb{A}}^{V,W}_{\lambda_0,\nu_0,\pm} \neq 0$, we prove a *localness theorem* asserting that all symmetry breaking operators are differential operators (Proposition 6.16).

We obtain a condition for the (non) vanishing of $\widetilde{\mathbb{A}}^{V,W}_{\lambda,\nu,\pm}$ as follows. Using the same notation as in [42, Chap. 1], we define the following two subsets in $\mathbb{Z}^2$:

$$L_{\mathrm{even}} := \{\, (-i, -j) : 0 \leq j \leq i \text{ and } i \equiv j \qquad \mod 2 \,\}, \tag{3.10}$$

$$L_{\mathrm{odd}} := \{\, (-i, -j) : 0 \leq j \leq i \text{ and } i \equiv j+1 \mod 2 \,\}. \tag{3.11}$$

Theorem 3.15. *Let $(\sigma, V) \in \widehat{O(n)}$ and $(\tau, W) \in \widehat{O(n-1)}$ with $[V:W] \neq 0$.*

(1) There exists $N(\sigma) \in \mathbb{N}$ such that

$$\widetilde{\mathbb{A}}^{V,W}_{\lambda,\nu,+} = 0 \qquad \text{if } (\lambda, \nu) \in L_{\mathrm{even}} \text{ and } \nu \leq -N(\sigma),$$

$$\widetilde{\mathbb{A}}^{V,W}_{\lambda,\nu,-} = 0 \qquad \text{if } (\lambda, \nu) \in L_{\mathrm{odd}} \text{ and } \nu \leq -N(\sigma).$$

(2) If $\widetilde{\mathbb{A}}^{V,W}_{\lambda,\nu,+} = 0$ then $\nu - \lambda \in 2\mathbb{N}$; if $\widetilde{\mathbb{A}}^{V,W}_{\lambda,\nu,-} = 0$ then $\nu - \lambda \in 2\mathbb{N}+1$.

Remark 3.16. We shall show in Lemma 5.35 that $N(\sigma)$ can be taken to be $\ell(\sigma)$, as defined in (2.22).

Theorem 3.15 (2) is a part of Theorem 3.13 (2), and will be proved in Section 5.8.

Combining Theorems 3.13 and 3.15, we see that there exist infinitely many $(\lambda, \nu) \in \mathbb{C}^2$ such that the multiplicity $m(I_\delta(V, \lambda), J_\varepsilon(W, \nu)) > 1$ as follows:

Corollary 3.17. *Let $(\sigma, V) \in \widehat{O(n)}$ and $(\tau, W) \in \widehat{O(n-1)}$ satisfy $[V:W] \neq 0$. If*

$$(\lambda, \nu) \in \begin{cases} L_{\mathrm{even}} \cap \{\nu \leq -N(\sigma)\} & \text{for } \delta\varepsilon = +, \\ L_{\mathrm{odd}} \cap \{\nu \leq -N(\sigma)\} & \text{for } \delta\varepsilon = -, \end{cases}$$

then we have

$$\dim_{\mathbb{C}} \mathrm{Hom}_{G'}(I_\delta(V, \lambda)|_{G'}, J_\varepsilon(W, \nu)) > 1.$$

By Theorem 3.15, we get readily the following corollary, to which we shall return in Chapter 13 (see Example 13.32).

Corollary 3.18. *Suppose that* $\widetilde{\mathbb{A}}^{V,W}_{\lambda,\nu,\delta} = 0$. *Then* $\widetilde{\mathbb{A}}^{V,W}_{n-\lambda,n-1-\nu,\delta} \neq 0$.

Theorem 3.15 means that

$$L_{\text{even}} \cap \{\nu \leq -N(\sigma)\} \subset \{(\lambda,\nu) \in \mathbb{C}^2 : \widetilde{\mathbb{A}}^{V,W}_{\lambda,\nu,+} = 0\}$$

$$\subset \{(\lambda,\nu) \in \mathbb{C}^2 : \nu - \lambda \in 2\mathbb{N}\},$$

$$L_{\text{odd}} \cap \{\nu \leq -N(\sigma)\} \subset \{(\lambda,\nu) \in \mathbb{C}^2 : \widetilde{\mathbb{A}}^{V,W}_{\lambda,\nu,-} = 0\}$$

$$\subset \{(\lambda,\nu) \in \mathbb{C}^2 : \nu - \lambda \in 2\mathbb{N}+1\}.$$

We shall determine in Theorem 3.19 the set $\{(\lambda,\nu) \in \mathbb{C}^2 : \widetilde{\mathbb{A}}^{V,W}_{\lambda,\nu,\gamma} = 0\}$ for $\gamma = \pm$ in the special case where $(V,W) = (\bigwedge^i(\mathbb{C}^n), \bigwedge^j(\mathbb{C}^{n-1}))$. If σ is the i-th exterior representation $\sigma^{(i)}$ on $\bigwedge^i(\mathbb{C}^n)$, then we can take $N(\sigma)$ to be 0 if $i = 0$ or n; to be 1 if $1 \leq i \leq n-1$. In this case, the left inclusion is almost a bijection. On the other hand, concerning the right inclusions, we refer to Theorem 3.13 (2-b), which will be proved in Section 6.8, see Proposition 6.16.

3.5   The Classification of Symmetry Breaking Operators for Differential Forms

Let $(G,G') = (O(n+1,1), O(n,1))$ with $n \geq 3$ as before. We consider the special case

$$(V,W) = (\textstyle\bigwedge^i(\mathbb{C}^n), \bigwedge^j(\mathbb{C}^{n-1})).$$

Then the corresponding principal series representations $I_\delta(V,\lambda)$ of G and $J_\varepsilon(W,\nu)$ of the subgroup G' are denoted by $I_\delta(i,\lambda)$ and $J_\varepsilon(j,\nu)$, respectively. In this section we summarize the complete classification of symmetry breaking operators from the G-module $I_\delta(i,\lambda)$ to the G'-module $J_\varepsilon(j,\nu)$. The main results are stated in Theorems 3.25 and 3.26. Our results rely on the vanishing condition of the normalized regular symmetry breaking operators $\widetilde{\mathbb{A}}^{i,j}_{\lambda,\nu,\gamma}$ (Theorem 3.19) and the classification of differential symmetry breaking operators (Fact 3.22).

3.5.1 Vanishing Condition for the Regular Symmetry Breaking Operators $\widetilde{\mathbb{A}}^{i,j}_{\lambda,\nu,\gamma}$

We apply the general construction of the (normalized) symmetry breaking operators $\widetilde{\mathbb{A}}^{V,W}_{\lambda,\nu,\gamma}$ in Theorem 3.10 to the pair of representations $(V, W) = (\bigwedge^i(\mathbb{C}^n), \bigwedge^j(\mathbb{C}^{n-1}))$. Then we obtain (normalized) symmetry breaking operators, to be denoted by $\widetilde{\mathbb{A}}^{i,j}_{\lambda,\nu,\gamma}$, that depend holomorphically on (λ, ν) in the entire complex plane $\mathbb{C}^2$ if $j \in \{i-1, i\}$ and $\gamma \in \{\pm\}$, see Theorem 9.2.

We determine the zero set of $\widetilde{\mathbb{A}}^{i,j}_{\lambda,\nu,\gamma}$ explicitly as follows:

Theorem 3.19 (zeros of regular symmetry breaking operators $\widetilde{\mathbb{A}}^{i,j}_{\lambda,\nu,\pm}$).

(1) *For* $0 \le i \le n-1$,

$$\{(\lambda, \nu) \in \mathbb{C}^2 : \widetilde{\mathbb{A}}^{i,i}_{\lambda,\nu,+} = 0\}$$

$$= \begin{cases} L_{\mathrm{even}} & \text{if } i = 0, \\ (L_{\mathrm{even}} - \{\nu = 0\}) \cup \{(i,i)\} & \text{if } 1 \le i \le n-1. \end{cases}$$

(2) *For* $1 \le i \le n$,

$$\{(\lambda, \nu) \in \mathbb{C}^2 : \widetilde{\mathbb{A}}^{i,i-1}_{\lambda,\nu,+} = 0\}$$

$$= \begin{cases} (L_{\mathrm{even}} - \{\nu = 0\}) \cup \{(n-i, n-i)\} & \text{if } 1 \le i \le n-1, \\ L_{\mathrm{even}} & \text{if } i = n. \end{cases}$$

(3) *For* $0 \le i \le n-1$,

$$\{(\lambda, \nu) \in \mathbb{C}^2 : \widetilde{\mathbb{A}}^{i,i}_{\lambda,\nu,-} = 0\}$$

$$= \begin{cases} L_{\mathrm{odd}} & \text{if } i = 0, \\ L_{\mathrm{odd}} - \{\nu = 0\} & \text{if } 1 \le i \le n-1. \end{cases}$$

(4) *For* $1 \le i \le n$,

$$\{(\lambda, \nu) \in \mathbb{C}^2 : \widetilde{\mathbb{A}}^{i,i-1}_{\lambda,\nu,-} = 0\}$$

$$= \begin{cases} L_{\mathrm{odd}} - \{\nu = 0\} & \text{if } 1 \le i \le n-1, \\ L_{\mathrm{odd}} & \text{if } i = n. \end{cases}$$

Theorem 3.19 will be proved in Section 9.2 by using the residue formula of $\widetilde{\mathbb{A}}^{i,j}_{\lambda,\nu,\pm}$ [34].

A special case of Theorem 3.19 includes the following.

Example 3.20.

(1) For $0 \le i \le n$, $\widetilde{\mathbb{A}}^{i,i}_{i,i,+} = 0$ and $\widetilde{\mathbb{A}}^{i,i}_{n-i,n-i-1,+} \ne 0$.

(2) For $0 \le i \le n-1$, $\widetilde{\mathbb{A}}^{n-i,n-i-1}_{i,i,+} = 0$ and $\widetilde{\mathbb{A}}^{n-i,n-i-1}_{n-i,n-i-1,+} \ne 0$.

Remark 3.21. In the case $i = 0$, $\widetilde{\mathbb{A}}^{i,i}_{\lambda,\nu,+}$ is the scalar-valued symmetry breaking operator induced from the scalar-valued distribution $\widetilde{A}_{\lambda,\nu,+}$, as we recall from (5.40). Thus the case $i = 0$ in (1) was proved in [42, Thm. 8.1].

3.5.2 Differential Symmetry Breaking Operators

We review from [35] the notation of conformally equivariant *differential* operators $\mathcal{E}^i(S^n) \to \mathcal{E}^j(S^{n-1})$, namely, *differential* symmetry breaking operators $I_\delta(V, \lambda)|_{G'} \to J_\varepsilon(W, \nu)$ with $(V, W) = (\bigwedge^i(\mathbb{C}^n), \bigwedge^j(\mathbb{C}^{n-1}))$. The complete classification of *differential* symmetry breaking operators was recently accomplished in [35, Thm. 2.8] based on the F-method [30].

Fact 3.22 (classification of differential symmetry breaking operators). *Let $n \ge 3$. Suppose $0 \le i \le n$, $0 \le j \le n-1$, $\lambda, \nu \in \mathbb{C}$, and $\delta, \varepsilon \in \{\pm\}$. Then the following three conditions on 6-tuple $(i, j, \lambda, \nu, \delta, \varepsilon)$ are equivalent.*

(i) $\mathrm{Diff}_{G'}(I_\delta(i, \lambda)|_{G'}, J_\varepsilon(j, \nu)) \ne \{0\}$.
(ii) $\dim_\mathbb{C} \mathrm{Diff}_{G'}(I_\delta(i, \lambda)|_{G'}, J_\varepsilon(j, \nu)) = 1$.
(iii) $\nu - \lambda \in \mathbb{N}$, $(-1)^{\nu-\lambda} = \delta\varepsilon$, *and one of the following conditions holds:*

    (a) $j = i - 2$, $2 \le i \le n-1$, $(\lambda, \nu) = (n-i, n-i+1)$;
    (a') $(i, j) = (n, n-2)$, $-\lambda \in \mathbb{N}$, $\nu = 1$;
    (b) $j = i - 1$, $1 \le i \le n$;
    (c) $j = i$, $0 \le i \le n-1$;
    (d) $j = i + 1$, $1 \le i \le n-2$, $(\lambda, \nu) = (i, i+1)$;
    (d') $(i, j) = (0, 1)$, $-\lambda \in \mathbb{N}$, $\nu = 1$.

The generators are explicitly constructed in [35, (2.24)–(2.32)] (see [21, 38, 42] for the $i = 0$ case), which we review quickly. Let $\widetilde{C}^\alpha_\ell(z)$ be the Gegenbauer polynomial of degree ℓ, normalized by

$$\widetilde{C}^\alpha_\ell(z) := \frac{1}{\Gamma(\alpha + [\frac{\ell+1}{2}])} \sum_{k=0}^{[\frac{\ell}{2}]} (-1)^k \frac{\Gamma(\ell - k + \alpha)}{k!(\ell - 2k)!} (2z)^{\ell - 2k} \qquad (3.12)$$

as in [35, (14.3)]. Then $\widetilde{C}^\alpha_\ell(z) \not\equiv 0$ for all $\alpha \in \mathbb{C}$ and $\ell \in \mathbb{N}$.

For $\ell \in \mathbb{N}$, we inflate $\widetilde{C}_\ell^\alpha(z)$ to a polynomial of two variables x and y:

$$\widetilde{C}_\ell^\alpha(x, y) := x^{\frac{\ell}{2}} \widetilde{C}_\ell^\alpha\left(\frac{y}{\sqrt{x}}\right)$$

$$= \sum_{k=0}^{[\frac{\ell}{2}]} \frac{(-1)^k \Gamma(\ell - k + \alpha)}{\Gamma(\alpha + [\frac{\ell+1}{2}]) \Gamma(\ell - 2k + 1) k!} (2y)^{\ell - 2k} x^k. \tag{3.13}$$

For instance, $\widetilde{C}_0^\alpha(x, y) = 1$, $\widetilde{C}_1^\alpha(x, y) = 2y$, $\widetilde{C}_2^\alpha(x, y) = 2(\alpha + 1)y^2 - x$, etc. Notice that $\widetilde{C}_\ell^\alpha(x^2, y)$ is a homogeneous polynomial of x and y of degree ℓ.

For $\nu - \lambda \in \mathbb{N}$, we set a scalar-valued differential operator $\widetilde{\mathbb{C}}_{\lambda,\nu} : C^\infty(\mathbb{R}^n) \to C^\infty(\mathbb{R}^{n-1})$ by

$$\widetilde{\mathbb{C}}_{\lambda,\nu} := \mathrm{Rest}_{x_n=0} \circ \widetilde{C}_{\nu-\lambda}^{\lambda - \frac{n-1}{2}} \left(-\Delta_{\mathbb{R}^{n-1}}, \frac{\partial}{\partial x_n}\right). \tag{3.14}$$

For $\mu \in \mathbb{C}$ and $a \in \mathbb{N}$, we set

$$\gamma(\mu, a) := \begin{cases} 1 & \text{if } a \text{ is odd,} \\ \mu + \frac{a}{2} & \text{if } a \text{ is even.} \end{cases} \tag{3.15}$$

We are ready to define matrix-valued differential operators

$$\widetilde{\mathbb{C}}_{\lambda,\nu}^{i,j} : \mathcal{E}^i(\mathbb{R}^n) \to \mathcal{E}^j(\mathbb{R}^{n-1})$$

which were introduced in [35, (2.24) and (2.26)] by the following formulæ:

$$\mathbb{C}_{\lambda,\nu}^{i,i} := \widetilde{\mathbb{C}}_{\lambda+1,\nu-1} d_{\mathbb{R}^n} d_{\mathbb{R}^n}^* - \gamma\left(\lambda - \frac{n}{2}, \nu - \lambda\right) \widetilde{\mathbb{C}}_{\lambda,\nu-1} d_{\mathbb{R}^n} \iota_{\frac{\partial}{\partial x_n}} + \frac{1}{2}(\nu - i) \widetilde{\mathbb{C}}_{\lambda,\nu}, \tag{3.16}$$

$$\mathbb{C}_{\lambda,\nu}^{i,i-1} := -\widetilde{\mathbb{C}}_{\lambda+1,\nu-1} d_{\mathbb{R}^n} d_{\mathbb{R}^n}^* \iota_{\frac{\partial}{\partial x_n}}$$
$$- \gamma\left(\lambda - \frac{n-1}{2}, \nu - \lambda\right) \widetilde{\mathbb{C}}_{\lambda+1,\nu} d_{\mathbb{R}^n}^* + \frac{1}{2}(\lambda + i - n) \widetilde{\mathbb{C}}_{\lambda,\nu} \iota_{\frac{\partial}{\partial x_n}}. \tag{3.17}$$

Here $\iota_Z : \mathcal{E}^i(\mathbb{R}^n) \to \mathcal{E}^{i-1}(\mathbb{R}^n)$ stands for the interior product which is defined to be the contraction with a vector field Z.

We note that

$$\mathbb{C}_{\lambda,\nu}^{0,0} = \frac{1}{2} \nu \widetilde{\mathbb{C}}_{\lambda,\nu}, \qquad\qquad \mathbb{C}_{\nu,\nu}^{i,i} = \frac{1}{2}(\nu - i) \mathrm{Rest}_{x_n=0},$$

$$\mathbb{C}_{\lambda,\lambda}^{i,i-1} = \frac{1}{2}(\lambda + i - n) \mathrm{Rest}_{x_n=0} \circ \iota_{\frac{\partial}{\partial x_n}}, \qquad \mathbb{C}_{\lambda,\nu}^{n,n-1} = \frac{1}{2} \nu \widetilde{\mathbb{C}}_{\lambda,\nu} \circ \iota_{\frac{\partial}{\partial x_n}}.$$

The operators $\mathbb{C}^{i,j}_{\lambda,\nu}$ vanish for the following special values of (λ,ν):

$$\mathbb{C}^{i,i}_{\lambda,\nu} = 0 \qquad \text{if and only if } \lambda = \nu = i \qquad \text{or } \nu = i = 0,$$

$$\mathbb{C}^{i,i-1}_{\lambda,\nu} = 0 \qquad \text{if and only if } \lambda = \nu = n-i \ \text{ or } \nu = n-i = 0.$$

In order to provide *nonzero* operators, following the notation as in [35, (2.30)], we renormalize $\mathbb{C}^{i,j}_{\lambda,\nu}$ as

$$\widetilde{\mathbb{C}}^{i,i}_{\lambda,\nu} := \begin{cases} \text{Rest}_{x_n=0} & \text{if } \lambda = \nu, \\ \widetilde{\mathbb{C}}_{\lambda,\nu} & \text{if } i = 0, \\ \mathbb{C}^{i,i}_{\lambda,\nu} & \text{otherwise,} \end{cases} \tag{3.18}$$

$$\widetilde{\mathbb{C}}^{i,i-1}_{\lambda,\nu} := \begin{cases} \text{Rest}_{x_n=0} \circ \iota_{\frac{\partial}{\partial x_n}} & \text{if } \lambda = \nu, \\ \widetilde{\mathbb{C}}_{\lambda,\nu} \circ \iota_{\frac{\partial}{\partial x_n}} & \text{if } i = n, \\ \mathbb{C}^{i,i-1}_{\lambda,\nu} & \text{otherwise.} \end{cases} \tag{3.19}$$

For $j = i-2$ or $i+1$, we also set

$$\widetilde{\mathbb{C}}^{i,i-2}_{\lambda,n-i+1} := \begin{cases} -d^*_{\mathbb{R}^{n-1}} \circ \widetilde{\mathbb{C}}^{n,n-1}_{\lambda,0} & \text{if } i = n, \lambda \in -\mathbb{N}, \\ \text{Rest}_{x_n=0} \circ \iota_{\frac{\partial}{\partial x_n}} d^*_{\mathbb{R}^n} & \text{if } 2 \le i \le n-1, \lambda = n-i. \end{cases}$$

$$\widetilde{\mathbb{C}}^{i,i+1}_{\lambda,i+1} := \begin{cases} d_{\mathbb{R}^{n-1}} \circ \widetilde{\mathbb{C}}_{\lambda,0} & \text{if } i = 0, \lambda \in -\mathbb{N}, \\ \text{Rest}_{x_n=0} \circ d_{\mathbb{R}^n} & \text{if } 1 \le i \le n-2, \lambda = i. \end{cases}$$

With the notation as above, we can describe explicit generators of the space $\text{Diff}_{G'}(I_\delta(i,\lambda)|_{G'}, J_\varepsilon(j,\varepsilon))$ of differential symmetry breaking operators:

Fact 3.23 (basis, [35, Thm. 2.9]). *Suppose that 6-tuple $(i,j,\lambda,\nu,\delta,\varepsilon)$ is one of the six cases in Fact 3.22 (iii). Then the differential symmetry breaking operators $I_\delta(i,\lambda) \to J_\varepsilon(j,\nu)$ are proportional to*

$$j = i-2 : \widetilde{\mathbb{C}}^{i,i-2}_{n-i,n-i+1} \ (2 \le i \le n-1); \quad \widetilde{\mathbb{C}}^{n,n-2}_{\lambda,1} \ (i = n),$$

$$j = i-1 : \widetilde{\mathbb{C}}^{i,i-1}_{\lambda,\nu},$$

$$j = i : \quad \widetilde{\mathbb{C}}^{i,i}_{\lambda,\nu},$$

$$j = i+1 : \widetilde{\mathbb{C}}^{i,i+1}_{i,i+1} \ (1 \le i \le n-2); \quad \widetilde{\mathbb{C}}^{0,1}_{\lambda,1} \ (i = 0).$$

Remark 3.24. The scalar case ($i = j = 0$) was classified in Juhl [21] for $n \ge 3$. See also [38] for a different approach using the F-method. The case $n=2$ (and $i = j = 0$)

is essentially equivalent to find differential symmetry breaking operators from the tensor product of two principal series representations to another principal series representation for $SL(2, \mathbb{R})$. In this case, generic (but not all) operators are given by the Rankin–Cohen brackets, and the complete classification was accomplished in [41, Thms. 9.1 and 9.2]. We note that the dimension of differential symmetry breaking operators may jump to two at some singular parameters where $n = 2$.

3.5.3 Formula of the Dimension of $\mathrm{Hom}_{G'}(I_\delta(i, \lambda)|_{G'}, J_\varepsilon(j, \nu))$

For admissible smooth representations Π of G and π of the subgroup G', we set

$$m(\Pi, \pi) := \dim_{\mathbb{C}} \mathrm{Hom}_{G'}(\Pi|_{G'}, \pi).$$

In this subsection we give a formula of the multiplicity $m(\Pi, \pi)$ for $\Pi = I_\delta(i, \lambda)$ and $\pi = J_\varepsilon(j, \nu)$.

Theorem 3.25 (multiplicity formula). *Let* $(G, G') = (O(n + 1, 1), O(n, 1))$ *with* $n \geq 3$. *Suppose* $\Pi = I_\delta(i, \lambda)$ *and* $\pi = J_\varepsilon(j, \nu)$ *for* $0 \leq i \leq n$, $0 \leq j \leq n - 1$, $\delta, \varepsilon \in \{\pm\}$, *and* $\lambda, \nu \in \mathbb{C}$. *Then we have the following.*

(1)

$$
\begin{aligned}
m(\Pi, \pi) &\in \{1, 2\} && \text{if } j = i - 1 \text{ or } i, \\
m(\Pi, \pi) &\in \{0, 1\} && \text{if } j = i - 2 \text{ or } i + 1, \\
m(\Pi, \pi) &= 0 && \text{otherwise.}
\end{aligned}
$$

(2) *Suppose* $j = i - 1$ *or* i. *Then* $m(\Pi, \pi) = 1$ *except for the countable set described as below.*

  (a) *Case* $1 \leq i \leq n - 1$. *Then* $m(I_\delta(i, \lambda), J_\varepsilon(i, \nu)) = 2$ *if and only if*

$$
\begin{aligned}
j = i, \quad & \delta\varepsilon = +, \quad (\lambda, \nu) \in L_{\mathrm{even}} - \{\nu = 0\} \cup \{(i, i)\}, \\
j = i, \quad & \delta\varepsilon = -, \quad (\lambda, \nu) \in L_{\mathrm{odd}} - \{\nu = 0\}, \\
j = i - 1, \quad & \delta\varepsilon = +, \quad (\lambda, \nu) \in L_{\mathrm{even}} - \{\nu = 0\} \cup \{(n - i, n - i)\},
\end{aligned}
$$

  or

$$
j = i - 1, \quad \delta\varepsilon = -, \quad (\lambda, \nu) \in L_{\mathrm{odd}} - \{\nu = 0\}.
$$

(b) *Case $i = 0$. Then $m(I_\delta(0, \lambda), J_\varepsilon(0, \nu)) = 2$ if $\delta\varepsilon = +, (\lambda, \nu) \in L_{\mathrm{even}}$ or $\delta\varepsilon = -, (\lambda, \nu) \in L_{\mathrm{odd}}$.*

(c) *Case $i = n$. Then $m(I_\delta(n, \lambda), J_\varepsilon(n - 1, \nu)) = 2$ if*
   $$\delta\varepsilon = +, (\lambda, \nu) \in L_{\mathrm{even}} \text{ or } \delta\varepsilon = -, (\lambda, \nu) \in L_{\mathrm{odd}}.$$

(3) *Suppose $j = i - 2$ or $i + 1$. Then $m(\Pi, \pi) = 1$ if one of the following conditions (d)–(g) is satisfied, and $m(\Pi, \pi) = 0$ otherwise.*

   (d) *Case $j = i - 2$, $2 \le i \le n - 1$, $(\lambda, \nu) = (n - i, n - i + 1)$, $\delta\varepsilon = -1$.*
   (e) *Case $(i, j) = (n, n - 2)$, $-\lambda \in \mathbb{N}$, $\nu = 1$, $\delta\varepsilon = (-1)^{\lambda+1}$.*
   (f) *Case $j = i + 1$, $1 \le i \le n - 2$, $(\lambda, \nu) = (i, i + 1)$, $\delta\varepsilon = -1$.*
   (g) *Case $(i, j) = (0, 1)$, $-\lambda \in \mathbb{N}$, $\nu = 1$, $\delta\varepsilon = (-1)^{\lambda+1}$.*

The proof of Theorem 3.25 will be given right after Theorem 3.26, by using Fact 3.22 and Theorems 3.13 and 3.19, whose proofs are deferred at later chapters.

3.5.4 Classification of Symmetry Breaking Operators $I_\delta(i, \lambda) \rightarrow J_\varepsilon(j, \nu)$

In this subsection, we give explicit generators of

$$\mathrm{Hom}_{G'}(I_\delta(i, \lambda)|_{G'}, J_\varepsilon(j, \nu)),$$

of which the dimension is determined in Theorem 3.25. For most of the cases, the regular symmetry breaking operators $\widetilde{\mathbb{A}}^{i,j}_{\lambda,\nu,\pm}$ and the differential symmetry breaking operators $\widetilde{\mathbb{C}}^{i,j}_{\lambda,\nu}$ give the generators. However, for the exceptional discrete set classified in Theorem 3.19, we need more operators which are defined as follows: for $(\lambda_0, \nu_0) \in \mathbb{C}^2$ such that $\widetilde{\mathbb{A}}^{i,j}_{\lambda_0,\nu_0,\pm} = 0$, we renormalize the regular symmetry breaking operators $\widetilde{\mathbb{A}}^{i,j}_{\lambda,\nu,\pm}$ as follows (see Section 9.9). For $j = i$ of $i - 1$, we set

$$\widetilde{\widetilde{\mathbb{A}}}^{i,j}_{\lambda_0,\nu_0,+} := \lim_{\lambda \to \lambda_0} \Gamma(\frac{\lambda - \nu_0}{2}) \widetilde{\mathbb{A}}^{i,j}_{\lambda,\nu_0,+}, \tag{3.20}$$

$$\widetilde{\widetilde{\mathbb{A}}}^{i,j}_{\lambda_0,\nu_0,-} := \lim_{\lambda \to \lambda_0} \Gamma(\frac{\lambda - \nu_0 + 1}{2}) \widetilde{\mathbb{A}}^{i,j}_{\lambda,\nu_0,-}. \tag{3.21}$$

Then $\widetilde{\widetilde{\mathbb{A}}}^{i,j}_{\lambda,\nu,\pm}$ are well-defined and nonzero symmetry breaking operators (Theorem 5.45).

For $j \in \{i - 1, i\}$ and $\gamma \in \{\pm\}$, the set

$$\{(\lambda, \nu) \in \mathbb{C}^2 : \widetilde{\mathbb{A}}^{i,j}_{\lambda,\nu,\gamma} = 0\}$$

is classified in Theorem 3.19. Then we are ready to give an explicit basis of symmetry breaking operators:

Theorem 3.26 (generators). *Suppose $j = i$ or $i - 1$.*

(1) $m(I_\delta(i, \lambda), J_\varepsilon(j, v)) = 1$ *if and only if* $\widetilde{\mathbb{A}}^{i,j}_{\lambda, v, \delta\varepsilon} \neq 0$. *In this case*

$$\mathrm{Hom}_{G'}(I_\delta(i, \lambda)|_{G'}, J_\varepsilon(j, v)) = \mathbb{C}\widetilde{\mathbb{A}}^{i,j}_{\lambda, v, \delta\varepsilon}.$$

(2) $m(I_\delta(i, \lambda), J_\varepsilon(j, v)) = 2$ *if and only if* $\widetilde{\mathbb{A}}^{i,j}_{\lambda, v, \delta\varepsilon} = 0$. *In this case*

$$\mathrm{Hom}_{G'}(I_\delta(i, \lambda)|_{G'}, J_\varepsilon(j, v)) = \mathbb{C}\widetilde{\widetilde{\mathbb{A}}}^{i,j}_{\lambda, v, \delta\varepsilon} \oplus \mathbb{C}\widetilde{\mathbb{C}}^{i,j}_{\lambda, v}.$$

See Theorem 3.19 for the necessary and sufficient condition on $(i, j, \lambda, v, \gamma)$ for $\widetilde{\mathbb{A}}^{i,j}_{\lambda, v, \gamma}$ to vanish.

Remark 3.27. For $j = i + 1$ or $i - 2$, all symmetry breaking operators are differential operators by the localness theorem (Theorem 3.6), and the generators are given in Fact 3.23.

Proof of Theorems 3.25 and 3.26. We apply the general scheme of symmetry breaking operators (Theorem 3.13) to the special setting:

$$V = \bigwedge\nolimits^i(\mathbb{C}^n) \quad \text{and} \quad W = \bigwedge\nolimits^j(\mathbb{C}^{n-1}).$$

Then the theorem follow from the explicit description of the zero sets of the (normalized) regular symmetry breaking operators $\widetilde{\mathbb{A}}^{i,j}_{\lambda, v, \gamma}$ (Theorem 3.19) and the classification of *differential* symmetry breaking operators (Fact 3.22). □

Remark 3.28. The first statement (*i.e.*, $\delta\varepsilon = +$ case) of Theorem 3.25 (2) (b) was established in [42, Thm. 1.1], and the second statement (*i.e.*, $\delta\varepsilon = -$ case) of (b) can be proved similarly. In this article, we take another approach for the latter case: we deduce results for all the matrix-valued cases (including the scalar-valued case with $\delta\varepsilon = -$) from the scalar valued case with $\delta\varepsilon = +$.

3.6　Consequences of Main Theorems in Sections 3.3 and 3.5

In this section we discuss symmetry breaking from principal series representations $\Pi = I_\delta(V, \lambda)$ of G to $\pi = J_\varepsilon(W, v)$ of the subgroup G' in the case where Π and π are *unitarizable*. Unitary principal series representations are treated in Section 3.6.1, and complementary series representations are treated in Sections 3.6.2 and 3.6.3. We note that Π and π are irreducible in these cases. On the other hand, if λ (resp. v) is integral, then Π (resp. π) may be reducible. We shall discuss symmetry breaking operators for the subquotients in the next chapter in detail when they have the trivial infinitesimal character ρ.

3.6.1 Tempered Representations

We recall the concept of tempered unitary representations of locally compact groups.

Definition 3.29 (tempered unitary representation). A unitary representation of a unimodular group G is called *tempered* if it is weakly contained in the regular representations on $L^2(G)$. By a little abuse of notation, we also say the smooth representation Π^∞ is *tempered*.

Returning to our setting where $(G, G') = (O(n + 1, 1), O(n, 1))$, we see that the principal series representations $I_\delta(V, \lambda)$ and $J_\varepsilon(W, \nu)$ are tempered if and only if $\lambda \in \sqrt{-1}\mathbb{R} + \frac{n}{2}$ and $\nu \in \sqrt{-1}\mathbb{R} + \frac{1}{2}(n - 1)$, respectively. We refer to them as *tempered principal series representations*.

We recall $[V : W] = \dim_{\mathbb{C}} \operatorname{Hom}_{O(n-1)}(V|_{O(n-1)}, W)$. Then Theorem 3.13 implies the following:

Theorem 3.30 (tempered principal series representations). *Let $(\sigma, V) \in \widehat{O(n)}$, $(\tau, W) \in \widehat{O(n-1)}$, $\delta, \varepsilon \in \{\pm\}$, and $\lambda \in \sqrt{-1}\mathbb{R} + \frac{n}{2}$, $\nu \in \sqrt{-1}\mathbb{R} + \frac{1}{2}(n - 1)$ so that $I_\delta(V, \lambda)$ and $J_\varepsilon(W, \nu)$ are tempered principal series representations. Then the following four conditions are equivalent:*

- (i) $[V : W] \neq 0$;
- (i′) $[V : W] = 1$;
- (ii) $\operatorname{Hom}_{G'}(I_\delta(V, \lambda)|_{G'}, J_\varepsilon(W, \nu)) \neq \{0\}$;
- (ii′) $\dim_{\mathbb{C}} \operatorname{Hom}_{G'}(I_\delta(V, \lambda)|_{G'}, J_\varepsilon(W, \nu)) = 1$.

Applying Theorem 3.30 to the exterior tensor representations $V = \bigwedge^i(\mathbb{C}^n)$ of $O(n)$ and $W = \bigwedge^j(\mathbb{C}^{n-1})$ of $O(n - 1)$, we get:

Corollary 3.31. *Suppose $\lambda \in \sqrt{-1}\mathbb{R} + \frac{n}{2}$, and $\nu \in \sqrt{-1}\mathbb{R} + \frac{1}{2}(n - 1)$. Then*

$$
\dim_{\mathbb{C}} \operatorname{Hom}_{G'}(I_\delta(i, \lambda)|_{G'}, J_\varepsilon(j, \nu)) = \begin{cases} 1 & \text{if } i = j \text{ or } j = i - 1, \\ 0 & \text{otherwise.} \end{cases}
$$

3.6.2 Complementary Series Representations

We say that $I_\delta(V, \lambda)$ is a (smooth) *complementary series representation* if it has a Hilbert completion to a unitary complementary series representation. If the irreducible $O(n)$-module (σ, V) is of type X (see Definition 2.6), *i.e.*, the last digit of the highest weight of V is not zero, then the principal series representation $I_\delta(V, \lambda)$ is irreducible at $\lambda = \frac{n}{2}$, and consequently, there exist complementary series representations $I_\delta(V, \lambda)$ for some interval $\lambda \in (\frac{n}{2} - a, \frac{n}{2} + a)$ with $a > 0$.

Example 3.32. Suppose (σ, V) is the i-th exterior tensor representation $\bigwedge^i(\mathbb{C}^n)$. We assume that this representation is of type X, equivalently, $n \neq 2i$ (see Example 2.8). The first reduction point of the principal series representation of $I_\delta(i, \lambda)$ is given by $\lambda = i$ or $n - i$ (see Proposition 2.18). Therefore $I_\delta(i, \lambda) \equiv I_\delta(\bigwedge^i(\mathbb{C}^n), \lambda)$ is a complementary series representation if

$$\min(i, n - i) < \lambda < \max(i, n - i).$$

In the category of unitary representations, the restriction of a tempered representation of G to a reductive subgroup G' decomposes into the direct integral of irreducible unitary tempered representations of a reductive subgroup G' because it is weakly contained in the regular representation. In particular, complementary series representations of the subgroup G' do not appear in the *unitary* branching law of the restriction of a unitary tempered principal series representation $I_\delta(V, \lambda)$, whereas Theorem 3.13 in the category of admissible *smooth* representations shows that there are nontrivial symmetry breaking operators

$$\widetilde{\mathbb{A}}^{V,W}_{\lambda, \nu, \delta\varepsilon} : I_\delta(V, \lambda) \to J_\varepsilon(W, \nu)$$

to all complementary series representations $J_\delta(W, \nu)$ of the subgroup G' if $[V : W] \neq 0$.

   Moreover, Theorem 3.13 (2) implies also that there are nontrivial symmetry breaking operators from any (smooth) complementary series representation $I_\delta(V, \lambda)$ of G to all (smooth) tempered principal series representations $J_\varepsilon(W, \nu)$ of the subgroup G' as far as $[V : W] \neq 0$.

3.6.3 Singular Complementary Series Representations

We consider the complementary series representations $I_\delta(i, s)$ for $i < s < \frac{n}{2}$ with an additional assumption that s is an integer. These representations are irreducible and have *singular* integral infinitesimal characters. We may describe the underlying $(\mathfrak{g}, K)$-modules of these singular complementary series representations in terms of cohomological parabolic induction $A_{\mathfrak{q}}(\lambda)$ where the parameter λ wanders outside the good range relative to the θ-stable parabolic subalgebra $\mathfrak{q}$ (see [24, Def. 0.49] for the definition).

   For $0 \le r \le \frac{n+1}{2}$, we denote by $\mathfrak{q}_r$ the θ-stable parabolic subalgebra of $\mathfrak{g}_\mathbb{C} = \mathfrak{o}(n+2, \mathbb{C})$ with Levi factor $SO(2)^r \times O(n - 2r + 1, 1)$ in $G = O(n+1, 1)$ (see Definition 14.37).

Lemma 3.33. *Let $0 \le i \le [\frac{n}{2}] - 1$. For $s \in \{i+1, i+2, \cdots, [\frac{n}{2}]\}$, we have an iso-morphism as $(\mathfrak{g}, K)$-modules:*

$$I_+(i,s)_K \simeq A_{\mathfrak{q}_{i+1}}(0, \cdots, 0, s - i).$$

See Remark 14.43 in Appendix I for the normalization of the $(\mathfrak{g}, K)$-module $A_{\mathfrak{q}}(\lambda)$ and Theorem 14.53 for more details about Lemma 3.33. See also [27, Thm. 3] for some more general cases. The restriction of these representations to the special orthogonal group $SO(n+1, 1)$ stays irreducible (see Lemma 15.3 in Appendix II). Bergeron and Clozel proved that there are automorphic square integrable representations, whose component at infinity is isomorphic to a representation $I_\delta(i,s)|_{SO(n+1,1)}$ (see [5, 10]).

A special case of Theorem 3.25 includes:

Proposition 3.34. *Suppose $s \in \mathbb{N}$ and $i < s \le [\frac{n}{2}]$. Let $\delta, \varepsilon \in \{\pm\}$.*

(1) *For $i < r \le [\frac{n-1}{2}]$,*

$$\mathrm{Hom}_{G'}(I_\delta(i,s)|_{G'}, J_\varepsilon(i,r)) = \mathbb{C}.$$

(2) *For $0 \le i - 1 < r \le [\frac{n-1}{2}]$,*

$$\mathrm{Hom}_{G'}(I_\delta(i,s)|_{G'}, J_\varepsilon(i-1,r)) = \mathbb{C}.$$

Remark 3.35. Proposition 3.34 may be viewed as symmetry breaking operators from the Casselman–Wallach globalization of the irreducible $(\mathfrak{g}, K)$-module $A_{\mathfrak{q}}(\lambda)$ to that of the irreducible $(\mathfrak{g}', K')$-module $A_{\mathfrak{q}'}(\nu)$ in some special cases where both λ and ν are outside the good range of parameters relative to the θ-stable parabolic subalgebras.

In the next chapter, we treat the case with trivial infinitesimal character ρ, and thus the parameters stay in the good range relative to the θ-stable parabolic subalgebras. In particular, we shall determine a necessary and sufficient condition for a pair $(\mathfrak{q}, \mathfrak{q}')$ of θ-stable parabolic subalgebras $\mathfrak{q}$ of $\mathfrak{g}_\mathbb{C}$ and $\mathfrak{q}'$ of its subalgebra $\mathfrak{g}'_\mathbb{C}$ such that

$$\mathrm{Hom}_{G'}(\Pi|_{G'}, \pi) \ne \{0\},$$

when the underlying $(\mathfrak{g}, K)$-module Π_K of $\Pi \in \mathrm{Irr}(G)$ is isomorphic to $(A_{\mathfrak{q}})_{\pm\pm}$ and the underlying $(\mathfrak{g}', K')$-module of $\pi \in \mathrm{Irr}(G')$ is $(A_{\mathfrak{q}'})_{\pm\pm}$, see Theorems 4.1 and 4.2 for the multiplicity-formula, and Proposition 14.44 in Appendix I for the description of Π_K in terms of $(A_{\mathfrak{q}})_{\pm\pm}$. In contrast to the case of Proposition 3.34, the irreducible G-module Π and G'-module π do not coincide with principal series representations, but appear as their subquotients in this case, see Theorem 2.20 (1).

3.7   Actions of $(G/G_0)\hat{\ } \times (G'/G'_0)\hat{\ }$ on Symmetry Breaking Operators

In this section we discuss the action of the character group of $G \times G'$ on the set

$$\{\mathrm{Hom}_{G'}(\Pi|_{G'}, \pi)\}$$

of the spaces of symmetry breaking operators where admissible smooth representations Π of G and those π of the subgroup G' vary. Actual computations for the pair $(G, G') = (O(n+1, 1), O(n, 1))$ are carried out by using Lemma 2.14 for principal series representations and Theorem 2.20 (5) for their irreducible subquotients.

3.7.1   Generalities: The Action of Character Group of $G \times G'$ on $\{\mathrm{Hom}_{G'}(\Pi|_{G'}, \pi)\}$ in the General Case

Let $G \supset G'$ be a pair of real reductive Lie groups. Then the character group of $G \times G'$ acts on the set of vector spaces $\{\mathrm{Hom}_{G'}(\Pi|_{G'}, \pi)\}$ where Π runs over admissible smooth representations of G, and π runs over those of the subgroup G'. Here the action is given by

$$\mathrm{Hom}_{G'}(\Pi|_{G'}, \pi) \mapsto \mathrm{Hom}_{G'}((\Pi \otimes \chi^{-1})|_{G'}, \pi \otimes \chi')$$

for a character χ of G and χ' of the subgroup G'.

In what follows, we regard a character of G as a character of G' by restriction, and use the same letter to denote its restriction to the subgroup G'. Then for all characters χ and χ' of G, we have the following isomorphisms:

$$\mathrm{Hom}_{G'}((\Pi \otimes \chi)|_{G'}, \pi \otimes \chi') \simeq \mathrm{Hom}_{G'}(\Pi|_{G'}, \pi \otimes \chi^{-1} \otimes \chi')$$

$$\simeq \mathrm{Hom}_{G'}((\Pi \otimes (\chi')^{-1})|_{G'}, \pi \otimes \chi^{-1})$$

$$\simeq \mathrm{Hom}_{G'}((\Pi \otimes \chi \otimes (\chi')^{-1})|_{G'}, \pi). \qquad (3.22)$$

The above isomorphisms define an equivalence relation on the set

$$\{\mathrm{Hom}_{G'}(\Pi|_{G'}, \pi)\}$$

of the spaces of symmetry breaking operators where Π and π vary.

3.7.2 Actions of the Character Group of the Component Group on $\{\mathrm{Hom}_{G'}(I_\delta(i, \lambda)|_{G'}, J_\varepsilon(j, \nu))\}$

We apply the above idea to our setting

$$(G, G') = (O(n+1, 1), O(n, 1)).$$

Then the component groups of G and G' are a finite abelian group given by

$$G'/G'_0 \simeq G/G_0 \simeq \mathbb{Z}/2\mathbb{Z} \times \mathbb{Z}/2\mathbb{Z}. \tag{3.23}$$

We recall from (2.13) that the set of their one-dimensional representations is parametrized by

$$(G'/G'_0)\hat{\ } \simeq (G/G_0)\hat{\ } = \{\chi_{ab} : a, b \in \{\pm\}\}.$$

By abuse of notation, we shall use the same letters χ_{ab} to denote the corresponding one-dimensional representations of G, G', G/G_0, and G'/G'_0.

The action of the character group (*Pontrjagin dual*) $(G/G_0)\hat{\ }$ on the set of principal series representations can be computed by using Lemma 2.14. To describe the action of the Pontrjagin dual $(G/G_0)\hat{\ } \simeq (G'/G'_0)\hat{\ }$ on the parameter set of the principal series representations $I_\delta(i, \lambda)$ of G and $J_\varepsilon(j, \nu)$ of the subgroup G', we define

$$S := \{0, 1, \cdots, n\} \times \mathbb{C} \times \mathbb{Z}/2\mathbb{Z}, \qquad I(s) := I_\delta(i, \lambda) \quad \text{for } s = (i, \lambda, \delta) \in S,$$

$$T := \{0, 1, \cdots, n-1\} \times \mathbb{C} \times \mathbb{Z}/2\mathbb{Z}, \quad J(t) := J_\varepsilon(j, \nu) \quad \text{for } t = (j, \nu, \varepsilon) \in T.$$

We let the character group $(G/G_0)\hat{\ }$ act on S by the following formula:

$$\chi_{++} \cdot (i, \lambda, \delta) := (i, \lambda, \delta), \qquad \chi_{+-} \cdot (i, \lambda, \delta) := (i, \lambda, -\delta),$$

$$\chi_{-+} \cdot (i, \lambda, \delta) := (\tilde{i}, \lambda, -\delta), \qquad \chi_{--} \cdot (i, \lambda, \delta) := (\tilde{i}, \lambda, \delta),$$

where $\tilde{i} := n - i$. The action of $(G'/G'_0)\hat{\ }$ on the set T is defined similarly, with obvious modification

$$\tilde{j} := n - 1 - j$$

when we discuss representations of the subgroup $G' = O(n, 1)$. By Lemma 2.14 and by the $O(n)$-isomorphism $\bigwedge^i(\mathbb{C}^n) \simeq \bigwedge^{n-i}(\mathbb{C}^n) \otimes \det$, we obtain the following.

Lemma 3.36. *For all* $\chi \in (G/G_0)\hat{\ } \simeq (G'/G'_0)\hat{\ }$ *and for* $s \in S, t \in T$, *we have the following isomorphisms as G-modules and G'-modules, respectively:*

$$I(s) \otimes \chi \simeq I(\chi \cdot s),$$

$$J(t) \otimes \chi \simeq J(\chi \cdot t).$$

Then the equivalence defined by the isomorphisms (3.22) implies that it suffices to consider symmetry breaking operators for $(\delta, \varepsilon) = (+, +)$ and $(\delta, \varepsilon) = (+, -)$. To be more precise, we obtain the following.

Proposition 3.37. *Let $\lambda, \nu \in \mathbb{C}$. Then every symmetry breaking operator in*

$$\bigcup_{\delta, \varepsilon \in \{\pm\}} \bigcup_{0 \le i \le n} \bigcup_{0 \le j \le n-1} \mathrm{Hom}_{G'}(I_\delta(i, \lambda)|_{G'}, J_\varepsilon(j, \nu))$$

is equivalent to a symmetry breaking operator in

$$\bigcup_{0 \le i \le [\frac{n}{2}]} \bigcup_{0 \le j \le n-1} (\mathrm{Hom}_{G'}(I_+(i, \lambda)|_{G'}, J_+(j, \nu)) \cup \mathrm{Hom}_{G'}(I_+(i, \lambda)|_{G'}, J_-(j, \nu))).$$

Proof. We use a graph to prove this. We set

$$(\delta, \varepsilon) := \mathrm{Hom}_{G'}(I_\delta(i, \lambda)|_{G'}, J_\varepsilon(j, \nu)),$$

$$\begin{pmatrix} \delta \\ \varepsilon \end{pmatrix} := \mathrm{Hom}_{G'}(I_\delta(n - i, \lambda)|_{G'}, J_\varepsilon(n - j - 1, \nu)).$$

In the following graph the nodes are indexed by (δ, ε) in first row and by $\begin{pmatrix} \delta \\ \varepsilon \end{pmatrix}$ in the second row. The nodes are connected by a line if they are equivalent. By Lemma 3.36, we obtain the graph by taking $\chi = \chi' = \chi_{+-}$ in (3.22) for horizontal equivalence, and $\chi = \chi' = \chi_{-+}$ in (3.22) for crossing equivalence (we omit here lines in the graph corresponding to $\chi = \chi' = \chi_{--}$ in (3.22) for vertical equivalence):

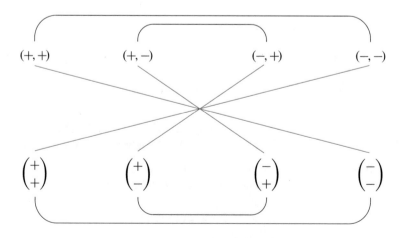

We observe that there are exactly two connected components of the graph, and that $\mathrm{Hom}_{G'}(I_+(i,\lambda)|_{G'}, J_+(j,v))$ and $\mathrm{Hom}_{G'}(I_+(i,\lambda)|_{G'}, J_-(j,v))$ are in a different connected component. Moreover, we may choose i or $n-i$ in the same equivalence classes, and thus we may take $0 \le i \le \frac{n}{2}$ as a representative.   □

Example 3.38.

(1) Suppose $n = 2m$ and $i = m$. Applying the isomorphism (3.22) to $(\Pi, \pi) = (I_\delta(m, \lambda), J_\varepsilon(m, v))$ with $\chi = \chi' = \chi_{--}$, we obtain a natural bijection:

$$\mathrm{Hom}_{G'}(I_\delta(m, \lambda)|_{G'}, J_\varepsilon(m, v)) \simeq \mathrm{Hom}_{G'}(I_\delta(m, \lambda)|_{G'}, J_\varepsilon(m-1, v)).$$

We note that the G-module $I_\delta(m, \lambda)$ at $\lambda = m$ splits into the direct sum of two irreducible smooth tempered representations (Theorem 2.20 (1) and (8)).

(2) Suppose $n = 2m + 1$ and $i = m$. Similarly to the first statement, we have a natural bijection:

$$\mathrm{Hom}_{G'}(I_\delta(m, \lambda)|_{G'}, J_\varepsilon(m, v)) \simeq \mathrm{Hom}_{G'}(I_\delta(m+1, \lambda)|_{G'}, J_\varepsilon(m, v)).$$

In this case, the G'-module $J_\varepsilon(m, v)$ at $v = m$ splits into the direct sum of two irreducible smooth tempered representations.

3.7.3   Actions of Characters of the Component Group on $\mathrm{Hom}_{G'}(\Pi_{i,\delta}|_{G'}, \pi_{j,\varepsilon})$

In the next chapter, we discuss

$$\mathrm{Hom}_{G'}(\Pi|_{G'}, \pi)$$

for $\Pi \in \mathrm{Irr}(G)_\rho$ and $\pi \in \mathrm{Irr}(G')_\rho$. In this case, (3.22) implies the following:

Proposition 3.39 (duality for symmetry breaking operators). *There are natural isomorphisms*

$$\mathrm{Hom}_{G'}(\Pi_{i,\delta}|_{G'}, \pi_{j,\varepsilon}) \simeq \mathrm{Hom}_{G'}(\Pi_{n+1-i,\delta}|_{G'}, \pi_{n-j,\varepsilon})$$

$$\simeq \mathrm{Hom}_{G'}(\Pi_{i,-\delta}|_{G'}, \pi_{j,-\varepsilon})$$

$$\simeq \mathrm{Hom}_{G'}(\Pi_{n+1-i,-\delta}|_{G'}, \pi_{n-j,-\varepsilon}).$$

Proof. By Theorem 2.20 (5), we have a natural G-isomorphism $\Pi_{i,\delta} \otimes \chi_{-+} \simeq \Pi_{n+1-i,\delta}$ and a G'-isomorphism $\pi_{j,\varepsilon} \otimes \chi_{-+} \simeq \pi_{n-j,\varepsilon}$. Hence the first isomorphism is derived from (3.22). By taking the tensor product with χ_{+-}, we get the last two isomorphisms again by Theorem 2.20 (5).   □

Chapter 4
Symmetry Breaking for Irreducible Representations with Infinitesimal Character ρ

In this chapter, we focus on symmetry breaking operators from *irreducible* representations Π of $G = O(n+1, 1)$ with $\mathfrak{Z}_G(\mathfrak{g})$-infinitesimal character ρ_G to *irreducible* representations π of the subgroup $G' = O(n, 1)$ with $\mathfrak{Z}_{G'}(\mathfrak{g}')$-infinitesimal character $\rho_{G'}$. The main results are Theorems 4.1 and 4.2, where we determine the multiplicity $\dim_{\mathbb{C}} \mathrm{Hom}_{G'}(\Pi|_{G'}, \pi)$ for all pairs (Π, π). A diagrammatic formulation of the main results is given in Theorem 4.3.

The proof uses basic properties of the normalized symmetry breaking operators for principal series representations of G and G',

$$\widetilde{\mathbb{A}}^{i,j}_{\lambda, \nu, \delta\varepsilon} : I_\delta(i, \lambda) \to J_\varepsilon(j, \nu),$$

in particular, the (K, K')-spectrum on basic K-types (Theorem 9.8) and their functional equations (Theorems 9.24 and 9.25).

4.1 Main Theorems

We recall from Theorem 2.20 that irreducible admissible smooth representations of G with trivial $\mathfrak{Z}_G(\mathfrak{g})$-infinitesimal character ρ_G are classified as

$$\mathrm{Irr}(G)_\rho = \{\Pi_{i,\delta} : 0 \le i \le n+1, \delta = \pm\}.$$

Similarly, irreducible admissible smooth representations of the subgroup $G' = O(n, 1)$ with trivial $\mathfrak{Z}_{G'}(\mathfrak{g}')$-infinitesimal character $\rho_{G'}$ are classified as

$$\mathrm{Irr}(G')_\rho = \{\pi_{j,\varepsilon} : 0 \le j \le n, \varepsilon = \pm\},$$

© Springer Nature Singapore Pte Ltd. 2018
T. Kobayashi, B. Speh, *Symmetry Breaking for Representations of Rank One Orthogonal Groups II*, Lecture Notes in Mathematics 2234,
https://doi.org/10.1007/978-981-13-2901-2_4

where we have used lowercase letters π for the subgroup G' instead of Π. We also recall that the representation $\Pi_{i,\delta}$ of $G = O(n+1,1)$ is

- one-dimensional if and only if $i = 0$ or $n+1$;
- the smooth representation of a discrete series representation if $i = \frac{n+1}{2}$ (n: odd);
- that of a tempered representation if $i = \frac{n}{2}$ (n: even).

The following two theorem determine the dimension of

$$\mathrm{Hom}_{G'}(\Pi|_{G'}, \pi) \quad \text{for } \Pi \in \mathrm{Irr}(G)_\rho \text{ and } \pi \in \mathrm{Irr}(G')_\rho.$$

Theorem 4.1 (vanishing). *Suppose* $0 \leq i \leq n+1, 0 \leq j \leq n, \delta, \varepsilon \in \{\pm\}$.

(1) *If* $j \neq i, i-1$ *then* $\mathrm{Hom}_{G'}(\Pi_{i,\delta}|_{G'}, \pi_{j,\varepsilon}) = \{0\}$.
(2) *If* $\delta\varepsilon = -$, *then* $\mathrm{Hom}_{G'}(\Pi_{i,\delta}|_{G'}, \pi_{j,\varepsilon}) = \{0\}$.

Theorem 4.2 (multiplicity-one). *Suppose* $0 \leq i \leq n+1, 0 \leq j \leq n$ *and* $\delta, \varepsilon \in \{\pm\}$. *If* $j = i-1$ *or* i *and if* $\delta\varepsilon = +$, *then*

$$\dim_{\mathbb{C}} \mathrm{Hom}_{G'}(\Pi_{i,\delta}|_{G'}, \pi_{j,\varepsilon}) = 1.$$

The proof of Theorems 4.1 and 4.2 will be given in Chapter 10. The nonzero symmetry breaking operators from $\Pi_{i,+}$ to $\pi_{j,+}$ ($j \in \{i-1,i\}$) will be applied to construct periods in Chapter 12 (see Theorem 12.6 for example).

4.2 Graphic Description of the Multiplicity for Irreducible Representations with Infinitesimal Character ρ

Using the action of the Pontrjagin dual of the component group $(G/G_0)^\widehat{} \times (G'/G'_0)^\widehat{}$ on $\mathrm{Hom}_{G'}(\Pi_{i,\delta}|_{G'}, \pi_{j,\varepsilon})$, see Proposition 3.39, we see that Theorems 4.1 and 4.2 are equivalent to their special case where $i \leq \frac{n+1}{2}$ and $\delta = +$. Furthermore, taking the vanishing result (Theorem 4.1) into account, we focus on the case $j \leq \frac{n}{2}$ and $\varepsilon = +$. We then describe Theorems 4.1 and 4.2 graphically in this setting.

We suppress the subscript, and write Π_i for $\Pi_{i,+}$, and π_j for $\pi_{j,+}$. Then Π_i $(0 \leq i \leq \frac{n+1}{2})$ and π_j $(0 \leq j \leq \frac{n}{2})$ are the standard sequence of representations with infinitesimal character ρ of G, respectively G' starting with the trivial one-dimensional representation (Definition 2.21). In the diagrams below, the first row are representations of G, the second row are representations of the subgroup G'. Arrows mean that there exist nonzero symmetry breaking operators.

Theorem 4.3. *Symmetry breaking for the standard sequence of irreducible representations starting at the trivial one-dimensional representations are represented graphically in Diagrams 4.1 and 4.2.*

$$\Pi_0 \quad \Pi_1 \quad \cdots \quad \Pi_{m-1} \quad \Pi_m$$
$$\downarrow \swarrow \downarrow \swarrow \quad \swarrow \downarrow \swarrow \downarrow$$
$$\pi_0 \quad \pi_1 \quad \cdots \quad \pi_{m-1} \quad \pi_m$$

Diagram. 4.1 Symmetry breaking for $O(2m+1,1) \downarrow O(2m,1)$

$$\Pi_0 \quad \Pi_1 \quad \cdots \quad \Pi_{m-1} \quad \Pi_m \quad \Pi_{m+1}$$
$$\downarrow \swarrow \downarrow \swarrow \quad \swarrow \downarrow \swarrow \downarrow \swarrow$$
$$\pi_0 \quad \pi_1 \quad \cdots \quad \pi_{m-1} \quad \pi_m$$

Diagram. 4.2 Symmetry breaking for $O(2m+2,1) \downarrow O(2m+1,1)$

Chapter 5
Regular Symmetry Breaking Operators

Let $I_\delta(V, \lambda)$ be a principal series representation of $G = O(n + 1, 1)$ realized in the Fréchet space $C^\infty(G/P, \mathcal{V}_{\lambda,\delta})$, and $J_\varepsilon(W, \nu)$ that of $G' = O(n, 1)$ realized in $C^\infty(G'/P', \mathcal{W}_{\nu,\varepsilon})$ as in Section 2.3.1. In this chapter we apply the general result in [42, Chap. 3] to construct a "matrix-valued regular symmetry breaking operators" $\widetilde{\mathbb{A}}^{V,W}_{\lambda,\nu,\pm} : I_\delta(V, \lambda) \to J_{\pm\delta}(W, \nu)$ that depend holomorphically on $(\lambda, \nu) \in \mathbb{C}^2$. We shall prove that the normalization (3.7) and (3.8) is optimal in the sense that the zeros of the operator $\widetilde{\mathbb{A}}^{V,W}_{\lambda,\nu,\pm}$ are of codimension > 1 in the parameter space of (λ, ν), that is, discrete in $\mathbb{C}^2$ in our setting. A key idea of the proof is a reduction to the scalar case.

5.1 Generalities

We recall from the general theory [42, Chap. 3] on the distribution kernels of symmetry breaking operators, which will be the basic tool in this chapter. Furthermore, we discuss some subtle questions on the underlying topology of representation spaces for symmetry breaking, see Theorem 5.4.

5.1.1 Distribution Kernels of Symmetry Breaking Operators

Throughout this monograph, we shall regard distributions as the dual of compactly supported smooth densities rather than that of compactly supported smooth functions. Thus we treat distributions as "generalized functions", and write their pairing with test functions by using the integral symbol, as if they were ordinary functions (with densities).

© Springer Nature Singapore Pte Ltd. 2018   63
T. Kobayashi, B. Speh, *Symmetry Breaking for Representations*
of Rank One Orthogonal Groups II, Lecture Notes in Mathematics 2234,
https://doi.org/10.1007/978-981-13-2901-2_5

Let $G \supset G'$ be a pair of real reductive Lie groups, and P, P' their parabolic subgroups. We do not require an inclusive relation $P \supset P'$ in this subsection. Let $(\widetilde{\sigma}, V)$ be a finite-dimensional representation of P, and $(\widetilde{\tau}, W)$ that of the subgroup P'. We form homogeneous vector bundles over flag manifolds by

$$\mathcal{V} := G \times_P V \quad \to G/P,$$
$$\mathcal{W} := G' \times_{P'} W \to G'/P'.$$

We write $\mathrm{Ind}_P^G(\widetilde{\sigma})$ for the admissible smooth representation of G on the Fréchet space $C^\infty(G/P, \mathcal{V})$, and $\mathrm{Ind}_{P'}^{G'}(\widetilde{\tau})$ for that of the subgroup G' on $C^\infty(G'/P', \mathcal{W})$.

We denote by $\mathcal{V}^*$ the dualizing bundle of $\mathcal{V}$, which is a G-homogeneous vector bundle over G/P associated to the representation

$$V^* := V^\vee \otimes |\det(\mathrm{Ad}_{\mathfrak{g}/\mathfrak{p}})|^{-1}$$

of the group P, where $V^\vee$ denotes the contragredient representation of $(\widetilde{\sigma}, V)$. Then the regular representation of G on the space $\mathcal{D}'(G/P, \mathcal{V}^*)$ of $\mathcal{V}^*$-valued distribution sections is the dual of the representation on $C^\infty(G/P, \mathcal{V})$.

The Schwartz kernel theorem guarantees that any symmetry breaking operator can be expressed by using a distribution kernel. Conversely, distributions that give rise to symmetry breaking operators are characterized as follows.

Fact 5.1 ([42, Prop. 3.2]). *There are natural linear bijections:*

$$\mathrm{Hom}_{G'}(C^\infty(G/P, \mathcal{V})|_{G'}, C^\infty(G'/P', \mathcal{W})) \simeq \mathcal{D}'(G/P \times G'/P', \mathcal{V}^* \boxtimes \mathcal{W})^{\Delta(G')}.$$

Here $\mathcal{V}^* \boxtimes \mathcal{W}$ denotes the outer tensor product bundle over the direct product manifold $G/P \times G'/P'$.

We note that the multiplication map

$$m \colon G \times G' \to G, \quad (x, y) \mapsto y^{-1}x$$

induces a linear bijection

$$\mathcal{D}'(G/P \times G'/P', \mathcal{V}^* \boxtimes \mathcal{W})^{\Delta(G')} \xleftarrow[m^*]{\sim} (\mathcal{D}'(G/P, \mathcal{V}^*) \otimes W)^{\Delta(P')},$$

where the right-hand side stands for the space of P'-invariant vectors under the diagonal action on the tensor product of the G-module $\mathcal{D}'(G/P, \mathcal{V}^*)$ and the P'-module W.

Thus Fact 5.1 may be reformulated as the following linear bijection

$$\mathrm{Hom}_{G'}(C^\infty(G/P, \mathcal{V})|_{G'}, C^\infty(G'/P', \mathcal{W})) \simeq (\mathcal{D}'(G/P, \mathcal{V}^*) \otimes W)^{\Delta(P')}. \qquad (5.1)$$

The point of Fact 5.1 is that the map

$$C^\infty(G/P, \mathcal{V}) \to \mathcal{D}'(G'/P', \mathcal{W}), \qquad f \mapsto \int_X K(x, y) f(x)$$

to the space $\mathcal{D}'(G'/P', \mathcal{W})$ of *distribution* sections becomes automatically a continuous map to the space $C^\infty(G'/P', \mathcal{W})$ of *smooth* sections for any $K \in \mathcal{D}'(G/P \times G'/P', \mathcal{V}^* \boxtimes \mathcal{W})^{\Delta(G')}$. This observation leads us to the proof of the isomorphism (5.3) in Theorem 5.4.

5.1.2 Invariant Bilinear Forms on Admissible Smooth Representations and Symmetry Breaking Operators

We retain the setting of the previous subsection, in particular, we suppose that $G \supset G'$ are a pair of real reductive Lie groups.

Let (Π, U) and (π, U') be admissible smooth representations of G and G', respectively. We recall that the underlying topological vector space of any admissible smooth representation is a nuclear Fréchet space. We define $\Pi \boxtimes \pi$ to be the natural representation of the direct product group $G \times G'$ on the space $U \widehat{\otimes} U'$. In this subsection, we study the space $\mathrm{Hom}_{G'}(\Pi \boxtimes \pi, \mathbb{C})$ of continuous functionals that are invariant under the diagonal action of the subgroup G'.

For an admissible smooth representation (Π, U) of G, we denote by $\Pi^\vee$ the contragredient representation of Π in the category of admissible smooth representations, namely, the Casselman–Wallach minimal globalization of $(\Pi^\vee)_K$ [62, Chap. 11]. The topological dual $U^\vee$ of U is the space of distribution vectors, on which we can define a continuous representation of G. This is the maximal globalization of $(\Pi^\vee)_K$ in the sense of Casselman–Wallach, which we refer to $(\Pi^\vee)^{-\infty}$. Thus we have

$$(\Pi^\vee)_K \subset \Pi^\vee \subset (\Pi^\vee)^{-\infty}.$$

We shall use these symbols for a representation π of the subgroup G' below.

Example 5.2. Let $\widetilde{\tau}$ be a finite-dimensional representation of a parabolic subgroup P' of G', and $\pi := \mathrm{Ind}_{P'}^{G'}(\widetilde{\tau})$ the representation on $C^\infty(G'/P', \mathcal{W})$. The dualizing bundle $\mathcal{W}^*$ is given as the G'-homogeneous vector bundle over G'/P' associated to $\tau^* := \widetilde{\tau}^\vee \otimes |\det(\mathrm{Ad}|_{\mathfrak{g}'/\mathfrak{p}'})|^{-1}$, where $\widetilde{\tau}^\vee$ is the contragredient representation of $\widetilde{\tau}$. Then the smooth admissible representation $\pi^\vee$ of G' is given as a representation $\mathrm{Ind}_{P'}^{G'}(\tau^*)$ on $C^\infty(G'/P', \mathcal{W}^*)$, whereas $(\pi^\vee)^{-\infty}$ is given as a representation on $\mathcal{D}'(G'/P', \mathcal{W}^*)$.

Any symmetry breaking operator $T\colon \Pi|_{G'} \to \pi^{\vee}$ induces a continuous bilinear form

$$\Pi \boxtimes \pi \to \mathbb{C}, \qquad u \otimes v \mapsto \langle Tu, v \rangle,$$

and we have a natural embedding

$$\operatorname{Hom}_{G'}(\Pi|_{G'}, \pi^{\vee}) \hookrightarrow \operatorname{Hom}_{G'}(\Pi \boxtimes \pi, \mathbb{C}) \simeq \operatorname{Hom}_{G'}(\Pi|_{G'}, (\pi^{\vee})^{-\infty}). \qquad (5.2)$$

Here the second isomorphism follows from the natural bijections for nuclear Fréchet spaces [57, Prop. 50.7]:

$$\operatorname{Hom}_{\mathbb{C}}(U \otimes U', \mathbb{C}) \simeq \operatorname{Hom}_{\mathbb{C}}(U, (U')^{\vee}),$$

where $\operatorname{Hom}_{\mathbb{C}}$ denotes the space of continuous linear maps.

As an immediate consequence of Fact 5.1, we have the following:

Proposition 5.3. *Suppose $\widetilde{\sigma}$ and $\widetilde{\tau}$ are finite-dimensional representations of parabolic subgroups P and P', respectively. Let $\Pi = \operatorname{Ind}_P^G(\widetilde{\sigma})$ and $\pi = \operatorname{Ind}_{P'}^{G'}(\widetilde{\tau})$ be admissible smooth representations of G and G', respectively. Then the embedding in (5.2) is an isomorphism.*

Proof. We recall that $\operatorname{Hom}_{\mathbb{C}}(\cdot, \mathbb{C})$ denotes the space of (continuous) functionals. Then $\operatorname{Hom}_{G'}(\Pi \boxtimes \pi, \mathbb{C})$ is naturally isomorphic to the spaces of G'-invariant elements of the following vector spaces

$$\operatorname{Hom}_{\mathbb{C}}(C^{\infty}(G/P \times G'/P', \mathcal{V} \boxtimes \mathcal{W}), \mathbb{C}) \simeq \mathcal{D}'(G/P \times G'/P', \mathcal{V}^* \boxtimes \mathcal{W}^*),$$

and so we have

$$\operatorname{Hom}_{G'}(\Pi \boxtimes \pi, \mathbb{C}) \simeq \mathcal{D}'(G/P \times G'/P', \mathcal{V}^* \boxtimes \mathcal{W}^*)^{\Delta(G')}.$$

Since $\tau^{**} \simeq \tau$, the right-hand side is canonically isomorphic to

$$\operatorname{Hom}_{G'}(C^{\infty}(G/P, \mathcal{V})|_{G'}, C^{\infty}(G'/P', \mathcal{W}^*)) \simeq \operatorname{Hom}_{G'}(\Pi|_{G'}, \pi^{\vee})$$

by Fact 5.1 and Example 5.2. Hence Proposition 5.3 is proved. □

More generally, we obtain the following.

Theorem 5.4. *Let $G \supset G'$ be a pair of real reductive Lie groups. For any $\Pi \in \operatorname{Irr}(G)$ and $\pi \in \operatorname{Irr}(G')$, we have a canonical bijection:*

$$\operatorname{Hom}_{G'}(\Pi|_{G'}, \pi^{\vee}) \xrightarrow{\sim} \operatorname{Hom}_{G'}(\Pi \boxtimes \pi, \mathbb{C}). \qquad (5.3)$$

By the second isomorphism (5.2), Theorem 5.4 is deduced from the following proposition, where we change the notation from $\pi^\vee$ to π for simplicity.

Proposition 5.5. *Suppose $\Pi \in \mathrm{Irr}(G)$ and $\pi \in \mathrm{Irr}(G')$, Let $\pi^{-\infty}$ be the representation of G' on distribution vectors. Then the natural embedding*

$$\mathrm{Hom}_{G'}(\Pi|_{G'}, \pi) \hookrightarrow \mathrm{Hom}_{G'}(\Pi|_{G'}, \pi^{-\infty})$$

is a bijection.

Proof of Proposition 5.5. We take P and P' to be minimal parabolic subgroups of G and G', respectively. By Casselman's subrepresentation theorem (or equivalently, "quotient theorem"), see [62, Chap. 3, Sect. 8] for instance, for any $\Pi \in \mathrm{Irr}(G)$, there exists an irreducible finite-dimensional representation $(\widetilde{\sigma}, V)$ of P such that Π_K is obtained as a quotient of $\mathrm{Ind}_P^G(\widetilde{\sigma})_K$, and therefore, there is a surjective continuous G-homomorphism $p \colon C^\infty(G/P, V) \to \Pi$ by the automatic continuity theorem [62, Chap. 11, Sect. 4]. Likewise, for any $\pi \in \mathrm{Irr}(G')$, there exists an irreducible finite-dimensional representation $(\widetilde{\tau}, W)$ of P' such that $\pi_{K'}$ is a subrepresentation of $\mathrm{Ind}_{P'}^{G'}(\widetilde{\tau})_{K'}$, and therefore, there is an injective continuous G'-homomorphism $\iota \colon \pi^{-\infty} \hookrightarrow \mathcal{D}'(G'/P', W)$ by the dual of the automatic continuity theorem. If $T \colon \Pi \to \pi^{-\infty}$ is a continuous G'-homomorphism, then T induces a continuous G'-homomorphism

$$\iota \circ T \circ p \colon C^\infty(G/P, V) \to \mathcal{D}'(G'/P', W).$$

By Proposition 5.3, $\iota \circ T \circ p$ is actually a continuous G'-homomorphism,

$$C^\infty(G/P, V) \to C^\infty(G'/P', W).$$

Hence the image of T is contained in the admissible smooth representation π. Since the topology of the admissible smooth representation π coincides with the relative topology of $C^\infty(G'/P', W)$, T is actually a G'-homomorphism $\Pi|_{G'} \to \pi$. □

Remark 5.6.

(1) In [2, Lem. A.0.8], the authors proved the injectivity of the map (5.2).
(2) Theorem 5.3 simplifies part of the proof of [32, Thm. 4.1] on twelve equivalence conditions including the finiteness criterion for the dimension of continuous invariant bilinear forms.

5.2   Distribution Kernels of Symmetry Breaking Operators for $G = O(n+1, 1)$

We analyze the distribution kernels of symmetry breaking operators in coordinates. For this, we set up some structural results for $G = O(n+1, 1)$.

5.2.1   Bruhat and Iwasawa Decompositions for $G = O(n+1, 1)$

We recall from (3.4) that the map $\psi_n : \mathbb{R}^n - \{0\} \to O(n)$, $x \mapsto \psi_n(x)$ is defined as the reflection with respect to the hyperplane orthogonal to x. By using $\psi_n(x)$, we give an explicit formula of the Bruhat decomposition $G = N_+ w M A N_+ \cup M A N_+$ and the Iwasawa decomposition $G = K A N_+$ for an element of the subgroup N_- for $G = O(n+1, 1)$. Here we set

$$w := \operatorname{diag}(1, \cdots, 1, -1) \in N_K(\mathfrak{a}). \tag{5.4}$$

Retain the notation as in Section 2.1.1. In particular, we recall from (2.4) and (2.5) the definition of the diffeomorphisms $n_+ : \mathbb{R}^n \xrightarrow{\sim} N_+$ and $n_- : \mathbb{R}^n \xrightarrow{\sim} N_-$, respectively.

Lemma 5.7 (Bruhat decomposition).   *For $b \in \mathbb{R}^n - \{0\}$,*

$$n_-(b) = n_+(a) \begin{pmatrix} -1 & & \\ & \psi_n(b) & \\ & & 1 \end{pmatrix} e^{tH} n,$$

where $a \in \mathbb{R}^n$ and $t \in \mathbb{R}$ are given uniquely by $a = -\dfrac{b}{|b|^2}$ and $e^t = |b|^2$, respectively, and $n \in N_+$.

Proof. Suppose that $a \in \mathbb{R}^n$, $\varepsilon = \pm 1$, $B \in O(n)$, $t \in \mathbb{R}$ and $n \in N_+$ satisfies

$$n_-(b) = n_+(a) w \begin{pmatrix} \varepsilon & & \\ & B & \\ & & \varepsilon \end{pmatrix} e^{tH} n. \tag{5.5}$$

Applying (5.5) to the vector $p_+ = {}^t(1, 0, \cdots, 0, 1) \in \Xi$ (see (2.8)), we have

$$\begin{pmatrix} 1 - |b|^2 \\ 2b \\ 1 + |b|^2 \end{pmatrix} = \varepsilon e^t \begin{pmatrix} 1 - |a|^2 \\ 2a \\ -1 - |a|^2 \end{pmatrix}.$$

Hence $\varepsilon = -1$, $e^t = \frac{1}{|a|^2}$, and $a = -|a|^2 b$. Thus $|a|\,|b| = 1$. In turn, (5.5) amounts to

$$n_+(a)^{-1}n_-(b) = \begin{pmatrix} -1 \\ & B \\ & & 1 \end{pmatrix} e^{tH} n,$$

whence $B = I_n + 2a^t b = I_n - \frac{2b^t b}{|b|^2} = \psi_n(b)$.   □

For $b \in \mathbb{R}^n$, we define $k(b) \in SO(n+1)$ by

$$k(b) := I_{n+1} + \frac{1}{1+|b|^2}\begin{pmatrix} -2|b|^2 & -2\,^t b \\ 2b & -2b^t b \end{pmatrix} = \psi_{n+1}(1,b)\begin{pmatrix} -1 \\ & I_n \end{pmatrix}. \quad (5.6)$$

Lemma 5.8 (Iwasawa decomposition).   *For any $b \in \mathbb{R}^n$, we have*

$$n_-(b) = k(b)e^{tH}n_+(a) \in KAN_+, \quad (5.7)$$

where $a \in \mathbb{R}^n$ and $t \in \mathbb{R}$ are given by $a = \frac{-b}{1+|b|^2}$ and $e^t = 1 + |b|^2$.

Proof. We shall prove that $k(b)$ in (5.7) is given by the formula (5.6). Since $n_-(b)$ is contained in the connected component of G, $k(b) = (k(b)_{ij})_{0 \le i,j \le n}$ in (5.7) belongs to the connected group $SO(n+1)$. We write $k(b) = (k(b)_0, k'(b))$ where $k(b)_0 \in \mathbb{R}^{n+1}$ and $k'(b) := (k(b)_{ij})_{\substack{0 \le i \le n \\ 1 \le j \le n}} \in M(n+1,n;\mathbb{R})$. Applying (5.7) to the vector $p_+ = {}^t(1,0,\cdots,0,1)$, we have

$$\begin{pmatrix} 1 - |b|^2 \\ 2b \\ 1 + |b|^2 \end{pmatrix} = e^t\begin{pmatrix} k(b)_0 \\ 1 \end{pmatrix}.$$

The last component shows $e^t = 1 + |b|^2$. In turn, we get the first column vector $k(b)_0$ of $k(b)$. On the other hand, we observe

$$k(b)_{ij} = (n_-(b)n_+(a)^{-1}e^{-tH})_{ij} = (n_-(b)n_+(a)^{-1})_{ij}$$

for $0 \le i \le n+1$ and $1 \le j \le n$. Hence we get

$$\begin{pmatrix} k'(b) \\ 0\cdots 0 \end{pmatrix} = \begin{pmatrix} (1-|b|^2)^t a - {}^t b \\ I_n + 2b^t a \\ (1+|b|^2)^t a + {}^t b \end{pmatrix},$$

which implies

$$a = -\frac{b}{1+|b|^2} \quad \text{and} \quad k'(b) = \begin{pmatrix} -\frac{1-|b|^2}{1+|b|^2}{}^t b - {}^t b \\ I_n - \frac{2b^t b}{1+|b|^2} \end{pmatrix} = \begin{pmatrix} \frac{-2}{1+|b|^2}{}^t b \\ I_n - \frac{2b^t b}{1+|b|^2} \end{pmatrix}.$$

In particular, we have shown that $k(b)$ in (5.7) is given by the formula (5.6). □

5.2.2 Distribution Kernels for Symmetry Breaking Operators

We apply Fact 5.1 to the pair $(G, G') = (O(n+1,1), O(n,1))$ and a pair of the minimal parabolic subgroups P and P'. With the notation of Fact 5.1, we shall take

$$\tilde{\sigma} = V \otimes \delta \otimes \mathbb{C}_\lambda \quad \text{on } V_{\lambda,\delta}$$

$$\tilde{\tau} = W \otimes \varepsilon \otimes \mathbb{C}_\nu \quad \text{on } W_{\varepsilon,\nu}$$

as (irreducible) representations of P and P', respectively, for $(\sigma, V) \in \widehat{O(n)}$, $\delta \in \{\pm\}$, and $\lambda \in \mathbb{C}$ and $(\tau, W) \in \widehat{O(n-1)}$, $\varepsilon \in \{\pm\}$, and $\nu \in \mathbb{C}$. We recall from (2.27) that $\mathcal{V}_{\lambda,\delta} = G \times_P V_{\lambda,\delta}$ is a homogeneous vector bundle over the real flag variety G/P. The dualizing bundle $\mathcal{V}_{\lambda,\delta}^*$ of $\mathcal{V}_{\lambda,\delta}$, is given by a G-homogeneous vector bundle over G/P associated to the representation of $P/N_+ \simeq MA \simeq O(n) \times \mathbb{Z}/2\mathbb{Z} \times \mathbb{R}$:

$$V_{\lambda,\delta}^* := (V_{\lambda,\delta})^\vee \otimes \mathbb{C}_{2\rho} \simeq V^\vee \boxtimes \delta \boxtimes \mathbb{C}_{n-\lambda},$$

where $V^\vee$ denotes the contragredient representation of (σ, V). Then the regular representation of G on the space $\mathcal{D}'(G/P, \mathcal{V}_{\lambda,\delta}^*)$ of $\mathcal{V}_{\lambda,\delta}^*$-valued distribution sections is the dual of the representation $I_\delta(V, \lambda)$ of G on $C^\infty(G/P, \mathcal{V}_{\lambda,\delta})$ as we discussed in Example 5.2.

In this special setting, Fact 5.1 amounts to the following.

Fact 5.9. *There is a natural bijective map:*

$$\mathrm{Hom}_{G'}(I_\delta(V, \lambda)|_{G'}, J_\varepsilon(W, \nu)) \xrightarrow{\sim} (\mathcal{D}'(G/P, \mathcal{V}_{\lambda,\delta}^*) \otimes W_{\nu,\varepsilon})^{\Delta(P')}, \quad T \mapsto K_T.$$

$$(5.8)$$

In [42, Def. 3.3], we defined regular symmetry breaking operators in the general setting. In our special setting, there is only one open P'-orbit in the real flag manifold G/P, and thus the definition is reduced to the following.

Definition 5.10 (regular symmetry breaking operator). A symmetry breaking operator $T: I_\delta(V, \lambda) \to J_\varepsilon(W, \nu)$ is *regular* if the support of the distribution kernel K_T is G/P.

5.2.3 Distribution Sections for Dualizing Bundle $\mathcal{V}^*_{\lambda, \delta}$ over G/P

This section provides a concrete description of the right-hand side of (5.8) in the coordinates on the open Bruhat cell.

We begin with a description of the G- and $\mathfrak{g}$-action on $\mathcal{D}'(G/P, \mathcal{V}^*_{\lambda,\delta})$ in the coordinates. We identify $\mathcal{D}'(G/P, \mathcal{V}^*_{\lambda,\delta})$ with a subspace of $V^\vee$-valued distribution on G via the following map:

$$\mathcal{D}'(G/P, \mathcal{V}^*_{\lambda,\delta}) \simeq (\mathcal{D}'(G) \otimes V^*_{\lambda,\delta})^{\Delta(P)} \subset \mathcal{D}'(G) \otimes V^\vee.$$

We recall that the Bruhat decomposition of G is given by $G = N_+ w P \cup P$ where $w = \mathrm{diag}(1, \cdots, 1, -1) \in G$, see (5.4). Since the real flag manifold G/P is covered by the two open subsets $N_+ w P/P$ and $N_- P/P$, distribution sections on G/P are determined uniquely by the restriction to these two open sets:

$$\mathcal{D}'(G/P, \mathcal{V}^*_{\lambda,\delta}) \hookrightarrow \mathcal{D}'(N_+ w P/P, \mathcal{V}^*_{\lambda,\delta}|_{N_+ w P/P}) \oplus \mathcal{D}'(N_- P/P, \mathcal{V}^*_{\lambda,\delta}|_{N_- P/P}).$$

$$(5.9)$$

By a little abuse of notation, we use the letters n_+ and n_- to denote the induced diffeomorphisms $\mathbb{R}^n \xrightarrow{\sim} N_+ w P/P$ and $\mathbb{R}^n \xrightarrow{\sim} N_- P/P$, respectively. Via the following trivialization of the two restricted bundles:

$$
\begin{array}{ccccccccc}
\mathbb{R}^n \times V^\vee & \xrightarrow{\sim} & \mathcal{V}^*_{\lambda,\delta}|_{N_+ w P/P} & \subset & \mathcal{V}^*_{\lambda,\delta} & \supset & \mathcal{V}^*_{\lambda,\delta}|_{N_- P/P} & \xleftarrow{\sim} & \mathbb{R}^n \times V^\vee \\
\downarrow & & \downarrow & & \downarrow & & \downarrow & & \downarrow \\
\mathbb{R}^n & \xrightarrow[n_+]{\sim} & N_+ w P/P & \subset & G/P & \supset & N_- P/P & \xleftarrow[n_-]{\sim} & \mathbb{R}^n,
\end{array}
$$

the injection (5.9) is restated as the following map:

$$\mathcal{D}'(G/P, \mathcal{V}^*_{\lambda,\delta}) \hookrightarrow (\mathcal{D}'(\mathbb{R}^n) \otimes V^\vee) \oplus (\mathcal{D}'(\mathbb{R}^n) \otimes V^\vee), \qquad f \mapsto (F_\infty, F)$$

$$(5.10)$$

where

$$F_\infty(a) := f(n_+(a)w), \qquad F(b) := f(n_-(b)).$$

Lemma 5.11. *Let $\psi_n \colon \mathbb{R}^n - \{0\} \to O(n)$ be the map taking the reflection defined in* (3.4).

(1) *The image of the injective map* (5.10) *is characterized by the following identity in $\mathcal{D}'(\mathbb{R}^n - \{0\}) \otimes V^\vee$:*

$$F(b) = \delta\sigma^\vee(\psi_n(b)^{-1})|b|^{2\lambda - 2n} F_\infty(-\frac{b}{|b|^2}) \qquad on \ \mathbb{R}^n - \{0\}. \qquad (5.11)$$

(2) (first projection) $f \in \mathcal{D}'(G/P, V_{\lambda,\delta}^*)$ *is supported at the singleton $\{[p_+]\} = \{eP/P\}$ if and only if $F_\infty = 0$.*

(3) (second projection) *The second projection $f \mapsto F$ is injective.*

Proof.

(1) The image of the map (5.9) is characterized by the compatibility condition on the intersection $(N_+ w P \cap N_- P)/P$, namely, the pair (F_∞, F) in (5.10) should satisfy:

$$F(b) = \sigma_{\lambda,\delta}^*(p)^{-1} F_\infty(a)$$

for all $(a, b, p) \in \mathbb{R}^n \times \mathbb{R}^n \times P$ such that $n_+(a)wp = n_-(b)$. In this case, $b \neq 0$ because $N_+ w P \not\ni e$. By Lemma 5.7, we have

$$a = -\frac{b}{|b|^2}, \quad p = \begin{pmatrix} -1 & & \\ & \psi_n(b) & \\ & & -1 \end{pmatrix} e^{tH},$$

where $e^t = |b|^2$. Then

$$F(b) = f(n_-(b))$$
$$= \sigma_{\lambda,\delta}^*(p^{-1}) f(n_+(a)w)$$
$$= \delta|b|^{2\lambda - 2n}\sigma^\vee(\psi_n(b)) F_\infty(a).$$

(2) Clear from $G - N_+ w P = P$.

(3) Since $P' N_- P = G$ [42, Cor. 5.5], the third statement follows from [42, Thm. 3.16]. $\qquad\qquad\qquad\qquad\qquad\qquad\qquad\qquad\qquad\qquad\square$

The regular representation of G on $\mathcal{D}'(G/P, V_{\lambda,\delta}^*)$ induces an action on the pairs (F_∞, F) of $V^\vee$-valued distributions through Lemma 5.11 (1). We need an explicit formula of the action of the parabolic subgroup $P = MAN_+$ or its Lie algebra $\mathfrak{p} = \mathfrak{m} + \mathfrak{a} + \mathfrak{n}_+$, which is given in the following two elementary lemmas.

We begin with the first projection $f \mapsto F_\infty$ in (5.10). Since the action of P on G/P leaves the open subset $N_+ w P/P = PwP/P$ invariant, we can define the geometric action of the group P on $\mathcal{D}'(N_+ w P/P, V_{\lambda,\delta}^*)$ as follows. We recall

$M = O(n) \times \{1, m_-\}$ (see (2.7)). We collect some basic formulæ for the coordinates $n_\varepsilon \colon \mathbb{R}^n \xrightarrow{\sim} N_\varepsilon$: for $\varepsilon = \pm$ (by abuse of notation, we also write as $\varepsilon = \pm 1$),

$$n_\varepsilon(Bb) = \begin{pmatrix} 1 & \\ & B \\ & & 1 \end{pmatrix} n_\varepsilon(b) \begin{pmatrix} 1 & \\ & B^{-1} \\ & & 1 \end{pmatrix} \quad \text{for } B \in O(n), \tag{5.12}$$

$$n_\varepsilon(-b) = m_- n_\varepsilon(b) m_-^{-1}, \tag{5.13}$$

$$n_\varepsilon(e^{\varepsilon t} b) = e^{tH} n_\varepsilon(b) e^{-tH}. \tag{5.14}$$

Lemma 5.12. *We let $P = MAN_+$ act on $\mathcal{D}'(\mathbb{R}^n) \otimes V^\vee$ by*

$$\left(\pi \begin{pmatrix} 1 & \\ & B \\ & & 1 \end{pmatrix} F_\infty \right)(a) = \sigma^\vee(B) F_\infty(B^{-1} a) \quad \text{for } B \in O(n), \tag{5.15}$$

$$(\pi(m_-) F_\infty)(a) \quad = \delta F_\infty(-a), \tag{5.16}$$

$$(\pi(e^{tH}) F_\infty)(a) \quad = e^{(\lambda - n)t} F_\infty(e^{-t} a) \quad \text{for all } t \in \mathbb{R}, \tag{5.17}$$

$$(\pi(n_+(c)) F_\infty)(a) \quad = F_\infty(a - c) \quad \text{for all } c \in \mathbb{R}^n. \tag{5.18}$$

Then the first projection $f \mapsto F_\infty$ in (5.10) is a P-homomorphism.

Proof. We give a proof for (5.17) on the action of the split abelian group A. Let $t \in \mathbb{R}$. By (5.14) and $e^{-tH} w = w e^{tH}$, we have

$$f(e^{-tH} n_+(a) w) = f(n_+(e^{-t} a) e^{-tH} w) = e^{(\lambda - n)t} f(n_+(e^{-t} a) w)$$
$$= e^{(\lambda - n)t} F_\infty(e^{-t} a),$$

whence we get the desired formula. The proof for the actions of M and N_+ is similar. □

Next, we consider the second projection $f \mapsto F$ in (5.10). In this case, the group N_+ does not preserve the open subset $N_- P/P$ in G/P, and therefore we shall use the action of the Lie algebra $\mathfrak{n}_+$ instead (see (5.22) below). We denote by E the Euler homogeneity operator $\sum_{\ell=1}^n x_\ell \frac{\partial}{\partial x_\ell}$.

Lemma 5.13. *We let the group MA and the Lie algebra $\mathfrak{n}_+$ act on $\mathcal{D}'(\mathbb{R}^n) \otimes V^\vee$ by*

$$\left(\pi \begin{pmatrix} 1 & \\ & B \\ & & 1 \end{pmatrix} F \right)(b) = \sigma^\vee(B) F(B^{-1} b) \quad \text{for } B \in O(n), \tag{5.19}$$

$$(\pi(m_-) F)(b) \quad = \delta F(-b), \tag{5.20}$$

$$(\pi(e^{tH})F)(b) \qquad = e^{(n-\lambda)t}F(e^t b) \quad \textit{for all } t \in \mathbb{R}, \tag{5.21}$$

$$d\pi(N_j^+)F(b) \qquad = \left((\lambda-n)b_j - b_j E + \frac{1}{2}|b|^2\frac{\partial}{\partial b_j}\right)F \quad \textit{for } 1 \le j \le n. \tag{5.22}$$

Here $b = (b_1, \cdots, b_n)$. *Then the second projection* $f \mapsto F$ *in* (5.10) *is an* $(MA, \mathfrak{n}_+)$-*homomorphism.*

Proof. See [42, Prop. 6.4] for (5.22). The other formulæ are easy, and we omit the proof.   □

5.2.4   Pair of Distribution Kernels for Symmetry Breaking Operators

We extend Lemma 5.11 to give a local expression of the distribution kernels of symmetry breaking operators via the isomorphism (5.8). Suppose $(\tau, W) \in \widehat{O(n-1)}$, $\nu \in \mathbb{C}$, and $\varepsilon \in \{\pm\}$. We define

$$(\mathcal{D}'(\mathbb{R}^n) \otimes \mathrm{Hom}_{\mathbb{C}}(V, W))^{\Delta(P')} \equiv (\mathcal{D}'(\mathbb{R}^n) \otimes \mathrm{Hom}_{\mathbb{C}}(V_{\lambda,\delta}, W_{\nu,\varepsilon}))^{\Delta(P')} \tag{5.23}$$

to be the space of $\mathrm{Hom}_{\mathbb{C}}(V, W)$-valued distributions $\mathcal{T}_{\infty}$ on $\mathbb{R}^n$ satisfying the following four conditions:

$$\tau(B) \circ \mathcal{T}_{\infty}(B^{-1}y, y_n) \circ \sigma^{-1}(B) = \mathcal{T}_{\infty}(y, y_n) \qquad \text{for all } B \in O(n-1), \tag{5.24}$$

$$\mathcal{T}_{\infty}(-y, -y_n) = \delta\varepsilon \mathcal{T}_{\infty}(y, y_n), \tag{5.25}$$

$$\mathcal{T}_{\infty}(e^t y, e^t y_n) = e^{(\lambda+\nu-n)t}\mathcal{T}_{\infty}(y, y_n) \qquad \text{for all } t \in \mathbb{R}, \tag{5.26}$$

$$\mathcal{T}_{\infty}(y - z, y_n) = \mathcal{T}_{\infty}(y, y_n) \qquad \text{for all } z \in \mathbb{R}^{n-1}. \tag{5.27}$$

For the open Bruhat cell $N_- P \subset G$, we consider the following.

Definition 5.14. We define $Sol(\mathbb{R}^n; V_{\lambda,\delta}, W_{\nu,\varepsilon}) \subset \mathcal{D}'(\mathbb{R}^n) \otimes \mathrm{Hom}_{\mathbb{C}}(V, W)$ to be the space of $\mathrm{Hom}_{\mathbb{C}}(V, W)$-valued distributions $\mathcal{T}$ on $\mathbb{R}^n$ satisfying the following invariance conditions under the action of the Lie algebras $\mathfrak{a}$, $\mathfrak{n}'_+$, and the group $M' \simeq O(n-1) \times O(1)$:

$$(E - (\lambda - \nu - n))\mathcal{T} = 0, \tag{5.28}$$

$$\left((\lambda-n)x_j - x_j E + \frac{1}{2}(|x|^2 + x_n^2)\frac{\partial}{\partial x_j}\right)\mathcal{T} = 0 \quad (1 \le j \le n-1), \tag{5.29}$$

$$\tau(m) \circ \mathcal{T}(m^{-1}b) \circ \sigma(m^{-1}) = \mathcal{T}(b) \qquad \text{for all } m \in O(n-1), \tag{5.30}$$

$$\mathcal{T}(-b) = \delta\varepsilon\mathcal{T}(b). \tag{5.31}$$

Applying Lemma 5.11 to the right-hand side of (5.8), we have the following:

Proposition 5.15. *Let $(\sigma, V) \in \widehat{M}$, $(\tau, W) \in \widehat{M'}$, $\delta, \varepsilon \in \{\pm\}$, and $\lambda, \nu \in \mathbb{C}$.*

(1) There is a one-to-one correspondence between a symmetry breaking operator

$$\mathbb{T} \in \mathrm{Hom}_{G'}(I_\delta(V, \lambda)|_{G'}, J_\varepsilon(W, \nu))$$

and a pair $(\mathcal{T}_\infty, \mathcal{T})$ of $\mathrm{Hom}_{\mathbb{C}}(V, W)$-valued distributions on $\mathbb{R}^n$ subject to the following three conditions:

$$\mathcal{T}_\infty \in (\mathcal{D}'(\mathbb{R}^n) \otimes \mathrm{Hom}_{\mathbb{C}}(V_{\lambda,\delta}, W_{\nu,\varepsilon}))^{\Delta(P')}, \tag{5.32}$$

$$\mathcal{T} \in Sol(\mathbb{R}^n; V_{\lambda,\delta}, W_{\nu,\varepsilon}), \tag{5.33}$$

$$\mathcal{T}(b) = \delta Q(b)^{\lambda-n} \mathcal{T}_\infty \left(-\frac{b}{|b|^2}\right) \circ \sigma(\psi_n(b)) \quad on \ \mathbb{R}^n - \{0\}. \tag{5.34}$$

(2) $\mathcal{T}$ determines $\mathbb{T}$ uniquely.

(3) Suppose that $\mathbb{T} \leftrightarrow (\mathcal{T}_\infty, \mathcal{T})$ is the correspondence in (1). Then the following three conditions are equivalent:

(i) $\mathcal{T}_\infty = 0$.
(ii) $\mathrm{Supp}\,\mathcal{T} \subset \{0\}$.
(iii) $\mathbb{T}$ is a differential operator (see Definition 6.3).

Proof. The first statement follows from Fact 5.9, Lemmas 5.11 (1), 5.12 and 5.13. The second statement is immediate from Lemma 5.11 (3). The third one is proved in [40], see Section 6.1 for more details about differential operators between two manifolds. □

Remark 5.16. The advantage of using $\mathcal{T}$ is that the second projection

$$\mathrm{Hom}_{G'}(I_\delta(V, \lambda)|_{G'}, J_\varepsilon(W, \nu)) \overset{\sim}{\to} Sol(\mathbb{R}^n; \sigma_{\lambda,\delta}, \tau_{\nu,\varepsilon}), \quad \mathbb{T} \mapsto \mathcal{T}$$

is bijective, and therefore, it is sufficient to use $\mathcal{T}$ in order to describe a symmetry breaking operator $\mathbb{T}$. This was the approach that we took in [42]. In this monograph, we shall use both $\mathcal{T}_\infty$ and $\mathcal{T}$. The advantage of using $\mathcal{T}_\infty$ is that the group P' leaves N_+wP/P invariant, and consequently, we can easily determine $\mathcal{T}_\infty$ (see Proposition 5.20 below), although the first projection

$$\mathrm{Hom}_{G'}(I_\delta(V, \lambda)|_{G'}, J_\varepsilon(W, \nu)) \to (\mathcal{D}'(\mathbb{R}^n) \otimes \mathrm{Hom}_{\mathbb{C}}(V, W))^{\Delta(P')}, \quad \mathbb{T} \mapsto \mathcal{T}_\infty$$

is neither injective nor surjective. We shall return to this point in Section 5.6.

5.3   Distribution Kernels near Infinity

Let $(\mathcal{T}_\infty, \mathcal{T})$ be as in Proposition 5.15. This section determines $\mathcal{T}_\infty$ up to scalar multiplication. The main result is Proposition 5.20, which also determines uniquely the restriction of $\mathcal{T}$ to $\mathbb{R}^n - \{0\}$ up to scalar multiplication.

Example 5.17. For $\sigma = 1$, $\tau = 1$, $\delta = +1$, and

$$\mathcal{T}_\infty(y, y_n) = |y_n|^{\lambda + \nu - n},$$

we have from (5.34)

$$\mathcal{T}(x, x_n) = (|x|^2 + x_n^2)^{-\nu} |x_n|^{\lambda + \nu - n}.$$

We begin with the following classical result on homogeneous distributions of one variable:

Lemma 5.18.

(1) *Both* $\frac{1}{\Gamma(\frac{\mu}{2})}|t|^{\mu-1}$ *and* $\frac{1}{\Gamma(\frac{\mu+1}{2})}|t|^{\mu-1}\operatorname{sgn}t$ *are nonzero distributions on* $\mathbb{R}$ *that depend holomorphically on* μ *in the entire complex plane* $\mathbb{C}$.

(2) *Suppose* $k \in \mathbb{N}$. *Then*

$$\frac{|t|^{\mu-1}}{\Gamma(\frac{\mu}{2})} = \frac{(-1)^k}{2^k(2k-1)!!}\delta^{(2k)}(t) \qquad\qquad if\ \mu = -2k,$$

$$\frac{|t|^{\mu-1}\operatorname{sgn}t}{\Gamma(\frac{\mu+1}{2})} = \frac{(-1)^k(k-1)!}{(2k-1)!}\delta^{(2k-1)}(t) \qquad if\ \mu = -2k - 1.$$

(3) *Suppose* $\mu \in \mathbb{C}$ *and* $\gamma = \pm 1$. *Then any distribution* $g(t)$ *on* $\mathbb{R}$ *satisfying the homogeneity condition*

$$g(at) = a^{\mu-1}g(t)\ for\ all\ a > 0,\ and\ g(-t) = \gamma g(t)$$

is a scalar multiple of $\frac{1}{\Gamma(\frac{\mu}{2})}|t|^{\mu-1}$ $(\gamma = 1)$, *or of* $\frac{1}{\Gamma(\frac{\mu+1}{2})}|t|^{\mu-1}\operatorname{sgn}t$ $(\gamma = -1)$.

For $(\sigma, V) \in \widehat{O(n)}$ and $(\tau, W) \in \widehat{O(n-1)}$, we recall that $[V : W]$ is the dimension of $\operatorname{Hom}_{O(n-1)}(V|_{O(n-1)}, W)$. Suppose $[V : W] \neq 0$, or equivalently, $[V : W] = 1$. We fix a generator

$$\operatorname{pr}_{V \to W} \in \operatorname{Hom}_{O(n-1)}(V|_{O(n-1)}, W)$$

which is unique up to nonzero scalar multiplication by Schur's lemma. In light of the Γ-factors in Lemma 5.18, we introduce $\operatorname{Hom}_{\mathbb{C}}(V, W)$-valued distributions

$(\tilde{\tilde{\mathcal{A}}}^{V,W}_{\lambda,\nu,\pm})_\infty$ on $\mathbb{R}^n$ that depend holomorphically on $(\lambda,\nu) \in \mathbb{C}^2$ by

$$(\tilde{\tilde{\mathcal{A}}}^{V,W}_{\lambda,\nu,+})_\infty(x,x_n) := \frac{1}{\Gamma(\frac{\lambda+\nu-n+1}{2})}|x_n|^{\lambda+\nu-n}\,\mathrm{pr}_{V\to W}, \tag{5.35}$$

$$(\tilde{\tilde{\mathcal{A}}}^{V,W}_{\lambda,\nu,-})_\infty(x,x_n) := \frac{1}{\Gamma(\frac{\lambda+\nu-n+2}{2})}|x_n|^{\lambda+\nu-n}\,\mathrm{sgn}\,x_n\,\mathrm{pr}_{V\to W}. \tag{5.36}$$

We regard $\mathrm{pr}_{V\to W} = 0$ if $[V:W] = 0$.

Remark 5.19. The notation $(\tilde{\tilde{\mathcal{A}}}^{V,W}_{\lambda,\nu,\gamma})_\infty$ with double tildes is used here because it will be compatible with the *renormalization* $\tilde{\tilde{\mathbb{A}}}^{V,W}_{\lambda,\nu,\gamma}$ of the normalized symmetry breaking operator $\widetilde{\mathbb{A}}^{V,W}_{\lambda,\nu,\gamma}$ which we will introduce in the next sections.

Let $\gamma = \delta\varepsilon$. If there exists $\mathcal{T}_\gamma \in Sol(\mathbb{R}^n; V_{\lambda,\delta}, W_{\nu,\varepsilon})$ such that the pair $((\tilde{\tilde{\mathcal{A}}}^{V,W}_{\lambda,\nu,\varepsilon})_\infty, \mathcal{T}_\gamma)$ satisfies the compatibility condition (5.34), then the restriction $\mathcal{T}_\gamma|_{\mathbb{R}^n-\{0\}}$ must be of the form $(\tilde{\tilde{\mathcal{A}}}^{V,W}_{\lambda,\nu,\gamma})' \in \mathcal{D}'(\mathbb{R}^n - \{0\}) \otimes \mathrm{Hom}_{\mathbb{C}}(V, W)$ where we set

$$(\tilde{\tilde{\mathcal{A}}}^{V,W}_{\lambda,\nu,+})' := \frac{1}{\Gamma(\frac{\lambda+\nu-n+1}{2})}(|x|^2 + x_n^2)^{-\nu}|x_n|^{\lambda+\nu-n}R^{V,W}(x,x_n), \tag{5.37}$$

$$(\tilde{\tilde{\mathcal{A}}}^{V,W}_{\lambda,\nu,-})' := \frac{1}{\Gamma(\frac{\lambda+\nu-n+2}{2})}(|x|^2 + x_n^2)^{-\nu}|x_n|^{\lambda+\nu-n}\,\mathrm{sgn}\,x_n\,R^{V,W}(x,x_n), \tag{5.38}$$

with $R^{V,W} = \mathrm{pr}_{V\to W} \circ \sigma \circ \psi_n$ (see (3.6)). We have used the notation $(\tilde{\tilde{\mathcal{A}}}^{V,W}_{\lambda,\nu,\gamma})'$ instead of $\tilde{\tilde{\mathcal{A}}}^{V,W}_{\lambda,\nu,\gamma}$ because it is defined only on $\mathbb{R}^n - \{0\}$ and may not extend to $\mathbb{R}^n$ (see Proposition 6.19 below).

Then we have:

Proposition 5.20.

(1) *For any $(\sigma, V) \in \widehat{O(n)}$, $(\tau, W) \in \widehat{O(n-1)}$, $\delta, \varepsilon \in \{\pm\}$, and $\lambda, \nu \in \mathbb{C}$, we have*

$$(\mathcal{D}'(\mathbb{R}^n) \otimes \mathrm{Hom}_{\mathbb{C}}(V_{\lambda,\delta}, W_{\nu,\varepsilon}))^{\Delta(P')} = \mathbb{C}(\tilde{\tilde{\mathcal{A}}}^{V,W}_{\lambda,\nu,\delta\varepsilon})_\infty.$$

(2) *If $[V:W] \neq 0$ then $(\tilde{\tilde{\mathcal{A}}}^{V,W}_{\lambda,\nu,+})' \neq 0$ for all $\lambda, \nu \in \mathbb{C}$.*

(3) *If $\mathcal{T} \in Sol(\mathbb{R}^n; V_{\lambda,\delta}, W_{\nu,\varepsilon})$, then $\mathcal{T}|_{\mathbb{R}^n-\{0\}}$ is a scalar multiple of $(\tilde{\tilde{\mathcal{A}}}^{V,W}_{\lambda,\nu,\delta\varepsilon})'$.*

Proof. Suppose $F \in (\mathcal{D}'(\mathbb{R}^n) \otimes \mathrm{Hom}_{\mathbb{C}}(V_{\lambda,\delta}, W_{\nu,\varepsilon}))^{\Delta(P')}$.

(1) Let $p_n \colon \mathbb{R}^n \to \mathbb{R}$ be the n-th projection, and $p_n^* \colon \mathcal{D}'(\mathbb{R}) \to \mathcal{D}'(\mathbb{R}^n)$ the pull-back of distributions. By the N'_+-invariance (5.27), F depends only on the last

coordinate, namely, F is of the form $p_n^* f$ for some $f \in \mathcal{D}'(\mathbb{R}) \otimes \mathrm{Hom}_\mathbb{C}(V, W)$. In turn, the $O(n-1)$-invariance (5.24) implies

$$f \in \mathcal{D}'(\mathbb{R}) \otimes \mathrm{Hom}_{O(n-1)}(V|_{O(n-1)}, W).$$

In particular, $F = 0$ if $[V : W] = 0$.

From now, we assume $[V : W] \neq 0$. Then f is of the form $h(y_n) \mathrm{pr}_{V \to W}$ for some $h(t) \in \mathcal{D}'(\mathbb{R})$. By (5.25) and (5.26), h is a homogeneous distribution of degree $\lambda + \nu - n$ and of parity $\delta\varepsilon$. Then $h(t)$ is determined by Lemma 5.18, and we get the desired result.

(2) The assertion follows from the nonvanishing statement for the distribution of one-variable (see Lemma 5.18 (1)).

(3) The third statement follows from the first assertion and Proposition 5.15. □

5.4   Vanishing Condition of Differential Symmetry Breaking Operators: Proof of Theorem 3.12 (1)

In this section, we prove a necessary condition for the existence of nonzero differential symmetry breaking operators as stated in Theorem 3.12 (1):

Theorem 5.21 (vanishing of differential symmetry breaking operators). *Suppose that V and W are finite-dimensional representations of $O(n)$ and $O(n-1)$, respectively, δ, $\varepsilon \in \{\pm\}$, and $(\lambda, \nu) \in \mathbb{C}^2$. If $(\lambda, \nu, \delta, \varepsilon)$ satisfies the generic parameter condition (3.2), namely, $\nu - \lambda \notin 2\mathbb{N}$ for $\delta\varepsilon = +$, or $\nu - \lambda \notin 2\mathbb{N} + 1$ for $\delta\varepsilon = -$, then*

$$\mathrm{Diff}_{G'}(I_\delta(V, \lambda)|_{G'}, J_\varepsilon(W, \nu)) = \{0\}.$$

Remark 5.22. In the above theorem, we do not impose any assumption on V and W. In Chapter 6, we give a converse implication under the assumption $[V : W] \neq 0$, see Theorem 6.1.

For the proof of Theorem 5.21, we use the following properties of distributions supported at the origin:

Lemma 5.23. *Let F be any $\mathrm{Hom}_\mathbb{C}(V, W)$-valued distribution on $\mathbb{R}^n$ supported at the origin and satisfying the Euler homogeneity differential equation (5.28).*

(1) *Assume $\nu - \lambda \notin \mathbb{N}$. Then F must be zero.*

(2) *Assume $\nu - \lambda \in \mathbb{N}$. Then $F(-x) = (-1)^{\nu-\lambda} F(x)$.*

Proof. Let $\delta(x) \equiv \delta(x_1, \cdots, x_n)$ be the Dirac delta function on $\mathbb{R}^n$. For a multi-index $\alpha = (\alpha_1, \cdots, \alpha_n) \in \mathbb{N}^n$, we define another distribution by

$$\delta^{(\alpha)}(x_1, \cdots, x_n) := \frac{\partial^{|\alpha|}}{\partial x_1^{\alpha_1} \cdots \partial x_n^{\alpha_n}} \delta(x_1, \cdots, x_n)$$

where $|\alpha| = \alpha_1 + \cdots + \alpha_n$. By the structural theory of distributions, F must be of the following form

$$F = \sum_{\alpha \in \mathbb{N}^n} a_\alpha \delta^{(\alpha)}(x_1, \cdots, x_n) \qquad \text{(finite sum)}$$

with some $a_\alpha \in \mathrm{Hom}_{\mathbb{C}}(V, W)$ for $\alpha \in \mathbb{N}^n$. Since $\delta^{(\alpha)}(x_1, \cdots, x_n)$ is a homogeneous distribution of degree $-n - |\alpha|$, the Euler homogeneity operator E acts as the scalar multiplication by $-(n + |\alpha|)$, and thus

$$EF = -\sum_{\alpha \in \mathbb{N}^n} (n + |\alpha|) a_\alpha \delta^{(\alpha)}(x_1, \cdots, x_n).$$

Since $\{\delta^{(\alpha)}(x_1, \cdots, x_n)\}_{\alpha \in \mathbb{N}^n}$ are linearly independent distributions, the differential equation (5.28), namely, $EF = (\lambda - \nu - n)F$ implies that

$$a_\alpha = 0 \qquad \text{whenever} \ -n - |\alpha| \neq \lambda - \nu - n.$$

Thus we conclude:

(1) If $\nu - \lambda \notin \mathbb{N}$, we get $a_\alpha = 0$ for all $\alpha \in \mathbb{N}^n$, whence $F = 0$.
(2) If $\nu - \lambda \in \mathbb{N}$, then a_α can survive only when $|\alpha| = \nu - \lambda$. Then $F(-x) = (-1)^{|\alpha|}F(x) = (-1)^{\nu - \lambda}F(x)$ because $\delta(x) = \delta(-x)$.

Therefore Lemma 5.23 is proved. □

Proof of Theorem 5.21. Immediate from the characterization of differential symmetry breaking operators (Proposition 5.15 (3)) and from Lemma 5.23. □

5.5 Upper Estimate of the Multiplicities

We recall from the general theory [39] that there exists a constant $C > 0$ such that

$$\dim_{\mathbb{C}} \mathrm{Hom}_{G'}(I_\delta(V, \lambda)|_{G'}, J_\varepsilon(W, \nu)) \leq C \qquad (5.39)$$

for any $(\sigma, V) \in \widehat{O(n)}$, $(\tau, W) \in \widehat{O(n-1)}$, $\delta, \varepsilon \in \{\pm\}$, and $(\lambda, \nu) \in \mathbb{C}^2$. Moreover, we also know that the left-hand side of (5.39) is either 0 or 1 if both the G-module $I_\delta(V, \lambda)$ and the G'-module $J_\varepsilon(W, \nu)$ are irreducible [55]. In this section, we give a more precise upper estimate of the dimension of (continuous) symmetry breaking operators by that of *differential* symmetry breaking operators. Owing to the "duality theorem" (see [40, Thm. 2.9], see also Fact 6.5 in the next chapter), the latter object can be studied algebraically as a branching problem for generalized Verma modules, and is completely classified in [35] when $(V, W) = (\bigwedge^i (\mathbb{C}^n), \bigwedge^j (\mathbb{C}^{n-1}))$. The proof for the upper estimate leads us to complete the proof of a localness theorem (Theorem 3.6), namely, a sufficient condition for all symmetry breaking operators to be differential operators.

Theorem 5.24 (upper estimate of dimension). *For any* $V \in \widehat{O(n)}$, $W \in \widehat{O(n-1)}$, $\delta, \varepsilon \in \{\pm\}$, *and* $(\lambda, \nu) \in \mathbb{C}^2$, *we have*

$$\dim_{\mathbb{C}} \mathrm{Hom}_{G'}(I_\delta(V, \lambda)|_{G'}, J_\varepsilon(W, \nu)) \le 1 + \dim_{\mathbb{C}} \mathrm{Diff}_{G'}(I_\delta(V, \lambda)|_{G'}, J_\varepsilon(W, \nu)).$$

Proof. Let $(\mathcal{T}_\infty, T)$ be the pair of distribution kernels of a symmetry breaking operator $\mathbb{T}$ as in Proposition 5.15. Then the first projection $\mathbb{T} \mapsto \mathcal{T}_\infty$ induces an exact sequence:

$$0 \to \mathrm{Diff}_{G'}(I_\delta(V, \lambda)|_{G'}, J_\varepsilon(W, \nu)) \to \mathrm{Hom}_{G'}(I_\delta(V, \lambda)|_{G'}, J_\varepsilon(W, \nu))$$

$$\to \mathbb{C}(\tilde{\mathcal{A}}_{\lambda, \nu, \delta\varepsilon}^{V, W})_\infty,$$

by Propositions 5.15 (3) and 5.20. Thus Theorem 5.24 is proved.   □

We are ready to prove a localness theorem stated in Theorem 3.6.

Proof of Theorem 3.6. If $[V : W] = 0$ then $(\mathcal{D}'(\mathbb{R}^n) \otimes \mathrm{Hom}_{\mathbb{C}}(V_{\lambda,\delta}, W_{\nu,\varepsilon}))^{P'} = \{0\}$ by Proposition 5.20 because $\mathrm{pr}_{V \to W} = 0$. Hence we get Theorem 3.6 by the exact sequence in the above proof.   □

We also prove a part of Theorem 3.3, a generic uniqueness result.

Corollary 5.25. *Suppose* $(\sigma, V) \in \widehat{O(n)}$, $(\tau, W) \in \widehat{O(n-1)}$, $\delta, \varepsilon \in \{\pm\}$, *and* $(\lambda, \nu) \in \mathbb{C}^2$. *If* $(\lambda, \nu, \delta, \varepsilon)$ *satisfies the generic parameter condition* (3.2), *namely, if* $\nu - \lambda \notin 2\mathbb{N}$ *for* $\delta\varepsilon = +$, *or* $\nu - \lambda \notin 2\mathbb{N} + 1$ *for* $\delta\varepsilon = -$, *then*

$$\dim_{\mathbb{C}} \mathrm{Hom}_{G'}(I_\delta(V, \lambda)|_{G'}, J_\varepsilon(W, \nu)) \le 1.$$

Proof of Corollary 5.25. Owing to Theorem 5.24, we obtain Corollary 5.25 by Theorem 5.21.   □

We shall see that the inequality in Corollary 5.25 is actually the equality by showing the lower estimate of the multiplicities in Theorem 5.42 below.

5.6   Proof of Theorem 3.10: Analytic Continuation of Symmetry Breaking Operators $\widetilde{\mathbb{A}}^{V,W}_{\lambda,\nu,\pm}$

The goal of this section is to complete the proof of Theorem 3.10 about the analytic continuation of $\widetilde{\mathbb{A}}^{V,W}_{\lambda,\nu,\pm}$. For $(\sigma, V) \in \widehat{O(n)}$ and $(\tau, W) \in \widehat{O(n-1)}$ such that $[V : W] \neq 0$ and for $\delta, \varepsilon \in \{\pm\}$, we set $\gamma = \delta\varepsilon$ and construct a family of matrix-valued symmetry breaking operators, to be denoted by

$$\widetilde{\mathbb{A}}^{V,W}_{\lambda,\nu,\gamma} : I_\delta(V, \lambda) \to J_\varepsilon(W, \nu),$$

which are initially defined for $\mathrm{Re}\,\lambda \gg |\mathrm{Re}\,\nu|$ in Lemma 5.31. We show that they have a holomorphic continuation to the entire plane $(\lambda, \nu) \in \mathbb{C}^2$, and thus complete the proof of Theorem 3.10.

Here is a strategy.

Step 0.   (distribution kernel near infinity)

We define $\mathrm{Hom}_\mathbb{C}(V, W)$-valued distributions $(\widetilde{\mathcal{A}}^{V,W}_{\lambda,\nu,\gamma})_\infty$ on $\mathbb{R}^n$ as a multiplication of $(\widetilde{\mathcal{A}}^{V,W}_{\lambda,\nu,\gamma})_\infty$ (see (5.35) and (5.36)) by appropriate holomorphic functions of λ and ν (Section 5.6.1). The distributions $(\widetilde{\mathcal{A}}^{V,W}_{\lambda,\nu,\gamma})_\infty$ depend holomorphically on (λ, ν) in the entire plane $\mathbb{C}^2$ (but may vanish at special (λ, ν)).

Step 1.   (very regular case)   For $\mathrm{Re}\,\lambda \gg |\mathrm{Re}\,\nu|$, we define $\mathrm{Hom}_\mathbb{C}(V, W)$-valued, locally integrable functions $\widetilde{\mathcal{A}}^{V,W}_{\lambda,\nu,\pm}$ on $\mathbb{R}^n$ such that the restriction $\widetilde{\mathcal{A}}^{V,W}_{\lambda,\nu,\pm}|_{\mathbb{R}^n - \{0\}}$ satisfies the compatibility condition (5.34). We then prove that the pair $((\widetilde{\mathcal{A}}^{V,W}_{\lambda,\nu,\gamma})_\infty, \widetilde{\mathcal{A}}^{V,W}_{\lambda,\nu,\gamma})$ belongs to $(\mathcal{D}'(G/P, \mathcal{V}^*_{\lambda,\delta}) \otimes W_{\nu,\varepsilon})^{\Delta(P')}$ for $\delta\varepsilon = \gamma$ if $\mathrm{Re}\,\lambda \gg |\mathrm{Re}\,\nu|$.

Step 2.   (meromorphic continuation and possible poles of $\widetilde{\mathcal{A}}^{V,W}_{\lambda,\nu,\pm}$)   We find polynomials $p^{V,W}_\gamma(\lambda, \nu)$ such that $p^{V,W}_\gamma(\lambda, \nu)\widetilde{\mathcal{A}}^{V,W}_{\lambda,\nu,\gamma}$ is a family of distributions on $\mathbb{R}^n$ that depend *holomorphically* on $(\lambda, \nu) \in \mathbb{C}^2$ (see Proposition 5.32).

Step 3.   (holomorphic continuation of $\widetilde{\mathcal{A}}^{V,W}_{\lambda,\nu,\pm}$)   We prove that there are actually no poles of the distributions $\widetilde{\mathcal{A}}^{V,W}_{\lambda,\nu,\gamma}$ by inspecting the residue formula of the *scalar-valued* symmetry breaking operators and the zeros of the polynomials $p^{V,W}_\gamma(\lambda, \nu)$. Thus $\widetilde{\mathcal{A}}^{V,W}_{\lambda,\nu,\gamma}$ are distributions on $\mathbb{R}^n$ that depend holomorphically on $(\lambda, \nu) \in \mathbb{C}^2$.

Thus the pair $((\widetilde{\mathcal{A}}^{V,W}_{\lambda,\nu,\gamma})_\infty, \widetilde{\mathcal{A}}^{V,W}_{\lambda,\nu,\gamma})$ gives an element of $\mathcal{D}'(G/P, \mathcal{V}^*_{\lambda,\delta}) \otimes W_{\nu,\varepsilon}$ for $\delta\varepsilon = \gamma$ which is invariant under the diagonal action of P', yielding a regular symmetry breaking operator $\widetilde{\mathbb{A}}^{V,W}_{\lambda,\nu,\gamma}$ that depends holomorphically on $(\lambda, \nu) \in \mathbb{C}^2$ by Proposition 5.15.

The key idea for Steps 1 and 2 is a reduction to *scalar-valued* symmetry breaking operators which will be discussed in Section 5.6.2 (Lemma 5.37).

5.6.1   Normalized Distributions $(\widetilde{\mathcal{A}}_{\lambda,\nu,\gamma}^{V,W})_\infty$ at Infinity

This is for Step 0. We note that the map $\mathbb{T} \mapsto \mathcal{T}_\infty$ in Proposition 5.15 is neither injective nor surjective in general. In particular, the nonzero distribution $(\widetilde{\tilde{\mathcal{A}}}_{\lambda,\nu,\pm}^{V,W})_\infty$ on $\mathbb{R}^n$ (see (5.35) and (5.36)) does not always extend to the compactification G/P as an element in $(\mathcal{D}'(G/P, \mathcal{V}_{\lambda,\delta}^*) \otimes W_{\nu,\varepsilon})^{\Delta(P')}$, see Proposition 6.19. However, we shall see in Section 5.6.6 that the following *renormalization* extends to a distribution on the compact manifold G/P for any $\lambda, \nu \in \mathbb{C}$.

$$(\widetilde{\mathcal{A}}_{\lambda,\nu,+}^{V,W})_\infty := \frac{1}{\Gamma(\frac{\lambda-\nu}{2})}(\widetilde{\tilde{\mathcal{A}}}_{\lambda,\nu,+}^{V,W})_\infty$$

$$= \frac{1}{\Gamma(\frac{\lambda-\nu}{2})\Gamma(\frac{\lambda+\nu-n+1}{2})}|x_n|^{\lambda+\nu-n}\,\mathrm{pr}_{V\to W},$$

$$(\widetilde{\mathcal{A}}_{\lambda,\nu,-}^{V,W})_\infty := \frac{1}{\Gamma(\frac{\lambda-\nu+1}{2})}(\widetilde{\tilde{\mathcal{A}}}_{\lambda,\nu,-}^{V,W})_\infty$$

$$= \frac{1}{\Gamma(\frac{\lambda-\nu+1}{2})\Gamma(\frac{\lambda+\nu-n+2}{2})}|x_n|^{\lambda+\nu-n}\,\mathrm{sgn}\,x_n\,\mathrm{pr}_{V\to W}\,.$$

By definition, $(\widetilde{\mathcal{A}}_{\lambda,\nu,\pm}^{V,W})_\infty$ are distributions on $\mathbb{R}^n$ that depend holomorphically on (λ,ν) in the entire $\mathbb{C}^2$. Inspecting the poles of $\Gamma(\frac{\lambda-\nu}{2})$ and $\Gamma(\frac{\lambda-\nu+1}{2})$, we immediately have the following:

Lemma 5.26. *Suppose* $[V : W] \neq 0$. *Then,* $(\widetilde{\mathcal{A}}_{\lambda,\nu,+}^{V,W})_\infty = 0$ *if and only if* $\nu - \lambda \in 2\mathbb{N}$; $(\widetilde{\mathcal{A}}_{\lambda,\nu,-}^{V,W})_\infty = 0$ *if and only if* $\nu - \lambda \in 2\mathbb{N}+1$.

5.6.2   Preliminary Results in the Scalar-Valued Case

As we have seen in Section 5.6.1, the analytic continuation of the distribution $(\widetilde{\mathcal{A}}_{\lambda,\nu,\gamma}^{V,W})_\infty$ at infinity is easy. In order to deal with the nontrivial case, *i.e.*, the distribution kernel $\widetilde{\mathcal{A}}_{\lambda,\nu,\gamma}^{V,W}$ near the origin, we begin with some basic properties of the *scalar-valued* symmetry breaking operators. We recall from [42, (7.8)] that the (scalar-valued) distribution kernels $\widetilde{A}_{\lambda,\nu,\pm} \in \mathcal{D}'(\mathbb{R}^n)$ are initially defined as locally integrable functions on $\mathbb{R}^n$ by

$$\widetilde{A}_{\lambda,\nu,+}(x,x_n) = \frac{1}{\Gamma(\frac{\lambda+\nu-n+1}{2})\Gamma(\frac{\lambda-\nu}{2})}(|x|^2+x_n^2)^{-\nu}|x_n|^{\lambda+\nu-n}, \qquad (5.40)$$

$$\widetilde{A}_{\lambda,\nu,-}(x,x_n) = \frac{1}{\Gamma(\frac{\lambda+\nu-n+2}{2})\Gamma(\frac{\lambda-\nu+1}{2})}(|x|^2+x_n^2)^{-\nu}|x_n|^{\lambda+\nu-n}\,\mathrm{sgn}\,x_n, \qquad (5.41)$$

respectively for $\operatorname{Re}\lambda \gg |\operatorname{Re}\nu|$. (In [42], we used the notation $\widetilde{K}^A_{\lambda,\nu}$ for the scalar-valued distribution kernel $\widetilde{A}_{\lambda,\nu,+}$.) More precisely, we have:

Fact 5.27 ([42, Chap. 7]). $\widetilde{A}_{\lambda,\nu,\pm}$ *are locally integrable on* $\mathbb{R}^n$ *if* $\operatorname{Re}(\lambda - \nu) > 0$ *and* $\operatorname{Re}(\lambda + \nu) > n - 1$, *and extend   as distributions on* $\mathbb{R}^n$ *that depend holomorphically on* λ, ν *in the entire* $(\lambda, \nu) \in \mathbb{C}^2$.

The distributions $\widetilde{A}_{\lambda,\nu,+}$ were thoroughly studied in [42, Chap. 7], and analogous results for $\widetilde{A}_{\lambda,\nu,-}$ can be proved exactly in the same way.

We introduce polynomials $p_{\pm,N}(\lambda, \nu)$ of the two-variables λ and ν by

$$p_{+,N}(\lambda, \nu) := \prod_{j=1}^{N}(\lambda - \nu - 2j) \qquad\qquad \text{for } N \in \mathbb{N}_+, \qquad (5.42)$$

$$p_{-,N}(\lambda, \nu) := (\lambda + \nu - n)\prod_{j=0}^{N}(\lambda - \nu - 1 - 2j) \quad \text{for } N \in \mathbb{N}. \qquad (5.43)$$

We use a trick to raise the regularity of the distribution $\widetilde{A}_{\lambda,\nu,+}(x, x_n)$ at the origin by shifting the parameter. The resulting distributions are under control by the polynomials $p_{\pm,N}(\lambda, \nu)$ as follows:

Lemma 5.28.   *We have the following identities as distributions on* $\mathbb{R}^n$:

$$p_{+,N}(\lambda, \nu)\widetilde{A}_{\lambda,\nu,+}(x, x_n) = 2^N(|x|^2 + x_n^2)^N\widetilde{A}_{\lambda-N,\nu+N,+}(x, x_n),$$

$$p_{-,N}(\lambda, \nu)\widetilde{A}_{\lambda,\nu,-}(x, x_n) = 2^{N+2}(|x|^2 + x_n^2)^N x_n\widetilde{A}_{\lambda-N-1,\nu+N,+}(x, x_n).$$

Proof. For $\operatorname{Re}\lambda \gg |\operatorname{Re}\nu|$, we have from the definition (5.40),

$$(|x|^2 + x_n^2)^N\widetilde{A}_{\lambda-N,\nu+N,+}(x, x_n) = \frac{\Gamma(\frac{\lambda-\nu}{2})}{\Gamma(\frac{\lambda-\nu}{2} - N)}\widetilde{A}_{\lambda,\nu,+}(x, x_n)$$

$$= \frac{1}{2^N}p_{+,N}(\lambda, \nu)\widetilde{A}_{\lambda,\nu,+}(x, x_n).$$

Since both sides depend holomorphically on $(\lambda, \nu) \in \mathbb{C}^2$, we get the first assertion. The proof of the second assertion goes similarly. $\square$

Lemma 5.29.   *If* $(\lambda, \nu) \in \mathbb{C}^2$ *satisfies* $p_{+,N}(\lambda, \nu) = 0$, *then*

$$h(x, x_n)\widetilde{A}_{\lambda-N,\nu+N,+} = 0 \quad \text{in } \mathcal{D}'(\mathbb{R}^n)$$

for all homogeneous polynomials $h(x, x_n)$ *of degree* $2N$.

Proof. It follows from $p_{+,N}(\lambda, \nu) = 0$ that

$$(\nu + N) - (\lambda - N) \in \{0, 2, 4, \cdots, 2N - 2\}.$$

By the residue formula of the scalar-valued symmetry breaking operator $\widetilde{A}_{\lambda', \nu', +}$ (see [42, Thm. 12.2 (2)]), we have

$$\widetilde{A}_{\lambda-N, \nu+N, +} = q\, \widetilde{C}_{\lambda-N, \nu+N}$$

for some constant $q \equiv q_C^A(\lambda - N, \nu + N)$ depending on $\lambda - N$ and $\nu + N$. Since $\widetilde{C}_{\lambda-N, \nu+N}$ is a distribution of the form $D\delta(x_1, \cdots, x_n)$ where $D = \widetilde{C}_{2N-2j}^{\lambda-N-\frac{n-1}{2}}(-\Delta_{\mathbb{R}^{n-1}}, \frac{\partial}{\partial x_n})$ is a differential operator of homogeneous degree $2N - 2j$ $(< 2N)$, see (3.13), an iterated use of the Leibniz rule shows

$$h(x, x_n) D\delta(x_1, \cdots, x_n) = 0 \text{ in } \mathcal{D}'(\mathbb{R}^n)$$

for any homogeneous polynomial $h(x, x_n)$ of degree $2N$. □

Lemma 5.30. *If $(\lambda, \nu) \in \mathbb{C}^2$ satisfies $p_{-,N}(\lambda, \nu) = 0$, then*

$$x_n h(x, x_n) \widetilde{A}_{\lambda-N-1, \nu+N, +}(x, x_n) = 0 \quad \text{in } \mathcal{D}'(\mathbb{R}^n)$$

for all homogeneous polynomial $h(x, x_n)$ of degree $2N$.

Proof. It follows from $p_{-,N}(\lambda, \nu) = 0$ that $(\nu + N) - (\lambda - N - 1) \in \{0, 2, \cdots, 2N\}$ or $(\lambda - N - 1) + (\nu + N) = n - 1$. By using again the residue formula of the scalar-valued symmetry breaking operator $\widetilde{A}_{\lambda', \nu', +}$ in [42, Thm. 12.2], we see that the distribution kernel $\widetilde{A}_{\lambda-N-1, \nu+N, +}(x, x_n)$ is a scalar multiple of the following distributions:

$$\delta(x_n) \qquad\qquad \text{if } \lambda + \nu = n,$$

$$D\delta(x_1, \cdots, x_n) \qquad \text{if } \lambda - \nu = 2j + 1 \qquad (0 \le j \le N),$$

where $D = \widetilde{C}_{2N-2j}^{\lambda-N-1-\frac{n-1}{2}}(-\Delta_{\mathbb{R}^{n-1}}, \frac{\partial}{\partial x_n})$ is a differential operator of homogeneous degree $2N - 2j$ $(< 2N + 1)$. Then the multiplication by a homogeneous polynomial $x_n h(x, x_n)$ of degree $2N + 1$ annihilates these distributions. Hence the lemma follows. □

5.6.3 Step 1: Very Regular Case

We recall from (3.6) that $R^{V,W} = \mathrm{pr}_{V \to W} \circ \sigma \circ \psi_n \in C^\infty(\mathbb{R}^n - \{0\}) \otimes \mathrm{Hom}_\mathbb{C}(V, W)$. For $\mathrm{Re}\,\lambda \gg |\mathrm{Re}\,\nu|$, we define $\widetilde{\mathcal{A}}^{V,W}_{\lambda,\nu,\pm} \in C(\mathbb{R}^n - \{0\}) \otimes \mathrm{Hom}_\mathbb{C}(V, W)$ by

$$\widetilde{\mathcal{A}}^{V,W}_{\lambda,\nu,+} := \frac{1}{\Gamma(\frac{\lambda+\nu-n+1}{2})\Gamma(\frac{\lambda-\nu}{2})}(|x|^2 + x_n^2)^{-\nu}|x_n|^{\lambda+\nu-n}R^{V,W}(x, x_n),$$

$$\widetilde{\mathcal{A}}^{V,W}_{\lambda,\nu,-} := \frac{1}{\Gamma(\frac{\lambda+\nu-n+2}{2})\Gamma(\frac{\lambda-\nu+1}{2})}(|x|^2 + x_n^2)^{-\nu}|x_n|^{\lambda+\nu-n}\,\mathrm{sgn}\,x_n\,R^{V,W}(x, x_n),$$

(see (3.7) and (3.8)), respectively. The goal of this section is to prove the following lemma in the matrix-valued case for $\mathrm{Re}\,\lambda \gg |\mathrm{Re}\,\nu|$.

Lemma 5.31. Let $(\sigma, V) \in \widehat{O(n)}$, $(\tau, W) \in \widehat{O(n-1)}$ and δ, $\varepsilon \in \{\pm\}$. Suppose $\mathrm{Re}\,(\lambda - \nu) > 0$ and $\mathrm{Re}\,(\lambda + \nu) > n - 1$.

(1) $\widetilde{\mathcal{A}}^{V,W}_{\lambda,\nu,\pm}$ are $\mathrm{Hom}_\mathbb{C}(V, W)$-valued locally integrable functions on $\mathbb{R}^n$.

(2) The pair $((\widetilde{\mathcal{A}}^{V,W}_{\lambda,\nu,\delta\varepsilon})_\infty, \widetilde{\mathcal{A}}^{V,W}_{\lambda,\nu,\delta\varepsilon})$ defines an element of $(\mathcal{D}'(G/P, \mathcal{V}^*_{\lambda,\delta}) \otimes W_{\nu,\varepsilon})^{\Delta(P')}$, and thus yield a symmetry breaking operator $\widetilde{\mathbb{A}}^{V,W}_{\lambda,\nu,\delta\varepsilon} : I_\delta(V, \lambda) \to J_\varepsilon(W, \nu)$.

Proof. We fix inner products on V and W that are invariant by $O(n)$ and $O(n-1)$, respectively. Let $\|\cdot\|_{\mathrm{op}}$ denote the operator norm for linear maps between (finite-dimensional) Hilbert spaces. In view of the definition $R^{V,W} = \mathrm{pr}_{V \to W} \circ \sigma \circ \psi_n$ (see (3.6)), we have

$$\|R^{V,W}(x, x_n)\|_{\mathrm{op}} \leq \|\sigma \circ \psi_n(x, x_n)\|_{\mathrm{op}} = 1 \quad \text{for all } (x, x_n) \in \mathbb{R}^n - \{0\}.$$

Hence the first statement is reduced to the scalar case as stated in Fact 5.27.

The compatibility condition (5.34) can be verified readily from the definition of $(\widetilde{\mathcal{A}}^{V,W}_{\lambda,\nu,\pm})_\infty$ and $\widetilde{\mathcal{A}}^{V,W}_{\lambda,\nu,\pm}$. Hence the pair $((\widetilde{\mathcal{A}}^{V,W}_{\lambda,\nu,\delta\varepsilon})_\infty, \widetilde{\mathcal{A}}^{V,W}_{\lambda,\nu,\delta\varepsilon})$ defines an element of $\mathcal{D}'(G/P, \mathcal{V}^*_{\lambda,\delta}) \otimes W_{\nu,\varepsilon}$ by Lemma 5.11. The invariance under the diagonal action of P' follows from Proposition 5.20 for $(\widetilde{\mathcal{A}}^{V,W}_{\lambda,\nu,\delta\varepsilon})_\infty$ and from a direct computation for $\widetilde{\mathcal{A}}^{V,W}_{\lambda,\nu,\delta\varepsilon}$ when $\mathrm{Re}\,\lambda \gg |\mathrm{Re}\,\nu|$ because both $(\widetilde{\mathcal{A}}^{V,W}_{\lambda,\nu,\delta\varepsilon})_\infty$ and $\widetilde{\mathcal{A}}^{V,W}_{\lambda,\nu,\delta\varepsilon} \in L^1_{\mathrm{loc}}(\mathbb{R}^n)$. $\qquad\square$

5.6.4 Step 2: Reduction to the Scalar-Valued Case

We shall prove:

Proposition 5.32. Let $(\sigma, V) \in \widehat{O(n)}$ and $(\tau, W) \in \widehat{O(n-1)}$. Then the distributions $\widetilde{\mathcal{A}}^{V,W}_{\lambda,\nu,\pm}$, initially defined as an element of $L^1_{\mathrm{loc}}(\mathbb{R}^n) \otimes \mathrm{Hom}_\mathbb{C}(V, W)$ for $\mathrm{Re}\,\lambda \gg |\mathrm{Re}\,\nu|$ in Lemma 5.31, extend meromorphically in the entire plane $(\lambda, \nu) \in \mathbb{C}^2$.

In order to prove Proposition 5.32, we need to control the singularity of $\sigma \circ \psi_n \in C^\infty(\mathbb{R}^n - \{0\}) \otimes \mathrm{End}_\mathbb{C}(V)$ at the origin. We formulate a necessary lemma:

Lemma 5.33. *For any irreducible representation (σ, V) of $O(n)$, there exists $N \in \mathbb{N}$ such that*

$$g(x, x_n) := (|x|^2 + x_n^2)^N \sigma(\psi_n(x, x_n))$$

is an $\mathrm{End}(V)$-valued homogeneous polynomial of degree $2N$.

Definition 5.34. For $\sigma \in \widehat{O(n)}$, we denote by $N(\sigma)$ the smallest integer N satisfying the conclusion of Lemma 5.33.

We prove Lemma 5.33 by showing the following estimate of the integer $N(\sigma)$. Let $\ell(\sigma)$ be as defined in (2.21).

Lemma 5.35. $N(\sigma) \le \ell(\sigma)$ *for all $\sigma \in \widehat{O(n)}$.*

Proof of Lemma 5.35. Suppose $(\sigma_1, \cdots, \sigma_n) \in \Lambda^+(O(n))$, and let (σ, V) be the irreducible finite-dimensional representation $F^{O(n)}(\sigma_1, \cdots, \sigma_n)$ of $O(n)$ via the Cartan–Weyl isomorphism (2.20). It is convenient to set $\sigma_{n+1} = 0$. Since the exterior representations of $GL(n, \mathbb{C})$ on $\bigwedge^j(\mathbb{C}^n)$ have highest weights $(1, \cdots, 1, 0, \cdots, 0)$, and since

$$\sum_{j=1}^n (\sigma_j - \sigma_{j+1})(\underbrace{1, \cdots, 1}_{j}, 0, \cdots, 0) = (\sigma_1, \cdots, \sigma_n),$$

we can realize the irreducible representation of $GL(n, \mathbb{C})$ with highest weight $(\sigma_1, \cdots, \sigma_n)$ as a subrepresentation of the tensor product representation

$$\bigotimes_{j=1}^n (\bigwedge^j(\mathbb{C}^n))^{\sigma_j - \sigma_{j+1}}.$$

This is a polynomial representation of homogeneous degree

$$\sum_{j=1}^n j(\sigma_j - \sigma_{j+1}) = \sum_{j=1}^n \sigma_j.$$

We set $N := \sum_{j=1}^n \sigma_j$. Then the matrix coefficients of this $GL(n, \mathbb{C})$-module are given by homogeneous polynomials of degree N of z_{ij} ($1 \le i, j \le n$) where z_{ij} are the coordinates of $GL(n, \mathbb{C})$. Since the representation (σ, V) of $O(n)$ arises as a subrepresentation of this $GL(n, \mathbb{C})$-module, the formula (3.4) of ψ_n shows that the matrix coefficients of $\sigma(\psi_n(x, x_n))$ is a polynomial of x and x_n after multiplying $(|x|^2 + x_n^2)^N$.

We note that $\det \psi_n(x, x_n) = -1$ for all $(x, x_n) \in \mathbb{R}^n - \{0\}$ by (3.5). Therefore, we may assume that (σ, V) is of type I by (2.23), namely, $\sigma_{k+1} = \cdots = \sigma_n$ for some k with $2k \leq n$. In this case $N = \ell(\sigma)$ by the definition (2.21). By (2.22), we have shown the lemma. $\square$

The estimate in Lemma 5.35 is not optimal.

Example 5.36.

1) $N(\sigma) = 0$ if (σ, V) is a one-dimensional representation.
2) $N(\sigma) = 1$ if σ is the exterior representation on $V = \bigwedge^i(\mathbb{C}^n)$ $(1 \leq i \leq n-1)$.
   See (7.10) and Lemma 7.4 (2) for the proof.

Let $N \equiv N(\sigma) \in \mathbb{N}$ and $g \in \mathrm{Pol}[x_1, \cdots, x_n] \otimes \mathrm{End}_{\mathbb{C}}(V)$ be as in Lemma 5.33, and $\mathrm{pr}_{V \to W} : V \to W$ be a nonzero $O(n-1)$-homomorphism. We define $g^{V,W} \in \mathrm{Pol}[x_1, \cdots, x_n] \otimes \mathrm{Hom}_{\mathbb{C}}(V, W)$ by

$$g^{V,W} := \mathrm{pr}_{V \to W} \circ g. \tag{5.44}$$

With notation of $R^{V,W}$ as in (3.6), we have

$$g^{V,W}(x, x_n) = (|x|^2 + x_n^2)^N R^{V,W}(x, x_n) \tag{5.45}$$

$$= (|x|^2 + x_n^2)^N \mathrm{pr}_{V \to W} \circ \sigma(\psi_n(x, x_n)).$$

Then $g^{V,W}$ is a $\mathrm{Hom}_{\mathbb{C}}(V, W)$-valued polynomial of homogeneous degree $2N$.

The following lemma will imply that the singularity at the origin of the matrix-valued distributions $\widetilde{\mathcal{A}}^{V,W}_{\lambda, \nu, \pm}$ is under control by the scalar-valued case:

Lemma 5.37. *Suppose* $\mathrm{Re}\,\lambda \gg |\mathrm{Re}\,\nu|$. *Let* $p_{\pm}(\lambda, \nu)$ *be the polynomials of* λ *and* ν *defined in (5.42) and (5.43). Then,*

$$p_{+,N}(\lambda, \nu)\widetilde{\mathcal{A}}^{V,W}_{\lambda, \nu, +}(x, x_n) = 2^N \widetilde{A}_{\lambda-N, \nu+N, +}(x, x_n) g^{V,W}(x, x_n), \tag{5.46}$$

$$p_{-,N}(\lambda, \nu)\widetilde{\mathcal{A}}^{V,W}_{\lambda, \nu, -}(x, x_n) = 2^{N+2} x_n \widetilde{A}_{\lambda-N-1, \nu+N, +}(x, x_n) g^{V,W}(x, x_n). \tag{5.47}$$

Proof. For $\mathrm{Re}\,\lambda \gg |\mathrm{Re}\,\nu|$, both $\widetilde{\mathcal{A}}^{V,W}_{\lambda, \nu, \pm}$ and $\widetilde{A}_{\lambda, \nu, \pm}$ are locally integrable in $\mathbb{R}^n$. By definition, we have

$$(|x|^2 + x_n^2)^N \widetilde{\mathcal{A}}^{V,W}_{\lambda, \nu, \gamma} = \widetilde{A}_{\lambda, \nu, \gamma}(x, x_n) g^{V,W}(x, x_n)$$

for $\gamma = \pm$. By Lemma 5.28, we have

$$(|x|^2 + x_n^2)^N (p_{+,N}(\lambda, \nu)\widetilde{\mathcal{A}}^{V,W}_{\lambda, \nu, +}(x, x_n) - 2^N \widetilde{A}_{\lambda-N, \nu+N, +}(x, x_n) g^{V,W}(x, x_n)) = 0.$$

Hence we get the equality (5.46) as $\mathrm{Hom}_{\mathbb{C}}(V, W)$-valued locally integrable functions in $\mathbb{R}^n$. Similarly, we obtain

$$(|x|^2 + x_n^2)^N \widetilde{A}_{\lambda-N-1, \nu+N, +}(x, x_n)x_n = \frac{1}{2^{N+2}} p_{-,N}(\lambda, \nu)\widetilde{A}_{\lambda, \nu, -}(x, x_n).$$

Thus the second statement follows.   □

We are ready to prove the main result of this section.

Proof of Proposition 5.32. Since $g^{V,W}(x, x_n)$ is a polynomial of $(x, x_n) = (x_1, \cdots, x_n)$, the multiplication of any distributions on $\mathbb{R}^n$ by $g^{V,W}$ is well defined. Therefore, the right-hand sides of (5.46) and (5.47) make sense as distributions on $\mathbb{R}^n$ that depend holomorphically in $(\lambda, \nu) \in \mathbb{C}^2$.

Taking their quotients by the polynomials $p_{\pm, N}(\lambda, \nu)$, we set

$$\widetilde{\mathcal{A}}_{\lambda, \nu, +}^{V, W}(x, x_n) := \frac{2^N}{p_{+,N}(\lambda, \nu)} \widetilde{A}_{\lambda-N, \nu+N, +}(x, x_n)g^{V,W}(x, x_n), \qquad (5.48)$$

$$\widetilde{\mathcal{A}}_{\lambda, \nu, -}^{V, W}(x, x_n) := \frac{2^{N+2}}{p_{-,N}(\lambda, \nu)} \widetilde{A}_{\lambda-N-1, \nu+N, +}(x, x_n)x_n g^{V,W}(x, x_n). \qquad (5.49)$$

Then $\widetilde{\mathcal{A}}_{\lambda, \nu, \pm}^{V, W}$ are $\mathrm{Hom}_{\mathbb{C}}(V, W)$-valued distributions on $\mathbb{R}^n$ which depend meromorphically on $(\lambda, \nu) \in \mathbb{C}^2$ because $\widetilde{A}_{\lambda', \nu', +}(x, x_n)$ is a family of scalar-valued distributions on $\mathbb{R}^n$ that depend holomorphically on $(\lambda', \nu') \in \mathbb{C}^2$ (Fact 5.27) and $g^{V,W}(x, x_n)$ is a polynomial. By Lemma 5.37, they coincide locally integrable functions on $\mathbb{R}^n$ that are defined in (3.7) and (3.8), respectively, when $\mathrm{Re}\,\lambda \gg |\mathrm{Re}\,\nu|$. Thus Proposition 5.32 is proved.   □

5.6.5 Step 3: Proof of Holomorphic Continuation

In this section, we show that there are no poles of $\widetilde{\mathcal{A}}_{\lambda, \nu, \pm}^{V, W}$.

Lemma 5.38. $\widetilde{\mathcal{A}}_{\lambda, \nu, \pm}^{V, W}$ *are distributions on* $\mathbb{R}^n$ *that depend holomorphically on* $(\lambda, \nu) \in \mathbb{C}^2$.

Proof. By (5.48) and (5.49), the only possible places that the distribution $\widetilde{\mathcal{A}}_{\lambda, \nu, \gamma}^{V, W}$ may have poles are the zeros of the denominators, namely,

$$p_{+,N}(\lambda, \nu) = \prod_{j=1}^{N} (\lambda - \nu - 2j) \qquad\qquad \gamma = +,$$

$$p_{-,N}(\lambda, \nu) = (\lambda + \nu - n) \prod_{j=0}^{N} (\lambda - \nu - 1 - 2j) \qquad \gamma = -,$$

however, we have proved that they are not actually poles by Lemmas 5.29 and 5.30, respectively. Hence $\widetilde{\mathbb{A}}_{\lambda,\nu,\gamma}^{V,W}$ are distributions that depend holomorphically on $(\lambda, \nu) \in \mathbb{C}^2$.

$\square$

5.6.6 Proof of Theorem 3.10

We are ready to prove that the matrix-valued symmetry breaking operator $\widetilde{\mathbb{A}}_{\lambda,\nu,\pm}^{V,W}$ has a holomorphic continuation in the entire plane $(\lambda, \nu) \in \mathbb{C}^2$.

Proof of Theorem 3.10. Suppose $(\sigma, V) \in \widehat{O(n)}$. Let $N \equiv N(\sigma) \in \mathbb{N}$ be as in Lemma 5.33. We recall from (5.45) that the $\mathrm{Hom}_{\mathbb{C}}(V, W)$-valued function

$$g^{V,W}(x, x_n) = (|x|^2 + x_n^2)^N \, \mathrm{pr}_{V \to W} \circ \sigma(\psi_n(x, x_n))$$

is actually a $\mathrm{Hom}_{\mathbb{C}}(V, W)$-valued polynomial of homogeneous degree $2N$.

We know that the pair $((\widetilde{A}_{\lambda,\nu,\pm}^{V,W})_\infty, \widetilde{A}_{\lambda,\nu,\pm}^{V,W})$ satisfies the following properties:

(1) $(\widetilde{A}_{\lambda,\nu,\pm}^{V,W})_\infty$ is a $\mathrm{Hom}_{\mathbb{C}}(V, W)$-valued distribution on $\mathbb{R}^n$ satisfying (5.32) that depend holomorphically in $(\lambda, \nu) \in \mathbb{C}^2$.

(2) $\widetilde{A}_{\lambda,\nu,\pm}^{V,W}$ is a $\mathrm{Hom}_{\mathbb{C}}(V, W)$-valued distribution on $\mathbb{R}^n$ that depend holomorphically on $(\lambda, \nu) \in \mathbb{C}^2$.

(3) For $\delta, \varepsilon \in \{\pm\}$, $\widetilde{A}_{\lambda,\nu,\delta\varepsilon}^{V,W} \in Sol(\mathbb{R}^n; V_{\lambda,\delta}, W_{\nu,\varepsilon})$. Moreover, the conditions (5.33) and (5.34) are satisfied when $\mathrm{Re}\,\lambda \gg |\mathrm{Re}\,\nu|$.

All the equations concerning $Sol(\mathbb{R}^n; V_{\lambda,\delta}, W_{\nu,\varepsilon})$ depend holomorphically on (λ, ν) in the entire $\mathbb{C}^2$. On the other hand, for $\gamma \in \{\pm\}$, the properties (1) and (2) tell that the pair $((\widetilde{A}_{\lambda,\nu,\gamma}^{V,W})_\infty, \widetilde{A}_{\lambda,\nu,\gamma}^{V,W})$ depends holomorphically on (λ, ν) in the entire $\mathbb{C}^2$. Hence the property (3) holds in the entire $(\lambda, \nu) \in \mathbb{C}^2$ by analytic continuation. In turn, Proposition 5.15 implies that the pair $((\widetilde{A}_{\lambda,\nu,\gamma}^{V,W})_\infty, \widetilde{A}_{\lambda,\nu,\gamma}^{V,W})$ gives an element of $(\mathcal{D}'(G/P, V_{\lambda,\delta}^*) \otimes W_{\nu,\varepsilon})^{\Delta(P')}$ for all $(\lambda, \nu) \in \mathbb{C}^2$, and we have completed the proof of Theorem 3.10.

$\square$

5.7 Existence Condition for Regular Symmetry Breaking Operators: Proof of Theorem 3.9

In Theorem 3.10, we have assumed $[V : W] \neq 0$ for the construction of symmetry breaking operators. In this section we complete the proof of Theorem 3.9, which asserts that the condition $[V : W] \neq 0$ is necessary and sufficient for the existence of regular symmetry breaking operators.

Suppose $[V : W] \neq 0$. Let $\widetilde{\mathbb{A}}^{V,W}_{\lambda,\nu,\delta\varepsilon} : I_\delta(V,\lambda) \to J_\varepsilon(W,\nu)$ be the normalized symmetry breaking operator which is obtained by the analytic continuation of the integral operator in Section 5.6. We study the support of its distribution kernel $\widetilde{\mathcal{A}}^{V,W}_{\lambda,\nu,\delta\varepsilon}$. We define subsets U^{reg}_+ and U^{reg}_- in $\mathbb{C}^2$ by

$$U^{\mathrm{reg}}_+ := \{(\lambda,\nu) \in \mathbb{C}^2 : n - \lambda - \nu - 1 \notin 2\mathbb{N}, \nu - \lambda \notin 2\mathbb{N}\}, \tag{5.50}$$

$$U^{\mathrm{reg}}_- := \{(\lambda,\nu) \in \mathbb{C}^2 : n - \lambda - \nu - 2 \notin 2\mathbb{N}, \nu - \lambda - 1 \notin 2\mathbb{N}\}. \tag{5.51}$$

Proposition 5.39. *Suppose $V \in \widehat{O(n)}$ and $W \in \widehat{O(n-1)}$ satisfy $[V : W] \neq 0$. Let $\delta, \varepsilon \in \{\pm\}$. Then $\widetilde{\mathbb{A}}^{V,W}_{\lambda,\nu,\delta\varepsilon}$ is a nonzero regular symmetry breaking operator in the sense of Definition 5.10 for all $(\lambda,\nu) \in U^{\mathrm{reg}}_{\delta\varepsilon}$.*

Proof of Proposition 5.39. As in Proposition 5.15, the distribution kernel of the operator $\widetilde{\mathbb{A}}^{V,W}_{\lambda,\nu,\delta\varepsilon}$ can be expressed by a pair $((\widetilde{\mathcal{A}}^{V,W}_{\lambda,\nu,\delta\varepsilon})_\infty, \widetilde{\mathcal{A}}^{V,W}_{\lambda,\nu,\delta\varepsilon})$ of $\mathrm{Hom}_{\mathbb{C}}(V,W)$-valued distributions on $\mathbb{R}^n$ corresponding to the open covering $G/P = N_+ w P/P \cup N_- P/P$. Then it suffices to show $\mathrm{Supp}(\widetilde{\mathcal{A}}^{V,W}_{\lambda,\nu,\delta\varepsilon})_\infty = \mathbb{R}^n$ for $(\lambda,\nu) \in U^{\mathrm{reg}}_{\delta\varepsilon}$. If $(\lambda,\nu) \in U^{\mathrm{reg}}_{\delta\varepsilon}$, then $(\lambda,\nu,\delta,\varepsilon) \notin \Psi_{\mathrm{sp}}$, and therefore $(\widetilde{\mathcal{A}}^{V,W}_{\lambda,\nu,\delta\varepsilon})_\infty \neq 0$ by Lemma 5.26. Moreover, if $n - \lambda - \nu - 1 \notin 2\mathbb{N}$ for $\delta\varepsilon = +$ (or if $n - \lambda - \nu - 2 \notin 2\mathbb{N}$ for $\delta\varepsilon = -$), then we deduce $\mathrm{Supp}(\widetilde{\mathcal{A}}^{V,W}_{\lambda,\nu,\delta\varepsilon})_\infty = \mathbb{R}^n$ from Lemma 5.18 about the support of the Riesz distribution. Hence Proposition 5.39 is proved.   □

Definition 5.40 (normalized regular symmetry breaking operator). We shall say $\widetilde{\mathbb{A}}^{V,W}_{\lambda,\nu,\delta\varepsilon} : I_\delta(V,\lambda) \to J_\varepsilon(W,\nu)$ is a holomorphic family of the normalized *(generically) regular symmetry breaking operators*. For simplicity, we also call it a holomorphic family of the normalized regular symmetry breaking operators by a little abuse of terminology. We are ready to complete the proof of Theorem 3.9.

Proof of Theorem 3.9. The implication (i) $\Rightarrow$ (iii) follows from the explicit construction of the (normalized) regular symmetry breaking operators $\widetilde{\mathbb{A}}^{V,W}_{\lambda,\nu,\pm}$ in Theorem 3.10, and from Proposition 5.39.

(iii) $\Rightarrow$ (ii) Clear.

Let us prove the implication (ii) $\Rightarrow$ (i). We use the notation as in Section 2.1 which is adopted from [42, Chap. 5]. Then there exists a unique open orbit of P' on G/P, and the isotropy subgroup at $[q_+] = [{}^t(0, \cdots, 0, 1, 1)] \in \Xi/\mathbb{R}^\times \simeq G/P$ is given by

$$\left\{ \begin{pmatrix} 1 & & \\ & B & \\ & & 1 \\ & & & 1 \end{pmatrix} : B \in O(n-1) \right\} \simeq O(n-1).$$

Then the implication (ii) $\Rightarrow$ (i) follows from the necessary condition for the existence of regular symmetry breaking operators proved in [42, Prop. 3.5].

Thus Theorem 3.9 is proved.   □

5.8   Zeros of $\widetilde{\mathbb{A}}^{V,W}_{\lambda,\nu,\pm}$: Proof of Theorem 3.15

This section discusses the zeros of the analytic continuation of the symmetry breaking operator $\widetilde{\mathbb{A}}^{V,W}_{\lambda,\nu,\gamma} : I_\delta(V, \lambda) \to J_\varepsilon(W, \nu)$ with $\delta\varepsilon = \gamma$.

Proof of Theorem 3.15.

(1) Let $N := N(\sigma)$ as in Definition 5.34. We first observe that

$$(\lambda - N, \nu + N) \in L_{\text{even}} \quad \text{if } (\lambda, \nu) \in L_{\text{even}} \text{ and } \nu \leq -N,$$

$$(\lambda - N - 1, \nu + N) \in L_{\text{even}} \quad \text{if } (\lambda, \nu) \in L_{\text{odd}} \text{ and } \nu \leq -N.$$

Then the scalar-valued distributions $\widetilde{\mathcal{A}}_{\lambda-N,\nu+N,+}$ and $\widetilde{\mathcal{A}}_{\lambda-N-1,\nu+N,+}$ vanish, respectively by [42, Thm. 8.1]. By Lemma 5.37, the $\text{Hom}_{\mathbb{C}}(V, W)$-valued distributions $p_{+,N}(\lambda, \nu)\widetilde{\mathcal{A}}^{V,W}_{\lambda,\nu,+}$ and $p_{-,N}(\lambda, \nu)\widetilde{\mathcal{A}}^{V,W}_{\lambda,\nu,-}$ vanish, respectively, because the multiplication of distributions by the polynomial $g^{V,W}(x, x_n)$ is well-defined. Since $p_{+,N}(\lambda, \nu) \neq 0$ for $(\lambda, \nu) \in L_{\text{even}}$ and $p_{-,N}(\lambda, \nu) \neq 0$ for $(\lambda, \nu) \in L_{\text{odd}}$, the first assertion follows from Proposition 5.15 (2).

(2) If the symmetry breaking operator $\widetilde{\mathbb{A}}^{V,W}_{\lambda,\nu,\gamma}$ vanishes, then its distribution kernel is zero, and in particular, $(\widetilde{\mathcal{A}}^{V,W}_{\lambda,\nu,\gamma})_\infty = 0$ (see Proposition 5.15). This implies $\nu - \lambda \in 2\mathbb{N}$ for $\gamma = +$, and $\nu - \lambda \in 2\mathbb{N} + 1$ for $\gamma = -$, owing to Lemma 5.26. Hence Theorem 3.15 is proved.   □

5.9   Generic Multiplicity-one Theorem: Proof of Theorem 3.3

We recall from (3.3) the definition of "generic parameter" (3.2) that $(\lambda, \nu, \delta, \varepsilon) \notin \Psi_{\text{sp}}$ if and only if

$$\nu - \lambda \notin 2\mathbb{N} \text{ for } \delta\varepsilon = +; \quad \nu - \lambda \notin 2\mathbb{N} + 1 \text{ for } \delta\varepsilon = -.$$

We are ready to classify symmetry breaking operators for generic parameters. The main result of this section is Theorem 5.41, from which Theorem 3.3 follows.

Theorem 5.41 (generic multiplicity-one theorem).   *Suppose* $(\sigma, V) \in \widehat{O(n)}$, $(\tau, W) \in \widehat{O(n-1)}$ *with* $[V : W] \neq 0$. *Assume* $(\lambda, \nu) \in \mathbb{C}^2$ *and* $\delta, \varepsilon \in \{\pm\}$ *satisfy the generic parameter condition, namely,* $(\lambda, \nu, \delta, \varepsilon) \notin \Psi_{\text{sp}}$. *Then the normalized operator* $\widetilde{\mathbb{A}}^{V,W}_{\lambda,\nu,\delta\varepsilon}$ *is nonzero and is not a differential operator. Furthermore we have*

$$\text{Hom}_{G'}(I_\delta(V, \lambda)|_{G'}, J_\varepsilon(W, \nu)) = \mathbb{C}\widetilde{\mathbb{A}}^{V,W}_{\lambda,\nu,\delta\varepsilon}.$$

Proof. By Theorem 3.10, $\widetilde{\mathbb{A}}^{V,W}_{\lambda,\nu,\pm}$ is a symmetry breaking operator for all $\lambda, \nu \in \mathbb{C}$. The generic assumption on $(\lambda, \nu, \delta, \varepsilon)$ implies $\widetilde{\mathbb{A}}^{V,W}_{\lambda,\nu,\delta\varepsilon} \neq 0$ by Theorem 3.15 (2). On the other hand, by Theorem 5.24 and Corollary 5.25, we see that $\widetilde{\mathbb{A}}^{V,W}_{\lambda,\nu,\delta\varepsilon}$ is not a differential operator and $\dim_{\mathbb{C}} \mathrm{Hom}_{G'}(I_\delta(V,\lambda)|_{G'}, J_\varepsilon(W, \nu)) \leq 1$. Thus we have proved Theorem 5.41.   □

The generic multiplicity-one theorem given in Theorem 3.3 is the second statement of Theorem 5.41.

5.10 Lower Estimate of the Multiplicities

In this section we do not assume the generic parameter condition (Definition 3.2), and allow the case $(\lambda, \nu, \delta, \varepsilon) \in \Psi_{\mathrm{sp}}$. In this generality, we give a lower estimate of the dimension of the space of symmetry breaking operators.

Theorem 5.42. *Let $(\sigma, V) \in \widehat{O(n)}$ and $(\tau, W) \in \widehat{O(n-1)}$ satisfying $[V : W] \neq 0$. For any $\delta, \varepsilon \in \{\pm\}$ and $(\lambda, \nu) \in \mathbb{C}^2$, we have*

$$\dim_{\mathbb{C}} \mathrm{Hom}_{G'}(I_\delta(V,\lambda)|_{G'}, J_\varepsilon(W, \nu)) \geq 1.$$

We use a general technique from [42, Lem. 11.10] to prove that the multiplicity function is upper semicontinuous.

As before, we denote by $((\widetilde{\mathcal{A}}^{V,W}_{\lambda,\nu,\gamma})_\infty, \widetilde{\mathcal{A}}^{V,W}_{\lambda,\nu,\gamma})$ the pair of $\mathrm{Hom}_{\mathbb{C}}(V, W)$-valued distributions on $\mathbb{R}^n$ that represents the symmetry breaking operator $\widetilde{\mathbb{A}}^{V,W}_{\lambda,\nu,\gamma}$ via Proposition 5.15.

We fix $(\lambda_0, \nu_0) \in \mathbb{C}^2$, and define $\mathrm{Hom}_{\mathbb{C}}(V, W)$-valued distributions on $\mathbb{R}^n$ for $k, \ell \in \mathbb{N}$ as follows:

$$F_{k\ell} := \frac{\partial^{k+\ell}}{\partial\lambda^k \partial\nu^\ell}\bigg|_{\substack{\lambda=\lambda_0 \\ \nu=\nu_0}} \widetilde{\mathcal{A}}^{V,W}_{\lambda,\nu,\gamma},$$

$$(F_{k\ell})_\infty := \frac{\partial^{k+\ell}}{\partial\lambda^k \partial\nu^\ell}\bigg|_{\substack{\lambda=\lambda_0 \\ \nu=\nu_0}} (\widetilde{\mathcal{A}}^{V,W}_{\lambda,\nu,\gamma})_\infty.$$

Lemma 5.43. *Let $\gamma \in \{\pm\}$ and m a positive integer such that*

$$((F_{k\ell})_\infty, F_{k\ell}) = (0,0) \quad \text{for all } (k, \ell) \in \mathbb{N}^2 \text{ with } k + \ell < m.$$

Then for any (k, ℓ) with $k + \ell = m$, the pair $((F_{k\ell})_\infty, F_{k\ell})$ defines a symmetry breaking operator $I_\delta(V,\lambda) \to J_\varepsilon(W, \nu)$ for (δ, ε) with $\delta\varepsilon = \gamma$.

Proof. Since both Equations (5.32)–(5.34) and the pairs $((\widetilde{\mathcal{A}}^{V,W}_{\lambda,\nu,\gamma})_\infty, \widetilde{\mathcal{A}}^{V,W}_{\lambda,\nu,\gamma})$ satisfying (5.32)–(5.34) depend holomorphically on (λ, ν) in the entire $\mathbb{C}^2$, we can apply [42, Lem. 11.10] to conclude that the pair $((F_{k\ell})_\infty, F_{k\ell})$ satisfies (5.32)–(5.34) at $(\lambda, \nu) = (\lambda_0, \nu_0)$ for any $(k, \ell) \in \mathbb{N}^2$ with $k + \ell = m$. Then $((F_{k\ell})_\infty, F_{k\ell})$ gives an element in $\mathrm{Hom}_{G'}(I_\delta(V, \lambda_0)|_{G'}, J_\varepsilon(W, \nu_0))$ by Proposition 5.15. $\square$

Definition 5.44. Suppose we are in the setting of Lemma 5.43. For (k, ℓ) with $k + \ell = m$ and $\delta, \varepsilon \in \{\pm\}$ with $\delta\varepsilon = \gamma$, we denote by

$$\frac{\partial^{k+\ell}}{\partial\lambda^k\partial\nu^\ell}\bigg|_{\substack{\lambda=\lambda_0\\\nu=\nu_0}} \widetilde{\mathbb{A}}^{V,W}_{\lambda,\nu,\gamma} \in \mathrm{Hom}_{G'}(I_\delta(V, \lambda_0)|_{G'}, J_\varepsilon(W, \nu_0)),$$

the symmetry breaking operator associated to the pair $((F_{k\ell})_\infty, F_{k\ell})$.

Proof of Theorem 5.42. Set $\gamma := \delta\varepsilon$. Then the pair $((\widetilde{\mathcal{A}}^{V,W}_{\lambda,\nu,\gamma})_\infty, \widetilde{\mathcal{A}}^{V,W}_{\lambda,\nu,\gamma})$ of $\mathrm{Hom}_{\mathbb{C}}(V, W)$-valued distributions depends holomorphically on (λ, ν) in the entire $\mathbb{C}^2$ and satisfies (5.32)–(5.34) for all $(\lambda, \nu) \in \mathbb{C}^2$. Moreover, the pair $((\widetilde{\mathcal{A}}^{V,W}_{\lambda,\nu,\gamma})_\infty, \widetilde{\mathcal{A}}^{V,W}_{\lambda,\nu,\gamma})$ is nonzero as far as $\nu - \lambda \notin \mathbb{N}$ by Lemma 5.26. This implies that, given $(\lambda_0, \nu_0) \in \mathbb{C}^2$, there exists $(k, \ell) \in \mathbb{N}^2$ for which $((F_{k\ell})_\infty, F_{k\ell})$ is nonzero. Take $(k, \ell) \in \mathbb{N}^2$ such that $k + \ell$ attains the minimum among all (k, ℓ) for which the pair $((F_{k\ell})_\infty, F_{k\ell})$ is nonzero. By Lemma 5.43, $\frac{\partial^{k+\ell}}{\partial\lambda^k\partial\nu^\ell}\big|_{\substack{\lambda=\lambda_0\\\nu=\nu_0}} \widetilde{\mathbb{A}}^{V,W}_{\lambda,\nu,\gamma}$ is a symmetry breaking operator. $\square$

5.11 Renormalization of Symmetry Breaking Operators $\widetilde{\mathbb{A}}^{V,W}_{\lambda,\nu,\gamma}$

In this section we construct a nonzero symmetry breaking operator $\widetilde{\widetilde{\mathbb{A}}}^{V,W}_{\lambda_0,\nu_0,\gamma}$ by "renormalization" when $\widetilde{\mathbb{A}}^{V,W}_{\lambda_0,\nu_0,\gamma} = 0$. We shall also prove that the renormalized operator is *not* a differential operator. The main results are stated in Theorem 5.45.

5.11.1 Expansion of $\widetilde{\mathbb{A}}^{V,W}_{\lambda,\nu,\gamma}$ Along $\nu = $ Constant

We fix $\gamma \in \{\pm\}$ and $(\lambda_0, \nu_0) \in \mathbb{C}^2$ such that

$$\nu_0 - \lambda_0 = \begin{cases} 2\ell & \text{for } \gamma = +, \\ 2\ell + 1 & \text{for } \gamma = -, \end{cases}$$

with $\ell \in \mathbb{N}$. For every $(\sigma, V) \in \widehat{O(n)}$ and $(\tau, W) \in \widehat{O(n-1)}$, the distribution kernel $\widetilde{\mathbb{A}}^{V,W}_{\lambda,\nu,\gamma}$ of the symmetry breaking operator $\widetilde{\mathbb{A}}^{V,W}_{\lambda,\nu,\gamma}$ is a $\mathrm{Hom}_{\mathbb{C}}(V,W)$-valued distribution on $\mathbb{R}^n$ that depend holomorphically on $(\lambda, \nu) \in \mathbb{C}^2$ by Theorem 3.10. We fix $\nu = \nu_0$ and expand $\widetilde{\mathbb{A}}^{V,W}_{\lambda,\nu_0,\gamma}$ with respect to λ near $\lambda = \lambda_0$ as

$$\widetilde{\mathbb{A}}^{V,W}_{\lambda,\nu_0,\gamma} = F_0 + (\lambda - \lambda_0) F_1 + (\lambda - \lambda_0)^2 F_2 + \cdots \tag{5.52}$$

with $\mathrm{Hom}_{\mathbb{C}}(V,W)$-valued distributions $F_0, F_1, F_2, \cdots$ on $\mathbb{R}^n$. By definition,

$$\widetilde{\mathbb{A}}^{V,W}_{\lambda_0,\nu_0,\gamma} \neq 0 \text{ if and only if } F_0 \neq 0.$$

For the next term F_1, we have the following two equivalent expressions:

$$F_1 = \lim_{\lambda \to \lambda_0} \frac{1}{\lambda - \lambda_0} (\widetilde{\mathbb{A}}^{V,W}_{\lambda,\nu_0,\gamma} - \widetilde{\mathbb{A}}^{V,W}_{\lambda_0,\nu_0,\gamma}), \tag{5.53}$$

and

$$F_1 = \frac{\partial}{\partial \lambda}\bigg|_{\lambda=\lambda_0} \widetilde{\mathbb{A}}^{V,W}_{\lambda,\nu_0,\gamma}. \tag{5.54}$$

5.11.2   Renormalized Regular Symmetry Breaking Operator $\widetilde{\widetilde{\mathbb{A}}}^{V,W}_{\lambda,\nu,\gamma}$

We consider the following renormalized operators

$$\widetilde{\widetilde{\mathbb{A}}}^{V,W}_{\lambda,\nu,+} := \Gamma(\frac{\lambda-\nu}{2})\widetilde{\mathbb{A}}^{V,W}_{\lambda,\nu,+} \qquad \text{for } \nu - \lambda \notin 2\mathbb{N}, \tag{5.55}$$

$$\widetilde{\widetilde{\mathbb{A}}}^{V,W}_{\lambda,\nu,-} := \Gamma(\frac{\lambda-\nu+1}{2})\widetilde{\mathbb{A}}^{V,W}_{\lambda,\nu,-} \qquad \text{for } \nu - \lambda \notin 2\mathbb{N}+1. \tag{5.56}$$

Since $\widetilde{\mathbb{A}}^{V,W}_{\lambda,\nu,\gamma}$ depend holomorphically on (λ, ν) in $\mathbb{C}^2$, $\widetilde{\widetilde{\mathbb{A}}}^{V,W}_{\lambda,\nu,\gamma}$ are obviously well-defined as symmetry breaking operators $I_\delta(V,\lambda) \to J_\varepsilon(W,\nu)$ if $\gamma = \delta\varepsilon$, because the gamma factors do not have poles in the domain of definitions (5.55) and (5.56).

On the other hand, Theorem 3.15 (2) implies that the gamma factors in (5.55) or (5.56) have poles if $\widetilde{\mathbb{A}}^{V,W}_{\lambda_0,\nu_0,\gamma} = 0$. Nevertheless we shall see in Theorem 5.45 below that the renormalization $\widetilde{\widetilde{\mathbb{A}}}^{V,W}_{\lambda_0,\nu_0,\gamma}$ still makes sense if $\widetilde{\mathbb{A}}^{V,W}_{\lambda_0,\nu_0,\gamma} = 0$.

Theorem 5.45. *Suppose $[V:W] \neq 0$ and let $(\lambda_0, \nu_0) \in \mathbb{C}^2$ such that $\widetilde{\mathbb{A}}^{V,W}_{\lambda_0,\nu_0,\gamma} = 0$.*

(1) *There exists $\ell \in \mathbb{N}$ such that*

$$\nu_0 - \lambda_0 = \begin{cases} 2\ell & \text{when } \gamma = +, \\ 2\ell+1 & \text{when } \gamma = -. \end{cases}$$

(2) *We set*

$$\widetilde{\widetilde{\mathbb{A}}}^{V,W}_{\lambda_0,\nu_0,\gamma} := \frac{2(-1)^\ell}{\ell!} \left. \frac{\partial}{\partial\lambda} \right|_{\lambda=\lambda_0} \widetilde{\mathbb{A}}^{V,W}_{\lambda,\nu_0,\gamma}. \tag{5.57}$$

Then $\widetilde{\widetilde{\mathbb{A}}}^{V,W}_{\lambda_0,\nu_0,\gamma}$ gives a nonzero symmetry breaking operator from $I_\delta(V,\lambda)$ to $J_\varepsilon(W,\nu_0)$ with $\delta\varepsilon = \gamma$.

(3) *We fix $\nu = \nu_0$. Then $\widetilde{\widetilde{\mathbb{A}}}^{V,W}_{\lambda,\nu_0,\gamma}$ defined by (5.55) and (5.56) for $\lambda \neq \lambda_0$, and by (5.57) for $\lambda = \lambda_0$, is a family of symmetry breaking operators from $I_\delta(V,\lambda)$ to $J_\varepsilon(W,\nu_0)$ with $\delta\varepsilon = \gamma$ that depend holomorphically on λ in the entire complex plane $\mathbb{C}$. In particular, we have*

$$\widetilde{\widetilde{\mathbb{A}}}^{V,W}_{\lambda_0,\nu_0,\gamma} = \lim_{\lambda\to\lambda_0} \widetilde{\widetilde{\mathbb{A}}}^{V,W}_{\lambda,\nu_0,\gamma}. \tag{5.58}$$

(4) *$\widetilde{\widetilde{\mathbb{A}}}^{V,W}_{\lambda_0,\nu_0,\gamma}$ is not a differential operator.*

Proof.

(1) The assertion is already given in Theorem 3.15 (2).
(2) The assertion follows from Lemma 5.43.
(3) By the first statement, we see $(\lambda, \nu_0, \delta, \varepsilon)$ with $\delta\varepsilon = \gamma$ satisfies the generic parameter condition (3.2) if and only if $\lambda \neq \lambda_0$ and that

$$\widetilde{\widetilde{\mathbb{A}}}^{V,W}_{\lambda,\nu_0,\gamma} = \Gamma(\frac{\lambda-\lambda_0}{2} - \ell)\widetilde{\mathbb{A}}^{V,W}_{\lambda,\nu_0,\gamma} \quad \text{if } \lambda \neq \lambda_0.$$

We expand the distribution $\widetilde{\mathcal{A}}^{V,W}_{\lambda,\nu_0,\gamma}$ as in (5.52) near $\lambda = \lambda_0$. By the assumption that $\widetilde{\mathbb{A}}^{V,W}_{\lambda_0,\nu_0,\gamma} = 0$, it follows from the two expressions (5.53) and (5.54) of the second term F_1 that

$$F_1 = \lim_{\lambda\to\lambda_0} \frac{1}{\lambda-\lambda_0}\widetilde{\mathcal{A}}^{V,W}_{\lambda,\nu_0,\gamma} = \lim_{\lambda\to\lambda_0} \frac{1}{(\lambda-\lambda_0)\Gamma(\frac{\lambda-\lambda_0}{2}-\ell)}\widetilde{\mathcal{A}}^{V,W}_{\lambda,\nu_0,\gamma},$$

$$F_1 = \frac{(-1)^\ell\ell!}{2}\widetilde{\widetilde{\mathcal{A}}}^{V,W}_{\lambda_0,\nu_0,\gamma}.$$

In light that $\lim_{\mu \to 0} \mu \Gamma(\frac{\mu}{2} - \ell) = \frac{2(-1)^\ell}{\ell!}$, we obtain

$$\lim_{\lambda \to \lambda_0} \widetilde{\mathbb{A}}^{V,W}_{\lambda,\nu_0,\gamma} = \widetilde{\mathbb{A}}^{V,W}_{\lambda_0,\nu_0,\gamma}.$$

Since $\widetilde{\mathbb{A}}^{V,W}_{\lambda,\nu_0,\gamma}$ depends holomorphically on λ in $\mathbb{C} - \{\lambda_0\}$, and since it is continuous at $\lambda = \lambda_0$, $\widetilde{\mathbb{A}}^{V,W}_{\lambda,\nu_0,\gamma}$ is holomorphic in λ in the entire complex plane $\mathbb{C}$.

(4) Let $((\widetilde{\mathcal{A}}^{V,W}_{\lambda_0,\nu_0,\gamma})_\infty, \widetilde{\mathcal{A}}^{V,W}_{\lambda_0,\nu_0,\gamma})$ be the pair of the distribution kernels for $\widetilde{\mathbb{A}}^{V,W}_{\lambda_0,\nu_0,\gamma}$ via Proposition 5.15 (1). Then as in the above proof, we have

$$(\widetilde{\mathcal{A}}^{V,W}_{\lambda_0,\nu_0,\gamma})_\infty = \lim_{\lambda \to \lambda_0} (\widetilde{\mathcal{A}}^{V,W}_{\lambda,\nu_0,\gamma})_\infty.$$

By Proposition 5.20 (2), the right-hand side is not zero. Hence $\widetilde{\mathbb{A}}^{V,W}_{\lambda_0,\nu_0,\gamma}$ is not a differential operator by Proposition 5.15 (3). $\square$

We are ready to complete the proof of Theorem 3.13 (2-C).

Corollary 5.46. *Let* $\gamma \in \{\pm\}$. *Suppose* $\widetilde{\mathbb{A}}^{V,W}_{\lambda,\nu,\gamma} = 0$. *Then the following holds.*

$$\mathrm{Hom}_{G'}(I_\delta(V,\lambda)|_{G'}, J_\varepsilon(W,\nu)) = \mathbb{C}\widetilde{\mathbb{A}}^{V,W}_{\lambda,\nu,\delta\varepsilon} \oplus \mathrm{Diff}_{G'}(I_\delta(V,\lambda)|_{G'}, J_\varepsilon(W,\nu)). \tag{5.59}$$

Proof of Corollary 5.46. By Theorem 5.45, the renormalized operator $\widetilde{\mathbb{A}}^{V,W}_{\lambda,\nu,\delta\varepsilon}$ is well-defined and nonzero. Moreover, the right-hand side of (5.59) is a direct sum, and is contained in the left-hand side.

Conversely, take any $\mathbb{T} \in \mathrm{Hom}_{G'}(I_\delta(V,\lambda)|_{G'}, J_\varepsilon(W,\nu))$, and write $(\mathcal{T}_\infty, \mathcal{T})$ for the corresponding pair of distribution kernels for $\mathbb{T}$ via Proposition 5.15. Let $\gamma := \delta\varepsilon$. Then Proposition 5.20 tells that $\mathcal{T}_\infty$ must be proportional to $(\widetilde{\mathcal{A}}^{V,W}_{\lambda,\nu,\gamma})_\infty$, namely, $\mathcal{T}_\infty = C(\widetilde{\mathcal{A}}^{V,W}_{\lambda,\nu,\gamma})_\infty$ for some $C \in \mathbb{C}$. This implies that the distribution kernel $\mathcal{T} - C\widetilde{\mathcal{A}}^{V,W}_{\lambda,\nu,\gamma}$ of the symmetry breaking operator $\mathbb{T} - C\widetilde{\mathbb{A}}^{V,W}_{\lambda,\nu,\gamma}$ is supported at the origin, and consequently $\mathbb{T} - C\widetilde{\mathbb{A}}^{V,W}_{\lambda,\nu,\gamma}$ is a differential operator by Proposition 5.15. $\square$

Chapter 6
Differential Symmetry Breaking Operators

In this chapter, we analyze the space

$$\text{Diff}_{G'}(I_\delta(V, \lambda)|_{G'}, J_\varepsilon(W, \nu))$$

of differential symmetry breaking operators between principal series representations of $G = O(n+1, 1)$ and $G' = O(n, 1)$ for arbitrary $V \in \widehat{O(n)}$ and $W \in \widehat{O(n-1)}$ with $[V : W] \neq 0$.

The goal of this chapter is to prove Theorem 6.1 below. We recall from (1.3) that the set of "special parameters" is denoted by

$$\Psi_{\text{sp}} = \{(\lambda, \nu, \delta, \varepsilon) \in \mathbb{C}^2 \times \{\pm\}^2 : \nu - \lambda \in 2\mathbb{N}\,(\delta\varepsilon = +) \text{ or } \nu - \lambda \in 2\mathbb{N}+1\,(\delta\varepsilon = -)\}.$$

Theorem 6.1. *Let* $(G, G') = (O(n+1, 1), O(n, 1))$. *Suppose* $(\sigma, V) \in \widehat{O(n)}$ *and* $(\tau, W) \in \widehat{O(n-1)}$ *satisfy* $[V : W] \neq 0$.

(1) *The following two conditions on* $\lambda, \nu \in \mathbb{C}$ *and* $\delta, \varepsilon \in \{\pm\}$ *are equivalent:*

    (i) $(\lambda, \nu, \delta, \varepsilon) \in \Psi_{\text{sp}}$.
    (ii) $\text{Diff}_{G'}(I_\delta(V, \lambda)|_{G'}, J_\varepsilon(W, \nu)) \neq \{0\}$.

(2) *If* $2\lambda \notin \mathbb{Z}$ *then* (i) *(or equivalently,* (ii)*) implies*

    (ii)′ $\dim_{\mathbb{C}} \text{Diff}_{G'}(I_\delta(V, \lambda)|_{G'}, J_\varepsilon(W, \nu)) = 1$.

The implication (ii) $\Rightarrow$ (i) in Theorem 6.1 holds without the assumption $[V : W] \neq 0$ as we have seen in Theorem 5.21. Thus the remaining part is to show the opposite implication (i) $\Rightarrow$ (ii) and the second statement, which will be carried out in Sections 6.7 and 6.6, respectively.

© Springer Nature Singapore Pte Ltd. 2018   97
T. Kobayashi, B. Speh, *Symmetry Breaking for Representations*
of Rank One Orthogonal Groups II, Lecture Notes in Mathematics 2234,
https://doi.org/10.1007/978-981-13-2901-2_6

Remark 6.2. In the setting where $(V, W) = (\bigwedge^i (\mathbb{C}^n), \bigwedge^j (\mathbb{C}^{n-1}))$, an explicit construction and the complete classification of the space $\mathrm{Diff}_{G'}(I_\delta(V, \lambda)|_{G'}, J_\varepsilon(W, \nu))$ were carried out in [35] without the assumption $[V : W] \neq 0$, see Fact 3.23.

6.1 Differential Operators Between Two Manifolds

To give a rigorous definition of *differential* symmetry breaking operators, we need the notion of differential operators between two manifolds, which we now recall.

For any smooth vector bundle $\mathcal{V}$ over a smooth manifold X, there exists the unique (up to isomorphism) vector bundle $J^k \mathcal{V}$ over X (called the k-th *jet prolongation* of $\mathcal{V}$) together with the canonical differential operator

$$J^k \colon C^\infty(X, \mathcal{V}) \to C^\infty(X, J^k \mathcal{V})$$

of order k. We recall that a linear operator $D \colon C^\infty(X, \mathcal{V}) \to C^\infty(X, \mathcal{V}')$ between two smooth vector bundles $\mathcal{V}$ and $\mathcal{V}'$ over X is called a differential operator of order at most k, if there is a bundle morphism $Q \colon J^k \mathcal{V} \to \mathcal{V}'$ such that $D = Q_* \circ J^k$, where $Q_* \colon C^\infty(X, J^k \mathcal{V}) \to C^\infty(X, \mathcal{V}')$ is the induced homomorphism. We need a generalization of this classical definition to the case of linear operators acting between vector bundles over two *different* smooth manifolds.

Definition 6.3 (differential operators between two manifolds [38, 40]). Suppose that $p \colon Y \to X$ is a smooth map between two smooth manifolds Y and X. Let $\mathcal{V} \to X$ and $\mathcal{W} \to Y$ be two smooth vector bundles. A linear map $D \colon C^\infty(X, \mathcal{V}) \to C^\infty(Y, \mathcal{W})$ is said to be a *differential operator* of order at most k if there exists a bundle map $Q \colon p^*(J^k \mathcal{V}) \to \mathcal{W}$ such that

$$D = Q_* \circ p^* \circ J^k.$$

Alternatively, one can give the following equivalent definitions of differential operators acting between vector bundles over two manifolds Y and X with morphism p:

- based on local properties that generalize Peetre's theorem [52] in the $X = Y$ case [40, Def. 2.1];
- based on the Schwartz kernel theorem [40, Lem. 2.3];
- by local expression in coordinates [40, Ex. 2.4].

Here is a local expression in the case where p is an immersion:

Example 6.4 ([40, Ex. 2.4 (2)]). Suppose that $p \colon Y \hookrightarrow X$ is an immersion. Choose an atlas of local coordinates $\{(y_i, z_j)\}$ on X such that Y is given locally by $z_j = 0$ for all j. Then every differential operator $D \colon C^\infty(X, \mathcal{V}) \to C^\infty(Y, \mathcal{W})$ is locally of the form

$$D = \sum_{\alpha,\beta} g_{\alpha\beta}(y) \frac{\partial^{|\alpha|+|\beta|}}{\partial y^\alpha \partial z^\beta}\Big|_{z=0} \qquad \text{(finite sum)},$$

where $g_{\alpha\beta}(y)$ are $\mathrm{Hom}(V, W)$-valued smooth functions on Y.

Let X and Y be two smooth manifolds acted by G and its subgroup G', respectively, with a G'-equivariant smooth map $p: Y \to X$. When $V \to X$ is a G-equivariant vector bundle and $W \to Y$ is a G'-equivariant one, we denote by

$$\mathrm{Diff}_{G'}(C^\infty(X, V)|_{G'}, C^\infty(Y, W))$$

the space of differential symmetry breaking operators, namely, differential operators in the sense of Definition 6.3 that are also G'-homomorphisms.

6.2 Duality for Differential Symmetry Breaking Operators

We review briefly the duality theorem between differential symmetry breaking operators and morphisms for branching of generalized Verma modules. See [40, Sect. 2] for details.

Let G be a (real) Lie group. We denote by $U(\mathfrak{g})$ the universal enveloping algebra of the complexified Lie algebra $\mathfrak{g}_\mathbb{C} = \mathrm{Lie}(G) \otimes_\mathbb{R} \mathbb{C}$. Analogous notations will be applied to other Lie groups.

Let H be a (possibly disconnected) closed subgroup of G. Given a finite-dimensional representation F of H, we set

$$\mathrm{ind}_{\mathfrak{h}}^{\mathfrak{g}}(F) := U(\mathfrak{g}) \otimes_{U(\mathfrak{h})} F. \qquad (6.1)$$

The diagonal H-action on the tensor product $U(\mathfrak{g}) \otimes_\mathbb{C} F$ induces an action of H on $U(\mathfrak{g}) \otimes_{U(\mathfrak{h})} F$, and thus $\mathrm{ind}_{\mathfrak{h}}^{\mathfrak{g}}(F)$ is endowed with a $(\mathfrak{g}, H)$-module structure.

When X and Y are homogeneous spaces G/H and G'/H', respectively, with $G' \subset G$ and $H' \subset H \cap G'$, we have a natural G'-equivariant smooth map $G'/H' \to G/H$ induced from the inclusion map $G' \hookrightarrow G$. In this case, the following duality theorem ([40, Thm. 2.9], see also [38, Thm. 2.4]) is a generalization of the classical duality in the case where $G = G'$ are complex reductive Lie groups and $H = H'$ are Borel subgroups:

Fact 6.5 (duality theorem). *Let F and F' be finite-dimensional representations of H and H', respectively, and we define equivariant vector bundles $V = G \times_H F$ and $W = G' \times_{H'} F'$ over X and Y, respectively. Then there is a canonical linear* *isomorphism:*

$$\mathrm{Hom}_{\mathfrak{g}',H'}(\mathrm{ind}_{\mathfrak{h}'}^{\mathfrak{g}'}(F'^{\vee}), \mathrm{ind}_{\mathfrak{h}}^{\mathfrak{g}}(F^{\vee})|_{\mathfrak{g}',H'}) \simeq \mathrm{Diff}_{G'}(C^{\infty}(X,\mathcal{V})|_{G'}, C^{\infty}(Y,\mathcal{W})).$$
(6.2)

Applying Fact 6.5 to our special setting, we obtain the following:

Proposition 6.6. *Let* $(G,G') = (O(n+1,1), O(n,1))$, $V \in \widehat{O(n)}$, $W \in \widehat{O(n-1)}$, $\lambda, \nu \in \mathbb{C}$, *and* $\delta, \varepsilon \in \{\pm\}$. *Let* $V_{\lambda,\delta} = V \otimes \delta \otimes \mathbb{C}_{\lambda}$ *be the irreducible representation of* P *with trivial* N_+-*action as before, and* $V_{\lambda,\delta}^{\vee}$ *the contragredient representation. Similarly,* $W_{\nu,\varepsilon}^{\vee}$ *be the contragredient* P'-*module of* $W_{\nu,\varepsilon} = W \otimes \varepsilon \otimes \mathbb{C}_{\nu}$. *Then there is a canonical linear isomorphism:*

$$\mathrm{Hom}_{\mathfrak{g}',P'}(\mathrm{ind}_{\mathfrak{p}'}^{\mathfrak{g}'}(W_{\nu,\varepsilon}^{\vee}), \mathrm{ind}_{\mathfrak{p}}^{\mathfrak{g}}(V_{\lambda,\delta}^{\vee})|_{\mathfrak{g}',P'}) \simeq \mathrm{Diff}_{G'}(I_{\delta}(V,\lambda)|_{G'}, J_{\varepsilon}(W,\nu)).$$
(6.3)

6.3   Parabolic Subgroup Compatible with a Reductive Subgroup

In this section we treat the general setting where G is a real reductive Lie group and G' is a reductive subgroup, and study basic properties of differential symmetry breaking operators between principal series representation Π of G and π of the subgroup G'. We shall prove in Theorem 6.8 below that the image of any nonzero differential symmetry breaking operator is infinite-dimensional if Π is induced from a parabolic subgroup P which is compatible with the subgroup G' (see Definition 6.7).

Let us give a basic setup. Suppose that G is a real reductive Lie group with Lie algebra $\mathfrak{g}$. Take a hyperbolic element H of $\mathfrak{g}$, and we define the direct sum decomposition, referred sometimes to as the Gelfand–Naimark decomposition (*cf.* [13]):

$$\mathfrak{g} = \mathfrak{n}_- + \mathfrak{l} + \mathfrak{n}_+$$

where $\mathfrak{n}_-$, $\mathfrak{l}$, and $\mathfrak{n}_+$ are the sum of eigenspaces of $\mathrm{ad}(H)$ with negative, zero and positive eigenvalues, respectively. We define a parabolic subgroup $P \equiv P(H)$ of G by

$$P = LN_+ \qquad \text{(Levi decomposition)},$$

where $L = \{g \in G : \mathrm{Ad}(g)H = H\}$ and $N_+ = \exp(\mathfrak{n}_+)$. The following "compatibility" gives a sufficient condition for the "discrete decomposability" of the generalized Verma module $\mathrm{ind}_{\mathfrak{p}}^{\mathfrak{g}}(V^{\vee})$ when restricted to the subalgebra $\mathfrak{g}'$, which concerns with the left-hand side of the duality (6.2) (see [29, Thm. 4.1]):

Definition 6.7 (([29])). Suppose G' is a reductive subgroup of G with Lie algebra $\mathfrak{g}'$. A parabolic subgroup P of G is said to be G'-*compatible* if there exists a hyperbolic element H in $\mathfrak{g}'$ such that $P = P(H)$.

If P is G'-compatible, then $P' := P \cap G'$ is a parabolic subgroup of the reductive subgroup G' with Levi decomposition $P' = L'N'_+$ where $L' := L \cap G'$ and $N'_+ := N_+ \cap G'$.

Theorem 6.8. *Let G be a real reductive Lie group, P a parabolic subgroup which is compatible with a reductive subgroup G', and $P' := P \cap G'$. Suppose that V is a G-equivariant vector bundle of finite rank over the real flag manifold G/P, and that W is a G'-equivariant one over G'/P'. Then for any nonzero differential operator $D \colon C^\infty(G/P, V) \to C^\infty(G'/P', W)$, we have*

$$\dim_{\mathbb{C}} \operatorname{Image} D = \infty.$$

As we shall see in the proof below, Theorem 6.8 follows from the definition of differential operators (Definition 6.3) without the assumption that D intertwines the G'-action.

Proof of Theorem 6.8. We set $Y = G'/P'$ and $X = G/P$. Then $Y \subset X$ because $P' = P \cap G'$. There exist countably many disjoint open subsets $\{U_j\}$ of X such that $Y \cap U_j \neq \emptyset$. It suffices to show that for every j there exists $\varphi_j \in C^\infty(X, V)$ such that $\operatorname{Supp}(\varphi_j) \subset U_j$ and $D\varphi_j \neq 0$ because $\operatorname{Supp}(D\varphi_j) \subset U_j \cap Y$ and because $\{U_j \cap Y\}$ is a set of disjoint open sets of Y. We fix j, and write U simply for U_j. By shrinking U if necessary, we trivialize the bundles $V|_U$ and $W|_{U \cap Y}$. Then we see from Example 6.4 that D can be written locally as the matrix-valued operators:

$$D = \sum_{\alpha, \beta} g_{\alpha\beta}(y) \left. \frac{\partial^{|\alpha|+|\beta|}}{\partial y^\alpha \partial z^\beta} \right|_{z=0}.$$

Take a multi-index β such that $g_{\alpha\beta}(0) \neq 0$ on U for some α. We fix α such that $|\alpha| = \alpha_1 + \cdots + \alpha_{\dim Y}$ attains its maximum among all multi-indices α with $g_{\alpha\beta}(y) \neq 0$. Take v in the typical fiber V at $(y, z) = (0, 0)$ such that $g_{\alpha\beta}(0)v \neq 0$. By using a cut function, we can construct easily $\varphi \in C^\infty(X, V)$ such that $\operatorname{Supp}(\varphi) \subset U$ and that $\varphi(y, z) \equiv y^\alpha z^\beta v$ in a neighbourhood of $(y, z) = (0, 0)$. Then we have

$$D\varphi \neq 0.$$

Thus Theorem 6.8 is proved.   □

6.4 Character Identity for Branching in the Parabolic BGG Category

We retain the general setting as in Section 6.3, and discuss the duality theorem in Section 6.2. To study the left-hand side of (6.3), we use the results [29, 38] on the restriction of parabolic Verma modules $\mathrm{ind}_{\mathfrak{p}}^{\mathfrak{g}}(F)$ with respect to a reductive subalgebra $\mathfrak{g}'$ under the assumption that $\mathfrak{p}$ is compatible with $\mathfrak{g}'$. For later purpose, we need to formulate the results in [29, 38] in a slightly more general form as below, because a parabolic subgroup P of a real reductive Lie group is not always connected.

Suppose that $P = LN_+$ is a parabolic subgroup of G which is compatible with a reductive subgroup G'. We set $\mathfrak{n}'_- := \mathfrak{n}_- \cap \mathfrak{g}'$. Then the L'-module structure on the nilradical $\mathfrak{n}_-$ descends to the quotient $\mathfrak{n}_-/\mathfrak{n}'_-$, and extends to the (complex) symmetric tensor algebra $S((\mathfrak{n}_-/\mathfrak{n}'_-) \otimes_{\mathbb{R}} \mathbb{C})$.

For an irreducible L-module F and an irreducible L'-module F', we set

$$n(F, F') := \dim_{\mathbb{C}} \mathrm{Hom}_{L'}(F', F|_L \otimes S((\mathfrak{n}_-/\mathfrak{n}'_-) \otimes_{\mathbb{R}} \mathbb{C})). \tag{6.4}$$

Then we have the following branching rule in the Grothendieck group of the parabolic BGG category of $(\mathfrak{g}', P')$-modules ([29, Prop. 5.2],[38, Thm. 3.5]):

Fact 6.9 (character identity for branching to a reductive subalgebra). *Suppose that $P = LN_+$ is a G'-compatible parabolic subgroup of G (Definition 6.7). Let F be an irreducible finite-dimensional L-module.*

(1) *$n(F, F') < \infty$ for all irreducible finite-dimensional L'-modules F'.*
(2) *We inflate F to a P-module by letting N_+ act trivially, and form a $(\mathfrak{g}, P)$-module $\mathrm{ind}_{\mathfrak{p}}^{\mathfrak{g}}(F) = U(\mathfrak{g}) \otimes_{U(\mathfrak{p})} F$. Then we have the following identity in the Grothendieck group of the parabolic BGG category of $(\mathfrak{g}', P')$-modules:*

$$\mathrm{ind}_{\mathfrak{p}}^{\mathfrak{g}}(F)|_{\mathfrak{g}', P'} \simeq \bigoplus_{F'} n(F, F') \mathrm{ind}_{\mathfrak{p}'}^{\mathfrak{g}'}(F').$$

In the right-hand side, F' runs over all irreducible finite-dimensional P'-modules, or equivalently, all irreducible finite-dimensional L'-modules with trivial N'_+-actions.

Proof. The argument is parallel to the one in [38, Thm. 3.5] for $(\mathfrak{g}', \mathfrak{p}')$-modules, which is proved by using [29, Prop. 5.2]. □

6.5 Branching Laws for Generalized Verma Modules

In this section we refine the character identity (identity in the Grothendieck group) in Section 6.4 to obtain actual branching laws. The idea works in the general setting (*cf.* [38, Sect. 3]), however, we confine ourselves with the pair $(G, G') = (O(n+1, 1), O(n, 1))$ for actual computations below. In particular, under the assumption $2\lambda \notin \mathbb{Z}$, we give an explicit irreducible decomposition of the $(\mathfrak{g}, P)$-module $\mathrm{ind}_{\mathfrak{p}}^{\mathfrak{g}}(V_{\lambda,\delta}^{\vee})$ when we regard it as a $(\mathfrak{g}', P')$-module:

Theorem 6.10 (branching law for generalized Verma modules). *Let $V \in \widehat{O(n)}$, $\lambda \in \mathbb{C}$, and $\delta \in \{\pm\}$. Assume $2\lambda \notin \mathbb{Z}$. Then the $(\mathfrak{g}, P)$-module $\mathrm{ind}_{\mathfrak{p}}^{\mathfrak{g}}(V_{\lambda,\delta}^{\vee})$ decomposes into the multiplicity-free direct sum of irreducible $(\mathfrak{g}', P')$-modules as follows:*

$$\mathrm{ind}_{\mathfrak{p}}^{\mathfrak{g}}(V_{\lambda,\delta}^{\vee})|_{\mathfrak{g}', P'} \simeq \bigoplus_{a=0}^{\infty} \bigoplus_{[V:W] \neq 0} \mathrm{ind}_{\mathfrak{p}'}^{\mathfrak{g}'}((W_{\lambda+a,(-1)^a\delta})^{\vee}). \tag{6.5}$$

Here W runs over all irreducible $O(n-1)$-modules such that $[V : W] \neq 0$.

Proof of Theorem 6.10. The hyperbolic element H defined in (2.2) is contained in $\mathfrak{g}' = \mathfrak{o}(n, 1)$, and therefore, the parabolic subgroup P is compatible with the reductive subgroup $G' = O(n, 1)$ in the sense of Definition 6.7. We then apply Fact 6.9 to

$$(F, \mathfrak{n}_-, \mathfrak{n}'_-) = (V_{\lambda,\delta}^{\vee}, \sum_{j=1}^{n} \mathbb{R}N_j^-, \sum_{j=1}^{n-1} \mathbb{R}N_j^-).$$

Since $\mathfrak{n}_-/\mathfrak{n}'_- \simeq \mathbb{R}N_n^-$, the a-th symmetric tensor space amounts to

$$S^a((\mathfrak{n}_-/\mathfrak{n}'_-) \otimes_{\mathbb{R}} \mathbb{C}) \simeq \mathbf{1} \boxtimes (-1)^a \boxtimes \mathbb{C}_{-a}$$

as a module of $L' \simeq O(n-1) \times O(1) \times \mathbb{R}$. Therefore we have an L'-isomorphism:

$$F|_{L'} \otimes S^a((\mathfrak{n}_-/\mathfrak{n}'_-) \otimes_{\mathbb{R}} \mathbb{C}) \simeq \bigoplus_{\substack{W \in \widehat{O(n-1)} \\ [V:W] \neq 0}} W^{\vee} \boxtimes (-1)^a \delta \boxtimes \mathbb{C}_{-\lambda-a},$$

where we observe $[V^{\vee} : W^{\vee}] \neq 0$ if and only if $[V : W] \neq 0$. Thus the identity (6.5) in the level of the Grothendieck group of $(\mathfrak{g}', P')$-modules is deduced from Fact 6.9.

In order to prove the identity (6.5) as $(\mathfrak{g}', P')$-modules, we use the following two lemmas. □

Lemma 6.11. *Assume $2\lambda \notin \mathbb{Z}$. Then any $\mathfrak{z}(\mathfrak{g}')$-infinitesimal characters of the summands in (6.5) are all distinct.*

Lemma 6.12. *Assume* $2\lambda \notin \mathbb{Z}$. *Then any summand* $\mathrm{ind}_{\mathfrak{p}'}^{\mathfrak{g}'}((W_{\lambda+a,(-1)^a\delta})^{\vee})$ *in* (6.5) *is irreducible as a* $(\mathfrak{g}', P')$-*module.*

Proof of Lemma 6.11. Via the Cartan–Weyl bijection (2.20) for the disconnected group $O(N)$ ($N = n, n-1$), we write $V = F^{O(n)}(\mu)$ and $W = F^{O(n-1)}(\mu')$ for $\mu = (\mu_1, \cdots, \mu_n) \in \Lambda^+(O(n))$ and $\mu' = (\mu_1', \cdots, \mu_{n-1}') \in \Lambda^+(O(n-1))$. By the classical branching law for the restriction $O(n) \downarrow O(n-1)$ (Fact 2.12), $[V : W] \neq 0$ if and only if

$$\mu_1 \geq \mu_1' \geq \mu_2 \geq \cdots \geq \mu_{n-1}' \geq \mu_n. \tag{6.6}$$

Since any irreducible $O(N)$-module is self-dual, we have $W^{\vee} \simeq F^{O(n-1)}(\mu')$. Therefore, the $\mathfrak{Z}(\mathfrak{g}')$-infinitesimal character of the $\mathfrak{g}'$-module $\mathrm{ind}_{\mathfrak{p}'}^{\mathfrak{g}'}(W^{\vee} \otimes (-1)^a\delta \otimes \mathbb{C}_{-\lambda-a})$ is given by

$$\left(-\lambda - a + \frac{n-1}{2}, \mu_1' + \frac{n-3}{2}, \mu_2' + \frac{n-5}{2}, \cdots, \mu_{[\frac{n-1}{2}]}' + \frac{n-1}{2} - [\frac{n-1}{2}]\right)$$

modulo the Weyl group $\mathfrak{S}_m \ltimes (\mathbb{Z}/2\mathbb{Z})^m$ for the disconnected group $G' = O(n,1)$ where $m = [\frac{n+1}{2}]$. Hence, if $2\lambda \notin \mathbb{Z}$, they are all distinct when a runs over $\mathbb{N}$ and μ' runs over $\Lambda^+(O(n-1))$ subject to (6.6). Thus Lemma 6.11 is proved. $\quad\square$

Proof of Lemma 6.12. By the criterion of Conze-Berline and Duflo [7], the $\mathfrak{g}'$-module $\mathrm{ind}_{\mathfrak{p}'}^{\mathfrak{g}'}(\tau_\nu \otimes \mathbb{C}_{-\lambda-a})$ is irreducible if τ_ν is an irreducible $\mathfrak{so}(n-1)$-module with highest weight $(\nu_1, \cdots, \nu_{[\frac{n-1}{2}]})$ satisfying

$$\langle -\lambda - a + \frac{n-1}{2}, \nu_1 + \frac{n-3}{2}, \nu_2 + \frac{n-5}{2}, \cdots, \nu_{[\frac{n-1}{2}]} + \frac{n-1}{2} - [\frac{n-1}{2}], \beta^{\vee}\rangle \notin \mathbb{N}_+,$$

where $\beta^{\vee}$ is the coroot of β, and β runs over the set

$$\Delta^+(\mathfrak{g}_{\mathbb{C}}) - \Delta^+(\mathfrak{l}_{\mathbb{C}}) = \{e_1 \pm e_j : 2 \leq j \leq [\frac{n+1}{2}]\}(\cup\{e_1\}, \text{when } n \text{ is even}).$$

This condition is fulfilled if $2\lambda \notin \mathbb{Z}$ because $\nu_1, \cdots, \nu_{[\frac{n-1}{2}]} \in \frac{1}{2}\mathbb{Z}$ and $a \in \mathbb{N}$. Hence $\mathrm{ind}_{\mathfrak{p}'}^{\mathfrak{g}'}((W_{\lambda+a,(-1)^a\delta})^{\vee})$ is an irreducible $\mathfrak{g}'$-module if $W^{\vee}(\simeq W) \in O(n-1)$ is of type X (Definition 2.6), namely, if $W^{\vee}$ is irreducible as an $\mathfrak{so}(n-1)$-module. On the other hand, if $W^{\vee} \in O(n-1)$ is of type Y, then $\mathrm{ind}_{\mathfrak{p}'}^{\mathfrak{g}'}((W_{\lambda+a,(-1)^a\delta})^{\vee})$ splits into the direct sum of two irreducible $\mathfrak{g}'$-module according to the decomposition of $W^{\vee}$ into irreducible $\mathfrak{so}(n-1)$-modules. Since these two $\mathfrak{g}'$-submodules are not stable by the L'-action, we conclude that $\mathrm{ind}_{\mathfrak{p}'}^{\mathfrak{g}'}((W_{\lambda+a,(-1)^a\delta})^{\vee})$ is irreducible as a $(\mathfrak{g}', L')$-module, in particular, as a $(\mathfrak{g}', P')$-module. Thus Lemma 6.12 is proved. $\quad\square$

6.6 Multiplicity-one Theorem for Differential Symmetry Breaking Operators: Proof of Theorem 6.1 (2)

Combining Proposition 6.6 (duality theorem) with the branching law for generalized Verma modules (Theorem 6.10), we obtain a generic multiplicity-one theorem for differential symmetry breaking operators as follows:

Corollary 6.13. *Suppose* $V \in \widehat{O(n)}$ *and* $W \in \widehat{O(n-1)}$ *satisfy* $[V : W] \neq 0$. *Suppose that* $(\lambda, \nu, \delta, \varepsilon) \in \Psi_{\mathrm{sp}}$ *(see* (1.3)*). Assume further* $2\lambda \notin \mathbb{Z}$. *Then*

$$\dim_{\mathbb{C}} \mathrm{Diff}_{G'}(I_\delta(V, \lambda)|_{G'}, J_\varepsilon(W, \nu)) = 1.$$

This gives a proof of the second statement of Theorem 6.1.

6.7 Existence of Differential Symmetry Breaking Operators: Extension to Special Parameters

What remains to prove is the implication (i) $\Rightarrow$ (ii) in Theorem 6.1 for special parameters, namely, for $2\lambda \in \mathbb{Z}$. We shall use the general idea given in [42, Lem. 11.10] and deduce the implication (i) $\Rightarrow$ (ii) for the special parameters from Corollary 6.13 for the regular parameters, and thus complete the proof of Theorem 6.1 (1).

Let $\mathrm{Diff}^{\mathrm{const}}(\mathfrak{n}_-)$ denote the ring of holomorphic differential operators on $\mathfrak{n}_-$ with constant coefficients and $\langle \, , \, \rangle$ denote the natural pairing $\mathfrak{n}_- = \sum_{j=1}^n \mathbb{R}N_j^-$ and $\mathfrak{n}_+ = \sum_{j=1}^n \mathbb{R}N_j^+$. Then the symbol map

$$\mathrm{Symb} \colon \mathrm{Diff}^{\mathrm{const}}(\mathfrak{n}_-) \to \mathrm{Pol}(\mathfrak{n}_+), \quad D_z \mapsto Q(\zeta)$$

given by the characterization

$$D_z e^{\langle z, \zeta \rangle} = Q(\zeta) e^{\langle z, \zeta \rangle}$$

is a ring isomorphism between $\mathrm{Diff}^{\mathrm{const}}(\mathfrak{n}_-)$ and the polynomial ring $\mathrm{Pol}(\mathfrak{n}_+)$.

The F-method [40, Thm. 4.1] is a method to find differential symmetry breaking operators, which was inspired by the L^2-model of minimal representations [37]. It characterizes the "Fourier transform" of differential symmetry breaking operators by certain systems of differential equations. It tells that any element in $\mathrm{Diff}_{G'}(I_\delta(V, \lambda)|_{G'}, J_\varepsilon(W, \lambda + a))$ is given as a $\mathrm{Hom}_{\mathbb{C}}(V, W)$-valued differential operator D on the Bruhat cell $N_- \simeq \mathbb{R}^n$ as

$$D = \mathrm{Rest}_{x_n=0} \circ (\mathrm{Symb}^{-1} \otimes \mathrm{id})(\psi),$$

where $\psi(\zeta_1, \cdots, \zeta_n)$ is a $\mathrm{Hom}_{\mathbb{C}}(V, W)$-valued homogeneous polynomial of degree a satisfying a system of linear (differential) equations (cf. [40, (4.3) and (4.4)]) that depend holomorphically on $\lambda \in \mathbb{C}$.

If we write the solution $\psi(\zeta)$ as

$$\psi(\zeta) = \sum_{\beta_1 + \cdots + \beta_n = a} \varphi(\beta) \zeta_1^{\beta_1} \cdots \zeta_n^{\beta_n},$$

then the system of differential equations for $\psi(\zeta)$ in the F-method amounts to a system of linear (homogeneous) equations for the coefficients $\{\varphi(\beta) : |\beta| = a\}$. We regard $\varphi = (\varphi(\beta)) \in \mathbb{C}^k$ where $k := \#\{\beta \in \mathbb{N}^n : |\beta| = \alpha\}$, and use the following elementary lemma on the global basis of solutions:

Lemma 6.14. *Let $Q_\lambda \varphi = 0$ be a system of linear homogeneous equations of $\varphi \in \mathbb{C}^k$ such that Q_λ depends holomorphically on $\lambda \in \mathbb{C}$. Assume that there exists a nonempty open subset U of $\mathbb{C}$ such that the space of solutions to $Q_\lambda \varphi = 0$ is one-dimensional for every λ in U. Then there exists $\varphi_\lambda \in \mathbb{C}^k$ that depend holomorphically on λ in the entire $\mathbb{C}$ such that $Q_\lambda \varphi_\lambda = 0$ for all $\lambda \in \mathbb{C}$.*

Proof. We may regard the equation $Q_\lambda \varphi = 0$ as a matrix equation where Q_λ is an l by k matrix ($l \geq k$) whose entries are holomorphic functions of $\lambda \in \mathbb{C}$. By assumption, we have

$$\mathrm{rank}\, Q_\lambda = k - 1 \qquad \text{for all } \lambda \in U.$$

We can choose a nonempty open subset U' of U and k row vectors in Q_λ such that the corresponding square submatrix P_λ is of rank $k - 1$, provided λ belongs to U'. Then at least one of row vectors in the cofactor of P_λ is nonzero, which we choose and denote by φ_λ. Clearly, φ_λ depends holomorphically on the entire $\lambda \in \mathbb{C}$, and $Q_\lambda \varphi_\lambda = 0$ for all $\lambda \in U'$.

Since both Q_λ and φ_λ depend holomorphically on λ in the entire $\mathbb{C}$, the equation $Q_\lambda \varphi_\lambda = 0$ holds for all $\lambda \in \mathbb{C}$. $\qquad\qquad\square$

We note that the solution φ_λ in Lemma 6.14 may vanish for some $\lambda \in \mathbb{C}$. However, the following nonvanishing result holds for all $\lambda \in \mathbb{C}$.

Proposition 6.15. *Suppose we are in the setting of Lemma 6.14. Then*

$$\dim_{\mathbb{C}} \{\varphi \in \mathbb{C}^k : Q_\lambda \varphi = 0\} \geq 1 \qquad \text{for all } \lambda \in \mathbb{C}. \tag{6.7}$$

Proof. Let φ_λ be as in Lemma 6.14. Then it suffices to show (6.7) for λ belonging to the discrete set $\{\lambda \in \mathbb{C} : \varphi_\lambda = 0\}$. Take any λ_0 such that $\varphi_{\lambda_0} = 0$. Let k be the smallest positive integer such that

$$\psi_{\lambda_0} := \left.\frac{\partial^k}{\partial \lambda^k}\right|_{\lambda = \lambda_0} \varphi_\lambda \neq 0 \quad \text{and} \quad \left.\frac{\partial^j}{\partial \lambda^j}\right|_{\lambda = \lambda_0} \varphi_\lambda = 0 \quad \text{for } 0 \leq j \leq k - 1.$$

By the Leibniz rule, $\frac{\partial^k}{\partial\lambda^k}\big|_{\lambda=\lambda_0}(Q_\lambda\varphi_\lambda) = 0$ yields $Q_{\lambda_0}\psi_{\lambda_0} = 0$, because $\frac{\partial^j}{\partial\lambda^j}\big|_{\lambda=\lambda_0}\varphi_\lambda$ $= 0$ for all $0 \le j \le k-1$. Therefore ψ_{λ_0} is a nonzero solution to $Q_{\lambda_0}\varphi = 0$, showing (6.7) for $\lambda = \lambda_0$. Hence Proposition 6.15 is proved. $\qquad\qquad\square$

As in the proof of Theorem 5.42, the implication (i) $\Rightarrow$ (ii) in Theorem 6.1 follows from Corollary 6.13 (generic parameters) and the extension result to special parameters (Proposition 6.15). Thus we have completed a proof of Theorem 6.1, and in particular, of Theorem 3.12 (2).

6.8   Proof of Theorem 3.13 (2-b)

In this section, we give a proof of Theorem 3.13 (2-b), namely, we prove the following proposition.

Proposition 6.16 (localness theorem). *Suppose* $[V : W] \ne 0$. *Suppose that* $(\lambda, \nu, \delta, \varepsilon) \in \Psi_{\mathrm{sp}}$, *namely,* $(\lambda, \nu) \in \mathbb{C}^2$ *and* $\delta, \varepsilon \in \{\pm\}$ *satisfy*

$$\nu - \lambda \in 2\mathbb{N} \text{ when } \delta\varepsilon = +; \nu - \lambda \in 2\mathbb{N}+1 \text{ when } \delta\varepsilon = -.$$

Assume further that $\widetilde{\mathbb{A}}_{\lambda,\nu,\delta\varepsilon}^{V,W} \ne 0$. *The we have*

$$\mathrm{Hom}_{G'}(I_\delta(V,\lambda)|_{G'}, J_\varepsilon(W,\nu)) = \mathrm{Diff}_{G'}(I_\delta(V,\lambda)|_{G'}, J_\varepsilon(W,\nu)).$$

We need two lemmas from [42].

Lemma 6.17 ([42, Lem. 11.10]). *Suppose* D_μ *is a differential operator with holomorphic parameter* μ, *and* F_μ *is a distribution on* $\mathbb{R}^n$ *that depends holomorphically on* μ *having the following expansions:*

$$D_\mu = D_0 + \mu D_1 + \mu^2 D_2 + \cdots,$$

$$F_\mu = F_0 + \mu F_1 + \mu^2 F_2 + \cdots,$$

where D_j *are differential operators and* F_i *are distributions on* $\mathbb{R}^n$. *Assume that there exists* $\varepsilon > 0$ *such that* $D_\mu F_\mu = 0$ *for any complex number* μ *with* $0 < |\mu| < \varepsilon$. *Then the distributions* F_0 *and* F_1 *satisfy the following differential equations:*

$$D_0 F_0 = 0 \quad and \quad D_0 F_1 + D_1 F_0 = 0.$$

Lemma 6.18 ([42, Lem. 11.11]). *Suppose* $h \in \mathcal{D}'(\mathbb{R}^n)$ *is supported at the origin. Let* E *be the Euler homogeneity operator* $\sum_{\ell=1}^n x_\ell \frac{\partial}{\partial x_\ell}$ *as before. If* $(E + A)^2 h = 0$ *for some* $A \in \mathbb{Z}$ *then* $(E + A)h = 0$.

The argument below is partly similar to the one in Section 5.11.2, however, we note that the renormalization $\widetilde{\widetilde{\mathbb{A}}}_{\lambda_0,\nu_0,\gamma}^{V,W}$ in Theorem 5.45 is not defined under our assumption that $\widetilde{\mathbb{A}}_{\lambda_0,\nu_0,\gamma}^{V,W} \neq 0$ and $(\lambda_0,\nu_0,\delta,\varepsilon) \in \Psi_{\mathrm{sp}}$. Instead, we shall use the distribution $(\widetilde{\mathcal{A}}_{\lambda,\nu,\gamma}^{V,W})'$ on $\mathbb{R}^n - \{0\}$, of which we recall (5.37) and (5.38) for the definition.

Proof of Proposition 6.16. Take any symmetry breaking operator

$$\mathbb{T} \in \mathrm{Hom}_{G'}(I_\delta(V,\lambda_0)|_{G'}, J_\varepsilon(W,\nu_0)).$$

We write $(\mathcal{T}_\infty, \mathcal{T})$ for the pair of distribution kernels of $\mathbb{T}$ as in Proposition 5.15. We set $\gamma := \delta\varepsilon$.

It follows from Proposition 5.20 (3) that $\mathcal{T}|_{\mathbb{R}^n - \{0\}} = c'(\widetilde{\mathcal{A}}_{\lambda_0,\nu_0,\gamma}^{V,W})'$ for some $c' \in \mathbb{C}$.

Suppose $\widetilde{\mathbb{A}}_{\lambda_0,\nu_0,\gamma}^{V,W} \neq 0$ and $\nu_0 - \lambda_0 \in 2\mathbb{N}$ ($\gamma = +$) or $\in 2\mathbb{N}+1$ ($\gamma = -$). As in (5.52), we expand $\widetilde{\mathcal{A}}_{\lambda,\nu_0,\gamma}^{V,W}$ near $\lambda = \lambda_0$:

$$\widetilde{\mathcal{A}}_{\lambda,\nu_0,\gamma}^{V,W} = F_0 + (\lambda - \lambda_0)F_1 + (\lambda - \lambda_0)^2 F_2 + \cdots,$$

where $F_j \in \mathcal{D}'(\mathbb{R}^n) \otimes \mathrm{Hom}_\mathbb{C}(V,W)$. We note that $F_0 \neq 0$ because $\widetilde{\mathbb{A}}_{\lambda_0,\nu_0,\gamma}^{V,W} \neq 0$. We define a nonzero constant c by

$$c := \lim_{\mu \to 0} \mu \Gamma(\frac{\mu}{2} - l) = \frac{2(-1)^l}{l!}. \tag{6.8}$$

In view of the relation

$$\widetilde{\mathcal{A}}_{\lambda,\nu,+}^{V,W}|_{\mathbb{R}^n - \{0\}} = \frac{1}{\Gamma(\frac{\lambda-\nu}{2})}(\widetilde{\mathcal{A}}_{\lambda,\nu,+}^{V,W})', \quad \widetilde{\mathcal{A}}_{\lambda,\nu,-}^{V,W}|_{\mathbb{R}^n - \{0\}} = \frac{1}{\Gamma(\frac{\lambda-\nu+1}{2})}(\widetilde{\mathcal{A}}_{\lambda,\nu,-}^{V,W})',$$

we get

$$cF_1|_{\mathbb{R}^n - \{0\}} = (\widetilde{\mathcal{A}}_{\lambda_0,\nu_0,\gamma}^{V,W})',$$

as in the proof of Theorem 5.45 (3). We set

$$D_0 := E - \lambda_0 + \nu_0 + n = \sum_{j=1}^n x_j \frac{\partial}{\partial x_j} - \lambda_0 + \nu_0 + n.$$

Applying Lemma 6.17 to the differential equation (5.28):

$$(E - \lambda + \nu_0 + n)\widetilde{\mathcal{A}}_{\lambda,\nu_0,\gamma}^{V,W} = (D_0 - (\lambda - \lambda_0))\widetilde{\mathcal{A}}_{\lambda,\nu_0,\gamma}^{V,W} = 0,$$

we get

$$D_0 F_0 = 0, \quad D_0 F_1 - F_0 = 0. \tag{6.9}$$

We set

$$h := \mathcal{T} - cc' F_1 \in \mathcal{D}'(\mathbb{R}^n) \otimes \mathrm{Hom}_{\mathbb{C}}(V, W).$$

Then $\mathrm{Supp}\, h \subset \{0\}$. Moreover, $D_0^2 h = 0$ by $D_0 \mathcal{T} = 0$ and (6.9).

Applying Lemma 6.18, we get $D_0 h = 0$. It turn, $cc' F_0 = 0$ again by $D_0 \mathcal{T} = 0$ and (6.9). Therefore, if $\tilde{\mathcal{A}}_{\lambda_0, \nu_0, \gamma}^{V, W} \neq 0$, or equivalently, if $\tilde{\mathbb{A}}_{\lambda_0, \nu_0, \gamma}^{V, W} \neq 0$, then we conclude $c' = 0$ because $F_0 \neq 0$. Thus $\mathcal{T}$ is supported at the origin, and therefore $\mathbb{T}$ is a differential operator (see Proposition 5.15 (3)).

Hence Proposition 6.16 is proved. $\qquad\qquad\qquad\qquad\qquad\qquad\qquad\qquad\square$

The above proof implies that the distribution

$$(\tilde{\mathcal{A}}_{\lambda, \nu, \gamma}^{V, W})' \in \mathcal{D}'(\mathbb{R}^n - \{0\}) \otimes \mathrm{Hom}_{\mathbb{C}}(V, W)$$

in (5.37) and (5.38) does not always extend to an element of $\mathcal{S}ol(\mathbb{R}^n; V_{\lambda, \delta}, W_{\nu, \varepsilon})$ ($\gamma = \delta\varepsilon$):

Proposition 6.19. *Let* $\gamma \in \{\pm\}$. *Suppose* $(\lambda, \nu) \in \mathbb{C}^2$ *satisfies*

$$\nu - \lambda \in 2\mathbb{N} \text{ when } \gamma = +; \ \nu - \lambda \in 2\mathbb{N} + 1 \text{ when } \gamma = -.$$

If $\tilde{\mathbb{A}}_{\lambda, \nu, \gamma}^{V, W} \neq 0$, *then for* $\delta, \varepsilon \in \{\pm\}$ *with* $\delta\varepsilon = \gamma$, *the restriction map*

$$\mathcal{S}ol(\mathbb{R}^n; V_{\lambda, \delta}, W_{\nu, \varepsilon}) \to \mathcal{D}'(\mathbb{R}^n - \{0\}) \otimes \mathrm{Hom}_{\mathbb{C}}(V, W)$$

is identically zero.

Chapter 7
Minor Summation Formulæ Related to Exterior Tensor $\bigwedge^i(\mathbb{C}^n)$

This chapter collects some combinatorial formulæ, which will be used in later chapters to compute the (K, K')-spectrum for symmetry breaking operators between differential forms on spheres S^n and S^{n-1}, namely, between principal series representations $I_\delta(V, \lambda)$ of G and $J_\varepsilon(W, \nu)$ of its subgroup G' in the setting where $(V, W) = (\bigwedge^i(\mathbb{C}^n), \bigwedge^j(\mathbb{C}^{n-1}))$.

7.1 Some Notation on Index Sets

Let n be a positive integer. We shall use the following convention of index sets:

$$\mathfrak{I}_{n,i} := \{I \subset \{1, \cdots, n\} : \#I = i\}. \tag{7.1}$$

Convention 7.1. *We use calligraphic uppercase letters $\mathcal{I}$, $\mathcal{J}$ instead of Roman uppercase letters I, J if the index set may contain 0. That is, if we write $\mathcal{I} \in \mathfrak{I}_{n+1,i}$, then*

$$\mathcal{I} \subset \{0, 1, \cdots, n\} \text{ with } \#\mathcal{I} = i.$$

In later applications for symmetry breaking with respect to $(G, G') = (O(n+1, 1), O(n, 1))$, the notation $\mathfrak{I}_{n+1,i}$ for subsets of $\{0, 1, \cdots, n\}$ will be used when we describe the basis of the basic K-types and K'-types, whereas the notation $\mathfrak{I}_{n,i}$, $\mathfrak{I}_{n-1,i}$ will be used when we discuss representations of M and M', respectively.

7.1.1   Exterior Tensors $\bigwedge^i(\mathbb{C}^n)$

Let $\{e_1, \cdots, e_n\}$ be the standard basis of $\mathbb{C}^n$. For $I = \{k_1, k_2, \cdots, k_i\} \in \mathfrak{I}_{n,i}$ with $k_1 < k_2 < \cdots < k_i$, we set

$$e_I := e_{k_1} \wedge \cdots \wedge e_{k_i} \in \bigwedge^i(\mathbb{C}^n).$$

Then $\{e_I : I \in \mathfrak{I}_{n,i}\}$ forms a basis of the exterior tensor space $\bigwedge^i(\mathbb{C}^n)$. We define linear maps

$$\mathrm{pr}_{i \to j} : \bigwedge^i(\mathbb{C}^n) \to \bigwedge^j(\mathbb{C}^{n-1}), \quad (j = i - 1, i)$$

by

$$\mathrm{pr}_{i \to i}(e_I) = \begin{cases} e_I & \text{if } n \notin I, \\ 0 & \text{if } n \in I, \end{cases} \tag{7.2}$$

$$\mathrm{pr}_{i \to i-1}(e_I) = \begin{cases} 0 & \text{if } n \notin I, \\ (-1)^{i-1} e_{I - \{n\}} & \text{if } n \in I. \end{cases} \tag{7.3}$$

Then we have the direct sum decomposition

$$\bigwedge^i(\mathbb{C}^n) \simeq \bigwedge^i(\mathbb{C}^{n-1}) \oplus \bigwedge^{i-1}(\mathbb{C}^{n-1}). \tag{7.4}$$

7.1.2   Signatures for Index Sets

Let $N \in \mathbb{N}_+$. In later sections, N will be $n - 1, n$ or $n + 1$.

For a subset $I \subset \{1, \cdots, N\}$, we define a signature $\varepsilon_I(k)$ by

$$\varepsilon_I(k) := \begin{cases} 1 & \text{if } k \in I, \\ -1 & \text{if } k \notin I, \end{cases}$$

and a quadratic polynomial $Q_I(y)$ by

$$Q_I(y) := \sum_{\ell \in I} y_\ell^2 \qquad \text{for } y = (y_1, \cdots, y_N) \in \mathbb{R}^N. \tag{7.5}$$

We note that

$$2Q_I(y) - |y|^2 = \sum_{k=1}^{N} \varepsilon_I(k) y_k^2.$$

For $I, J \subset \mathfrak{I}_{N,i}$, we set

$$|I - J| := \#I - \#(I \cap J) = \#J - \#(I \cap J).$$

By definition, $|I - J| = 0$ if and only if $I = J$; $|I - J| = 1$ if and only if there exist $K \in \mathfrak{I}_{N,i-1}$ and $p, q \notin K$ with $p \neq q$ such that $I = K \cup \{p\}$ and $J = K \cup \{q\}$.

Definition 7.2. For $I \subset \{1, 2, \cdots, n\}$ and $p, q \in \mathbb{N}$, we set

$$\operatorname{sgn}(I; p) := (-1)^{\#\{r \in I : r < p\}},$$

$$\operatorname{sgn}(I; p, q) := (-1)^{\#\{r \in I : \min(p,q) < r < \max(p,q)\}}.$$

The following lemma is readily seen from the definition.

Lemma 7.3. *For $I \subset \{1, 2, \cdots, n\}$ and $p, q \in \mathbb{N}$, we have*

$$\operatorname{sgn}(I; p) \operatorname{sgn}(I; q) = \begin{cases} \operatorname{sgn}(I; p, q) & \text{if } \min(p, q) \notin I, \\ -\operatorname{sgn}(I; p, q) & \text{if } \min(p, q) \in I. \end{cases}$$

For $y = (y_1, \cdots, y_N) \in \mathbb{R}^N$, we define quadratic polynomials $S_{IJ}(y)$ by

$$S_{IJ}(y) := \begin{cases} \sum_{k=1}^{N} \varepsilon_I(k) y_k^2 & \text{if } I = J, \\ 2 \operatorname{sgn}(K; p, q) y_p y_q & \text{if } I = K \cup \{p\}, J = K \cup \{q\}, \\ 0 & \text{if } |I - J| \geq 2, \end{cases} \tag{7.6}$$

where we write $I = K \cup \{p\}$ and $J = K \cup \{q\}$ $(p \neq q)$ when $|I - J| = 1$.
It is convenient to set

$$S_{\emptyset\emptyset}(y) = -\sum_{k=1}^{N} y_k^2. \tag{7.7}$$

7.2 Minor Determinant for $\psi : \mathbb{R}^N - \{0\} \to O(N)$

We introduce the following map:

$$\psi_N : \mathbb{R}^N \times \mathbb{C} \to M(N, \mathbb{C}), \quad (y; \lambda) \mapsto I_N - \lambda\, y^t y. \tag{7.8}$$

Here we have used a similar notation to the map $\psi_N(y)$ defined in (3.4). In fact, the map (7.8) may be thought of as an extension of the previous one, since its special value at $\lambda = \dfrac{2}{|y|^2}$ recovers (3.4) by

$$\psi_N(y) = \psi_N(y; \frac{2}{|y|^2}) \qquad \text{for } y \in \mathbb{R}^N - \{0\}. \tag{7.9}$$

For $I, J \subset \{1, 2, \cdots, N\}$ with $\#I = \#J$, the minor determinant of $A = (A_{ij})_{1 \le i, j \le N}$ $\in M(N, \mathbb{R})$ is denoted by

$$\det A_{IJ} := \det(A_{ij})_{\substack{i \in I \\ j \in J}}.$$

Then the exterior representation

$$\sigma : O(N) \to GL_{\mathbb{C}}(\textstyle\bigwedge^k(\mathbb{C}^N))$$

is given by

$$\sigma(A)e_J = \sum_{J' \in \mathfrak{I}_{N,k}} (\det A)_{J'J} e_{J'}. \tag{7.10}$$

It follows from (7.10) that for $A, B \in O(N)$ we have

$$\det(AB)_{JJ'} = \sum_{J'' \in \mathfrak{I}_{N,j}} (\det A)_{JJ''} (\det B)_{J''J'} \tag{7.11}$$

Lemma 7.4. *Suppose* $I, J \subset \{1, \cdots, N\}$ *with* $\#I = \#J$.

(1) *For* $(y; \lambda) \in \mathbb{R}^N \times \mathbb{C}$,

$$\det \psi_N(y; \lambda)_{IJ} = \begin{cases} 1 - \lambda Q_I(y) & \text{if } I = J, \\ -\lambda \operatorname{sgn}(K; p, q) y_p y_q & \text{if } I = K \cup \{p\}, J = K \cup \{q\}, \\ 0 & \text{if } |I - J| \ge 2. \end{cases}$$

(2) *For* $y \in \mathbb{R}^N - \{0\}$,

$$\det \psi_N(y)_{IJ} = -\frac{1}{|y|^2} S_{IJ}(y)$$

$$= \frac{1}{|y|^2} \times \begin{cases} -\sum_{l=1}^{N} \varepsilon_I(l) y_l^2 & \text{if } I = J, \\ -2 \operatorname{sgn}(K; p, q) y_p y_q & \text{if } I = K \cup \{p\}, J = K \cup \{q\}, \\ 0 & \text{if } |I - J| \ge 2. \end{cases}$$

Proof.

(1) Suppose $I = J$. Since the symmetric matrix $y\,{}^t y$ is of rank 1, its characteristic polynomial has zeros of order $N - 1$:

$$\det(\mu I_N - y\,{}^t y) = \mu^N - \mu^{N-1}(\text{Trace } y\,{}^t y) = \mu^N - \mu^{N-1} \sum_{j=1}^{N} y_j^2,$$

and therefore

$$\det(I_N - \lambda y\,{}^t y) = 1 - \lambda \sum_{j=1}^{N} y_j^2.$$

Applying this to the principal minor of size $\#I$, we get the first formula.

Next suppose $|I - J| = 1$. We may write as $I = K \cup \{p\}$, $J = K \cup \{q\}$. Then the q-th column vector of the minor matrix $(I_N - \lambda y\,{}^t y)_{IJ}$ is of the form $-\lambda y_q (y_i)_{i \in I}$. Adding this vector multiplied by the scalar $(-y_j/y_q)$ to the j-th column vector for $j \in J - \{q\}$, we get

$$\det \psi_N(y; \lambda)_{IJ} = \text{sgn}(K; q) \det(-\lambda y_q (y_i)_{i \in I}, (\delta_{ij})_{\substack{i \in I \\ j \in K}})$$

$$= -\lambda \, \text{sgn}(K; p) \, \text{sgn}(K; q) y_p y_q.$$

Hence the second formula follows from Lemma 7.3. The third one is proved similarly.

(2) Substitute $\lambda = \frac{2}{|y|^2}$. □

As a special case of Lemma 7.4 (2) with $N = n + 1$, we have the following:

Lemma 7.5. *For $\mathcal{I}, \mathcal{J} \in \mathfrak{I}_{n+1,i}$ and $b \in \mathbb{R}^n$, we have*

$$\det \psi_{n+1}(1, b)_{\mathcal{I}\mathcal{J}} = \frac{-1}{1 + |b|^2} S_{\mathcal{I}\mathcal{J}}(1, b).$$

Here $(1, b) := (1, b_1, \cdots, b_n) \in \mathbb{R}^{n+1}$.

7.3 Minor Summation Formulæ

We collect minor summation formulæ that we shall need in computing the (K, K')-spectrum of symmetry breaking operators for "basic K-types".

We recall from (7.5) that $Q_I(b) = \sum_{k \in I} b_k^2$.

Lemma 7.6. *Suppose $I \in \mathfrak{I}_{n,i}$. For $b \in \mathbb{R}^n$ and $s, t \in \mathbb{C}$, we have:*

(1) $\displaystyle\sum_{J \in \mathfrak{I}_{n,i}} \det \psi_n(b; s)_{IJ} \det \psi_n(b; t)_{IJ} = 1 - (s+t)Q_I(b) + st|b|^2 Q_I(b).$

$$(7.12)$$

(2) $\displaystyle\sum_{J \in \mathfrak{I}_{n,i}} \det \psi_{n+1}(1, b; s)_{I \cup \{0\}, J \cup \{0\}} \det \psi_n(b; t)_{IJ}$

$$= 1 - s - (s + t - st)Q_I(b) + st|b|^2 Q_I(b). \qquad (7.13)$$

(3) $\displaystyle\sum_{J \in \mathfrak{I}_{n,i}} \det \psi_{n+1}(1, b; s)_{IJ} \det \psi_n(b; t)_{IJ} = 1 - (s+t)Q_I(b) + st|b|^2 Q_I(b).$

$$(7.14)$$

Proof.

(1) By Lemma 7.4, the left-hand side is equal to

$$(1 - sQ_I(b))(1 - tQ_I(b)) + \sum_{\substack{k \in I \\ \ell \notin I}} st b_k^2 b_\ell^2$$

$$= 1 - (s+t)Q_I(b) + st Q_I(b)^2 + st Q_I(b)(|b|^2 - Q_I(b)),$$

whence the equation (7.12).

(2) By Lemma 7.4, the left-hand side is equal to

$$(1 - s(1 + Q_I(b)))(1 - tQ_I(b)) + st \sum_{\substack{k \in I \\ \ell \notin I}} b_k^2 b_\ell^2,$$

whence the equation (7.13).

(3) By Lemma 7.4, the left-hand side is equal to

$$(1 - sQ_I(b))(1 - tQ_I(b)) + st \sum_{\substack{k \in I \\ \ell \notin I}} b_k^2 b_\ell^2$$

$$= 1 - (s+t)Q_I(b) + st Q_I(b)^2 + st Q_I(b)(|b|^2 - Q_I(b)),$$

whence the equation (7.14). $\square$

The following proposition will be used in obtaining the closed formulæ of the (K, K')-spectrum of the Knapp–Stein intertwining operators (Proposition 8.9) and the ones of the regular symmetry breaking operators (Theorem 9.8).

Proposition 7.7. *For $I \in \mathfrak{I}_{n,i}$, we have:*

(1) $\displaystyle\sum_{J \in \mathfrak{I}_{n,i}} \det \psi_n(b; \frac{2}{1+|b|^2})_{IJ} \det \psi_n(b)_{IJ} = 1 - \frac{2Q_I(b)}{(1+|b|^2)|b|^2}.$

(2) $\displaystyle\sum_{J \in \mathfrak{I}_{n,i}} \det \psi_{n+1}(1,b)_{I\cup\{0\}, J\cup\{0\}} \det \psi_n(b)_{IJ} = \frac{-1+|b|^2}{1+|b|^2} + \frac{2Q_I(b)}{(1+|b|^2)|b|^2}.$

(3) $\displaystyle\sum_{J \in \mathfrak{I}_{n,i}} \left(\det \psi_n(b; \frac{2}{1+|b|^2})_{IJ} + \det \psi_{n+1}(1,b)_{I\cup\{0\}, J\cup\{0\}} \right) \det \psi_n(b)_{IJ}$

$\displaystyle = \frac{2|b|^2}{1+|b|^2}.$

(4) $\displaystyle\sum_{J \in \mathfrak{I}_{n,i}} \det \psi_{n+1}(1,b)_{IJ} \det \psi_n(b)_{IJ} = 1 - \frac{2Q_I(b)}{(1+|b|^2)|b|^2}.$ \hfill (7.15)

Proof. The assertions (1), (2), and (4) are special cases of Lemma 7.6 (1), (2), and (3), respectively, with $s = \dfrac{2}{1+|b|^2}$ and $t = \dfrac{2}{|b|^2}$. The third one follows from the first two. $\qquad\square$

Lemma 7.8. *For $I \in \mathfrak{I}_{n-1,i-1}$,*

$$\sum_{J \in \mathfrak{I}_{n,i}} \det \psi_{n+1}(1,b)_{I\cup\{0\}, J} \det \psi_n(b)_{I\cup\{n\}, J} = \frac{2(-1)^{i+1} b_n}{1+|b|^2}.$$

Proof. Since $0 \notin J$, the summand vanishes except for the following two cases:

Case 1. $J = I \cup \{n\}$.
Case 2. $J = I \cup \{p\}$ for some $p \in \{1, 2, \cdots, n-1\} - I$.

By Lemma 7.4, we get

$(1+|b|^2)|b|^2 \displaystyle\sum_{J \in \mathfrak{I}_{n,i}} \det \psi_{n+1}(1,b)_{I\cup\{0\}, J} \det \psi_n(b)_{I\cup\{n\}, J}$

$\displaystyle = (-2\operatorname{sgn}(I; 0, n) b_n)(|b|^2 - 2Q_I(b) - 2b_n^2)$

$\displaystyle \quad + \sum_{p \in \{1,2,\cdots,n-1\}-I} (-2\operatorname{sgn}(I; 0, p) b_p)(-2\operatorname{sgn}(I; p, n) b_p b_n)$

$\displaystyle = 2(-1)^{i+1} b_n (2Q_I(b) + 2b_n^2 - |b|^2) + 4(-1)^{i+1} b_n (|b|^2 - Q_I(b) - b_n^2)$

$\displaystyle = 2(-1)^{i+1} |b|^2 b_n.$

Hence Lemma 7.8 is proved. $\qquad\square$

Lemma 7.9. *For $I \in \mathfrak{I}_{n-1,i}$,*

$$\sum_{J \in \mathfrak{I}_{n,i}} \det \psi_{n+1}(1,b)_{I \cup \{n\}, J \cup \{0\}} \det \psi_n(b)_{IJ} = \frac{2(-1)^{i+1}b_n}{1+|b|^2}.$$

Proof. Since $0 \notin I$, $|(I \cup \{n\}) - (J \cup \{0\})| \leq 1$ holds in the following two cases:

Case 1. $I = J$.
Case 2. $I = K \cup \{p\}$ and $J = K \cup \{n\}$ for some $K \in \mathfrak{I}_{n-1,i-1}$.

In Case 1,

$$\det \psi_{n+1}(1,b)_{I \cup \{n\}, J \cup \{0\}} \det \psi_n(b)_{II} = \frac{2(-1)^{i+1}b_n}{1+|b|^2} \times (1 - \frac{2Q_I(b)}{|b|^2}).$$

In Case 2,

$$\det \psi_{n+1}(1,b)_{K \cup \{p,n\}, K \cup \{0,n\}} \det \psi_n(b)_{K \cup \{p\}, K \cup \{n\}}$$

$$= \frac{-2\operatorname{sgn}(K \cup \{n\}; 0, p)b_p}{1+|b|^2} \times \frac{-2\operatorname{sgn}(K; p, n)b_p b_n}{|b|^2}$$

$$= (-1)^{i-1} \frac{4}{(1+|b|^2)|b|^2} b_p^2 b_n.$$

Adding the term in Case 1 and taking the summation of the terms over $p \in I$ in Case 2, we get the lemma. $\qquad\square$

Chapter 8
The Knapp–Stein Intertwining Operators Revisited: Renormalization and K-spectrum

In this chapter, we discuss the classical Knapp–Stein operators, which may be viewed as a baby case of symmetry breaking operators (*i.e.*, $G = G'$ case). We determine the (K, K)-spectrum (K-spectrum, for short) of the matrix-valued Knapp–Stein operators $\widetilde{\mathbb{T}}^V_{\lambda, n-\lambda} \colon I_\delta(V, \lambda) \to I_\delta(V, n - \lambda)$, see (8.13) in the case where $V = \bigwedge^i (\mathbb{C}^n)$. We also study the renormalization of the operator $\widetilde{\mathbb{T}}^V_{\lambda, n-\lambda}$ when it vanishes, see Section 8.4.

8.1 Basic K-types in the Compact Picture

Let (μ, U) be an irreducible representation of a compact Lie group K, and (σ, V) that of a subgroup M. The classical Frobenius reciprocity tells that μ occurs in the induced representation $\mathrm{Ind}^K_M \sigma$ if and only if $\mathrm{Hom}_M(\mu|_M, \sigma) \neq \{0\}$. In this section we provide a concrete realization of (μ, U) in the space $C^\infty(K/M, \mathcal{V})$ of global sections for the K-equivariant vector bundle $\mathcal{V} = K \times_M V$ which we will use later.

Lemma 8.1.

(1) *Let (μ, U) be a finite-dimensional representation of a compact Lie group K. The left regular representation on $C^\infty(K, U)$ is defined by $f(\cdot) \mapsto f(\ell^{-1} \cdot)$ for $f \in C^\infty(K, U)$ and $\ell \in K$, where we regard U just as a vector space. By assigning to $u \in U$, the function $f_u \colon K \to U$ is defined by $f_u(k) := \mu(k)^{-1} u$. Then the K-module U can be embedded as a submodule of the left regular representation $C^\infty(K, U)$ by*

$$U \to C^\infty(K, U), \qquad u \mapsto f_u.$$

© Springer Nature Singapore Pte Ltd. 2018
T. Kobayashi, B. Speh, *Symmetry Breaking for Representations of Rank One Orthogonal Groups II*, Lecture Notes in Mathematics 2234,
https://doi.org/10.1007/978-981-13-2901-2_8

(2) *Let V be a vector space over $\mathbb{C}$, and $\mathrm{pr}_{U \to V} : U \to V$ a linear map. Then we have a K-homomorphism*

$$U \to C^\infty(K, V), \qquad u \mapsto \mathrm{pr}_{U \to V} \circ f_u.$$

(3) *Suppose that $\sigma : M \to GL_\mathbb{C}(V)$ is a representation of a subgroup M of K and that $\mathrm{pr}_{U \to V}$ is an M-homomorphism. Then we have a well-defined K-homomorphism*

$$U \to C^\infty(K/M, \mathcal{V}), \qquad u \mapsto \mathrm{pr}_{U \to V} \circ f_u,$$

where we identify the space of smooth sections for $\mathcal{V} := K \times_M V$ over K/M with the space of M-invariant elements

$$C^\infty(K, V)^M := \{ F \in C^\infty(K, V) : F(\cdot m) = \sigma(m)^{-1} F(\cdot) \quad \text{for all } m \in M \}.$$

Proof. The detailed formulation of each statement gives a proof by itself.   □

Applying Lemma 8.1 to differential forms on the sphere, we obtain:

Example 8.2. Let $K := O(n+1)$, and σ be the i-th exterior tensor representation of the subgroup $M := O(n)$ on $V := \bigwedge^i(\mathbb{C}^n)$. Then the vector bundle $\mathcal{V} = K \times_M V$ is identified with the i-th exterior tensor of the cotangent bundle of the n-sphere $S^n \simeq K/M$, and we may identify $C^\infty(K, V)^M \simeq C^\infty(K/M, \mathcal{V})$ with the space $\mathcal{E}^i(S^n)$ of differential i-forms on S^n. Suppose that μ is the k-th exterior tensor representation of $K = O(n+1)$ on $U := \bigwedge^k(\mathbb{C}^{n+1})$. For $k = i$ or $i+1$, the projection $\mathrm{pr}_{k \to i} : \bigwedge^k(\mathbb{C}^{n+1}) \to \bigwedge^i(\mathbb{C}^n)$, see (7.2) and (7.3), is an M-homomorphism, and therefore, Lemma 8.1 gives a concrete realization of the K-module $U = \bigwedge^k(\mathbb{C}^{n+1})$ in $\mathcal{E}^i(S^n) \simeq C^\infty(K, V)^M$ as below. Let $\{e_0, e_1, \cdots, e_n\}$ be the standard basis of $\mathbb{C}^{n+1}$, and $\{e_\mathcal{I} : \mathcal{I} \in \mathfrak{I}_{n+1,k}\}$ the standard basis of $\bigwedge^k(\mathbb{C}^{n+1})$.

We treat the cases $k = i$ and $i+1$, separately. In what follows, we use Convention 7.1 for the index set $\mathfrak{I}_{n+1,k}$. See also Section 7.2 for minor determinant $(\det A)_{IJ}$ of $A \in M(N, \mathbb{R})$.

Case 1.   Suppose $k = i$. Then $\mathbf{1}^\mathcal{I} := \mathrm{pr}_{i \to i} \circ f_{e_\mathcal{I}}$ is a map given by

$$O(n+1) \to \bigwedge^i(\mathbb{C}^n), \qquad k \mapsto \mathbf{1}^\mathcal{I}(k) = \sum_{J \in \mathfrak{I}_{n,i}} (\det k)_{\mathcal{I}J} e_J. \tag{8.1}$$

Thus $\mathbf{1}^\mathcal{I}$ is regarded as an element of $C^\infty(O(n+1), \bigwedge^i(\mathbb{C}^n))^{O(n)} \simeq \mathcal{E}^i(S^n)$.

Case 2.   Suppose $k = i+1$. Then $h^\mathcal{I} := (-1)^i \, \mathrm{pr}_{i+1 \to i} \circ f_{e_\mathcal{I}}$ is a map given by

$$O(n+1) \to \bigwedge^i(\mathbb{C}^n), \quad k \mapsto h^\mathcal{I}(k) = \sum_{J \in \mathfrak{I}_{n,i}} (\det k)_{\mathcal{I}, J \cup \{0\}} e_J, \tag{8.2}$$

which is again regarded as an element of $\mathcal{E}^i(S^n)$. We remark that the projection

$$\mathrm{pr}_{i+1 \to j} : \bigwedge\nolimits^{i+1}(\mathbb{C}^{n+1}) \to \bigwedge\nolimits^i(\mathbb{C}^n)$$

is given by "removing" e_0, whereas the projection in (7.3) was by "removing" e_n.

By Lemma 8.1, we obtain injective $O(n+1)$-homomorphisms

$$\bigwedge\nolimits^i(\mathbb{C}^{n+1}) \quad \to \mathcal{E}^i(S^n), \qquad e_{\mathcal{I}} \mapsto \mathbf{1}^{\mathcal{I}},$$
$$\bigwedge\nolimits^{i+1}(\mathbb{C}^{n+1}) \to \mathcal{E}^i(S^n), \qquad e_{\mathcal{I}} \mapsto h^{\mathcal{I}}.$$

8.2   K-picture and N-picture of Principal Series Representations

Let $(\sigma, V) \in \widehat{O(n)}$, $\delta \in \{\pm 1\}$, and $\lambda \in \mathbb{C}$. We recall from Section 2.3.1 that the principal series representation

$$I_\delta(V, \lambda) = \mathrm{Ind}_P^G(V \otimes \delta \otimes \mathbb{C}_\lambda)$$

of $G = O(n+1, 1)$ is realized on the Fréchet space $C^\infty(G/P, \mathcal{V}_{\lambda,\delta})$ of smooth sections for the homogeneous vector bundle $G \times_P \mathcal{V}_{\lambda,\delta}$ over the real flag manifold G/P, see (2.27).

8.2.1   Explicit K-finite Vectors in the N-picture

In this subsection we review the K-picture and N-picture of the principal series representation $I_\delta(V, \lambda)$, and provide a concrete formula connecting the two pictures. As we saw in (2.28), the noncompact picture (N-picture) of $I_\delta(V, \lambda)$ is given by

$$\iota_N^* : I_\delta(V, \lambda) \hookrightarrow C^\infty(\mathbb{R}^n) \otimes V, \qquad F \mapsto f(b) := F(n_-(b)),$$

as the pull-back of sections via the coordinate map of the open Bruhat cell $\iota_N : \mathbb{R}^n \hookrightarrow G/P, b \mapsto n_-(b) \cdot o$, where $n_- : \mathbb{R}^n \xrightarrow{\sim} N_-$ is defined in (2.5).

Next, let V_δ denote the outer tensor product representation $V \boxtimes \delta$ of $M = O(n) \times O(1)$. Then the diffeomorphism $\iota_K : K/M \xrightarrow{\sim} G/P$ induces an isomorphism $\iota_K^*(\mathcal{V}_{\lambda,\delta}) \simeq K \times_M V_\delta$ as K-equivariant vector bundles over K/M, and hence K-isomorphisms between the space of sections:

$$\iota_K^* : I_\delta(V, \lambda) \xrightarrow{\sim} C^\infty(K/M, K \times_M V_\delta) \simeq (C^\infty(K) \otimes V_\delta)^M,$$

which is referred to as the K-picture of $I_\delta(V, \lambda)$.

The transform from the K-picture to the N-picture is given by

$$\iota_\lambda^* := \iota_N^* \circ (\iota_K^*)^{-1} : (C^\infty(K) \otimes V_\delta)^M \hookrightarrow C^\infty(\mathbb{R}^n) \otimes V. \qquad (8.3)$$

Then the three realizations of the principal series representation $I_\delta(V, \lambda)$ of G are summarized as below.

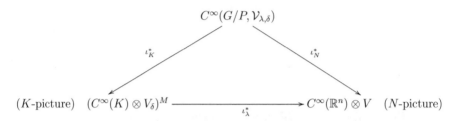

$C^\infty(G/P, \mathcal{V}_{\lambda,\delta})$

(K-picture) $(C^\infty(K) \otimes V_\delta)^M \xrightarrow{\quad\iota_\lambda^*\quad} C^\infty(\mathbb{R}^n) \otimes V$ (N-picture)

To compute ι_λ^*, we recall from Lemma 5.8 that the map

$$k : \mathbb{R}^n \to SO(n+1) \subset K = O(n+1) \times O(1),$$

see (5.6), induces the following commutative diagram:

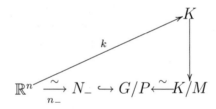

Lemma 8.3. *Suppose* $F \in (C^\infty(K) \otimes V_\delta)^M$. *Then we have*

$$(\iota_\lambda^* F)(b) = (1 + |b|^2)^{-\lambda} F(k(b)) \quad \text{for all } b \in \mathbb{R}^n. \qquad (8.4)$$

Here $k(b) \in SO(n+1)$ *is viewed as an element of* K *on the right-hand side.*

Proof. We define $t \in \mathbb{R}$ by $e^t = 1 + |b|^2$. It follows from Lemma 5.8 that

$$(\iota_\lambda^* F)(b) = (\iota_K^{*\,-1} F)(n_-(b)) = (\iota_K^{*\,-1} F) \left(\begin{pmatrix} k(b) & 0 \\ 0 & 1 \end{pmatrix} e^{tH} n_+ \left(\frac{-b}{1 + |b|^2} \right) \right)$$

$$= (1 + |b|^2)^{-\lambda} F \left(\begin{pmatrix} k(b) & 0 \\ 0 & 1 \end{pmatrix} \right).$$

Hence the lemma is verified. $\square$

8.2.2   Basic K-types in the N-picture

We recall from the K-type formula (Lemma 2.16) that the principal series representation $I_\delta(i,\lambda)$ of $G = O(n+1,1)$ contains two "basic K-types" $\mu^\flat(i,\delta) = \bigwedge^i(\mathbb{C}^{n+1}) \boxtimes \delta$ and $\mu^\sharp(i,\delta) = \bigwedge^{i+1}(\mathbb{C}^{n+1}) \boxtimes (-\delta)$ for $0 \leq i \leq n$.

   In this section, we write down explicit K-finite vectors belonging to $\mu^\flat(i,\delta)$ and $\mu^\sharp(i,\delta)$ in the noncompact picture.

   Let $\mathbf{1}^{\mathcal{I}}$ and $h^{\mathcal{I}}$ be the elements in $\mathcal{E}^i(S^n) \simeq C^\infty(O(n+1),V)^{O(n)}$ constructed in Example 8.2, where we take V to be $\bigwedge^i(\mathbb{C}^n)$. We note that the pair

$$(K,M) = (O(n+1) \times O(1), O(n) \times O(1))$$

is not exactly the same with the pair $(O(n+1), O(n))$ in Example 8.2, however, the diffeomorphism $O(n+1)/O(n) \xrightarrow{\sim} K/M$ induces the following isomorphisms

$$\mathcal{E}^i(S^n) \simeq C^\infty(O(n+1) \otimes V)^{O(n)} \xleftarrow{\sim} C^\infty(K \otimes V_\delta)^M.$$

Thus we may regard that $\{\mathbf{1}^{\mathcal{I}} : \mathcal{I} \in \mathfrak{I}_{n+1,i}\}$ is a basis of $\mu^\flat(i,\delta)$ and $\{h^{\mathcal{I}} : \mathcal{I} \in \mathfrak{I}_{n+1,i+1}\}$ is a basis of $\mu^\sharp(i,\delta)$. Applying the map $\iota_\lambda^*: C^\infty(K \otimes V_\delta)^M \hookrightarrow C^\infty(\mathbb{R}^n) \otimes V$ (see (8.4)), we set

$$\mathbf{1}_\lambda^{\mathcal{I}} := \iota_\lambda^* \mathbf{1}^{\mathcal{I}} \qquad \text{for } \mathcal{I} \in \mathfrak{I}_{n+1,i},$$

$$h_\lambda^{\mathcal{I}} := \iota_\lambda^* h^{\mathcal{I}} \qquad \text{for } \mathcal{I} \in \mathfrak{I}_{n+1,i+1}.$$

By Lemma 8.1 and Example 8.2, we have shown the following.

Proposition 8.4 (basic K-type $\mu^\flat$ and $\mu^\sharp$). *We define linear maps by*

$$\bigwedge^i(\mathbb{C}^{n+1}) \to C^\infty(\mathbb{R}^n, \bigwedge^i(\mathbb{C}^n)), \quad e_{\mathcal{I}} \mapsto \mathbf{1}_\lambda^{\mathcal{I}} \quad \text{for } \mathcal{I} \in \mathfrak{I}_{n+1,i}, \tag{8.5}$$

$$\bigwedge^{i+1}(\mathbb{C}^{n+1}) \to C^\infty(\mathbb{R}^n, \bigwedge^i(\mathbb{C}^n)), \quad e_{\mathcal{I}} \mapsto h_\lambda^{\mathcal{I}} \quad \text{for } \mathcal{I} \in \mathfrak{I}_{n+1,i+1}.$$

Then, for $\delta = \pm$, the images give the unique K-types $\mu^\flat(i,\delta) = \bigwedge^i(\mathbb{C}^{n+1}) \boxtimes \delta$ and $\mu^\sharp(i,\delta) = \bigwedge^{i+1}(\mathbb{C}^{n+1}) \boxtimes (-\delta)$ respectively, of the principal series representation $I_\delta(i,\lambda) = \mathrm{Ind}_P^G(\bigwedge^i(\mathbb{C}^n) \otimes \delta \otimes \mathbb{C}_\lambda)$ of G in the N-picture.

An explicit formula for $\mathbf{1}_\lambda^{\mathcal{I}}$ and $h_\lambda^{\mathcal{I}}$ is given as follows.

Lemma 8.5. *Let $S_{\mathcal{I}\mathcal{J}}(b)$ be the quadratic polynomial of $b = (b_1, \cdots, b_n)$ defined in (7.6).*

(1) *Let $0 \leq i \leq n$. For $\mathcal{I} \in \mathfrak{I}_{n+1,i}$ and $\lambda \in \mathbb{C}$, we have*

$$\mathbf{1}_{\lambda}^{\mathcal{I}}(b) = (1+|b|^2)^{-\lambda} \sum_{J \in \mathfrak{I}_{n,i}} \det \psi_{n+1}(1,b)_{\mathcal{I}J} e_J$$

$$= -(1+|b|^2)^{-\lambda-1} \sum_{J \in \mathfrak{I}_{n,i}} S_{\mathcal{I}J}(1,b) e_J. \qquad (8.6)$$

If $i = 0$, we regard $\mathcal{I} = \emptyset$ and $\mathbf{1}_{\lambda}^{\emptyset} = (1+|b|^2)^{-\lambda}$ (see (7.7)).
(2) *Let $0 \leq i \leq n$. For $\mathcal{I} \in \mathfrak{I}_{n+1,i+1}$ and $\lambda \in \mathbb{C}$, we have*

$$h_{\lambda}^{\mathcal{I}}(b) = -(1+|b|^2)^{-\lambda} \sum_{J \in \mathfrak{I}_{n,i}} \det \psi_{n+1}(1,b)_{\mathcal{I},J \cup \{0\}} e_J$$

$$= (1+|b|^2)^{-\lambda-1} \sum_{J \in \mathfrak{I}_{n,i}} S_{\mathcal{I},J \cup \{0\}}(1,b) e_J. \qquad (8.7)$$

We note that Lemma 8.5 implies

$$\mathbf{1}_{\lambda}^{\mathcal{I}}(0) = \begin{cases} e_{\mathcal{I}} & 0 \notin \mathcal{I}, \\ 0 & 0 \in \mathcal{I}, \end{cases} \qquad (8.8)$$

$$h_{\lambda}^{\mathcal{I}}(0) = \begin{cases} e_{\mathcal{I}-\{0\}} & 0 \in \mathcal{I}, \\ 0 & 0 \notin \mathcal{I}. \end{cases} \qquad (8.9)$$

Proof of Lemma 8.5. Suppose $b \in \mathbb{R}^n$, and let $k(b) \in SO(n+1)$ be as defined in (5.6). By (8.1) and (8.2), respectively, the formula (8.4) of ι_{λ}^* tells that

$$\mathbf{1}_{\lambda}^{\mathcal{I}}(b) = (\iota_{\lambda}^* \mathbf{1}^{\mathcal{I}})(b) = (1+|b|^2)^{-\lambda} \mathbf{1}^{\mathcal{I}}(k(b))$$

$$= (1+|b|^2)^{-\lambda} \sum_{J \in \mathfrak{I}_{n,i}} (\det k(b))_{\mathcal{I}J} e_J,$$

$$h_{\lambda}^{\mathcal{I}}(b) = (\iota_{\lambda}^* h^{\mathcal{I}})(b) = (1+|b|^2)^{-\lambda} h^{\mathcal{I}}(k(b))$$

$$= (1+|b|^2)^{-\lambda} \sum_{J \in \mathfrak{I}_{n,i}} (\det k(b))_{\mathcal{I},J \cup \{0\}} e_J.$$

It follows from Lemma 7.4 (2) that, for $\mathcal{I}, \mathcal{J} \subset \{0, 1, \cdots, n\}$ with $\#\mathcal{I} = \#\mathcal{J} = i$, the minor determinant of $k(b)$ is given by

$$(\det k(b))_{\mathcal{I}\mathcal{J}} = -\varepsilon_{\mathcal{J}}(0)(\det \psi_{n+1}(1,b))_{\mathcal{I}\mathcal{J}} = \varepsilon_{\mathcal{J}}(0) \frac{S_{\mathcal{I}\mathcal{J}}(1,b)}{1+|b|^2}, \qquad (8.10)$$

where we set $\varepsilon_{\mathcal{J}}(0) = -1$ for $0 \notin \mathcal{J}$ and $\varepsilon_{\mathcal{J}}(0) = 1$ for $0 \in \mathcal{J}$.
Now the second formulæ in Lemma 8.5 are also shown. $\square$

8.3 Knapp–Stein Intertwining Operator

In this section we summarize some basic results on the matrix-valued Knapp–Stein intertwining operators, see [22, 23]. In the general framework of symmetry breaking operators for the restriction $G \downarrow G'$, this classical case may be thought of as a special case where $G = G'$, and the proof is much easier than the general case $G \supsetneq G'$. Nevertheless, we sketch a proof of results which we need in other chapters.

8.3.1 Knapp–Stein Intertwining Operator

For $(\sigma, V) \in \widehat{O(n)}$, $\delta, \varepsilon \in \{\pm\}$ and $\lambda, \nu \in \mathbb{C}$, we consider intertwining operators between two principal series representations $I_\delta(V, \lambda)$ and $I_\varepsilon(V, \nu)$ of $G = O(n + 1, 1)$. They are determined by distribution kernels, and Fact 5.9 (see [42, Prop. 3.2]) with $G = G'$ and $V = W$ gives a linear isomorphism

$$\operatorname{Hom}_G(I_\delta(V, \lambda), I_\varepsilon(V, \nu)) \simeq (\mathcal{D}'(G/P, \mathcal{V}^*_{\lambda, \delta}) \otimes V_{\nu, \varepsilon})^{\Delta(P)}, \tag{8.11}$$

where P acts diagonally on the $(G \times P)$-module $\mathcal{D}'(G/P, \mathcal{V}^*_{\lambda, \delta}) \otimes V_{\nu, \varepsilon}$. As in Proposition 5.15 (2), the restriction to the open Bruhat cell determines invariant distributions in the right-hand side, and thus we have an injective homomorphism

$$(\mathcal{D}'(G/P, \mathcal{V}^*_{\lambda, \delta}) \otimes V_{\nu, \varepsilon})^{\Delta(P)} \hookrightarrow \mathcal{D}'(\mathbb{R}^n) \otimes \operatorname{End}_{\mathbb{C}}(V), \quad f \mapsto F(x) := f(n_-(x)),$$

where we have used the canonical isomorphism $V^\vee \otimes V \simeq \operatorname{End}_{\mathbb{C}}(V)$. Different from the case $G \supsetneq G'$ for symmetry breaking operators, there are strong constraints on the parameter for the existence of nonzero elements in (8.11). In fact, it follows readily from the P-invariance that $F|_{\mathbb{R}^n - \{0\}}$ is nonzero only if $\nu = n - \lambda$, and in this case it is proportional to $|x|^{2\lambda - 2n} \sigma(\psi_n(x))$, where we recall from (3.4) the definition of $\psi_n \colon \mathbb{R}^n - \{0\} \to O(n)$. We normalize as

$$\widetilde{\mathcal{T}}^V_{\lambda, n-\lambda}(x) := \frac{1}{\Gamma(\lambda - \frac{n}{2})} |x|^{2\lambda - 2n} \sigma(\psi_n(x)). \tag{8.12}$$

Remark 8.6. The normalization of the Knapp–Stein operator is not unique, and different choices are useful for different purposes. See for example Knapp–Stein [22] or Langlands [49].

With the normalization (8.12), we now review the Knapp–Stein intertwining operators in this setting as follows.

Lemma 8.7 (normalized Knapp–Stein operator). *The distribution (8.12) belongs to $L^1_{\mathrm{loc}}(\mathbb{R}^n) \otimes \operatorname{End}_{\mathbb{C}}(V)$ if $\operatorname{Re} \lambda \gg 0$, and extends to an element of $(\mathcal{D}'(G/P, \mathcal{V}^*_{\lambda, \delta}) \otimes V_{n-\lambda, \delta})^{\Delta(P)}$. Furthermore, it has an analytic continuation to the entire $\lambda \in \mathbb{C}$.*

By definition, the (normalized) Knapp–Stein intertwining operator

$$\widetilde{\mathbb{T}}^V_{\lambda,n-\lambda} : I_\delta(V,\lambda) \to I_\delta(V,n-\lambda) \tag{8.13}$$

is defined in the N-picture of the principal series representation by the formula

$$(\widetilde{\mathbb{T}}^V_{\lambda,n-\lambda} f)(x) = \int_{\mathbb{R}^n} \widetilde{\mathcal{T}}^V_{\lambda,n-\lambda}(x-y) f(y) dy.$$

When (σ, V) is the i-th exterior representation on $\bigwedge^i(\mathbb{C}^n)$, we write simply $\widetilde{\mathbb{T}}^i_{\lambda,n-\lambda}$ and $\widetilde{\mathcal{T}}^i_{\lambda,n-\lambda}$ for the operator $\widetilde{\mathbb{T}}^V_{\lambda,n-\lambda}$ and the distribution $\widetilde{\mathcal{T}}^V_{\lambda,n-\lambda}$, respectively.

The Knapp–Stein operator (8.13) gives a continuous G-homomorphism $I_\delta(i,\lambda) \to I_\varepsilon(j,\nu)$ when $j = i$ (and $\delta = \varepsilon$, $\nu = n - \lambda$). On the other hand, there exist G-intertwining operators $I_\delta(i,\lambda) \to I_\varepsilon(j,\nu)$ also when $i \neq j$ for special parameters. Like sporadic symmetry breaking operators (*cf.* Theorem 3.6), they are given by differential operators as follows.

Fact 8.8. *Suppose that $0 \leq i \leq n-1$.*

(1) *We can identify $I_{(-1)^i}(i,i)$ with the space $\mathcal{E}^i(S^n)$ of differential i-forms endowed with the natural action of the conformal group $G = O(n+1,1)$.*
(2) *The exterior derivative $d: \mathcal{E}^i(S^n) \to \mathcal{E}^{i+1}(S^n)$ induces a G-intertwining operator*

$$D_i : I_{(-1)^i}(i,i) \to I_{(-1)^{i+1}}(i+1,i+1).$$

The kernel of D_i is $\Pi_{i,(-1)^i}$, and the image is $\Pi_{i+1,(-1)^{i+1}}$.

This follows from [35, Thm. 12.2]. We note that the existence of such an intertwining operator is assured *a priori* by the composition series of the principal series representation (Theorem 2.20), see also [11].

8.3.2   K-spectrum of the Knapp–Stein Intertwining Operator

This section gives an explicit formula for the eigenvalues of the (normalized) Knapp–Stein intertwining operator

$$\widetilde{\mathbb{T}}^i_{\lambda,n-\lambda} : I_\delta(i,\lambda) \to I_\delta(i,n-\lambda) \tag{8.14}$$

on the basic K-types $\mu^\flat(i,\delta)$ and $\mu^\sharp(i,\delta)$ (see (2.30) and (2.31), respectively). For $0 \leq i \leq n$ and $\lambda \in \mathbb{C}$, we set

$$c^\flat(i,\lambda) = \frac{\pi^{\frac{n}{2}}}{\Gamma(\lambda+1)} \times \begin{cases} \lambda - i & \text{if } \natural = \flat, \\ n - i - \lambda & \text{if } \natural = \sharp. \end{cases} \tag{8.15}$$

Proposition 8.9. *Suppose $0 \le i \le n$, $\lambda \in \mathbb{C}$ and $\delta \in \{\pm\}$. Then the (normalized) Knapp–Stein intertwining operator*

$$\widetilde{\mathbb{T}}^i_{\lambda, n-\lambda} : I_\delta(i, \lambda) \to I_\delta(i, n-\lambda)$$

acts on the basic K-types $\mu^\flat(i, \delta) = \bigwedge^i(\mathbb{C}^{n+1}) \boxtimes \delta$ and $\mu^\sharp(i, \delta) = \bigwedge^{i+1}(\mathbb{C}^{n+1}) \boxtimes (-\delta)$ as the scalar multiplication:

$$\widetilde{\mathbb{T}}^i_{\lambda, n-\lambda} \circ \iota^*_\lambda = c^\natural(i, \lambda) \iota^*_{n-\lambda} \quad on \ \mu^\natural(i, \delta) \ for \ \natural = \flat \ or \ \sharp.$$

In other words, we have

$$\widetilde{\mathbb{T}}^i_{\lambda, n-\lambda}(\mathbf{1}^{\mathcal{I}}_\lambda) = \frac{(\lambda - i)\pi^{\frac{n}{2}}}{\Gamma(\lambda + 1)} \mathbf{1}^{\mathcal{I}}_{n-\lambda} \qquad for \ all \ \mathcal{I} \in \mathfrak{I}_{n+1,i},$$

$$\widetilde{\mathbb{T}}^i_{\lambda, n-\lambda}(h^{\mathcal{I}}_\lambda) = \frac{(n - i - \lambda)\pi^{\frac{n}{2}}}{\Gamma(\lambda + 1)} h^{\mathcal{I}}_{n-\lambda} \qquad for \ all \ \mathcal{I} \in \mathfrak{I}_{n+1,i+1}.$$

Remark 8.10. Proposition 8.9 in the $i = 0$ case for $\mu^\flat(i, \delta)$ was proved in [42, Prop. 4.6].

We will give a proof of Proposition 8.9 in Section 8.3.4.

We recall from Theorem 2.20 that the composition series of $I_\delta(i, i)$ and $I_\delta(i, n - i)$ are described by the following exact sequences of G-modules:

$$0 \to \Pi_{i,\delta} \to I_\delta(i, i) \to \Pi_{i+1,-\delta} \to 0,$$

$$0 \to \Pi_{i+1,-\delta} \to I_\delta(i, n-i) \to \Pi_{i,\delta} \to 0,$$

which do not split if $i \ne \frac{n}{2}$. Thus Proposition 8.9 implies:

Proposition 8.11. *Suppose $G = O(n + 1, 1)$ and $i \ne \frac{n}{2}$. Then the kernels and the images of the G-homomorphisms $\widetilde{\mathbb{T}}^i_{\lambda, n-\lambda} : I_\delta(i, \lambda) \to I_\delta(i, n-\lambda)$ for $\lambda = i$, $n - i$ are given by*

$$\mathrm{Ker}(\widetilde{\mathbb{T}}^i_{i, n-i}) \simeq \Pi_{i,\delta} \quad \simeq \mathrm{Image}(\widetilde{\mathbb{T}}^i_{n-i, i})$$

$$\mathrm{Image}(\widetilde{\mathbb{T}}^i_{i, n-i}) \simeq \Pi_{i+1,-\delta} \simeq \mathrm{Ker}(\widetilde{\mathbb{T}}^i_{n-i, i}).$$

8.3.3 Vanishing of the Knapp–Stein Operator

There are a few exceptional parameters (i, λ) for which $\widetilde{\mathbb{T}}^i_{\lambda, n-\lambda}$ vanishes:

Proposition 8.12. *Suppose $G = O(n + 1, 1)$, $0 \le i \le n$, and $\lambda \in \mathbb{C}$. Then the normalized Knapp–Stein intertwining operator $\widetilde{\mathbb{T}}^i_{\lambda, n-\lambda}$ is zero if and only if $\lambda = i = \frac{n}{2}$.*

Proof. See [34].   □

A renormalization of the Knapp–Stein intertwining operator $\widetilde{\mathbb{T}}^i_{\lambda,n-\lambda}$ for $n = 2i$ will be discussed in Section 8.4.

8.3.4   Integration Formula for the (K, K)-spectrum

In this subsection, we give a proof of Proposition 8.9. Let $\natural = \flat$ or $\sharp$. Since the multiplicity of the K-type $\mu^\natural(i, \delta)$ in the principal series representation $I_\delta(i, \lambda)$ is one, there exists a constant $c^\natural(i, \lambda)$ depending on i and λ such that

$$\widetilde{\mathbb{T}}^i_{\lambda,n-\lambda} \circ \iota^*_\lambda = c^\natural(i, \lambda)\iota^*_{n-\lambda} \quad \text{on } \mu^\natural(i, \delta). \tag{8.16}$$

We shall show that the constants $c^\natural(i, \lambda)$ in the equation (8.16) are given by the formulæ (8.15). The first step is to give an integral formula for the constants $c^\natural(i, \lambda)$ for $\natural = \flat$ and $\sharp$:

Lemma 8.13. *Suppose $0 \le i \le n$ and $\lambda \in \mathbb{C}$ with $\mathrm{Re}\,\lambda \gg 0$. Then we have*

$$c^\flat(i, \lambda) = \frac{1}{\Gamma(\lambda - \frac{n}{2})} \int_{\mathbb{R}^n} |b|^{2\lambda-2n}(1 + |b|^2)^{-\lambda}$$

$$\times \left(1 - \frac{2}{|b|^2(1 + |b|^2)} \sum_{k=1}^{i} b_k^2 \right) db,$$

$$c^\flat(i, \lambda) - c^\sharp(i, \lambda) = \frac{2}{\Gamma(\lambda - \frac{n}{2})} \int_{\mathbb{R}^n} |b|^{2\lambda-2n+2}(1 + |b|^2)^{-\lambda-1} db.$$

Proof of Lemma 8.13. We first consider (8.16) for $\natural = \flat$. Then we have

$$\widetilde{\mathbb{T}}^i_{\lambda,n-\lambda}(\mathbf{1}^\mathcal{I}_\lambda) = c^\flat(i, \lambda)\mathbf{1}^\mathcal{I}_{n-\lambda} \qquad \text{for all } \mathcal{I} \in \mathfrak{I}_{n+1,i}.$$

Take $\mathcal{I} \in \mathfrak{I}_{n+1,i}$ such that $0 \notin \mathcal{I}$. Then (8.8) tells that

$$(\widetilde{\mathbb{T}}^i_{\lambda,n-\lambda}\mathbf{1}^\mathcal{I}_\lambda)(0) = c^\flat(i, \lambda)e_\mathcal{I}. \tag{8.17}$$

Let us compute the left-hand side. In view of the distribution kernel (8.12) of the normalized Knapp–Stein operator $\widetilde{\mathbb{T}}^i_{\lambda,n-\lambda}$, for $\mathrm{Re}\,\lambda \gg 0$ we have

$$(\widetilde{\mathbb{T}}^i_{\lambda,n-\lambda}\mathbf{1}^\mathcal{I}_\lambda)(0) = \frac{1}{\Gamma(\lambda - \frac{n}{2})} \int_{\mathbb{R}^n} |-b|^{2\lambda-2n}\sigma(\psi_n(-b))\mathbf{1}^\mathcal{I}_\lambda(b)db.$$

By (7.10) and the formula (8.6) of $\mathbf{1}_{\lambda}^{\mathcal{I}}(b)$, the integrand amounts to

$$\sum_{J, J' \in \mathfrak{I}_{n,i}} |b|^{2\lambda - 2n}(1 + |b|^2)^{-\lambda}(\det \psi_{n+1}(1, b))_{\mathcal{I}J}(\det \psi_n(b))_{J'J} e_{J'}.$$

Comparing the coefficients of $e_{\mathcal{I}}$ in the both sides of (8.17), we get

$$c^{\flat}(i, \lambda) = \frac{1}{\Gamma(\lambda - \frac{n}{2})} \int_{\mathbb{R}^n} |b|^{2\lambda - 2n}(1 + |b|^2)^{-\lambda} g_{\mathcal{I}}(b) db,$$

where we set

$$g_{\mathcal{I}}(b) := \sum_{J \in \mathfrak{I}_{n,i}} (\det \psi_{n+1}(1, b))_{\mathcal{I}J}(\det \psi_n(b))_{\mathcal{I}J} = 1 - \frac{2Q_{\mathcal{I}}(b)}{(1 + |b|^2)|b|^2}. \tag{8.18}$$

The second equality was proved as the minor summation formula in Proposition 7.7 (4), where we recall $Q_{\mathcal{I}}(b) = \sum_{l \in \mathcal{I}} b_l^2$. Therefore, by taking $\mathcal{I} = \{1, 2, \cdots, n\}$, we get the first assertion of Lemma 8.13.

Next, we consider (8.16) for $\natural = \sharp$. Then we have

$$\widetilde{\mathbb{T}}_{\lambda, n-\lambda}^i(h_{\lambda}^{\mathcal{I}}) = c^{\sharp}(i, \lambda) h_{n-\lambda}^{\mathcal{I}} \quad \text{for all } \mathcal{I} \in \mathfrak{I}_{n+1, i+1}.$$

Take $I \in \mathfrak{I}_{n,i}$, and set $\mathcal{I} := I \cup \{0\} \in \mathfrak{I}_{n+1, i+1}$. By (8.9), we have

$$\widetilde{\mathbb{T}}_{\lambda, n-\lambda}^i(h_{\lambda}^{\mathcal{I}})(0) = c^{\sharp}(i, \lambda) e_I. \tag{8.19}$$

By (8.12), we have

$$\widetilde{\mathbb{T}}_{\lambda, n-\lambda}^i(h_{\lambda}^{\mathcal{I}})(0) = \frac{1}{\Gamma(\lambda - \frac{n}{2})} \int_{\mathbb{R}^n} |-b|^{2\lambda - 2n} \sigma(\psi_n(-b)) h_{\lambda}^{\mathcal{I}}(b) db.$$

Comparing the coefficients of e_I in the both sides of the equation (8.19), we get from (8.7) and (7.10)

$$c^{\sharp}(i, \lambda) = -\frac{1}{\Gamma(\lambda - \frac{n}{2})} \int_{\mathbb{R}^n} |b|^{2\lambda - 2n}(1 + |b|^2)^{-\lambda} g'_I(b) db,$$

where we set

$$g'_I(b) := \sum_{J \in \mathfrak{I}_{n,i}} (\det \psi_{n+1}(1, b))_{I \cup \{0\}, J \cup \{0\}}(\det \psi_n(b))_{IJ}.$$

We note that

$$\det \psi_{n+1}(1, b)_{IJ} = \det \psi_n(b; \frac{2}{1 + |b|^2})_{IJ}$$

if $I, J \in \mathfrak{I}_{n,i}$ is regarded as elements of $\mathfrak{I}_{n+1,i}$ in the left-hand side. Then we have

$$g_I(b) + g_I'(b) = \frac{2|b|^2}{1+|b|^2}$$

from Proposition 7.7 (3), and thus we get

$$c^\flat(i,\lambda) - c^\sharp(i,\lambda) = \frac{2}{\Gamma(\lambda - \frac{n}{2})} \int_{\mathbb{R}^n} |b|^{2\lambda+2-2n}(1+|b|^2)^{-\lambda-1}db.$$

Now Lemma 8.13 is proved.   □

The second step is to compute the integrals in Lemma 8.13.

Lemma 8.14. *For* $\operatorname{Re}\lambda \gg 0$, $c^\flat(i,\lambda)$ *and* $c^\sharp(i,\lambda)$ *take the form* (8.15).

Proof. Let $B(\lambda, \nu)$ denote the Beta function. By the change of variables $r^2 = \frac{x}{1-x}$, we have

$$\int_0^\infty r^a(1+r^2)^b dr = \frac{1}{2} \int_0^1 x^{\lambda-1}(1-x)^{\nu-1}dx = \frac{1}{2}B(\lambda, \nu), \qquad (8.20)$$

where $a = 2\lambda - 1$ and $b = -\lambda - \nu$. Then Lemma 8.13 in the polar coordinates tells that

$$c^\flat(i,\lambda) = \frac{\operatorname{vol}(S^{n-1})}{2\Gamma(\lambda - \frac{n}{2})} (B(\lambda - \frac{n}{2}, \frac{n}{2}) - \frac{2i}{n} B(\lambda - \frac{n}{2}, \frac{n}{2}+1))$$

by (8.20) and by the following observation:

$$\int_{S^{n-1}} |\omega_i|^2 d\omega = \frac{1}{n} \operatorname{vol}(S^{n-1}) \quad (1 \le i \le n).$$

Since $\operatorname{vol}(S^{n-1}) = \frac{2\pi^{\frac{n}{2}}}{\Gamma(\frac{n}{2})}$, we get the first statement.

By the second formula of Lemma 8.13, we have

$$c^\flat(i,\lambda) - c^\sharp(i,\lambda) = \frac{1}{\Gamma(\lambda - \frac{n}{2})} \operatorname{vol}(S^{n-1}) B(\lambda - \frac{n}{2}, \frac{n}{2}+1)$$

$$= \frac{(2\lambda - n)\pi^{\frac{n}{2}}}{\Gamma(\lambda + 1)}.$$

Thus the closed formula (8.15) for $c^\sharp(i,\lambda)$ is also proved.   □

Proof of Proposition 8.9. The assertion follows from Lemmas 8.13 and 8.14 for $\operatorname{Re}\lambda \gg 0$. For general $\lambda \in \mathbb{C}$, Proposition holds by the analytic continuation.   □

8.4 Renormalization of the Knapp–Stein Intertwining Operator

Because of the vanishing of the normalized Knapp–Stein intertwining operators in the middle degree when n is even (Proposition 8.12), intertwining operators from $I_\delta(\frac{n}{2}, \lambda)$ to $I_\delta(\frac{n}{2}, n - \lambda)$ require special attention. In this case, we set $n = 2m$ and renormalize the Knapp–Stein intertwining operator of $G = O(2m + 1, 1)$ at the middle degree by

$$\widetilde{\widetilde{\mathbb{T}}}^m_{\lambda, 2m-\lambda} := \frac{1}{\lambda - m} \widetilde{\mathbb{T}}^m_{\lambda, 2m-\lambda}. \tag{8.21}$$

Then $\widetilde{\widetilde{\mathbb{T}}}^m_{\lambda, 2m-\lambda} \colon I_\delta(m, \lambda) \to I_\delta(m, 2m - \lambda)$ depends holomorphically in the entire $\lambda \in \mathbb{C}$, and is vanishing nowhere.

If $\lambda = m$, then $\widetilde{\widetilde{\mathbb{T}}}^m_{\lambda, 2m-\lambda}$ acts as an endomorphism of $I_\delta(m, m)$. On the other hand, we know from Theorem 2.20 (1) that the principal series representation $I_\delta(m, m)$ decomposes into the direct sum of two irreducible tempered representations of G as follows:

$$I_\delta(m, m) \simeq I_\delta(m)^\flat \oplus I_\delta(m)^\sharp \equiv \Pi_{m,\delta} \oplus \Pi_{m+1,-\delta}.$$

Lemma 8.15. *Let $n = 2m$ and $G = O(2m + 1, 1)$. Then the renormalized Knapp–Stein operator $\widetilde{\widetilde{\mathbb{T}}}^m_{m,m}$ acts on $I_\delta(m, m) \simeq \Pi_{m,\delta} \oplus \Pi_{m+1,-\delta}$ as*

$$\frac{\pi^m}{m!} (\mathrm{id}_{\Pi_{m,\delta}} \oplus (-\mathrm{id})_{\Pi_{m+1,-\delta}}).$$

Proof. Since the irreducible G-module $\Pi_{m,\delta}$ is not isomorphic to the irreducible G-module $\Pi_{m+1,-\delta}$, the renormalized Knapp–Stein intertwining operator $\widetilde{\widetilde{\mathbb{T}}}^m_{m,m}$ acts on each irreducible summand by scalar multiplication. Therefore, it is sufficient to find the scalars on specific K-types occurring in each summand. By Proposition 8.9, the renormalized Knapp–Stein intertwining operator $\widetilde{\widetilde{\mathbb{T}}}^m_{\lambda, 2m-\lambda}$ acts on vectors that belong to the K-types $\mu^\flat(m, \delta)(\subset \Pi_{m,\delta})$ and $\mu^\sharp(m, \delta)(\subset \Pi_{m+1,-\delta})$ by the scalars

$$\frac{1}{\lambda - m} \frac{(\lambda - m)\pi^m}{\Gamma(\lambda + 1)} \quad \text{and} \quad \frac{1}{\lambda - m} \frac{(2m - m - \lambda)\pi^m}{\Gamma(\lambda + 1)},$$

respectively. Taking the limit as λ tends to m, we get the lemma. $\qquad\square$

8.5   Kernel of the Knapp–Stein Operator

In this section, we discuss the proper submodules of the principal series representation $I_\delta(i, \lambda)$ of $G = O(n+1, 1)$ at reducible points (see (2.33)).

We consider the composition of the Knapp–Stein operators, $\widetilde{\mathbb{T}}^i_{n-\lambda, \lambda} \circ \widetilde{\mathbb{T}}^i_{\lambda, n-\lambda} \in \mathrm{End}_G(I_\delta(i, \lambda))$. By Proposition 8.9, its K-spectrum on the basic K-type $\mu^\flat(i, \delta)$ is given as

$$\widetilde{\mathbb{T}}^i_{n-\lambda, \lambda} \circ \widetilde{\mathbb{T}}^i_{\lambda, n-\lambda}(1^{\mathcal{I}}_\lambda) = \frac{(\lambda - i)(n - \lambda - i)\pi^n}{\Gamma(\lambda + 1)\Gamma(n - \lambda + 1)}(1^{\mathcal{I}}_\lambda) \quad \text{for all } \mathcal{I} \in \mathfrak{I}_{n+1, i}.$$

Since the principal series representation $I_\delta(i, \lambda)$ is generically irreducible, we conclude

$$\widetilde{\mathbb{T}}^i_{n-\lambda, \lambda} \circ \widetilde{\mathbb{T}}^i_{\lambda, n-\lambda} = \frac{(\lambda - i)(n - \lambda - i)\pi^n}{\Gamma(\lambda + 1)\Gamma(n - \lambda + 1)}\mathrm{id} \quad \text{on } I_\delta(i, \lambda) \tag{8.22}$$

for generic λ by Schur's lemma, and then for all $\lambda \in \mathbb{C}$ by analytic continuation.

Lemma 8.16. *Let $G = O(n+1, 1)$, $0 \le i \le n$, and $\delta \in \{\pm\}$. Assume*

$$\lambda \in \{i, n - i\} \cup (-\mathbb{N}_+) \cup (n + \mathbb{N}_+).$$

Then $I_\delta(i, \lambda)$ is reducible.

Proof. If $(n, \lambda) = (2i, i)$, we already know that $I_\delta(i, \lambda)$ is reducible, see Lemma 8.15.

Assume now $(n, \lambda) \ne (2i, i)$. Then Proposition 8.12 tells that neither $\widetilde{\mathbb{T}}^i_{n-\lambda, \lambda}$ nor $\widetilde{\mathbb{T}}^i_{\lambda, n-\lambda}$ vanishes. On the other hand, by (8.22), the assumption on λ implies

$$\widetilde{\mathbb{T}}^i_{n-\lambda, \lambda} \circ \widetilde{\mathbb{T}}^i_{\lambda, n-\lambda} = 0,$$

which shows that at least one of the G-modules $I_\delta(i, \lambda)$ or $I_\delta(i, n - \lambda)$ is reducible. By Lemma 3.36, we conclude that both $I_\delta(i, \lambda)$ and its contragredient representation $I_\delta(i, n - \lambda)$ are reducible. $\qquad\square$

Lemma 8.16 gives an alternative proof for the "if part" of Proposition 2.18 (1).

Proposition 8.17. *Let $G = O(n+1, 1)$, $0 \le i \le n$, $\delta \in \{\pm\}$, and $\lambda \in \mathbb{C}$. Assume further that $I_\delta(i, \lambda)$ is reducible, namely,*

$$\lambda \in \{i, n - i\} \cup (-\mathbb{N}_+) \cup (n + \mathbb{N}_+).$$

(1) *Suppose $(n, \lambda) \ne (2i, i)$. Then the unique proper submodule of $I_\delta(i, \lambda)$ is given as the kernel of the Knapp–Stein operator $\widetilde{\mathbb{T}}^i_{\lambda, n-\lambda}: I_\delta(i, \lambda) \to I_\delta(i, n - \lambda)$.*

(2) *Suppose $(n, \lambda) = (2i, i)$. Then $\widetilde{\mathbb{T}}^i_{\lambda, n-i} = 0$, and there are two proper submodules of $I_\delta(i, \lambda)$, which are given as the kernel of $\widetilde{\widetilde{\mathbb{T}}}^i_{i,i} \pm \frac{\pi^i}{i!} \mathrm{id} \in \mathrm{End}_G(I_\delta(i, i))$ where $\widetilde{\widetilde{\mathbb{T}}}^i_{i,i}$ is the renormalized Knapp–Stein operator.*

Proof.

(1) There is a unique irreducible submodule of $I_\delta(i, \lambda)$ for the parameter λ under consideration. Hence $\mathrm{Ker}(\widetilde{\mathbb{T}}^i_{\lambda, n-\lambda})$ is the unique irreducible submodule by the proof of Lemma 8.16.

(2) This is already proved in Lemma 8.15. $\square$

Chapter 9
Regular Symmetry Breaking Operators $\widetilde{\mathbb{A}}^{i,j}_{\lambda,\nu,\delta\varepsilon}$ from $I_\delta(i,\lambda)$ to $J_\varepsilon(j,\nu)$

In this chapter we apply the general results developed in Chapter 5 on the analytic continuation of integral symmetry breaking operators $\widetilde{\mathbb{A}}^{V,W}_{\lambda,\nu,\delta\varepsilon}: I_\delta(V,\lambda) \to J_\varepsilon(W,\nu)$ to the special setting where

$$(V,W) = (\textstyle\bigwedge^i(\mathbb{C}^n), \bigwedge^j(\mathbb{C}^{n-1})), \tag{9.1}$$

and construct a holomorphic family of (normalized) regular symmetry breaking operators

$$\widetilde{\mathbb{A}}^{i,j}_{\lambda,\nu,\delta\varepsilon}: I_\delta(i,\lambda) \to J_\varepsilon(j,\nu),$$

which exist if and only if $j = i - 1$ or i (Theorems 9.1 and 9.2). Then the goal of this chapter is to determine

- the parameter (λ,ν) for which $\widetilde{\mathbb{A}}^{i,j}_{\lambda,\nu,\pm}$ vanishes (Section 9.2);
- the (K,K')-spectrum of $\widetilde{\mathbb{A}}^{i,j}_{\lambda,\nu,\pm}$ (Sections 9.3–9.7);
- functional equations of $\widetilde{\mathbb{A}}^{i,j}_{\lambda,\nu,\pm}$ (Sections 9.8–9.9).

Thus we will complete the proof of Theorem 3.19 that determines the zeros of the normalized operators $\widetilde{\mathbb{A}}^{i,j}_{\lambda,\nu,\delta\varepsilon}$. This is the last missing piece in the classification scheme (Theorem 3.13), and thus we complete the proof of the classification of the space $\mathrm{Hom}_{G'}(I_\delta(i,\lambda)|_{G'}, J_\varepsilon(j,\nu))$ of *all* symmetry breaking operators as stated in Theorems 3.25 and 3.26.

The (K,K')-spectrum resembles eigenvalues of a symmetry breaking operator (Definition 9.7), for which we find an integral expression and determine the explicit formula for basic K- and K'-types (Theorem 9.8).

The matrix-valued functional equations among various intertwining operators are determined explicitly in Theorems 9.24 and 9.25 by using the formula of the

© Springer Nature Singapore Pte Ltd. 2018
T. Kobayashi, B. Speh, *Symmetry Breaking for Representations of Rank One Orthogonal Groups II*, Lecture Notes in Mathematics 2234,
https://doi.org/10.1007/978-981-13-2901-2_9

(K, K')-spectrum, which in turn will play a crucial role in analyzing the behavior of the symmetry breaking operators at reducible places (Chapter 10).

Degenerate cases where the normalized operators $\widetilde{\mathbb{A}}^{i,j}_{\lambda,\nu,\pm}$ vanish will be discussed in Sections 9.9 and 9.10.

As an application of the matrix-valued functional equations (Theorems 9.24 and 9.25) and the residue formulæ of $\widetilde{\mathbb{A}}^{i,j}_{\lambda,\nu,\pm}$ (Fact 9.3), we determine when the *differential* symmetry breaking operators $\widetilde{\mathbb{C}}^{i,j}_{\lambda,\nu}$ $(j = i, i - 1)$ are surjective in Section 9.11.

9.1   Regular Symmetry Breaking Operators $\widetilde{\mathbb{A}}^{i,j}_{\lambda,\nu,\pm}$

In this section, we give the existence condition and an explicit construction of (generically) regular symmetry breaking operators from G-modules $I_\delta(V, \lambda)$ to G'-modules $J_\varepsilon(W, \nu)$ in the setting (9.1) by applying the general results of Chapters 3 and 5, in particular, Theorems 3.9 and 3.10 and their proofs.

9.1.1   Existence Condition for Regular Symmetry Breaking Operators

We recall from Definition 5.10 the notion of *regular* symmetry breaking operators. We also recall from (5.50) and (5.51) the definition of the open dense subsets $U^{\text{reg}}_\pm$ in $\mathbb{C}^2$. Then the existence condition of regular symmetry breaking operators in the setting (9.1) is stated as follows.

Theorem 9.1. *Suppose* $0 \leq i \leq n$ *and* $0 \leq j \leq n - 1$. *Then the following three conditions on the pair* (i, j) *are equivalent:*

(i) *there exists a nonzero regular symmetry breaking operator from the G-module $I_\delta(i, \lambda)$ to the G'-module $J_\varepsilon(j, \nu)$ for some $(\lambda, \nu, \delta, \varepsilon) \in \mathbb{C}^2 \times \{\pm\}^2$;*

(ii) *for any $(\delta, \varepsilon) \in \{\pm\}^2$, there exists a nonzero regular symmetry breaking operator from $I_\delta(i, \lambda)$ to $J_\varepsilon(j, \nu)$ for all $(\lambda, \nu) \in U^{\text{reg}}_{\delta\varepsilon}$;*

(iii) $j = i$ *or* $i - 1$.

Proof. As we have seen in the decomposition (7.4), $[V : W] \neq 0$ in the setting (9.1) if and only if $j = i - 1$ or i. Then Theorem 9.1 follows from Theorem 3.9 and Proposition 5.39. $\square$

9.1.2 Construction of $\widetilde{\mathbb{A}}^{i,j}_{\lambda,\nu,\pm}$ for $j \in \{i-1, i\}$

In this section we apply Theorem 3.10 about the construction of the (generically) regular symmetry breaking operators $\widetilde{\mathbb{A}}^{V,W}_{\lambda,\nu,\pm}$ in the setting (9.1) with $j = i - 1$ or i. In particular, we give concrete formulæ of the matrix-valued distribution kernels $\widetilde{\mathscr{A}}^{i,j}_{\lambda,\nu,\pm}$ for the operators.

Let $j = i - 1$ or i. We recall from (7.2) and (7.3) that the projection $\mathrm{pr}_{i \to j} : \bigwedge^i(\mathbb{C}^n) \to \bigwedge^j(\mathbb{C}^{n-1})$ defines an element of

$$\mathrm{Hom}_{O(n-1)}(V, W) = \mathrm{Hom}_{O(n-1)}(\textstyle\bigwedge^i(\mathbb{C}^n), \bigwedge^j(\mathbb{C}^{n-1})).$$

Denote by $\sigma \equiv \sigma^{(i)}$ the i-th exterior representation of $O(n)$ on $\bigwedge^i(\mathbb{C}^n)$. Then the matrix-valued function $R^{V,W}$ (see (3.6)) amounts to the following map

$$R^{i,j} : \mathbb{R}^n - \{0\} \to \mathrm{Hom}_{\mathbb{C}}(\textstyle\bigwedge^i(\mathbb{C}^n), \bigwedge^j(\mathbb{C}^{n-1}))$$

given by

$$R^{i,j} := \mathrm{pr}_{i \to j} \circ \sigma \circ \psi_n \tag{9.2}$$

where we recall from (3.4) that $\psi_n : \mathbb{R}^n - \{0\} \to O(n)$ is the map of taking "reflection".

Applying the general formulæ (3.7) and (3.8) of the distribution kernels $\widetilde{\mathscr{A}}^{V,W}_{\lambda,\nu,\pm}$ in the setting (9.1), we obtain $\mathrm{Hom}_{\mathbb{C}}(\bigwedge^i(\mathbb{C}^n), \bigwedge^j(\mathbb{C}^{n-1}))$-valued locally integrable functions on $\mathbb{R}^n$ for $\mathrm{Re}\,\lambda \gg |\mathrm{Re}\,\nu|$ as follows.

$$\widetilde{\mathscr{A}}^{i,j}_{\lambda,\nu,+} := \frac{1}{\Gamma(\frac{\lambda+\nu-n+1}{2})\Gamma(\frac{\lambda-\nu}{2})}(|x|^2 + x_n^2)^{-\nu}|x_n|^{\lambda+\nu-n} R^{i,j}(x, x_n), \tag{9.3}$$

$$\widetilde{\mathscr{A}}^{i,j}_{\lambda,\nu,-} := \frac{1}{\Gamma(\frac{\lambda+\nu-n+2}{2})\Gamma(\frac{\lambda-\nu+1}{2})}(|x|^2 + x_n^2)^{-\nu}|x_n|^{\lambda+\nu-n} \,\mathrm{sgn}\, x_n\, R^{i,j}(x, x_n). \tag{9.4}$$

Then, as a special case of Theorem 3.10, we obtain:

Theorem 9.2 (holomorphic continuation of integral operators). *Let (V, W) be as in (9.1) with $j = i, i - 1$, and $\delta, \varepsilon \in \{\pm\}$. Then the distributions $\widetilde{\mathscr{A}}^{i,j}_{\lambda,\nu,\delta\varepsilon}$, initially defined as $\mathrm{Hom}_{\mathbb{C}}(V, W)$-valued locally integrable functions on $\mathbb{R}^n$ for $\mathrm{Re}\,\lambda \gg |\mathrm{Re}\,\nu|$, extends to $(\mathcal{D}'(G/P, V^*_{\lambda,\delta}) \otimes W_{\nu,\varepsilon})^{\Delta(P')}$ that depends holomorphically on (λ, ν) in $\mathbb{C}^2$. Then the matrix-valued distribution kernels $\widetilde{\mathscr{A}}^{i,j}_{\lambda,\nu,\delta\varepsilon}$ induce a family of symmetry breaking operators*

$$\widetilde{\mathbb{A}}^{i,j}_{\lambda,\nu,\delta\varepsilon} : I_\delta(i, \lambda) \to J_\varepsilon(j, \nu) \tag{9.5}$$

for all $(\lambda, \nu) \in \mathbb{C}^2$.

Then $\widetilde{\mathbb{A}}^{i,j}_{\lambda,\nu,\pm}$ is the normalized (generically) regular symmetry breaking operator (Definition 5.40) in the sense that there exists an open dense subset U_γ in $\mathbb{C}^2$ for $\gamma \in \{\pm\}$ such that the support of the distribution kernel of $\widetilde{\mathbb{A}}^{i,j}_{\lambda,\nu,\gamma}$ equals the whole flag manifold G/P as far as $(\lambda, \nu) \in U_\gamma$, see Proposition 5.39. By a little abuse of terminology, we say that $\{\widetilde{\mathbb{A}}^{i,j}_{\lambda,\nu,\pm}\}$ is a family of *normalized regular symmetry breaking operators*.

9.2   Zeros of $\widetilde{\mathbb{A}}^{i,j}_{\lambda,\nu,\pm}$: Proof of Theorem 3.19

In this section we determine the exact place of the zeros of the normalized regular symmetry breaking operators $\widetilde{\mathbb{A}}^{i,j}_{\lambda,\nu,\delta\varepsilon}$, and thus give a proof of Theorem 3.19. In particular, we see that the Gamma factors in the normalization (9.3) and (9.4) are optimal in the sense that the zeros of $\widetilde{\mathbb{A}}^{i,j}_{\lambda,\nu,\gamma}$ are of codimension two in $\mathbb{C}^2$, namely, form a discrete subset of $\mathbb{C}^2$. The proof of Theorem 3.19 consists of the following steps.

Step 0.   (existence condition) Regular symmetry breaking operators from $I_\delta(i,\lambda)$ to $J_\varepsilon(j,\nu)$ exist if and only if $j \in \{i-1, i\}$ (Theorem 9.1).

Step 1.   (generically nonzero) If $\widetilde{\mathbb{A}}^{i,j}_{\lambda,\nu,\delta\varepsilon} = 0$, then $(\lambda, \nu, \delta, \varepsilon)$ belongs to the set Ψ_{sp} of special parameters (Theorem 5.41).

Step 2.   (residue formula) If $(\lambda, \nu, \delta, \varepsilon) \in \Psi_{sp}$, then $\widetilde{\mathbb{A}}^{i,j}_{\lambda,\nu,\delta\varepsilon}$ is proportional to the *differential* symmetry breaking operator $\mathbb{C}^{i,j}_{\lambda,\nu}$ with explicit proportional constant (Fact 9.3).

9.2.1   Residue Formula of the Regular Symmetry Breaking Operator $\widetilde{\mathbb{A}}^{i,j}_{\lambda,\nu,\pm}$

Generalizing the residue formula of the scalar-valued regular symmetry breaking operators $\widetilde{\mathbb{A}}^{0,0}_{\lambda,\nu,+}$ for spherical principal series representations given in [31] (see also [42, Thm. 12.2]), we determined the residue of the matrix-valued regular symmetry breaking operators $\widetilde{\mathbb{A}}^{i,j}_{\lambda,\nu,\pm}$ in [34], as follows:

Fact 9.3 (residue formula [34, Thm. 1.3]). *Let $\mathbb{C}^{i,j}_{\lambda,\nu}$ be the differential symmetry breaking operators defined in (3.16) and (3.17) for $j = i-1$ or i.*

(1) *Suppose $\nu - \lambda = 2\ell$ with $\ell \in \mathbb{N}$. Then,*

$$\widetilde{\mathbb{A}}^{i,j}_{\lambda,\nu,+} = \frac{(-1)^{i-j+\ell}\pi^{\frac{n-1}{2}}\ell!}{2^{2\ell-1}\Gamma(\nu+1)}\mathbb{C}^{i,j}_{\lambda,\nu}. \tag{9.6}$$

(2) *Suppose $\nu - \lambda = 2\ell + 1$ with $\ell \in \mathbb{N}$. Then,*

$$\widetilde{\mathbb{A}}^{i,j}_{\lambda,\nu,-} = \frac{(-1)^{i-j+\ell+1}\pi^{\frac{n-1}{2}}\ell!}{2^{2\ell+2}\Gamma(\nu+1)}\mathbb{C}^{i,j}_{\lambda,\nu}.$$

We may unify the two formulæ in Fact 9.3 into one formula: for $\nu - \lambda \in \mathbb{N}$ and $j \in \{i, i-1\}$,

$$\widetilde{\mathbb{A}}^{i,j}_{\lambda,\nu,(-1)^{\nu-\lambda}} = \frac{2(-1)^{i-j}\pi^{\frac{n-1}{2}}}{q(\nu-\lambda)\Gamma(\nu+1)}\mathbb{C}^{i,j}_{\lambda,\nu}, \tag{9.7}$$

where we set, for $m \in \mathbb{N}$,

$$q(m) := \begin{cases} \dfrac{(-1)^{\ell}2^{2\ell}}{\ell!} & \text{if } m = 2\ell, \\[4mm] \dfrac{(-1)^{\ell+1}2^{2\ell+3}}{\ell!} & \text{if } m = 2\ell + 1. \end{cases} \tag{9.8}$$

9.2.2 Zeros of $\widetilde{\mathbb{A}}^{i,j}_{\lambda,\nu,\pm}$

The zeros of the operators $\widetilde{\mathbb{A}}^{i,i}_{\lambda,\nu,\delta\varepsilon}$ for the special parameter in Ψ_{sp} (see (1.3) for the definition) were determined in [34] as a corollary of the residue formula (Fact 9.3), which we recall now.

Corollary 9.4 (zeros of $\widetilde{\mathbb{A}}^{i,i}_{\lambda,\nu,\pm}$ for Ψ_{sp}, [34, Thm. 8.1]).

(1) *Suppose $\nu - \lambda \in 2\mathbb{N}$.*
$\widetilde{\mathbb{A}}^{i,i}_{\lambda,\nu,+} = 0$ *if and only if*

$$(\lambda, \nu) \in \begin{cases} L_{\text{even}} & \text{for } i = 0, \\ (L_{\text{even}} - \{\nu = 0\}) \cup \{(i,i)\} & \text{for } 1 \leq i \leq n - 1. \end{cases}$$

$\widetilde{\mathbb{A}}^{i,i-1}_{\lambda,\nu,+} = 0$ *if and only if*

$$(\lambda, \nu) \in \begin{cases} (L_{\text{even}} - \{\nu = 0\}) \cup \{(n-i, n-i)\} & \text{for } 1 \leq i \leq n - 1, \\ L_{\text{even}} & \text{for } i = n. \end{cases}$$

(2) *Suppose $\nu - \lambda \in 2\mathbb{N} + 1$.*
   $\widetilde{\mathbb{A}}^{i,i}_{\lambda,\nu,-} = 0$ *if and only if*

$$(\lambda,\nu) \in \begin{cases} L_{\text{odd}} & \text{for } i = 0, \\ L_{\text{odd}} - \{\nu = 0\} & \text{for } 1 \le i \le n-1. \end{cases}$$

$\widetilde{\mathbb{A}}^{i,i-1}_{\lambda,\nu,-} = 0$ *if and only if*

$$(\lambda,\nu) \in \begin{cases} L_{\text{odd}} - \{\nu = 0\} & \text{for } 1 \le i \le n-1, \\ L_{\text{odd}} & \text{for } i = n. \end{cases}$$

We are ready to complete the proof of Theorem 3.19 on the zeros of the analytic continuation $\widetilde{\mathbb{A}}^{i,j}_{\lambda,\nu,\gamma}$ of regular symmetry breaking operators.

Proof of Theorem 3.19. We apply Theorem 5.41 to the exterior representations (9.1), and see that $\widetilde{\mathbb{A}}^{i,j}_{\lambda,\nu,\gamma} = 0$ only if

$$\nu - \lambda \in 2\mathbb{N} \quad (\gamma = +) \quad \text{or} \quad \nu - \lambda \in 2\mathbb{N}+1 \quad (\gamma = -). \tag{9.9}$$

Then Theorem 3.19 follows from Corollary 9.4.   □

9.3 (K, K')-spectrum for Symmetry Breaking Operators

The second goal of this chapter is to formulate the concept of the (K, K')-spectrum for symmetry breaking operators (Definition 9.7), and give an explicit formula of the (K, K')-spectrum

$$S(\widetilde{\mathbb{A}}^{i,j}_{\lambda,\nu,\varepsilon}) = \begin{pmatrix} a^{i,j}_\varepsilon(\lambda,\nu) & b^{i,j}_\varepsilon(\lambda,\nu) \\ c^{i,j}_\varepsilon(\lambda,\nu) & d^{i,j}_\varepsilon(\lambda,\nu) \end{pmatrix}, \tag{9.10}$$

(see (9.13)), for the regular symmetry breaking operator $\widetilde{\mathbb{A}}^{i,j}_{\lambda,\nu,\varepsilon} : I_\delta(i,\lambda) \to J_{\delta\varepsilon}(j,\nu)$ with respect to basic K-types $\mu^\natural(i,\delta)$ and K'-types $\mu^\natural(j,\delta\varepsilon)'$ (see (2.30) and (2.31)) for $\natural = \flat$ or $\sharp$. We will discuss the (K, K')-spectrum in Sections 9.3–9.7. The main results are Theorem 9.8 which will be proved in Proposition 9.9 (vanishing results) and Theorems 9.10 and 9.19.

    One of the algebraic clues that we introduced in the study of symmetry breaking operators A in [42] was an explicit formula of the "eigenvalues" of A on spherical vectors. In the setting of this article, there is no spherical vector in the principal series representation $I_\delta(i,\lambda)$ if $i > 0$ or $J_\varepsilon(j,\nu)$ if $j > 0$. In this section, we extend the idea of [42] to the (K, K')-*spectrum* for symmetry breaking operators with focus on basic K-types.

9.3.1 Generalities: (K, K')-spectrum of Symmetry Breaking Operators

We begin with a general setup. Let (G, G') be a pair of real reductive Lie groups. Suppose Π is a continuous representation of G, and π is that of the subgroup G'. We define a subset of $\widehat{K} \times \widehat{K'}$ by

$$\mathcal{D}(\Pi, \pi) := \{(\mu, \mu') \in \widehat{K} \times \widehat{K'} : [\Pi|_K : \mu], [\pi|_{K'} : \mu'], [\mu|_{K'} : \mu'] \in \{0, 1\}\}.$$

Here is a sufficient condition for $\mathcal{D}(\Pi, \pi)$ to be nonempty:

Proposition 9.5. *Let $P = LN$ and $P' = L'N'$ be parabolic subgroups of G and its subgroup G', respectively. Suppose that $\Pi = \mathrm{Ind}_P^G(\sigma \otimes \mathbb{C}_\lambda)$ and $\pi = \mathrm{Ind}_{P'}^{G'}(\tau \otimes \mathbb{C}_\nu)$ are the induced representations from irreducible finite-dimensional representations $\sigma \otimes \mathbb{C}_\lambda$ of $L \simeq P/N$ and $\tau \otimes \mathbb{C}_\nu$ of $L' \simeq P'/N'$, respectively.*

(1) *(spherical principal series) If σ and τ are the trivial one-dimensional representations, then $\mathcal{D}(\Pi, \pi) \ni (\mathbf{1}_K, \mathbf{1}_{K'})$.*
(2) *If $(K, L \cap K)$, $(K', L' \cap K')$ and (K, K') are strong Gelfand pairs, in particular, if they are symmetric pairs, then $\mathcal{D}(\Pi, \pi) = \widehat{K} \times \widehat{K'}$.*

Proof.

(1) Clear from the Frobenius reciprocity.
(2) Immediate from the multiplicity-free property for strong Gelfand pairs. □

The following is an example of Proposition 9.5 (2).

Example 9.6. Let $(G, G') = (O(n+1, 1), O(n, 1))$, and we consider $\Pi = I_\delta(V, \lambda)$, $\pi = J_\varepsilon(W, \nu)$ for any $(\sigma, V) \in \widehat{O(n)}$ and any $(\tau, W) \in \widehat{O(n-1)}$. Then $\mathcal{D}(\Pi, \pi) = \widehat{K} \times \widehat{K'}$.

Now we introduce a (K, K')-spectrum for symmetry breaking operators as follows.

Definition 9.7 $((K, K')$-spectrum$)$. Let $(\mu, \mu') \in \mathcal{D}(\Pi, \pi)$. Assume

$$[\Pi|_K : \mu] = [\pi|_{K'} : \mu'] = [\mu|_{K'} : \mu'] = 1.$$

Then we fix a nonzero K-homomorphism $\varphi : \mu \hookrightarrow \Pi$ and nonzero K'-homomorphisms $\varphi' : \mu' \hookrightarrow \pi$ and $\iota : \mu' \hookrightarrow \mu$ that are unique up to scalar multiplication. Suppose $A \in \mathrm{Hom}_{G'}(\Pi|_{G'}, \pi)$. Then by Schur's lemma, there exists a constant $S_{\mu,\mu'}(A) \in \mathbb{C}$ such that

$$A \circ \varphi \circ \iota = S_{\mu,\mu'}(A) \circ \varphi' \qquad \text{on } \mu'. \tag{9.11}$$

If one of $[\Pi|_K : \mu]$, $[\pi|_{K'} : \mu']$, or $[\mu|_{K'} : \mu']$ is 0, then we just set

$$S_{\mu,\mu'}(A) = 0 \quad \text{for any } A \in \mathrm{Hom}_{G'}(\Pi|_{G'}, \pi).$$

Thus we have defined a map

$$S\colon \mathrm{Hom}_{G'}(\Pi|_{G'},\pi) \times \mathcal{D}(\Pi,\pi) \to \mathbb{C}, \qquad (A,(\mu,\mu')) \mapsto S_{\mu,\mu'}(A). \qquad (9.12)$$

We say $S_{\mu,\mu'}(A)$ is the (K,K')-*spectrum* of the symmetry breaking operator A for $(\mu,\mu') \in \widehat{K} \times \widehat{K'}$. We note that it is independent of the choice of the normalizations of φ, φ', and ι whether $S_{\mu,\mu'}(A)$ vanishes or not.

9.4 Explicit Formula of (K,K')-spectrum on Basic K-types for Regular Symmetry Breaking Operators $\widetilde{\mathbb{A}}^{i,j}_{\lambda,\nu,\pm}$

We return to our setting where $(G,G') = (O(n+1,1), O(n,1))$, and thus

$$K = O(n+1) \times O(1) \supset K' = O(n) \times O(1).$$

We consider a pair of representations $\Pi = I_\delta(i,\lambda)$ of $G = O(n+1,1)$ and $\pi = J_\varepsilon(j,\nu)$ of the subgroup $G' = O(n,1)$. In this case $\mathcal{D}(\Pi,\pi) = \widehat{K} \times \widehat{K'}$ as we saw in Example 9.6, however, the following finite subset

$$\mathcal{D}^{\flat,\sharp} \equiv \mathcal{D}^{\flat,\sharp}(\Pi,\pi) := \{\mu^\flat(i,\delta), \mu^\sharp(i,\delta)\} \times \{\mu^\flat(j,\varepsilon)', \mu^\sharp(j,\varepsilon)'\} \subset \widehat{K} \times \widehat{K'}$$

will be sufficient for the later analysis of symmetry breaking operators. Here we recall from (2.30) and (2.31) that $\mu^\flat(i,\delta)$ and $\mu^\sharp(i,\delta)$ are "basic K-types" of the principal series representation $I_\delta(i,\lambda)$ of G and that $\mu^\flat(j,\varepsilon)'$ and $\mu^\sharp(j,\varepsilon)'$ are those for $J_\varepsilon(j,\nu)$ of the subgroup G'.

Then the (K,K')-spectrum restricted to the subset $\mathcal{D}^{\flat,\sharp}$ is described as a 2×2 matrix:

$$S\colon \mathrm{Hom}_{G'}(I_\delta(i,\lambda)|_{G'}, J_\varepsilon(j,\nu)) \to M(2,\mathbb{C}), \qquad A \mapsto \begin{pmatrix} a & b \\ c & d \end{pmatrix} \qquad (9.13)$$

by taking a, b, c, d to be $S_{\mu,\mu'}(A)$ as follows:

$S_{\mu,\mu'}(A)$	μ	μ'
a	$\mu^\flat(i,\delta) = \bigwedge^i(\mathbb{C}^{n+1}) \boxtimes \delta$	$\mu^\flat(j,\varepsilon)' = \bigwedge^j(\mathbb{C}^n) \boxtimes \varepsilon$
b	$\mu^\flat(i,\delta) = \bigwedge^i(\mathbb{C}^{n+1}) \boxtimes \delta$	$\mu^\sharp(j,\varepsilon)' = \bigwedge^{j+1}(\mathbb{C}^n) \boxtimes (-\varepsilon)$
c	$\mu^\sharp(i,\delta) = \bigwedge^{i+1}(\mathbb{C}^{n+1}) \boxtimes (-\delta)$	$\mu^\flat(j,\varepsilon)' = \bigwedge^j(\mathbb{C}^n) \boxtimes \varepsilon$
d	$\mu^\sharp(i,\delta) = \bigwedge^{i+1}(\mathbb{C}^{n+1}) \boxtimes (-\delta)$	$\mu^\sharp(j,\varepsilon)' = \bigwedge^{j+1}(\mathbb{C}^n) \boxtimes (-\varepsilon)$

To be more precise, we need a normalization of the map φ, φ' and ι in Definition 9.7 in this setting. For this, we realize the K-types $\mu^\flat(i,\delta) = \bigwedge^i(\mathbb{C}^{n+1}) \boxtimes \delta$ and $\mu^\sharp(i,\delta) = \bigwedge^{i+1}(\mathbb{C}^{n+1}) \boxtimes (-\delta)$ in $I_\delta(i,\lambda)$ as in Proposition 8.4. Similarly,

$\mu^\flat(j, \varepsilon)' = \bigwedge^j(\mathbb{C}^n) \boxtimes \varepsilon$ and $\mu^\sharp(j, \varepsilon)' = \bigwedge^{j+1}(\mathbb{C}^n) \boxtimes (-\varepsilon)$ are realized in $J_\varepsilon(j, \nu)$. When μ' and μ are representations on the exterior tensor spaces $\bigwedge^l(\mathbb{C}^n)$ and $\bigwedge^k(\mathbb{C}^{n+1})$ $(l = k$ or $k - 1)$ respectively, we normalize an $O(n)$-homomorphism

$$\iota_{l \to k}: \bigwedge^l(\mathbb{C}^n) \hookrightarrow \bigwedge^k(\mathbb{C}^{n+1})$$

such that $\mathrm{pr}_{k \to l} \circ \iota_{l \to k} = \mathrm{id}$, where the projection $\mathrm{pr}_{k \to l}: \bigwedge^k(\mathbb{C}^{n+1}) \to \bigwedge^l(\mathbb{C}^n)$ is defined in (7.2) and (7.3). With these normalizations, the map (9.13) is defined. We obtain the following closed formula of the (K, K')-spectrum for the normalized regular symmetry breaking operators $\widetilde{\mathbb{A}}^{i,j}_{\lambda,\nu,\pm}: I_\delta(i, \lambda) \to J_{\pm\delta}(j, \nu)$.

Theorem 9.8 $((K, K')$-spectrum for $\widetilde{\mathbb{A}}^{i,j}_{\lambda,\nu,\pm})$. *Suppose* $(\lambda, \nu) \in \mathbb{C}^2$. *Then the* (K, K')-*spectrum of the analytic continuation* $\widetilde{\mathbb{A}}^{i,j}_{\lambda,\nu,\pm}$ *of regular symmetry breaking operators takes the following form on basic K-types:*

$$S(\widetilde{\mathbb{A}}^{i,i}_{\lambda,\nu,+}) = \frac{\pi^{\frac{n-1}{2}}}{\Gamma(\lambda + 1)} \begin{pmatrix} \lambda - i & 0 \\ 0 & \nu - i \end{pmatrix} \qquad \text{for } 0 \leq i \leq n - 1;$$

$$S(\widetilde{\mathbb{A}}^{i,i}_{\lambda,\nu,-}) = \frac{\pi^{\frac{n-1}{2}}}{\Gamma(\lambda + 1)} \begin{pmatrix} 0 & 0 \\ 2(-1)^{i+1} & 0 \end{pmatrix} \qquad \text{for } 0 \leq i \leq n - 1;$$

$$S(\widetilde{\mathbb{A}}^{i,i-1}_{\lambda,\nu,+}) = \frac{\pi^{\frac{n-1}{2}}}{\Gamma(\lambda + 1)} \begin{pmatrix} n - \nu - i & 0 \\ 0 & \lambda - n + i \end{pmatrix} \quad \text{for } 1 \leq i \leq n;$$

$$S(\widetilde{\mathbb{A}}^{i,i-1}_{\lambda,\nu,-}) = \frac{\pi^{\frac{n-1}{2}}}{\Gamma(\lambda + 1)} \begin{pmatrix} 0 & -2 \\ 0 & 0 \end{pmatrix} \qquad \text{for } 1 \leq i \leq n.$$

The vanishing result (an easy part) of Theorem 9.8 will be shown in Proposition 9.9, and the remaining nontrivial part will be proved in Theorems 9.10 and 9.19.

9.5 Proof of Vanishing Results on (K, K')-spectrum

In this section, we formulate and prove vanishing results for (K, K')-spectrum that hold for general symmetry breaking operators.

Proposition 9.9. *Suppose that* $j \in \{i - 1, i\}$, $\delta, \varepsilon \in \{\pm\}$, *and* $\lambda, \nu \in \mathbb{C}$. *Let* $A: I_\delta(i, \lambda) \to J_\varepsilon(j, \nu)$ *be an arbitrary symmetry breaking operator. Then the* (K, K')-*spectrum* $S(A)$ *for basic K-types takes the following form:*

j	i	i	$i - 1$	$i - 1$
$\delta\varepsilon$	$+$	$-$	$+$	$-$
$S(A)$	$\begin{pmatrix} * & 0 \\ 0 & * \end{pmatrix}$	$\begin{pmatrix} 0 & 0 \\ * & 0 \end{pmatrix}$	$\begin{pmatrix} * & 0 \\ 0 & * \end{pmatrix}$	$\begin{pmatrix} 0 & * \\ 0 & 0 \end{pmatrix}$

Proof. Without loss of generality, we may assume $\delta = +$. The K-modules $\mu^\flat(i,+)$ and $\mu^\sharp(i,+)$ (see (2.30) and (2.31)) decompose into the sum of irreducible representations of the subgroup K':

$$\mu^\flat(i,+) = \textstyle\bigwedge^i(\mathbb{C}^{n+1}) \boxtimes \mathbf{1} \quad \simeq \textstyle\bigwedge^i(\mathbb{C}^n) \boxtimes \mathbf{1} \quad \oplus \textstyle\bigwedge^{i-1}(\mathbb{C}^n) \boxtimes \mathbf{1},$$

$$\mu^\sharp(i,+) = \textstyle\bigwedge^{i+1}(\mathbb{C}^{n+1}) \boxtimes \mathrm{sgn} \simeq \textstyle\bigwedge^{i+1}(\mathbb{C}^n) \boxtimes \mathrm{sgn} \oplus \textstyle\bigwedge^i(\mathbb{C}^n) \boxtimes \mathrm{sgn}.$$

Using the notion $\mu^\natural(j,\pm)'$ with $\natural = \flat$ or $\sharp$ for K'-types, we may rewrite these decompositions as

$$\mu^\flat(i,+)|_{K'} \simeq \mu^\flat(i,+)' \oplus \mu^\flat(i-1,+)' \tag{9.14}$$

$$\simeq \mu^\sharp(i-1,-)' \oplus \mu^\sharp(i-2,-)',$$

$$\mu^\sharp(i,+)|_{K'} \simeq \mu^\sharp(i,+)' \oplus \mu^\sharp(i-1,+)' \tag{9.15}$$

$$\simeq \mu^\flat(i+1,-)' \oplus \mu^\flat(i,-)'.$$

The second isomorphisms follow from (2.32).

For simplicity, we discuss the symmetry breaking operator $A: I_\delta(i,\lambda) \to J_\varepsilon(j,\nu)$ in the case $j = i$, $\delta = +$, and $\varepsilon = -$. Then the branching rule (9.14) tells that neither the K'-type $\mu^\flat(i,-)'$ nor $\mu^\sharp(i,-)'$ occurs in the K-type $\mu^\flat(i,+)$ of $I_+(i,\lambda)$. Likewise, (9.15) tells that the K'-type $\mu^\sharp(i,-)'$ does not occur in the K-type $\mu^\sharp(i,+)$. Hence the matrix $S(A)$ in (9.13) must be of the form $\begin{pmatrix} 0 & 0 \\ * & 0 \end{pmatrix}$.

The vanishing statements in the other cases are proved similarly. $\square$

9.6   Proof of Theorem 9.8 on (K, K')-spectrum for the Normalized Symmetry Breaking Operator $\widetilde{\mathbb{A}}^{i,j}_{\lambda,\nu,+}: I_\delta(i,\lambda) \to J_\delta(j,\nu)$

In this section, we determine the (K, K')-spectrum $a^{i,j}_\varepsilon(\lambda,\nu)$ and $d^{i,j}_\varepsilon(\lambda,\nu)$ for $j = i, i-1$ in (9.10) when $\varepsilon = +$. The case $\varepsilon = -$ will be discussed separately in Section 9.7. By definition (9.11), the constants $a^{i,j}_\varepsilon(\lambda,\nu)$ and $d^{i,j}_\varepsilon(\lambda,\nu)$ are characterized by the following equations:

$$\widetilde{\mathbb{A}}^{i,j}_{\lambda,\nu,+} \circ \iota^*_\lambda \circ \iota_{j \to i} = a^{i,j}_+(\lambda,\nu)\iota^*_\nu \quad \text{on } \textstyle\bigwedge^j(\mathbb{C}^n), \tag{9.16}$$

$$\widetilde{\mathbb{A}}^{i,j}_{\lambda,\nu,+} \circ \iota^*_\lambda \circ \iota_{j+1 \to i+1} = d^{i,j}_+(\lambda,\nu)\iota^*_\nu \quad \text{on } \textstyle\bigwedge^{j+1}(\mathbb{C}^n), \tag{9.17}$$

where $\widetilde{\mathbb{A}}^{i,j}_{\lambda,\nu,+} : I_\delta(i,\lambda) \to J_\delta(j,\nu)$ is the normalized symmetry breaking operator, ι^*_λ is the transform from the K-picture to the N-picture (see (8.3)), and $\iota_{j \to i} : \bigwedge^j(\mathbb{C}^n) \to \bigwedge^i(\mathbb{C}^{n+1})$ is the normalized injective $O(n)$-homomorphism such that $\mathrm{pr}_{i \to j} \circ \iota_{j \to i} = \mathrm{id}$. The main results of this section are part of Theorem 9.8, which is given as follows:

Theorem 9.10. *Suppose* $\lambda, \nu \in \mathbb{C}$.

$$a^{i,i}_+(\lambda, \nu) = \frac{\pi^{\frac{n-1}{2}}(\lambda - i)}{\Gamma(\lambda + 1)}.$$

$$a^{i,i-1}_+(\lambda, \nu) = \frac{\pi^{\frac{n-1}{2}}(n - \nu - i)}{\Gamma(\lambda + 1)}.$$

$$d^{i,i}_+(\lambda, \nu) = \frac{\pi^{\frac{n-1}{2}}(\nu - i)}{\Gamma(\lambda + 1)}. \tag{9.18}$$

$$d^{i,i-1}_+(\lambda, \nu) = \frac{\pi^{\frac{n-1}{2}}(\lambda - n + i)}{\Gamma(\lambda + 1)}. \tag{9.19}$$

Remark 9.11. Theorem 9.10 generalizes [42, Thm. 1.10] in the spherical case ($i = j = 0$ and $\delta = \varepsilon = +$).

The proof of Theorem 9.10 is divided into the following two steps:

- integral expression of $a^{i,j}_+(\lambda, \nu)$ and $d^{i,j}_+(\lambda, \nu)$ (Section 9.6.1);
- computation of the integral (Section 9.6.2).

9.6.1   Integral Expression of (K, K')-spectrum

As the first step of the proof, we give an integral expression of the (K, K')-spectrum $a^{i,j}_+(\lambda, \nu)$ and $d^{i,j}_+(\lambda, \nu)$. For $I \in \mathfrak{I}_{n,i}$, we recall from (7.5) that the quadratic form $Q_I(b)$ is defined to be $\sum_{k \in I} b_k^2$, and set

$$\alpha_I(b) := 1 - \frac{2Q_I(b)}{(1 + |b|^2)|b|^2}, \tag{9.20}$$

$$\delta_I(b) := 1 - \frac{2|b|^2}{1 + |b|^2} - \frac{2Q_I(b)}{(1 + |b|^2)|b|^2}. \tag{9.21}$$

Consider the following integrals:

$$A_I(\lambda,\nu) := \int_{\mathbb{R}^n} \widetilde{\mathcal{A}}_{\lambda,\nu,+}(b)(1+|b|^2)^{-\lambda}\alpha_I(b)db,$$

$$D_I(\lambda,\nu) := \int_{\mathbb{R}^n} \widetilde{\mathcal{A}}_{\lambda,\nu,+}(b)(1+|b|^2)^{-\lambda}\delta_I(b)db.$$

Then the (K,K')-spectrum $a_+^{i,j}(\lambda,\nu)$ and $d_+^{i,j}(\lambda,\nu)$ in (9.16) and (9.17), respectively, is given by the integrals $A_I(\lambda,\nu)$ and $D_I(\lambda,\nu)$ as follows:

Proposition 9.12 (integral expression of (K,K')-spectrum).

$$\begin{aligned}
a_+^{i,i}(\lambda,\nu) &= A_I(\lambda,\nu) & &\text{for any } I \in \mathfrak{I}_{n,i} \text{ with } n \notin I, \\
a_+^{i,i-1}(\lambda,\nu) &= A_I(\lambda,\nu) & &\text{for any } I \in \mathfrak{I}_{n,i} \text{ with } n \in I, \\
d_+^{i,i}(\lambda,\nu) &= D_I(\lambda,\nu) & &\text{for any } I \in \mathfrak{I}_{n,i} \text{ with } n \notin I, \\
d_+^{i,i-1}(\lambda,\nu) &= -D_I(\lambda,\nu) & &\text{for any } I \in \mathfrak{I}_{n,i} \text{ with } n \in I.
\end{aligned}$$

In order to prove Proposition 9.12, we use the N-picture of the principal series representations $I_\delta(i,\lambda)$ and $J_\varepsilon(j,\nu)$. By Proposition 8.4 for the vectors $\mathbf{1}^{\mathcal{I}}_\lambda$ and $h^{\mathcal{I}}_\lambda$ belonging to the basic K-types, the equation (9.16) means that for $\mathcal{I} \in \mathfrak{I}_{n+1,i}$

$$\widetilde{\mathbb{A}}^{i,i}_{\lambda,\nu,+}\mathbf{1}^{\mathcal{I}}_\lambda = a_+^{i,i}(\lambda,\nu)\mathbf{1}'^{\mathcal{I}}_\nu \qquad (n \notin \mathcal{I}),$$

$$\widetilde{\mathbb{A}}^{i,i-1}_{\lambda,\nu,+}\mathbf{1}^{\mathcal{I}}_\lambda = (-1)^{i-1}a_+^{i,i-1}(\lambda,\nu)\mathbf{1}'^{\mathcal{I}-\{n\}}_\nu \qquad (n \in \mathcal{I}).$$

The signature in the second formula arises from the definition (7.3) of the projection $\mathrm{pr}_{i\to i-1}$.

To compute the constants $a_+^{i,j}(\lambda,\nu)$, we take $I \in \mathfrak{I}_{n,i}$ and set $\mathcal{I} := I$, regarded as an element of $\mathfrak{I}_{n+1,i}$, where we recall Convention 7.1 of index sets. Since $0 \notin \mathcal{I}$, it follows from (8.8) that

$$\widetilde{\mathbb{A}}^{i,i}_{\lambda,\nu,+}(\mathbf{1}^I_\lambda)(0) = a_+^{i,i}(\lambda,\nu)e_I \qquad \text{if } n \notin I,$$

$$\widetilde{\mathbb{A}}^{i,i-1}_{\lambda,\nu,+}(\mathbf{1}^I_\lambda)(0) = (-1)^{i-1}a_+^{i,i-1}(\lambda,\nu)e_{I-\{n\}} \qquad \text{if } n \in I.$$

Likewise, the equation (9.17) means that for $\mathcal{I} \in \mathfrak{I}_{n+1,i+1}$

$$\widetilde{\mathbb{A}}^{i,i}_{\lambda,\nu,+}h^{\mathcal{I}}_\lambda = d_+^{i,i}(\lambda,\nu)h'^{\mathcal{I}}_\nu \qquad (n \notin \mathcal{I}),$$

$$\widetilde{\mathbb{A}}^{i,i-1}_{\lambda,\nu,+}h^{\mathcal{I}}_\lambda = (-1)^i d_+^{i,i-1}(\lambda,\nu)h'^{\mathcal{I}-\{n\}}_\nu \qquad (n \in \mathcal{I}).$$

In this case, we take $I \in \mathfrak{I}_{n,i}$ and set $\mathcal{I} := I \cup \{0\} \in \mathfrak{I}_{n+1,i+1}$. Then (8.9) implies

$$(\widetilde{\mathbb{A}}^{i,i}_{\lambda,\nu,+} h^{I\cup\{0\}}_\lambda)(0) = d^{i,i}_+(\lambda,\nu)e_I \qquad \text{if } n \notin I, \qquad (9.22)$$

$$(\widetilde{\mathbb{A}}^{i,i-1}_{\lambda,\nu,+} h^{I\cup\{0\}}_\lambda)(0) = (-1)^i d^{i,i-1}_+(\lambda,\nu)e_{I-\{n\}} \quad \text{if } n \in I.$$

Let us compute $\widetilde{\mathbb{A}}^{i,j}_{\lambda,\nu,+}(1^I_\lambda)(0)$ and $\widetilde{\mathbb{A}}^{i,j}_{\lambda,\nu,+}(h^{I\cup\{0\}}_\lambda)(0)$ for $j = i$ and $i - 1$. If $\operatorname{Re}\lambda \gg |\operatorname{Re}\nu|$, then the matrix-valued distribution kernel $\widetilde{A}^{i,j}_{\lambda,\nu,+}$ (see (9.3)) of the regular symmetry breaking operator $\widetilde{\mathbb{A}}^{i,j}_{\lambda,\nu,+}$ is decomposed as

$$\widetilde{A}^{i,j}_{\lambda,\nu,+} = \widetilde{A}_{\lambda,\nu,+} R^{i,j},$$

where $\widetilde{A}_{\lambda,\nu,+}$ is the scalar-valued, locally integrable function defined in (5.40) and the matrix-valued function $R^{i,j} \in C^\infty(\mathbb{R}^n - \{0\}) \otimes \operatorname{Hom}_{\mathbb{C}}(\bigwedge^i(\mathbb{C}^n), \bigwedge^j(\mathbb{C}^{n-1}))$ is defined in (9.2). Hence, we have

$$(\widetilde{\mathbb{A}}^{i,j}_{\lambda,\nu,+}\psi)(0) = \int_{\mathbb{R}^n} \widetilde{A}_{\lambda,\nu,+}(-b)R^{i,j}(-b)\psi(b)db$$

$$= \int_{\mathbb{R}^n} \widetilde{A}_{\lambda,\nu,+}(b)R^{i,j}(b)\psi(b)db$$

in the N-picture for any $\psi \in \iota^*_\lambda(\mathcal{E}^i(S^n)) \subset C^\infty(\mathbb{R}^n) \otimes \bigwedge^i(\mathbb{C}^n)$. Thus Proposition 9.12 is a consequence of the following two lemmas on the computation of $R^{i,j}(b)\psi(b) \in \bigwedge^j(\mathbb{C}^{n-1})$ for $\psi = 1^I_\lambda$ or $h^{I\cup\{0\}}_\lambda$ and for $j = i$ or $i - 1$.

Lemma 9.13. *Suppose $I \in \mathfrak{I}_{n,i}$.*

(1) *If $n \notin I$, then the coefficient of e_I in $R^{i,i}(b)1^I_\lambda(b)$ is given by*

$$(1 + |b|^2)^{-\lambda}\alpha_I(b) = (1 + |b|^2)^{-\lambda}(1 - \frac{2Q_I(b)}{(1 + |b|^2)|b|^2}),$$

where we recall $Q_I(b) = \sum_{l\in I} b_l^2$ from (7.5).

(2) *If $n \in I$, then the coefficient of $e_{I-\{n\}}$ in $R^{i,i-1}(b)1^I_\lambda(b)$ is given by*

$$(-1)^{i-1}(1 + |b|^2)^{-\lambda}\alpha_I(b) = (1 + |b|^2)^{-\lambda}(1 - \frac{2Q_I(b)}{(1 + |b|^2)|b|^2}).$$

Lemma 9.14. *Suppose $I \in \mathfrak{I}_{n,i}$.*

(1) *If $n \notin I$, then the coefficient of e_I in $R^{i,i}(b)h^{I\cup\{0\}}_\lambda(b)$ is given by*

$$(1 + |b|^2)^{-\lambda}\delta_I(b) = (1 + |b|^2)^{-\lambda-1}(1 - |b|^2 - \frac{2Q_I(b)}{|b|^2}).$$

(2) *If $n \in I$, then the coefficient of $e_{I-\{n\}}$ in $R^{i,i-1}(b)h_\lambda^{I\cup\{0\}}(b)$ is given by*

$$(-1)^{i-1}(1+|b|^2)^{-\lambda}\delta_I(b) = (-1)^{i-1}(1+|b|^2)^{-\lambda-1}(1-|b|^2-\frac{2Q_I(b)}{|b|^2}).$$

Proof of Lemma 9.13. Let σ be the i-th exterior representation on $\bigwedge^i(\mathbb{C}^n)$. We recall from (9.2) $R^{i,j} = \mathrm{pr}_{i\to j}\circ\sigma\circ\psi_n$. We identify $I \in \mathfrak{I}_{n,i}$ with $\mathcal{I} \in \mathfrak{I}_{n+1,i}$ such that $n \notin \mathcal{I}$ as usual, and apply the formula (8.6) of $\mathbf{1}_\lambda^I$. Then we have

$$\sigma(\psi_n(b))\mathbf{1}_\lambda^I(b) = (1+|b|^2)^{-\lambda}\sigma(\psi_n(b))\sum_{J\in\mathfrak{I}_{n,i}}(\det\psi_{n+1}(1,b))_{IJ}e_J.$$

By the formula (7.10) of the matrix coefficients of the exterior tensor representation, the coefficient of e_I in $\sigma(\psi_n(b))\mathbf{1}_\lambda^I(b)$ amounts to

$$(1+|b|^2)^{-\lambda}\sum_{J\in\mathfrak{I}_{n,i}}(\det\psi_{n+1}(1,b))_{IJ}(\det\psi_n(b))_{IJ},$$

which is equal to

$$(1+|b|^2)^{-\lambda}(1-\frac{2Q_I(b)}{(1+|b|^2)|b|^2}) = (1+|b|^2)^{-\lambda}\alpha_I(b)$$

by the minor summation formula (7.15) in Proposition 7.7. Hence the lemma follows from $\mathrm{pr}_{i\to i}(e_I) = e_I$ ($n \notin I$) and $\mathrm{pr}_{i\to i-1}(e_I) = (-1)^{i-1}e_{I-\{n\}}$ ($n \in I$) (see (7.2) and (7.3)). □

Proof of Lemma 9.14. The proof goes in parallel to that of Lemma 9.13. For the sake of completeness, we give a proof.

By (8.7) and (7.10), we have

$$\sigma(\psi_n(b))h_\lambda^{I\cup\{0\}}(b)$$

$$= -(1+|b|^2)^{-\lambda}\sigma(\psi_n(b))\sum_{J\in\mathfrak{I}_{n,i}}(\det\psi_{n+1}(1,b))_{I\cup\{0\},J\cup\{0\}}e_J$$

$$= -(1+|b|^2)^{-\lambda}\sum_{J\in\mathfrak{I}_{n,i}}\sum_{J'\in\mathfrak{I}_{n,i}}(\det\psi_{n+1}(1,b))_{I\cup\{0\},J\cup\{0\}}\det\psi_n(b)_{J'J}e_{J'}.$$

Hence the coefficient of e_I in $\sigma(\psi_n(b))h_\lambda^{I\cup\{0\}}(b)$ is equal to

$$-(1+|b|^2)^{-\lambda}\sum_{J\in\mathfrak{I}_{n,i}}(\det\psi_{n+1}(1,b))_{I\cup\{0\},J\cup\{0\}}\det\psi_n(b)_{IJ},$$

which amounts to

$$(1 + |b|^2)^{-\lambda-1}(1 - |b|^2 - \frac{2Q_I(b)}{|b|^2}) = (1 + |b|^2)^{-\lambda}\delta_I(b)$$

by the minor summation formula in Proposition 7.7 (2). Thus we have shown the lemma.   □

Therefore we have completed the proof of Proposition 9.12.

9.6.2   Integral Formula of the (K, K')-spectrum

As the second step, we compute the integrals $A_I(\lambda, \nu)$ and $D_I(\lambda, \nu)$ in Section 9.6.1. We begin with the following integral formulæ: Denote by $d\omega$ the standard measure on the unit sphere $S^{n-1} = \{\omega = (\omega_1, \cdots, \omega_n) \in \mathbb{R}^n : \sum_{j=1}^n \omega_j^2 = 1\}$.
For $a, b \in \mathbb{C}$ with $\operatorname{Re} a, \operatorname{Re} b > -1$, we set

$$S(a, b) \equiv S_n(a, b) := \int_{S^{n-1}} |\omega_n|^a |\omega_{n-1}|^b d\omega. \tag{9.23}$$

Then we have

$$S_n(a, 0) = \int_{S^{n-1}} |\omega_n|^a d\omega = \frac{2\pi^{\frac{n-1}{2}}\Gamma(\frac{a+1}{2})}{\Gamma(\frac{a+n}{2})}, \tag{9.24}$$

see [42, Lemma 7.6], for instance. More generally, we have the following.

Lemma 9.15. *Suppose* $\operatorname{Re} a > -1$ *and* $\operatorname{Re} b > -1$. *Then we have*

$$S(a, b) = \frac{2\pi^{\frac{n-2}{2}}\Gamma(\frac{a+1}{2})\Gamma(\frac{b+1}{2})}{\Gamma(\frac{a+b+n}{2})}. \tag{9.25}$$

It is convenient to write down the following recurrence relations that are derived readily from (9.25):

$$S(a, 2) = \frac{1}{a+n}S(a, 0), \tag{9.26}$$

$$S(a+2, 0) = \frac{a+1}{a+n}S(a, 0). \tag{9.27}$$

Proof of Lemma 9.15. For any $f \in C(S^{n-1})$, the polar coordinates give the following expression of the integral:

$$\int_{S^{n-1}} f(\omega)d\omega = \int_{-1}^1 \int_{S^{n-2}} f(\sqrt{1-t^2}\eta, t)(1 - t^2)^{\frac{n-3}{2}} d\eta dt. \tag{9.28}$$

Then we have

$$
S(a,b) = \int_{-1}^{1} \int_{S^{n-2}} |\sqrt{1-t^2}\eta_{n-1}|^b |t|^a (1-t^2)^{\frac{n-3}{2}} \, d\eta dt
$$

$$
= \int_{S^{n-2}} |\eta_{n-1}|^b d\eta \int_{-1}^{1} |t|^a (1-t^2)^{\frac{n+b-3}{2}} \, dt.
$$

The first term equals $S_{n-1}(b,0)$, see (9.24). The second term is given by the Beta function:

$$
\int_{0}^{1} t^{2A-1}(1-t^2)^{B-1} dt = \frac{\Gamma(A)\Gamma(B)}{2\Gamma(A+B)}. \tag{9.29}
$$

Here we get the lemma.   □

Lemma 9.16. *Let $\widetilde{\mathcal{A}}_{\lambda,\nu,+}$ be the (scalar-valued) locally integrable function on $\mathbb{R}^n$ defined in (5.40) for $\mathrm{Re}\,(\lambda-\nu) > 0$ and $\mathrm{Re}\,(\lambda+\nu) > n-1$.*

(1) *We have*

$$
\int_{\mathbb{R}^n} \widetilde{\mathcal{A}}_{\lambda,\nu,+}(b)(1+|b|^2)^{-\lambda} db = \frac{\pi^{\frac{n-1}{2}}}{\Gamma(\lambda)}.
$$

(2) *Let $\ell \in \{1,2,\cdots,n\}$. Then we have*

$$
\int_{\mathbb{R}^n} \widetilde{\mathcal{A}}_{\lambda,\nu,+}(b)(1+|b|^2)^{-\lambda} \frac{2b_\ell^2}{(1+|b|^2)|b|^2} db
$$

$$
= \frac{\pi^{\frac{n-1}{2}}}{\Gamma(\lambda+1)} \times \begin{cases} 1 & \text{if } 1 \leq \ell \leq n-1, \\ \lambda+\nu-n+1 & \text{if } \ell = n. \end{cases}
$$

Proof.

(1) This formula was given in [42, Prop. 7.4], but we give a proof here in order to illustrate our notation for later purpose. By (8.20), the left-hand side amounts to

$$
\frac{1}{\Gamma(\frac{\lambda+\nu-n+1}{2})\Gamma(\frac{\lambda-\nu}{2})} \int_{0}^{\infty} r^{\lambda-\nu-1}(1+r^2)^{-\lambda} dr \int_{S^{n-1}} |\omega_n|^{\lambda+\nu-n} d\omega
$$

$$
= \frac{1}{2\Gamma(\frac{\lambda+\nu-n+1}{2})\Gamma(\frac{\lambda-\nu}{2})} B(\frac{\lambda-\nu}{2},\frac{\lambda+\nu}{2}) S(\lambda+\nu-n,0),
$$

which equals $\frac{\pi^{\frac{n-1}{2}}}{\Gamma(\lambda)}$ by (9.25).

(2) By a similar computation as above, the ratio of the two integrals is given as

$$\frac{\text{the left-hand side of (2)}}{\text{the left-hand side of (1)}} = \frac{2\int_0^\infty r^{\lambda-\nu-1}(1+r^2)^{-\lambda-1}dr \int_{S^{n-1}} |\omega_n|^{\lambda+\nu-n}|\omega_\ell|^2 d\omega}{\int_0^\infty r^{\lambda-\nu-1}(1+r^2)^{-\lambda}dr \int_{S^{n-1}} |\omega_n|^{\lambda+\nu-n}d\omega}.$$

The right-hand side depends on whether $\ell = n$ or not. It amounts to

$$\frac{2B(\frac{\lambda-\nu}{2}, \frac{\lambda+\nu}{2}+1)}{B(\frac{\lambda-\nu}{2}, \frac{\lambda+\nu}{2})} \cdot \frac{1}{S(\lambda+\nu-n,0)} \times \begin{cases} S(\lambda+\nu-n,2) \\ S(\lambda+\nu-n+2,0) \end{cases}$$

$$= \frac{\lambda+\nu}{\lambda} \cdot \frac{1}{\lambda+\nu} \times \begin{cases} 1 & \text{if } 1 \le \ell \le n-1, \\ \lambda+\nu-n+1 & \text{if } \ell = n \end{cases}$$

by the recurrence relations (9.26) and (9.27).   □

Lemma 9.17. $A_I(\lambda, \nu) - D_I(\lambda, \nu) = \frac{\pi^{\frac{n-1}{2}}(\lambda-\nu)}{\Gamma(\lambda+1)}.$

Proof. By the definitions (9.20) and (9.21), we have $\alpha_I(b) - \delta_I(b) = \frac{2|b|^2}{1+|b|^2}$. Thus we have

$$A_I(\lambda, \nu) - D_I(\lambda, \nu) = 2\int_{\mathbb{R}^n} \widetilde{A}_{\lambda,\nu,+}(b)(1+|b|^2)^{-\lambda-1}|b|^2 db$$

$$= \frac{B(\frac{\lambda-\nu}{2}, \frac{\lambda+\nu}{2}+1)S(\lambda+\nu-n,0)}{\Gamma(\frac{\lambda-\nu}{2})\Gamma(\frac{\lambda+\nu-n+1}{2})},$$

as in the proof of Lemma 9.16 (1). Thus the lemma follows from (9.25).   □

Proof of Theorem 9.10. It follows from Lemma 9.16 that

$$A_I(\lambda, \nu) = \frac{\pi^{\frac{n-1}{2}}}{\Gamma(\lambda+1)} \times \begin{cases} \lambda-i & \text{if } n \notin I, \\ \lambda-(i-1)-(\lambda+\nu-n+1) & \text{if } n \in I, \end{cases}$$

whence the first two formulæ of Theorem 9.10 are proved by Proposition 9.12.
   By Lemma 9.17, we have

$$D_I(\lambda, \nu) = A_I(\lambda, \nu) - \frac{\pi^{\frac{n-1}{2}}(\lambda-\nu)}{\Gamma(\lambda+1)}$$

$$= \frac{\pi^{\frac{n-1}{2}}}{\Gamma(\lambda+1)} \times \begin{cases} (\lambda-i)-(\lambda-\nu) & \text{if } n \notin I, \\ (n-\nu-i)-(\lambda-\nu) & \text{if } n \in I, \end{cases}$$

whence the last two formulæ of Theorem 9.10 by Proposition 9.12.   □

Remark 9.18. Alternatively, one could derive the last two formulæ of Theorem 9.10 from the first two by using the duality theorem for symmetry breaking operators given in Proposition 3.39.

9.7 Proof of Theorem 9.8 on the (K,K')-spectrum for $\widetilde{\mathbb{A}}^{i,j}_{\lambda,\nu,-} : I_\delta(i,\lambda) \to J_{-\delta}(j,\nu)$

In this section, we determine the (K,K')-spectrum $b^{i,i-1}_-(\lambda,\nu)$ and $c^{i,i}_-(\lambda,\nu)$ in (9.10) for the normalized regular symmetry breaking operators $\widetilde{\mathbb{A}}^{i,j}_{\lambda,\nu,-} : I_\delta(i,\lambda) \to J_{-\delta}(j,\nu)$ with $j \in \{i-1,i\}$. By definition, these constants $b^{i,i-1}_-(\lambda,\nu)$ and $c^{i,i}_-(\lambda,\nu)$ are characterized by the following equations:

$$\widetilde{\mathbb{A}}^{i,i-1}_{\lambda,\nu,-} \circ \iota^*_\lambda = b^{i,i-1}_-(\lambda,\nu)\iota^*_\nu \circ \mathrm{pr}_{i\to i} \qquad \text{on } \bigwedge{}^i(\mathbb{C}^{n+1}), \tag{9.30}$$

$$\widetilde{\mathbb{A}}^{i,i}_{\lambda,\nu,-} \circ \iota^*_\lambda = (-1)^i c^{i,i}_-(\lambda,\nu) \circ \iota^*_\nu \circ \mathrm{pr}_{i+1\to i} \qquad \text{on } \bigwedge{}^{i+1}(\mathbb{C}^{n+1}). \tag{9.31}$$

The main results of this section are given as follows:

Theorem 9.19. *Suppose* $\lambda,\nu \in \mathbb{C}$. *Then we have*

$$b^{i,i-1}_-(\lambda,\nu) = -\frac{2\pi^{\frac{n-1}{2}}}{\Gamma(\lambda+1)}, \tag{9.32}$$

$$c^{i,i}_-(\lambda,\nu) = \frac{2(-1)^{i+1}\pi^{\frac{n-1}{2}}}{\Gamma(\lambda+1)}. \tag{9.33}$$

This is the remaining part of Theorem 9.8, and the proof of Theorem 9.8 will be complete when Theorem 9.19 is shown. The proof of Theorem 9.19 is parallel to that of Theorem 9.10, and thus will be discussed briefly. We begin with an integral expression of the constants $b^{i,i-1}_-(\lambda,\nu)$ and $c^{i,i}_-(\lambda,\nu)$ as follows.

Proposition 9.20 (integral expression of (K,K')-spectrum).

$$b^{i,i-1}_-(\lambda,\nu) = -2\int_{\mathbb{R}^n} \widetilde{A}_{\lambda,\nu,-}(b)(1+|b|^2)^{-\lambda-1}b_n db,$$

$$c^{i,i}_-(\lambda,\nu) = 2(-1)^{i+1}\int_{\mathbb{R}^n} \widetilde{A}_{\lambda,\nu,-}(b)(1+|b|^2)^{-\lambda-1}b_n db.$$

Admitting Proposition 9.20 for the time being, we complete the proof of Theorem 9.19.

Proof of Theorem 9.19. Theorem 9.19 is an immediate consequence of Proposition 9.20 and the following lemma.   □

Lemma 9.21.

$$\int_{\mathbb{R}^n} \widetilde{\mathcal{A}}_{\lambda,\nu,-}(b)(1+|b|^2)^{-\lambda-1}b_n db = \frac{\pi^{\frac{n-1}{2}}}{\Gamma(\lambda+1)}.$$

Proof. We use the identity

$$b_n \widetilde{\mathcal{A}}_{\lambda,\nu,-}(b) = \widetilde{\mathcal{A}}_{\lambda+1,\nu,+}(b).$$

Then the lemma follows from Lemma 9.16.   □

The rest of this section is devoted to the proof of Proposition 9.20. In the N-picture, the equation (9.30) amounts to

$$\widetilde{\mathbb{A}}^{i,i-1}_{\lambda,\nu,-}(\mathbf{1}^{\mathcal{I}}_\lambda) = \begin{cases} b^{i,i-1}_-(\lambda,\nu)h'^{\mathcal{I}}_\nu & \text{if } n \notin \mathcal{I}, \\ 0 & \text{if } n \in \mathcal{I}, \end{cases}$$

for all $\mathcal{I} \in \mathfrak{I}_{n+1,i}$, whereas (9.31) amounts to

$$\widetilde{\mathbb{A}}^{i,i}_{\lambda,\nu,-}h^{\mathcal{I}}_\lambda = \begin{cases} c^{i,i}_-(\lambda,\nu)\mathbf{1}'^{\mathcal{I}-\{n\}}_\nu & \text{if } n \in \mathcal{I}, \\ 0 & \text{if } n \notin \mathcal{I}, \end{cases}$$

for all $\mathcal{I} \in \mathfrak{I}_{n+1,i+1}$. In particular, we have

$$A^{i,i-1}_{\lambda,\nu,-}(\mathbf{1}^{I\cup\{0\}}_\lambda)(0) = b^{i,i-1}_-(\lambda,\nu)e_I \quad \text{for any } I \in \mathfrak{I}_{n-1,i-1}, \tag{9.34}$$

$$(\widetilde{\mathbb{A}}^{i,i}_{\lambda,\nu,-}h^{I\cup\{n\}}_\lambda)(0) = c^{i,i}_-(\lambda,\nu)e_I \quad \text{for any } I \in \mathfrak{I}_{n-1,i} \tag{9.35}$$

by (8.8) and (8.9) because $0 \notin I$.

The distribution kernel $\widetilde{\mathcal{A}}^{i,j}_{\lambda,\nu,-}$ of the regular symmetry breaking operator $\widetilde{\mathbb{A}}^{i,j}_{\lambda,\nu,-}$ is decomposed as

$$\widetilde{\mathcal{A}}^{i,j}_{\lambda,\nu,-} = \widetilde{\mathcal{A}}_{\lambda,\nu,-}R^{i,j},$$

where $\widetilde{\mathcal{A}}_{\lambda,\nu,-}$ is the scalar-valued, locally integrable function defined in (5.41) and the matrix-valued function $R^{i,j}$ is defined in (9.2). Then we have

$$(\widetilde{\mathbb{A}}^{i,j}_{\lambda,\nu,-}\psi)(0) = \int_{\mathbb{R}^n} \widetilde{\mathcal{A}}_{\lambda,\nu,-}(-b)R^{i,j}(-b)\psi(b)db$$

$$= -\int_{\mathbb{R}^n} \widetilde{\mathcal{A}}_{\lambda,\nu,-}(b)R^{i,j}(b)\psi(b)db$$

in the N-picture for any $\psi \in \iota^*_\lambda(\mathcal{E}^i(S^n))$. Hence Proposition 9.20 is a consequence of (9.34), (9.35), and of the following two lemmas.

Lemma 9.22. *Suppose $I \in \mathfrak{I}_{n-1,i}$. Then the coefficient of e_I in $R^{i,i-1}(b)\mathbf{1}_\lambda^{I\cup\{0\}}(b)$ is equal to*

$$2(1+|b|^2)^{-\lambda-1}b_n.$$

Proof. Using the formula (8.6) of $\mathbf{1}_\lambda^{\mathcal{I}}(b)$, we have for $I \in \mathfrak{I}_{n,i}$

$$(1+|b|^2)^\lambda R^{i,i-1}(b)\mathbf{1}_\lambda^{I\cup\{0\}}(b)$$

$$= R^{i,i-1}(b) \sum_{J\in\mathfrak{I}_{n,i}} \det\psi_{n+1}(1,b)_{I\cup\{0\},J}e_J$$

$$= \mathrm{pr}_{i\to i-1} \sum_{J\in\mathfrak{I}_{n,i}} \sum_{J'\in\mathfrak{I}_{n,i}} \det\psi_{n+1}(1,b)_{I\cup\{0\},J} \det\psi_n(b)_{J'J}e_{J'}$$

$$= (-1)^{i-1} \sum_{J\in\mathfrak{I}_{n,i}} \sum_{J'\in\mathfrak{I}_{n,i}-\mathfrak{I}_{n-1,i}} \det\psi_{n+1}(1,b)_{I\cup\{0\},J} \det\psi_n(b)_{J'J}e_{J'-\{n\}}.$$

Here, for $J' \in \mathfrak{I}_{n,i}$, we mean by $J' \notin \mathfrak{I}_{n-1,i}$ the condition that $n \notin J'$. Hence the coefficient of e_I in $R^{i,i-1}(b)\mathbf{1}_\lambda^{I\cup\{0\}}(b)$ amounts to

$$(-1)^{i-1} \sum_{J\in\mathfrak{I}_{n,i}} \det\psi_{n+1}(1,b)_{I\cup\{0\},J} \det\psi_n(b)_{I\cup\{n\},J}.$$

Now the lemma follows from Lemma 7.8.   □

Lemma 9.23. *Suppose $I \in \mathfrak{I}_{n-1,i}$. The coefficient of e_I in $R^{i,i}(b)h_\lambda^{I\cup\{n\}}(b)$ is given by*

$$2(-1)^i(1+|b|^2)^{-\lambda-1}b_n.$$

Proof. By (8.7) and (7.10), we have

$$\sigma(\psi_n(b))h_\lambda^{I\cup\{n\}}(b)$$

$$= -(1+|b|^2)^{-\lambda}\sigma(\psi_n(b)) \sum_{J\in\mathfrak{I}_{n,i}} \det\psi_{n+1}(1,b)_{I\cup\{n\},J\cup\{0\}}e_J$$

$$= -(1+|b|^2)^{-\lambda} \sum_{J\in\mathfrak{I}_{n,i}} \sum_{J'\in\mathfrak{I}_{n,i}} \det\psi_{n+1}(1,b)_{I\cup\{n\},J\cup\{0\}} \det\psi_n(b)_{J'J}e_{J'}.$$

Applying the projection $\mathrm{pr}_{i \to i} : \bigwedge^i(\mathbb{C}^n) \to \bigwedge^i(\mathbb{C}^{n-1})$ (see (7.2)), we find that the coefficient of e_I in $R^{i,i}(b)h_\lambda^{I \cup \{n\}}(b)$ is equal to

$$-(1+|b|^2)^{-\lambda} \sum_{J \in \mathfrak{I}_{n,i}} \det \psi_{n+1}(1,b)_{I \cup \{n\}, J \cup \{0\}} \det \psi_n(b)_{IJ}.$$

Hence the lemma follows from the minor summation formula in Lemma 7.9. $\qquad \square$

9.8 Matrix-Valued Functional Equations

The third goal of this chapter is to obtain explicit matrix-valued functional equations for the regular symmetry breaking operators $\widetilde{\mathbb{A}}_{\lambda,\nu,\pm}^{i,j}$. We retain the setting where $(G, G') = (O(n+1,1), O(n,1))$. By the generic multiplicity-one theorem (Theorem 3.3), two symmetry breaking operators from the G-module $I_\delta(V,\lambda)$ to the G'-module $J_\varepsilon(W,\nu)$ must be proportional to each other if $[V:W] \neq 0$ and $(\lambda, \nu, \delta, \varepsilon)$ does not belong to the set Ψ_{sp} of special parameters. In Sections 9.8 and 9.9, we consider the case

$$(V, W) = (\bigwedge^i(\mathbb{C}^n), \bigwedge^j(\mathbb{C}^{n-1})), \quad j \in \{i-1, i\},$$

and compare the (normalized) regular symmetry breaking operator $\widetilde{\mathbb{A}}_{\lambda,\nu,\gamma}^{i,j}$ with its composition of the Knapp–Stein intertwining operator for G or for the subgroup G' as in the following diagrams:

$$
\begin{array}{ccc}
I_\delta(i,\lambda) \xrightarrow{\;\widetilde{\mathbb{A}}_{\lambda,\nu,\gamma}^{i,j}\;} J_\varepsilon(j,\nu) & \qquad & I_\delta(i,\lambda) \\
\searrow_{\widetilde{\mathbb{A}}_{\lambda,n-1-\nu,\gamma}^{i,j}} \quad \downarrow^{\widetilde{\mathbb{T}}_{\nu,n-1-\nu}^{j}} & \qquad & {}^{\widetilde{\mathbb{T}}_{\lambda,n-\lambda}^{i}} \downarrow \quad \searrow^{\widetilde{\mathbb{A}}_{\lambda,\nu,\gamma}^{i,j}} \\
J_\varepsilon(j,n-1-\nu), & \qquad & I_\delta(i,n-\lambda) \xrightarrow{\;\widetilde{\mathbb{A}}_{n-\lambda,\nu,\gamma}^{i,j}\;} J_\varepsilon(j,\nu)
\end{array}
$$

where $\gamma = \delta\varepsilon$. We obtain closed formulæ of the proportional constants for the two operators in each diagram in Theorems 9.24 and 9.25. The zeros of the proportional constants provide us crucial information on the kernels and the images of the symmetry breaking operators $\widetilde{\mathbb{A}}_{\lambda,\nu,\delta\varepsilon}^{i,j} : I_\delta(i,\lambda) \to J_\varepsilon(j,\nu)$ at reducible places of the principal series representations, which will be investigated in Chapter 10.

9.8.1 Main Results: Functional Equations of $\widetilde{\mathbb{A}}^{i,j}_{\lambda,\nu,\varepsilon}$

Suppose $j \in \{i-1, i\}$. Let $\widetilde{\mathbb{A}}^{i,j}_{\lambda,\nu,\delta\varepsilon} : I_\delta(i,\lambda) \to J_\varepsilon(j,\nu)$ be the normalized symmetry breaking operators as defined in (9.5), and $\widetilde{\mathbb{T}}^j_{\nu,n-1-\nu} : J_\varepsilon(j,\nu) \to J_\varepsilon(j, n-1-\nu)$ be the normalized Knapp–Stein operators as defined in (8.14) for principal series representations of the subgroup G'. Then we obtain:

Theorem 9.24 (functional equation). *Suppose $(\lambda, \nu) \in \mathbb{C}^2$ and $\gamma \in \{\pm\}$. Then*

$$\widetilde{\mathbb{T}}^i_{\nu,n-1-\nu} \circ \widetilde{\mathbb{A}}^{i,i}_{\lambda,\nu,\gamma} = \frac{\pi^{\frac{n-1}{2}}(\nu - i)}{\Gamma(\nu+1)} \widetilde{\mathbb{A}}^{i,i}_{\lambda, n-1-\nu, \gamma} \qquad for\ 0 \le i \le n-1,$$

$$\widetilde{\mathbb{T}}^{i-1}_{\nu,n-1-\nu} \circ \widetilde{\mathbb{A}}^{i,i-1}_{\lambda,\nu,\gamma} = \frac{\pi^{\frac{n-1}{2}}(n - \nu - i)}{\Gamma(\nu+1)} \widetilde{\mathbb{A}}^{i,i-1}_{\lambda, n-1-\nu, \gamma} \qquad for\ 1 \le i \le n.$$

In the next theorem, we use the same letter $\widetilde{\mathbb{T}}^i_{\lambda, n-\lambda}$ to denote the normalized Knapp–Stein intertwining operators $\widetilde{\mathbb{T}}^i_{\lambda, n-\lambda} : I_\delta(i, \lambda) \to I_\delta(i, n-\lambda)$ for the group G. Then we obtain:

Theorem 9.25 (functional equation). *Suppose $(\lambda, \nu) \in \mathbb{C}^2$ and $\gamma \in \{\pm\}$. Then*

$$\widetilde{\mathbb{A}}^{i,i}_{n-\lambda,\nu,\gamma} \circ \widetilde{\mathbb{T}}^i_{\lambda,n-\lambda} = \frac{\pi^{\frac{n}{2}}(n - \lambda - i)}{\Gamma(n - \lambda + 1)} \widetilde{\mathbb{A}}^{i,i}_{\lambda,\nu,\gamma} \qquad for\ 0 \le i \le n-1,$$

$$\widetilde{\mathbb{A}}^{i,i-1}_{n-\lambda,\nu,\gamma} \circ \widetilde{\mathbb{T}}^i_{\lambda,n-\lambda} = \frac{\pi^{\frac{n}{2}}(\lambda - i)}{\Gamma(n - \lambda + 1)} \widetilde{\mathbb{A}}^{i,i-1}_{\lambda,\nu,\gamma} \qquad for\ 1 \le i \le n.$$

Remark 9.26. Theorems 9.24 and 9.25 generalize the functional equations which we proved in the scalar case [42, Thm. 8.5]. Matrix-valued functional identities (factorization identities) for *differential* symmetry breaking operators were recently proved explicitly in [35, Chap. 13]. Alternatively, we could deduce a large part of the identities [35, Chap. 13] from Theorems 9.24 and 9.25 by using the residue formula of the normalized symmetry breaking operators $\widetilde{\mathbb{A}}^{i,j}_{\lambda,\nu,\pm}$ given in Fact 9.3, see [34].

9.8.2 Proof of Functional Equations

In this section we give a proof of the functional equations that are stated in Theorems 9.24 and 9.25.

We apply Proposition 8.9 on the K'-spectrum of the Knapp–Stein intertwining operator to the subgroup $G' = O(n, 1)$. Then the K'-spectrum of the (normalized)

Knapp–Stein intertwining operator $\widetilde{\mathbb{T}}^j_{\nu,n-1-\nu} : J_\varepsilon(j,\nu) \to J_\varepsilon(j,n-1-\nu)$ of G' is given by

$$\widetilde{\mathbb{T}}^i_{\nu,n-1-\nu} \circ \iota^*_\nu = c^\natural(j,\nu)' \iota^*_{n-\nu} \qquad \text{on } \mu^\natural(j,\varepsilon)'$$

for $\natural = \flat$ or $\sharp$, where

$$c^\flat(j,\nu)' = \frac{(\nu-j)\pi^{\frac{n-1}{2}}}{\Gamma(\nu+1)}, \qquad c^\sharp(j,\nu)' = \frac{(n-1-j-\nu)\pi^{\frac{n-1}{2}}}{\Gamma(\nu+1)}.$$

Proof of Theorem 9.24. For $j = i$ or $j-1$ and for $(\lambda,\nu) \in \mathbb{C}^2$ with $\nu - \lambda \notin \mathbb{N}$, we recall from Theorem 3.26 and Corollary 5.25 that

$$\text{Hom}_{G'}(I_+(i,\lambda)|_{G'}, J_\varepsilon(j,n-1-\nu)) = \mathbb{C}\widetilde{\mathbb{A}}^{i,j}_{\lambda,n-1-\nu,\varepsilon}.$$

Hence, there exists a constant $p^{TA}_A(i,j,\varepsilon;\lambda,\nu) \in \mathbb{C}$ such that

$$\widetilde{\mathbb{T}}^j_{\nu,n-1-\nu} \circ \widetilde{\mathbb{A}}^{i,j}_{\lambda,\nu,\varepsilon} = p^{TA}_A(i,j,\varepsilon;\lambda,\nu)\widetilde{\mathbb{A}}^{i,j}_{\lambda,n-1-\nu,\varepsilon} \qquad (9.36)$$

if $n-1-\nu-\lambda \notin \mathbb{N}$. We compute $p^{TA}_A(i,j,\varepsilon;\lambda,\nu)$ by using the (K,K')-spectrum $S_{\mu,\mu'}$ (see Section 9.3) for (9.36) with an appropriate choice of basic K-types $\mu \in \widehat{K}$ and $\mu' \in \widehat{K'}$. We recall from Theorem 9.8 an explicit formula of the (K,K')-spectrum

$$S(\widetilde{\mathbb{A}}^{i,j}_{\lambda,\nu,\varepsilon}) = \begin{pmatrix} a^{i,j}_\varepsilon(\lambda,\nu) & b^{i,j}_\varepsilon(\lambda,\nu) \\ c^{i,j}_\varepsilon(\lambda,\nu) & d^{i,j}_\varepsilon(\lambda,\nu) \end{pmatrix}$$

for the regular symmetry breaking operator $\widetilde{\mathbb{A}}^{i,j}_{\lambda,\nu,\varepsilon} : I_+(i,\lambda) \to J_\varepsilon(j,\nu)$ with respect to basic K-types.

Case 1. $j = i$ and $\varepsilon = +$. Take $(\mu,\mu') = (\mu^\flat(i,+), \mu^\flat(i,+)')$. Then the computation of $S_{\mu,\mu'}$ on the both sides of (9.36) leads us to the following identity:

$$p^{TA}_A(i,i,+;\lambda,\nu) = c^\flat(i,\nu)' \cdot \frac{a^{i,i}_+(\lambda,\nu)}{a^{i,i}_+(\lambda,n-1-\nu)} = \frac{\pi^{\frac{n-1}{2}}(\nu-j)}{\Gamma(\nu+1)} \cdot 1.$$

Case 2. $j = i$ and $\varepsilon = -$. Take $(\mu,\mu') = (\mu^\sharp(i,+), \mu^\flat(i,-)')$. By the same argument as above, we have

$$p^{TA}_A(i,i,-;\lambda,\nu) = c^\flat(i,\nu)' \cdot \frac{c^{i,i}_-(\lambda,\nu)}{c^{i,i}_-(\lambda,n-1-\nu)} = \frac{\pi^{\frac{n-1}{2}}(\nu-j)}{\Gamma(\nu+1)} \cdot 1.$$

Case 3. $j = i - 1$ and $\varepsilon = +$. Take $(\mu, \mu') = (\mu^\sharp(i, +), \mu^\sharp(i - 1, +)')$.

$$p_A^{TA}(i, i - 1, +; \lambda, \nu) = c^\sharp(i - 1, \nu)' \cdot \frac{d_+^{i,i-1}(\lambda, \nu)}{d_+^{i,i-1}(\lambda, n - 1 - \nu)} = \frac{\pi^{\frac{n-1}{2}}(n - \nu - i)}{\Gamma(\nu + 1)} \cdot 1.$$

Case 4. $j = i - 1$ and $\varepsilon = -$. Take $(\mu, \mu') = (\mu^\flat(i, +), \mu^\sharp(i - 1, -)')$.

$$p_A^{TA}(i, i - 1, -; \lambda, \nu) = c^\sharp(i - 1, \nu)' \cdot \frac{b_-^{i,i-1}(\lambda, \nu)}{b_-^{i,i-1}(\lambda, n - 1 - \nu)} = \frac{\pi^{\frac{n-1}{2}}(n - \nu - i)}{\Gamma(\nu + 1)} \cdot 1.$$

Since both sides of (9.36) depend holomorphically in the entire $(\lambda, \nu) \in \mathbb{C}^2$, the identity (9.36) holds for all $(\lambda, \nu) \in \mathbb{C}^2$. Hence Theorem 9.24 is proved.   □

Proof of Theorem 9.25.   The proof of Theorem 9.25 goes similarly. Since $\widetilde{\mathbb{A}}^{i,j}_{n-\lambda,\nu,\varepsilon} \circ \widetilde{\mathbb{T}}^i_{\lambda,n-\lambda} \in \mathrm{Hom}_{G'}(I_+(i,\lambda)|_{G'}, J_\varepsilon(j, \nu))$, there exists a constant

$$p_A^{AT}(i, j, \varepsilon; \lambda, \nu) \in \mathbb{C}$$

such that

$$\widetilde{\mathbb{A}}^{i,j}_{n-\lambda,\nu,\varepsilon} \circ \widetilde{\mathbb{T}}^i_{\lambda,n-\lambda} = p_A^{AT}(i, j, \varepsilon; \lambda, \nu)\widetilde{\mathbb{A}}^{i,j}_{\lambda,\nu,\varepsilon} \tag{9.37}$$

by the generic multiplicity-one theorem (Theorem 5.41) for $j \in \{i - 1, i\}$, $\varepsilon \in \{\pm\}$, and $(\lambda, \nu) \in \mathbb{C}^2$ with $\nu - \lambda \notin \mathbb{N}$.

Case 1. $j = i$ and $\varepsilon = +$. Take $(\mu, \mu') = (\mu^\sharp(i, +), \mu^\sharp(i, +)')$. Applying both sides of (9.37) to the basic K'-type $\mu' = \mu^\sharp(i, +)'$ via the inclusion $\mu' \hookrightarrow \mu = \mu^\sharp(i, +)$, we get the following identities from Proposition 8.9 and Theorem 9.10:

$$p_A^{AT}(i, i, +; \lambda, \nu) = c^\sharp(i, \lambda) \cdot \frac{d_+^{i,i}(n - \lambda, \nu)}{d_+^{i,i}(\lambda, \nu)} = \frac{\pi^{\frac{n}{2}}(n - \lambda - i)}{\Gamma(\lambda + 1)} \cdot \frac{\Gamma(\lambda + 1)}{\Gamma(n - \lambda + 1)}.$$

The other three cases are proved similarly as below.
Case 2. $j = i$ and $\varepsilon = -$. Take $(\mu, \mu') = (\mu^\sharp(i, +), \mu^\flat(i, -)')$.

$$p_A^{AT}(i, i, -; \lambda, \nu) = c^\sharp(i, \lambda) \cdot \frac{c_-^{i,i}(n - \lambda, \nu)}{c_-^{i,i}(\lambda, \nu)} = \frac{\pi^{\frac{n}{2}}(n - \lambda - i)}{\Gamma(\lambda + 1)} \cdot \frac{\Gamma(\lambda + 1)}{\Gamma(n - \lambda + 1)}.$$

Case 3. $j = i - 1$ and $\varepsilon = +$. Take $(\mu, \mu') = (\mu^\flat(i, +), \mu^\flat(i - 1, +)')$.

$$p_A^{AT}(i, i - 1, +; \lambda, \nu) = c^\flat(i, \lambda) \cdot \frac{a_+^{i,i-1}(n - \lambda, \nu)}{a_+^{i,i-1}(\lambda, \nu)} = \frac{\pi^{\frac{n}{2}}(\lambda - i)}{\Gamma(\lambda + 1)} \cdot \frac{\Gamma(\lambda + 1)}{\Gamma(n - \lambda + 1)}.$$

Case 4. $j = i - 1$ and $\varepsilon = -$. Take $(\mu, \mu') = (\mu^{\flat}(i, +), \mu^{\sharp}(i - 1, -)')$.

$$p_A^{AT}(i, i - 1, -; \lambda, \nu) = c^{\flat}(i, \lambda) \cdot \frac{b_-^{i,i-1}(n - \lambda, \nu)}{b_-^{i,i-1}(\lambda, \nu)} = \frac{\pi^{\frac{n}{2}}(\lambda - i)}{\Gamma(\lambda + 1)} \cdot \frac{\Gamma(\lambda + 1)}{\Gamma(n - \lambda + 1)}.$$

Thus Theorem 9.25 is proved.   □

9.9   Renormalized Symmetry Breaking Operator $\widetilde{\mathbb{A}}^{i,j}_{\lambda,\nu,+}$

In Theorem 5.45, we constructed a *renormalized* symmetry breaking operator $\widetilde{\mathbb{A}}^{V,W}_{\lambda,\nu,\pm}$ when the normalized regular symmetry breaking operator $\widetilde{\mathbb{A}}^{V,W}_{\lambda,\nu,\pm}$ vanishes. We apply it to the special case $(V, W) = (\bigwedge^i(\mathbb{C}^n), \bigwedge^j(\mathbb{C}^{n-1}))$, and obtain for those (λ, ν) for which $\widetilde{\mathbb{A}}^{i,j}_{\lambda,\nu,\gamma} = 0$ the renormalized symmetry breaking operator $\widetilde{\mathbb{A}}^{i,j}_{\lambda,\nu,\gamma}$ as the analytic continuation of the following:

$$\widetilde{\widetilde{\mathbb{A}}}^{i,j}_{\lambda,\nu,\gamma} = \begin{cases} \Gamma(\frac{\lambda-\nu}{2})\widetilde{\mathbb{A}}^{i,j}_{\lambda,\nu,+} & \text{if } \gamma = +, \\ \Gamma(\frac{\lambda-\nu+1}{2})\widetilde{\mathbb{A}}^{i,j}_{\lambda,\nu,-} & \text{if } \gamma = -. \end{cases} \tag{9.38}$$

We recall that for $j \in \{i - 1, i\}$ and $\gamma \in \{\pm\}$, we have determined in Theorem 3.19 precisely the zero set

$$\{(\lambda, \nu) \in \mathbb{C}^2 : \widetilde{\mathbb{A}}^{i,j}_{\lambda,\nu,\gamma} = 0\}.$$

In this section, we discuss functional equations and (K, K')-spectrum of the renormalized operators $\widetilde{\widetilde{\mathbb{A}}}^{i,j}_{\lambda,\nu,\pm}$ only in the few cases that are necessary for later arguments.

9.9.1   *Functional Equations for the Renormalized Operator* $\widetilde{\widetilde{\mathbb{A}}}^{i,i}_{\lambda,i,+}$

In this subsection, we treat the case $j = i$. For $\nu = i(= j)$, $\widetilde{\mathbb{A}}^{i,i}_{\lambda,i,\gamma} = 0$ if and only if $\lambda = i \in \{0, 1, \cdots, n - 1\}$ and $\gamma = +$ by Theorem 3.19. Then the renormalized operator $\widetilde{\widetilde{\mathbb{A}}}^{i,i}_{\lambda,i,+} : I_{\delta}(i, \lambda) \to J_{\delta}(i, i)$ is the analytic continuation of the following:

$$\widetilde{\widetilde{\mathbb{A}}}^{i,i}_{\lambda,i,+} = \Gamma(\frac{\lambda - i}{2})\widetilde{\mathbb{A}}^{i,i}_{\lambda,i,+}. \tag{9.39}$$

Then $\widetilde{\widetilde{\mathbb{A}}}^{i,i}_{\lambda,i,+} : I_\delta(i,\lambda) \to J_\delta(i,i)$ is a G'-homomorphism that depends holomorphically on λ in the entire complex plane $\mathbb{C}$ by Theorem 5.45 (3).

We determine functional equations and (K,K')-spectrum $S(\widetilde{\widetilde{\mathbb{A}}}^{i,i}_{\lambda,i,+})$ (see (9.13)) on basic K- and K'-types for the renormalized operator $\widetilde{\widetilde{\mathbb{A}}}^{i,i}_{\lambda,i,+}$ as follows.

Lemma 9.27 (functional equations and the (K,K')-spectrum for $\widetilde{\widetilde{\mathbb{A}}}^{i,i}_{n-\lambda,i,+}$). *Suppose $0 \le i \le n-1$ and $\lambda \in \mathbb{C}$. Then we have*

$$\widetilde{\mathbb{T}}^i_{i,n-1-i} \circ \widetilde{\widetilde{\mathbb{A}}}^{i,i}_{\lambda,i,+} = 0, \tag{9.40}$$

$$\widetilde{\widetilde{\mathbb{A}}}^{i,i}_{n-\lambda,i,+} \circ \widetilde{\mathbb{T}}^i_{\lambda,n-\lambda} = \frac{2\pi^{\frac{n}{2}}\Gamma(\frac{n-\lambda-i}{2}+1)}{\Gamma(\frac{\lambda-i}{2})\Gamma(n-\lambda+1)} \widetilde{\widetilde{\mathbb{A}}}^{i,i}_{\lambda,i,+}, \tag{9.41}$$

$$S(\widetilde{\widetilde{\mathbb{A}}}^{i,i}_{\lambda,i,+}) = \frac{\pi^{\frac{n-1}{2}}}{\Gamma(\lambda+1)} \begin{pmatrix} 2 & 0 \\ 0 & 0 \end{pmatrix}. \tag{9.42}$$

Proof. Applying Theorem 9.24 with $\nu = i$ $(0 \le i \le n)$, we have

$$\widetilde{\mathbb{T}}^i_{i,n-1-i} \circ \widetilde{\mathbb{A}}^{i,i}_{\lambda,i,+} = 0 \quad \text{for all } \lambda \in \mathbb{C}.$$

Taking the limit as λ tends to i in the following equation:

$$\widetilde{\mathbb{T}}^i_{i,n-1-i} \circ \Gamma(\frac{\lambda-i}{2}) \widetilde{\mathbb{A}}^{i,i}_{\lambda,i,+} = 0,$$

we get the desired formula (9.40) by the definition (9.39) of the renormalization $\widetilde{\widetilde{\mathbb{A}}}^{i,i}_{\lambda,i,+}$.

Similarly, the formulæ (9.41) and (9.42) for the renormalized operator $\widetilde{\widetilde{\mathbb{A}}}^{i,i}_{\lambda,i,+}$ follow from the limit of the corresponding results for $\widetilde{\mathbb{A}}^{i,i}_{\lambda,i,+}$ given in Theorems 9.25 and 9.8, respectively. □

9.9.2 Functional Equations at Middle Degree for n Even

For n even (say, $n = 2m$), at the "middle degree" $i = \frac{n}{2}(=m)$, we observe that the Knapp–Stein operator $\widetilde{\mathbb{T}}^m_{\lambda,2m-\lambda} : I_+(m,\lambda) \to I_+(m,2m-\lambda)$ vanishes if $\lambda = m$ (see Proposition 8.12), and so the functional equation (9.41) is trivial. Instead we use the renormalized Knapp–Stein operator $\widetilde{\widetilde{\mathbb{T}}}^m_{\lambda,2m-\lambda}$ defined in (8.21) for another functional equation, see (9.43) below. We recall from Lemma 8.15 that $\widetilde{\widetilde{\mathbb{T}}}^m_{\lambda,2m-\lambda}$ is an endomorphism of $I_\delta(m,m)$ when $\lambda = m$, but is not proportional to the identity operator when $\lambda = m$.

Lemma 9.28 (functional equation for $\widetilde{\mathbb{A}}^{i,i}_{m,m,+}$). *Let* $(G,G') = (O(n+1,1), O(n,1))$ *with* $n = 2m$. *Then we have*

$$\widetilde{\widetilde{\mathbb{A}}}^{m,m}_{m,m,+} \circ \widetilde{\widetilde{\mathbb{T}}}^{m}_{m,m} = \frac{\pi^m}{m!} \widetilde{\widetilde{\mathbb{A}}}^{m,m}_{m,m,+}. \tag{9.43}$$

Proof. By Theorem 5.45 and (8.21),

$$\widetilde{\widetilde{\mathbb{A}}}^{m,m}_{m,m,+} \circ \widetilde{\widetilde{\mathbb{T}}}^{m}_{m,m} = \left(\lim_{\lambda \to m} \Gamma\left(\frac{(2m-\lambda)-m}{2} \right) \widetilde{\mathbb{A}}^{m,m}_{2m-\lambda,m,+} \right) \circ \left(\lim_{\lambda \to m} \frac{1}{\lambda - m} \widetilde{\mathbb{T}}^{m}_{\lambda,2m-\lambda} \right)$$

$$= \lim_{\lambda \to m} \frac{\Gamma(\frac{m-\lambda}{2})}{\lambda - m} \widetilde{\mathbb{A}}^{m,m}_{2m-\lambda,m,+} \circ \widetilde{\mathbb{T}}^{m}_{\lambda,2m-\lambda}.$$

In turn, the functional equation in Theorem 9.25 shows that the right-hand side amounts to

$$\lim_{\lambda \to m} \frac{\Gamma(\frac{m-\lambda}{2})}{\lambda - m} \frac{\pi^m (2m - \lambda - m)}{\Gamma(2m - \lambda + 1)} \widetilde{\mathbb{A}}^{m,m}_{\lambda,m,+} = \frac{-\pi^m}{\Gamma(m+1)} \left(\lim_{\lambda \to m} \frac{\Gamma(\frac{m-\lambda}{2})}{\Gamma(\frac{\lambda-m}{2})} \right) \widetilde{\widetilde{\mathbb{A}}}^{m,m}_{m,m,+}$$

$$= \frac{\pi^m}{\Gamma(m+1)} \widetilde{\widetilde{\mathbb{A}}}^{m,m}_{m,m,+}. \tag{9.44}$$

Hence the formula (9.43) is proved. □

In contrast to Lemma 9.28 where we needed to treat the renormalized operator $\widetilde{\widetilde{\mathbb{A}}}^{m,m}_{m,m,+}$ because $\widetilde{\mathbb{A}}^{m,m}_{m,m,+} = 0$, the normalized operator $\widetilde{\mathbb{A}}^{m,m}_{m,m,-}$ does not vanish (Theorem 3.19 (3)). In this case, the functional equations for $\widetilde{\mathbb{A}}^{m,m}_{m,m,-}$ are given as follows:

Lemma 9.29 (functional equation for $\widetilde{\mathbb{A}}^{m,m}_{m,m,-}$). *We retain the setting that* $(G,G') = (O(n+1,1), O(n,1))$ *with* $n = 2m$. *Then we have*

$$\widetilde{\mathbb{A}}^{m,m}_{m,m,-} \circ \widetilde{\widetilde{\mathbb{T}}}^{m}_{m,m} = -\frac{\pi^m}{m!} \widetilde{\mathbb{A}}^{m,m}_{m,m,-}, \tag{9.45}$$

$$\widetilde{\mathbb{T}}^{m}_{m,m-1} \circ \widetilde{\mathbb{A}}^{m,m}_{m,m,-} = 0. \tag{9.46}$$

Proof. By the definition of $\widetilde{\widetilde{\mathbb{T}}}^{m,m}_{\lambda,2m-\lambda}$ in (8.21) and the functional equation in Theorem 9.25, we have

$$\widetilde{\mathbb{A}}^{m,m}_{m,m,-} \circ \widetilde{\widetilde{\mathbb{T}}}^{m}_{m,m} = \lim_{\lambda \to m} \widetilde{\mathbb{A}}^{m,m}_{2m-\lambda,m,-} \circ \frac{1}{\lambda - m} \widetilde{\mathbb{T}}^{m}_{\lambda,2m-\lambda}$$

$$= \lim_{\lambda \to m} \frac{\pi^m (m - \lambda)}{(\lambda - m)\Gamma(2m - \lambda + 1)} \widetilde{\mathbb{A}}^{m,m}_{\lambda,m,-}$$

$$= \frac{-\pi^m}{m!} \widetilde{\mathbb{A}}^{m,m}_{m,m,-}.$$

Hence the first statement is verified. The second statement is a special case of Theorem 9.24.   □

9.9.3   Functional Equations for the Renormalized Operator $\widetilde{\mathbb{A}}^{i,i-1}_{\lambda,n-i,+}$

In this subsection, we treat the case $j = i - 1$. For $j = i - 1$ and $\nu = n - i$, $\widetilde{\mathbb{A}}^{i,j}_{\lambda,n-i,\gamma} = 0$ if and only if $\gamma = +$ and $\lambda = n - i$ by Theorem 3.19. In this case, the renormalized symmetry breaking operator $\widetilde{\mathbb{A}}^{i,i-1}_{\lambda,n-i,+} : I_\delta(i,\lambda) \to J_\delta(i-1, n-i)$ is obtained as the analytic continuation of the following:

$$\widetilde{\widetilde{\mathbb{A}}}^{i,i-1}_{\lambda,n-i,+} = \Gamma(\frac{\lambda - n + i}{2}) \widetilde{\mathbb{A}}^{i,i-1}_{\lambda,n-i,+},$$

see Theorem 5.45 (3).

We determine functional equations and (K, K')-spectrum $S(\widetilde{\widetilde{\mathbb{A}}}^{i,i-1}_{\lambda,n-i,+})$ (see (9.13)) on basic K- and K'-types for the renormalized operator $\widetilde{\widetilde{\mathbb{A}}}^{i,i-1}_{\lambda,n-i,+}$ as follows.

Lemma 9.30 (functional equations and the (K, K')-spectrum for $\widetilde{\widetilde{\mathbb{A}}}^{i,i-1}_{\lambda,n-i,+}$). *Suppose $1 \le i \le n$ and $\lambda \in \mathbb{C}$. Then we have*

$$\widetilde{\widetilde{\mathbb{A}}}^{i,i-1}_{n-\lambda,n-i,+} \circ \widetilde{\mathbb{T}}^i_{\lambda,n-\lambda} = -\frac{2\pi^{\frac{n}{2}}\Gamma(\frac{i-\lambda}{2} + 1)}{\Gamma(n - \lambda + 1)\Gamma(\frac{\lambda-n+i}{2})} \widetilde{\widetilde{\mathbb{A}}}^{i,i-1}_{\lambda,n-i,+}, \tag{9.47}$$

$$\widetilde{\mathbb{T}}^{i-1}_{n-i,i-1} \circ \widetilde{\widetilde{\mathbb{A}}}^{i,i-1}_{\lambda,n-i,+} = 0, \tag{9.48}$$

$$S(\widetilde{\widetilde{\mathbb{A}}}^{i,i-1}_{\lambda,n-i,+}) = \frac{\pi^{\frac{n-1}{2}}}{\Gamma(\lambda + 1)} \begin{pmatrix} 0 & 0 \\ 0 & 2 \end{pmatrix}.$$

Proof. The functional equations follow from Theorems 9.24 and 9.25. The formula for the (K, K')-spectrum is derived from Theorem 9.8.   □

9.9.4   Functional Equations at Middle Degree for n Odd

For n odd (say, $n = 2m + 1$), the Knapp–Stein operator $\widetilde{\mathbb{T}}^j_{\nu,n-\nu-1} : J_\varepsilon(j,\nu) \to J_\varepsilon(j, n - 1 - \nu)$ for the subgroup $G' = O(n, 1)$ vanishes at the middle degree $j = \frac{1}{2}(n-1)(= m)$ if $\nu = m$ by Proposition 8.12. We note that the exact sequence in Theorem 2.20 (1) for $G' = O(2m + 1, 1)$ splits, and we have a direct sum decomposition

$$J_\varepsilon(m, m) \simeq \pi_{m,\varepsilon} \oplus \pi_{m+1,-\varepsilon}$$

of two irreducible tempered representations of G'. In this case, the functional equations (9.40) in Lemma 9.27 and (9.48) in Lemma 9.30 are trivial, and we replace them by the following functional equations for the *renormalized* Knapp–Stein operator $\widetilde{\widetilde{\mathbb{T}}}^m_{m,m}$.

Lemma 9.31. *For $(G, G') = (O(n + 1, 1), O(n, 1))$ with $n = 2m + 1$ and for $\lambda \in \mathbb{C}$, we have*

$$\widetilde{\widetilde{\mathbb{T}}}^m_{m,m} \circ \widetilde{\widetilde{\mathbb{A}}}^{m,m}_{\lambda,m,+} = \frac{\pi^m}{m!} \widetilde{\widetilde{\mathbb{A}}}^{m,m}_{\lambda,m,+}, \tag{9.49}$$

$$\widetilde{\widetilde{\mathbb{T}}}^m_{m,m} \circ \widetilde{\widetilde{\mathbb{A}}}^{m+1,m}_{\lambda,m,+} = -\frac{\pi^m}{m!} \widetilde{\widetilde{\mathbb{A}}}^{m+1,m}_{\lambda,m,+}. \tag{9.50}$$

Lemma 9.31 tells that

$$\mathrm{Image}(\widetilde{\widetilde{\mathbb{A}}}^{m,m}_{\lambda,m,+} : I_\delta(m, \lambda) \to J_\delta(m, m)) \qquad \subset \pi_{m,\delta},$$

$$\mathrm{Image}(\widetilde{\widetilde{\mathbb{A}}}^{m+1,m}_{\lambda,m,+} : I_\delta(m+1, \lambda) \to J_\delta(m, m)) \subset \pi_{m+1,-\delta},$$

for all $\lambda \in \mathbb{C}$ by Lemma 8.15.

Proof. The functional equations in Theorem 9.24 tell that

$$\left(\frac{1}{\nu - m} \widetilde{\mathbb{T}}^m_{\nu,2m-\nu}\right) \circ \Gamma\left(\frac{\lambda - m}{2}\right) \widetilde{\mathbb{A}}^{m,m}_{\lambda,\nu,+} = \frac{\pi^m}{\Gamma(\nu + 1)} \Gamma\left(\frac{\lambda - m}{2}\right) \widetilde{\mathbb{A}}^{m,m}_{\lambda,2m-\nu,+},$$

$$\left(\frac{1}{\nu - m} \widetilde{\mathbb{T}}^m_{\nu,2m-\nu}\right) \circ \Gamma\left(\frac{\lambda - m}{2}\right) \widetilde{\mathbb{A}}^{m+1,m}_{\lambda,\nu,+} = -\frac{\pi^m}{\Gamma(\nu + 1)} \Gamma\left(\frac{\lambda - m}{2}\right) \widetilde{\mathbb{A}}^{m+1,m}_{\lambda,2m-\nu,+}.$$

Taking the limit as ν tends to m, we get Lemma 9.31.   □

9.10 Restriction Map $I_\delta(i, \lambda) \to J_\delta(i, \lambda)$

The restriction of (smooth) differential forms to a submanifold defines an obvious continuous map between Fréchet spaces, which intertwines the conformal representation (see [35, Lem. 8.9]). We end this section with the most elementary symmetry breaking operator, namely, the restriction map for the pair $(G'/P', G/P) \subset (S^{n-1}, S^n)$.

Lemma 9.32. *The restriction map from G/P to the submanifold G'/P' induces obvious symmetry breaking operators*

$$\mathrm{Rest}^{i,i}_{\lambda,\lambda,+} : I_\delta(i,\lambda) \to J_\delta(i,\lambda).$$

Then the (K,K')-spectrum for basic K- and K'-types (see (9.13)) is given by

$$S(\mathrm{Rest}^{i,i}_{\lambda,\lambda,+}) = \begin{pmatrix} 1 & 0 \\ 0 & 1 \end{pmatrix}. \tag{9.51}$$

Proof. We recall from Proposition 8.4 that

$$\{ \mathbf{1}^{\mathcal{I}}_\lambda : \mathcal{I} \in \mathfrak{I}_{n+1,i} \}$$

forms a basis of the basic K-type $\mu^\flat(i,\delta)$ of the principal series representation $I_\delta(i,\lambda)$ in the N-picture. Let $\mathcal{I} \in \mathfrak{I}_{n+1,i}$ and $(x,x_n) \in \mathbb{R}^{n-1} \oplus \mathbb{R} = \mathbb{R}^n$. By (8.6), we have

$$\mathbf{1}^{\mathcal{I}}_\lambda(x,x_n) = -(1+|x|^2+x_n^2)^{-\lambda-1} \sum_{J\in\mathfrak{I}_{n,i}} S_{\mathcal{I}J}(1,x,x_n)e_J.$$

Then an elementary computation by using (7.6) shows

$$\mathbf{1}^{\mathcal{I}}_\lambda(x,0) = \begin{cases} \mathbf{1}'^{\mathcal{I}}_\lambda(x) & \text{if } n \notin \mathcal{I} \in \mathfrak{I}_{n+1,i}, \\ \mathbf{1}'^{\mathcal{I}-\{n\}}_\lambda(x) \wedge e_n & \text{if } n \in \mathcal{I} \in \mathfrak{I}_{n+1,i}. \end{cases}$$

The case for the basic K-type $\mu^\sharp(i,\delta)$ is similar, where we recall from Proposition 8.4 that $\{h^{\mathcal{I}}_\lambda : \mathcal{I} \in \mathfrak{I}_{n+1,i+1}\}$ forms its basis, for which we can compute the restriction $x_n = 0$. Thus Lemma 9.32 is shown. $\square$

9.11   Image of the Differential Symmetry Breaking Operator $\widetilde{\mathbb{C}}^{i,j}_{\lambda,\nu}$

In Theorem 6.8, we have proved that the image of any nonzero differential symmetry breaking operator from principal series representation is infinite-dimensional. As an application of the functional equations of the (generically) regular symmetry breaking operators $\widetilde{\mathbb{A}}^{i,j}_{\lambda,\nu,\pm}$ (Theorems 9.24 and 9.25) and of the residue formulæ of $\widetilde{\mathbb{A}}^{i,j}_{\lambda,\nu,\pm}$ (Fact 9.3, see [34]), we end this chapter with a necessary and sufficient condition for the renormalized differential symmetry breaking operator $\widetilde{\mathbb{C}}^{i,j}_{\lambda,\nu}$ to be surjective when $j = i, i-1$, see Theorems 9.33 and 9.34.

9.11.1 Surjectivity Condition of $\widetilde{\mathbb{C}}_{\lambda,\nu}^{i,j}$

Suppose $j \in \{i, i-1\}$. We recall from (3.18) and (3.19) that the renormalized differential symmetry breaking operator $\widetilde{\mathbb{C}}_{\lambda,\nu}^{i,j} : I_\delta(i,\lambda) \to J_\varepsilon(j,\nu)$ is defined for $(\lambda,\nu) \in \mathbb{C}^2$ with $\nu - \lambda \in \mathbb{N}$ and $\delta\varepsilon = (-1)^{\nu-\lambda}$. Moreover, $\widetilde{\mathbb{C}}_{\lambda,\nu}^{i,j}$ is nonzero for any (i,j,λ,ν) with $j \in \{i, i-1\}$ and $\nu - \lambda \in \mathbb{N}$.

In what follows, we shall sometimes encounter the condition that $(\lambda, n-1-\nu) \in L_{\mathrm{even}} \cup L_{\mathrm{odd}}$, which is equivalent to

$$(\lambda,\nu) \in \mathbb{Z}^2 \quad \text{and} \quad \lambda + \nu \le n - 1 \le \nu. \tag{9.52}$$

Theorem 9.33. *Suppose* $0 \le i \le n-1$, $\nu - \lambda \in \mathbb{N}$, *and* δ, $\varepsilon \in \{\pm\}$ *with* $(-1)^{\nu-\lambda} = \delta\varepsilon$. *Then the following two conditions (i) and (ii) on* (i,λ,ν) *are equivalent:*

(i) $\widetilde{\mathbb{C}}_{\lambda,\nu}^{i,i} : I_\delta(i,\lambda) \to J_\varepsilon(i,\nu)$ *is not surjective.*
(ii) *One of the following conditions holds:*

   (ii-a) $1 \le i \le n-1$, $\nu = i$, *and* $\mathbb{Z} \ni \lambda < i$;
   (ii-b) n *is odd,* $i = 0$, *and* (9.52);
   (ii-c) n *is odd,* $1 \le i \le n-1$, (9.52), *and* $\nu \neq n-1$;
   (ii-d) n *is odd,* $1 \le i < \frac{1}{2}(n-1)$, $(\lambda,\nu) = (i, n-1-i)$.

Theorem 9.34. *Suppose* $1 \le i \le n$, $\nu - \lambda \in \mathbb{N}$, *and* δ, $\varepsilon \in \{\pm\}$ *with* $(-1)^{\nu-\lambda} = \delta\varepsilon$. *Then the following two conditions (i) and (ii) on* (i,λ,ν) *are equivalent:*

(i) $\widetilde{\mathbb{C}}_{\lambda,\nu}^{i,i-1} : I_\delta(i,\lambda) \to J_\varepsilon(i-1,\nu)$ *is not surjective.*
(ii) *One of the following conditions holds:*

   (ii-a) $1 \le i \le n-1$, $\nu = n-i$, *and* $\mathbb{Z} \ni \lambda < n-i$;
   (ii-b) n *is odd,* $1 \le i \le n-1$, (9.52), *and* $\nu \neq n-1$;
   (ii-c) n *is odd,* $i = n$, *and* (9.52);
   (ii-d) n *is odd,* $\frac{1}{2}(n+1) < i \le n-1$, *and* $(\lambda,\nu) = (n-i, i-1)$.

For the proof of Theorems 9.33 and 9.34, we first derive the functional equations for $\widetilde{\mathbb{C}}_{\lambda,\nu}^{i,j}$ in Theorem 9.35 from those for the regular symmetry breaking operators $\widetilde{\mathbb{A}}_{\lambda,\nu,\pm}^{i,j}$ in Chapter 9 and from the matrix-valued residue formulæ [34]. The results cover most of the cases where the Knapp–Stein intertwining operators $\widetilde{\mathbb{T}}_{\nu,n-1-\nu}^{j}$ do not vanish. A special attention is required when $\widetilde{\mathbb{T}}_{\nu,n-1-\nu}^{j} = 0$. In this case, the principal series representation $J_\varepsilon(j,\nu)$ splits into the direct sum of two irreducible representations of the subgroup $G' = O(n,1)$, and we shall treat this case separately in Section 9.11.3. The proof of Theorems 9.33 and 9.34 will be completed in Section 9.11.4.

9.11.2   Functional Equation for $\widetilde{\mathbb{C}}^{i,j}_{\lambda,\nu}$

Suppose $0 \le i \le n$, $0 \le j \le n-1$, and $j = i$ or $i-1$. We set $p^{i,j}(\lambda,\nu)$ by

$$p^{i,i}(\lambda,\nu) := \begin{cases} 1 & \text{if } i = 0 \text{ or } \lambda = \nu, \\ \frac{1}{2}(\nu - i) & \text{otherwise,} \end{cases} \tag{9.53}$$

$$p^{i,i-1}(\lambda,\nu) := \begin{cases} 1 & \text{if } i = n \text{ or } \lambda = \nu, \\ \frac{1}{2}(\nu + i - n) & \text{otherwise.} \end{cases}$$

Theorem 9.35 (functional equation for $\widetilde{\mathbb{C}}^{i,j}_{\lambda,\nu}$). *Suppose* $0 \le i \le n$, $0 \le j \le n-1$, *and* $j \in \{i, i-1\}$. *For* $(\lambda,\nu) \in \mathbb{C}^2$ *with* $\nu - \lambda \in \mathbb{N}$, *we have*

$$\widetilde{\mathbb{T}}^j_{\nu,n-1-\nu} \circ \widetilde{\mathbb{C}}^{i,j}_{\lambda,\nu} = q(\nu-\lambda)p^{i,j}(\lambda,\nu)\widetilde{\mathbb{A}}^{i,j}_{\lambda,n-1-\nu,(-1)^{\nu-\lambda}}, \tag{9.54}$$

where $q(m)$ *is a nonzero number defined in* (9.8).

Proof. We set

$$p_{i,j}(\nu) := \begin{cases} \frac{1}{2}(\nu - i) & \text{if } j = i, \\ \frac{1}{2}(n - \nu - i) & \text{if } j = i-1. \end{cases} \tag{9.55}$$

By the functional equation for the regular symmetry breaking operator $\widetilde{\mathbb{A}}^{i,j}_{\lambda,\nu,\pm}$ given in Theorem 9.24, we have for $\gamma \in \{\pm\}$

$$\widetilde{\mathbb{T}}^j_{\nu,n-1-\nu} \circ \widetilde{\mathbb{A}}^{i,j}_{\lambda,\nu,\gamma} = \frac{2\pi^{\frac{n-1}{2}} p_{i,j}(\nu)}{\Gamma(\nu+1)} \widetilde{\mathbb{A}}^{i,j}_{\lambda,n-1-\nu,\gamma}.$$

Suppose $\nu - \lambda \in \mathbb{N}$. Applying the residue formula (9.7) of $\widetilde{\mathbb{A}}^{i,j}_{\lambda,\nu,\pm}$ to the left-hand side, we get

$$\widetilde{\mathbb{T}}^j_{\nu,n-1-\nu} \circ \mathbb{C}^{i,j}_{\lambda,\nu} = (-1)^{i-j}q(\nu-\lambda)p_{i,j}(\nu)\widetilde{\mathbb{A}}^{i,j}_{\lambda,n-1-\nu,(-1)^{\nu-\lambda}}. \tag{9.56}$$

On the other hand, by using $p^{i,j}(\lambda,\nu)$ and $p_{i,j}(\nu)$, the relation between the unnormalized operators $\mathbb{C}^{i,j}_{\lambda,\nu}$ and the renormalized operators $\widetilde{\mathbb{C}}^{i,j}_{\lambda,\nu}$ defined in (3.18) and (3.19) are given as the following unified formula:

$$p^{i,j}(\lambda,\nu)\mathbb{C}^{i,j}_{\lambda,\nu} = (-1)^{i-j}p_{i,j}(\nu)\widetilde{\mathbb{C}}^{i,j}_{\lambda,\nu} \quad \text{for } j \in \{i, i-1\}.$$

Multiplying both sides of the equation (9.56) by $p^{i,j}(\lambda,\nu)$, we get the desired formula. $\square$

Proposition 9.36. *Suppose $v - \lambda \in \mathbb{N}$, and $\delta, \varepsilon \in \{\pm\}$ with $(-1)^{v-\lambda} = \delta\varepsilon$.*

(1) *Suppose $0 \le i \le n - 1$. Then the following two conditions on (i, λ, v) are equivalent:*

   (i) *The image of $\widetilde{\mathbb{C}}^{i,i}_{\lambda,v} : I_\delta(i, \lambda) \to J_\varepsilon(i, v)$ is contained in $\mathrm{Ker}\,(\widetilde{\mathbb{T}}^i_{v,n-1-v})$.*
   (ii) *One of the following conditions holds:*

      (ii-a) $1 \le i \le n - 1$, $v = i$, *and* $\mathbb{Z} \ni \lambda < i$;
      (ii-b) *n is odd, $i = 0$, and (9.52);*
      (ii-c) *n is odd, $1 \le i \le n - 1$, (9.52), and $v \ne n - 1$;*
      (ii-d) *n is odd, $1 \le i \le \frac{1}{2}(n - 1)$, and $(\lambda, v) = (i, n - 1 - i)$.*

(2) *Suppose $1 \le i \le n$. Then the following two conditions on (i, λ, v) are equivalent:*

   (iii) *The image of $\widetilde{\mathbb{C}}^{i,i-1}_{\lambda,v} : I_\delta(i, \lambda) \to J_\varepsilon(i - 1, v)$ is contained in $\mathrm{Ker}\,(\widetilde{\mathbb{T}}^{i-1}_{v,n-1-v})$.*
   (iv) *One of the following holds:*

      (iv-a) $1 \le i \le n - 1$, $v = n - i$, *and* $\mathbb{Z} \ni \lambda < n - i$;
      (iv-b) *n is odd, $1 \le i \le n - 1$, (9.52), and $v \ne n - 1$;*
      (iv-c) *n is odd, $i = n$, and (9.52);*
      (iv-d) *n is odd, $\frac{1}{2}(n + 1) \le i \le n - 1$, and $(\lambda, v) = (n - i, i - 1)$.*

The difference of this proposition from Theorems 9.33 and 9.34 is that the cases $i = \frac{1}{2}(n - 1)$ in (1) and $i = \frac{1}{2}(n + 1)$ in (2) are included in Proposition 9.36. In these cases, the Knapp–Stein intertwining operator $\widetilde{\mathbb{T}}^j_{v,n-1-v}$ vanishes where $j = i$ in (1) and $= i - 1$ in (2), and the conditions (i) and (iii) do not provide any information of $\mathrm{Image}(\widetilde{\mathbb{C}}^{i,j}_{\lambda,v})$. In these special cases, we shall study $\mathrm{Image}(\widetilde{\mathbb{C}}^{i,j}_{\lambda,v})$ separately in Section 9.11.3 by using the renormalized Knapp–Stein operators $\widetilde{\widetilde{\mathbb{T}}}^j_{v,n-1-v}$.

Proof. By the functional equation (9.54) in Theorem 9.35, we see that

$$\mathrm{Image}\,(\widetilde{\mathbb{C}}^{i,j}_{\lambda,v}) \subset \mathrm{Ker}\,(\widetilde{\mathbb{T}}^j_{v,n-1-v})$$

if and only if $p^{i,j}(\lambda, v) = 0$ or $\widetilde{\mathbb{A}}^{i,j}_{\lambda,n-1-v,(-1)^{v-\lambda}} = 0$. Suppose $0 \le i \le n$, $0 \le j \le n - 1$, and $j \in \{i, i - 1\}$. By definition (9.53),

$$p^{i,i}(\lambda, v) = 0 \Leftrightarrow \lambda \ne v = i \quad \text{and} \quad 1 \le i \le n - 1,$$

$$p^{i,i-1}(\lambda, v) = 0 \Leftrightarrow \lambda \ne v = n - i \quad \text{and} \quad 1 \le i \le n - 1.$$

On the other hand, we claim

$$\widetilde{\mathbb{A}}^{i,i}_{\lambda,n-1-v,(-1)^{v-\lambda}} = 0 \Leftrightarrow \text{(ii-b), (ii-c), or (ii-d) holds;}$$

$$\widetilde{\mathbb{A}}^{i,i-1}_{\lambda,n-1-v,(-1)^{v-\lambda}} = 0 \Leftrightarrow \text{(iv-b), (iv-c), or (iv-d) holds.}$$

Let us verify the first equivalence for $1 \le i \le n-1$. The vanishing condition of $\widetilde{\mathbb{A}}^{i,j}_{\lambda,\nu,\pm}$ given in Theorem 3.19 (1) and (3) shows that $\widetilde{\mathbb{A}}^{i,i}_{\lambda,n-1-\nu,(-1)^{\nu-\lambda}} = 0$ if and only if one of the following three conditions holds:

- $i = 0$, $(\lambda, n-1-\nu) \in \mathbb{Z}^2$, $(\nu+1-n) - \lambda \equiv \nu - \lambda \mod 2$, and $0 \le \nu + 1 - n \le -\lambda$;
- $i \ne 0$, $(\lambda, n-1-\nu) \in \mathbb{Z}^2$, $(\nu+1-n) - \lambda \equiv \nu - \lambda \mod 2$, and $0 < \nu + 1 - n \le -\lambda$;
- $i \ne 0$, $\nu - \lambda \in 2\mathbb{Z}$, and $(\lambda, n-1-\nu) = (i,i)$.

These conditions amount to (ii-b), (ii-c), and (ii-d) in Proposition 9.36 (1), respectively. The second equivalence is shown similarly. Hence Proposition 9.36 is proved. $\qquad\square$

Remark 9.37. For $\lambda = \nu$, the above conditions are fulfilled if and only if $(\lambda, \nu) = (\frac{1}{2}(n-1), \frac{1}{2}(n-1))$ and $i = \frac{1}{2}(n-1)$ in Proposition 9.36 (1) or $i = \frac{1}{2}(n+1)$ in Proposition 9.36 (2). This is exactly when $\widetilde{\mathbb{T}}^j_{\nu,n-1-\nu}$ $(j = i, i-1)$ vanishes.

9.11.3 The Case When $\widetilde{\mathbb{T}}^j_{\nu,n-1-\nu} = 0$

By Proposition 8.12, the Knapp–Stein operators $\widetilde{\mathbb{T}}^j_{\nu,n-1-\nu}$ for the subgroup $G' = O(n,1)$ vanishes if and only if n is odd and

$$\nu = j = \frac{n-1}{2}.$$

We note in this case that $\nu - i = 0$ for $i = j$ and $\nu + i - n = 0$ for $i = j+1$, and therefore the definition (9.53) tells

$$p^{i,j}(\lambda, \nu) = \begin{cases} 1 & \text{if } \lambda = \nu, \\ 0 & \text{if } \lambda \ne \nu. \end{cases}$$

When $n = 2m+1$ and $j = \frac{1}{2}(n-1)$ $(=m)$, we use the renormalized Knapp–Stein operator, see (8.21), given by

$$\widetilde{\widetilde{\mathbb{T}}}^m_{\nu,2m-\nu} = \frac{1}{\nu - m} \widetilde{\mathbb{T}}^m_{\nu,2m-\nu}.$$

Proposition 9.38. *Suppose* $(G, G') = (O(n+1,1), O(n,1))$ *with* $n = 2m+1$. *Let* $i = m$ *or* $m+1$. *Then the composition* $\widetilde{\widetilde{\mathbb{T}}}^m_{\nu,2m-\nu} \circ \widetilde{\mathbb{C}}^{i,m}_{\lambda,\nu} : I_\delta(i,\lambda) \to J_\varepsilon(m, 2m-\nu)$ *for* $\nu - \lambda \in \mathbb{N}$ *and* $\delta\varepsilon \in (-1)^{\nu-\lambda}$ *is given as follows.*

(1) *For $\nu - \lambda \in \mathbb{N}_+$,*

$$\widetilde{\widetilde{\mathbb{T}}}^m_{\nu,2m-\nu} \circ \widetilde{\mathbb{C}}^{i,m}_{\lambda,\nu} = \frac{1}{2}q(\nu - \lambda)\widetilde{\mathbb{A}}^{i,m}_{\lambda,2m-\nu,(-1)^{\nu-\lambda}}.$$

In particular, if $m - \lambda \in \mathbb{N}_+$, then

$$\widetilde{\widetilde{\mathbb{T}}}^m_{m,m} \circ \widetilde{\mathbb{C}}^{i,m}_{\lambda,m} = (-1)^{i-m}\frac{\pi^m}{m!}\widetilde{\mathbb{C}}^{i,m}_{\lambda,m}.$$

(2) *For $\nu = \lambda = m$,*

$$\widetilde{\widetilde{\mathbb{T}}}^m_{m,m} \circ \widetilde{\mathbb{C}}^{i,m}_{m,m} = \widetilde{\mathbb{A}}^{i,m}_{m,m,+} + (-1)^{i-m+1}\frac{\pi^m}{m!}\widetilde{\mathbb{C}}^{i,m}_{m,m}.$$

Proof.

(1) The functional equation (9.54) with $n = 2m + 1$ and $j = m$ shows

$$\widetilde{\widetilde{\mathbb{T}}}^m_{\nu,2m-\nu} \circ \widetilde{\mathbb{C}}^{i,m}_{\lambda,\nu} = q(\nu - \lambda)p^{i,m}(\lambda,\nu)\widetilde{\mathbb{A}}^{i,m}_{\lambda,2m-\nu,(-1)^{\nu-\lambda}}.$$

By (9.53), we have for $i \in \{m, m+1\}$ and $\lambda \neq \nu$,

$$p^{i,m}(\lambda,\nu) = \frac{1}{2}(\nu - m).$$

Hence the first equation is verified. For the second statement, we substitute $\nu = m$. Then the second equation follows from the residue formula (9.7) and from the fact $\widetilde{\mathbb{C}}^{i,m}_{\lambda,m} = \mathbb{C}^{i,m}_{\lambda,m}$ when $\lambda \neq m$.

(2) The case $i = m + 1$ will be shown in Lemma 10.25. The case $i = m$ is similar by using

$$\lim_{\nu \to m} \frac{1}{\nu - m}\widetilde{\mathbb{A}}^{m,m}_{\nu,2m-\nu,+} = \widetilde{\mathbb{A}}^{m,m}_{m,m,+} - \frac{\pi^m}{m!}\widetilde{\mathbb{C}}^{m,m}_{m,m}$$

as it will be explained in (10.20) of Chapter 10.   □

9.11.4   Proof of Theorems 9.33 and 9.34

Suppose $0 \leq j \leq n - 1$. Then the principal series representation $J_\delta(j,\nu)$ is reducible as a module of $G' = O(n,1)$ if and only if

$$\nu \in \{j, n-1-j\} \cup (-\mathbb{N}_+) \cup (n + \mathbb{N}), \tag{9.57}$$

see Proposition 2.18 (1). Suppose ν satisfies (9.57). Then the proper submodules of $J_\delta(j,\nu)$ are described as follows:

Case 1. $(n,\nu) \neq (2j+1,j)$, equivalently, $\widetilde{\mathbb{T}}^j_{\nu,n-1-\nu} \neq 0$.
In this case, the unique proper submodule of $J_\delta(j,\nu)$ is given as the kernel of the Knapp–Stein operator $\widetilde{\mathbb{T}}^j_{\nu,n-1-\nu} : J_\delta(j,\nu) \to J_\delta(j,n-1-\nu)$.

Case 2. $(n,\nu) = (2j+1,j)$, equivalently, $\widetilde{\mathbb{T}}^j_{\nu,n-1-\nu} = 0$.
In this case, there are two proper submodules of $J_\delta(j,\nu)$, which are given as the kernel of $\widetilde{\widetilde{\mathbb{T}}}^j_{j,j} \pm \frac{\pi^j}{j!}\mathrm{id} \in \mathrm{End}_{G'}(J_\delta(j,j))$, see Lemma 8.15.

Proof of Theorems 9.33 and 9.34.   Assume $(n,\nu) \neq (2j+1,j)$. This excludes the case where $\mathbb{Z} \ni \lambda \leq j$ from the conditions (ii) $(i=j)$ and (iv) $(i=j+1)$ in Proposition 9.36. In this case Theorems 9.33 and 9.34 are immediate consequences of Proposition 9.36.

Assume now $(n,\nu,j) = (2m+1,m,m)$ for some $m \in \mathbb{N}_+$. Then $\widetilde{\mathbb{C}}^{i,m}_{\lambda,m}$ is not surjective if $\lambda < m$, and is surjective if $\lambda = m$ by Proposition 9.38 (1) and (2), respectively. Thus Theorems 9.33 and 9.34 are proved.   □

Chapter 10
Symmetry Breaking Operators for Irreducible Representations with Infinitesimal Character ρ: Proof of Theorems 4.1 and 4.2

In the first half of this chapter, we give a proof of Theorems 4.1 and 4.2 that determine the dimension of the space of symmetry breaking operators from *irreducible* representations Π of $G = O(n+1,1)$ to *irreducible* representations π of the subgroup $G' = O(n,1)$ when both Π and π have the trivial infinitesimal characters ρ, or equivalently by Theorem 2.20 (2), when

$$\Pi \in \mathrm{Irr}(G)_\rho = \{\Pi_{i,\delta} : 0 \le i \le n+1, \, \delta \in \{\pm\}\},$$

$$\pi \in \mathrm{Irr}(G')_\rho = \{\pi_{j,\varepsilon} : 0 \le j \le n, \, \varepsilon \in \{\pm\}\}.$$

The proofs of Theorems 4.1 and 4.2 are completed in Sections 10.1 and 10.2.4, respectively. In the latter half of this chapter, we give a concrete construction of such symmetry breaking operators from Π to π. We pursue such constructions more than what we need for the proof for Theorems 4.1 and 4.2: some of the results will be used in calculating "periods" in Chapter 12. Our proof uses the symmetry breaking operators for *principal series representations* and their basic properties that we have developed in the previous chapters.

10.1 Proof of the Vanishing Result (Theorem 4.1)

This section gives a proof of the vanishing theorem of symmetry breaking operators (Theorem 4.1). In the same circle of the ideas, we also give a proof of multiplicity-free results (Proposition 10.7). In order to study symmetry breaking for irreducible representations $\Pi_{i,\delta}$ of G, we embed $\mathrm{Hom}_{G'}(\Pi_{i,\delta}|_{G'}, \pi_{j,\varepsilon})$ into the space of symmetry breaking operators between principal series representations as follows:

© Springer Nature Singapore Pte Ltd. 2018
T. Kobayashi, B. Speh, *Symmetry Breaking for Representations of Rank One Orthogonal Groups II*, Lecture Notes in Mathematics 2234,
https://doi.org/10.1007/978-981-13-2901-2_10

Lemma 10.1. *Let $\delta, \varepsilon \in \{\pm\}$. Then we have natural embeddings:*

(1) *for $0 \le i \le n$ and $0 \le j \le n - 1$,*

$$\mathrm{Hom}_{G'}(\Pi_{i,\delta}|_{G'}, \pi_{j,\varepsilon}) \subset \mathrm{Hom}_{G'}(I_\delta(i, n-i)|_{G'}, J_\varepsilon(j, j)); \qquad (10.1)$$

(2) *for $1 \le i \le n + 1$ and $0 \le j \le n - 1$,*

$$\mathrm{Hom}_{G'}(\Pi_{i,\delta}|_{G'}, \pi_{j,\varepsilon}) \subset \mathrm{Hom}_{G'}(I_{-\delta}(i-1, i-1)|_{G'}, J_\varepsilon(j, j)); \qquad (10.2)$$

(3) *for $0 \le i \le n$ and $1 \le j \le n$,*

$$\mathrm{Hom}_{G'}(\Pi_{i,\delta}|_{G'}, \pi_{j,\varepsilon}) \subset \mathrm{Hom}_{G'}(I_\delta(i, n-i)|_{G'}, J_{-\varepsilon}(j-1, n-j)); \qquad (10.3)$$

(4) *for $1 \le i \le n + 1$ and $1 \le j \le n$,*

$$\mathrm{Hom}_{G'}(\Pi_{i,\delta}|_{G'}, \pi_{j,\varepsilon}) \subset \mathrm{Hom}_{G'}(I_{-\delta}(i-1, i-1)|_{G'}, J_{-\varepsilon}(j-1, n-j)). \qquad (10.4)$$

Proof. We recall from Theorem 2.20 (1) that there are surjective G-homomorphisms

$$I \twoheadrightarrow \Pi_{i,\delta} \quad \text{for } I = I_\delta(i, n-i) \text{ or } I_{-\delta}(i-1, i-1)$$

and injective G'-homomorphisms

$$\pi_{j,\varepsilon} \hookrightarrow J \quad \text{for } J = J_\varepsilon(j, j) \text{ or } J_{-\varepsilon}(j-1, n-j).$$

Then the composition $I \twoheadrightarrow \Pi_{i,\delta} \to \pi_{j,\varepsilon} \hookrightarrow J_\varepsilon(j, j)$ yields the embeddings (10.1)–(10.4). $\qquad \square$

Proposition 10.2. *If $j \notin \{i-1, i\}$, then $\mathrm{Hom}_{G'}(\Pi_{i,\delta}|_{G'}, \pi_{j,\varepsilon}) = \{0\}$.*

Proof. Assume $\mathrm{Hom}_{G'}(\Pi_{i,\delta}|_{G'}, \pi_{j,\varepsilon}) \ne \{0\}$.

Suppose first $1 \le i \le n$. By Theorem 3.25 (1), we get $j \in \{i-3, i-2, i-1, i\}$ from (10.2), and $j \in \{i-1, i, i+1, i+2\}$ from (10.3). Hence we conclude that $j \in \{i-1, i\}$.

Suppose next $i = 0$ or $n+1$. By using Theorem 3.25 (1) again, we get $j \in \{0, 1\}$ from (10.1) for $i = 0$, and $j \in \{n-1, n\}$ from (10.2) for $i = n+1$. Since $\dim_{\mathbb{C}} \Pi_{0,\delta} = \dim_{\mathbb{C}} \Pi_{n+1,\delta} = 1$ whereas both $\pi_{1,\varepsilon}$ and $\pi_{n-1,\varepsilon}$ are infinite-dimensional irreducible representations of G' (Theorem 2.20 (4)), we have an obvious vanishing result:

$$\mathrm{Hom}_{G'}(\Pi_{i,\delta}|_{G'}, \pi_{j,\varepsilon}) = \{0\} \quad \text{if } (i, j) = (0, 1) \text{ or } (n+1, n-1).$$

Hence we conclude $j \in \{i-1, i\}$ for $i = 0$ or $n+1$, too. $\qquad \square$

Proposition 10.3. *If $\delta\varepsilon = -$, then $\mathrm{Hom}_{G'}(\Pi_{i,\delta}|_{G'}, \pi_{j,\varepsilon}) = \{0\}$.*

Proof. We have already proved the assertion in the case $j \notin \{i-1, i\}$ in Proposition 10.2. Therefore it suffices to prove the assertion in the case $j = i - 1$ and i. We begin with the case $j = i - 1$.

Suppose $2 \leq i \leq n$. Then by Theorem 3.25 (3),

$$\mathrm{Hom}_{G'}(I_\delta(i, n-i)|_{G'}, J_{-\varepsilon}(i-2, n-i+1)) = \{0\}$$

because $\delta(-\varepsilon) = +$. This implies $\mathrm{Hom}_{G'}(\Pi_{i,\delta}|_{G'}, \pi_{i-1,\varepsilon}) = \{0\}$ from (10.3).

For the case $(i, j) = (1, 0)$, we know from [42, Thm. 2.5 (1-a)] that

$$\mathrm{Hom}_{G'}(\Pi_{1,-}|_{G'}, \pi_{0,+}) = \{0\}.$$

($F(0) = \pi_{0,+}$ and $T(0) = \Pi_{1,-}$ with the notation therein.) It then follows from Proposition 3.39 that $\mathrm{Hom}_{G'}(\Pi_{1,+}|_{G'}, \pi_{0,-}) = \{0\}$.

For the case $(i, j) = (n+1, n)$, we use the fact that both $\Pi_{i,\delta}$ and $\pi_{j,\varepsilon}$ are one-dimensional. In fact, we have isomorphisms $\Pi_{n+1,\delta} \simeq \chi_{-,\delta}$ and $\pi_{n,\varepsilon} \simeq \chi_{-,\varepsilon}|_{G'}$ by Theorem 2.20 (4). Thus the vanishing assertion is straightforward for $j = i - 1$ ($1 \leq i \leq n+1$).

The case $j = i$ is derived from the case $j = i - 1$ by duality (see Proposition 3.39). $\qquad\square$

By Propositions 10.2 and 10.3, we have completed the proof of Theorem 4.1.

10.2 Construction of Symmetry Breaking Operators from $\Pi_{i,\delta}$ to $\pi_{i,\delta}$: Proof of Theorem 4.2

In this section we prove the existence and the uniqueness (up to scalar multiplication) of symmetry breaking operators from the irreducible G-module $\Pi_{i,\delta}$ to the irreducible G'-module $\pi_{j,\varepsilon}$ when $j \in \{i-1, i\}$ and $\delta\varepsilon = +$, and thus complete the proof of Theorem 4.2. Moreover, we investigate their (K, K')-spectrum for minimal K- and K'-types, and also give an explicit construction of such operators.

10.2.1 Generators of Symmetry Breaking Operators Between Principal Series Representations Having the Trivial Infinitesimal Character ρ

We have determined explicit generators of symmetry breaking operators $I_\delta(i, \lambda) \to J_\varepsilon(j, v)$ in Theorem 3.26. In this subsection, we extract some special cases which will be used for the proof of Theorem 4.2.

The following lemma is used for the proof of the multiplicity-free theorem (Proposition 10.7 below), and also for an explicit construction of nonzero symmetry breaking operators $\Pi \to \pi$ with $\Pi \in \mathrm{Irr}(G)_\rho$ and $\pi \in \mathrm{Irr}(G')_\rho$ (Proposition 10.13).

Lemma 10.4.

(1) *Suppose* $0 \leq i \leq n - 1$ *and* $\delta\varepsilon = +$. *Then*

$$\mathrm{Hom}_{G'}(I_\delta(i, n - i)|_{G'}, J_\delta(i, i)) \simeq \begin{cases} \mathbb{C}\widetilde{\mathrm{A}}^{i,i}_{n-i,i,+} & \text{if } 2i \neq n, \\ \mathbb{C}\widetilde{\mathrm{A}}^{i,i}_{n-i,i,+} \oplus \mathbb{C}\widetilde{\mathbb{C}}^{i,i}_{n-i,i} & \text{if } 2i = n. \end{cases}$$

(2) *Suppose* $1 \leq i \leq n - 1$ *and* $\delta\varepsilon = +$. *Then*

$$\mathrm{Hom}_{G'}(I_{-\delta}(i - 1, i - 1)|_{G'}, J_\varepsilon(i, i)) = \mathbb{C}\widetilde{\mathbb{C}}^{i-1,i}_{i-1,i}.$$

(3) *Suppose* $0 \leq i \leq n - 1$ *and* $\delta \in \{\pm\}$. *Then we have*

$$\mathrm{Hom}_{G'}(I_\delta(i, i)|_{G'}, J_\delta(i, i)) = \mathbb{C}\widetilde{\mathrm{A}}^{i,i}_{i,i,+} \oplus \mathbb{C}\widetilde{\mathbb{C}}^{i,i}_{i,i}.$$

(4) *Suppose* $1 \leq i \leq n$ *and* $\delta\varepsilon = +$. *Then*

$$\mathrm{Hom}_{G'}(I_{-\delta}(i - 1, i - 1)|_{G'}, J_{-\varepsilon}(i - 1, n - i))$$

$$\simeq \begin{cases} \mathbb{C}\widetilde{\mathrm{A}}^{i-1,i-1}_{i-1,n-i,+} & \text{if } n \neq 2i - 1, \\ \mathbb{C}\widetilde{\mathrm{A}}^{i-1,i-1}_{i-1,n-i,+} \oplus \mathbb{C}\widetilde{\mathbb{C}}^{i-1,i-1}_{i-1,n-i} & \text{if } n = 2i - 1. \end{cases}$$

Proof. We determined the dimension of the left-hand side by Theorem 3.25 (2) and (3). Then the lemma follows from Theorem 3.26 for (1), (3), (4); and from Fact 3.23 for (2). □

Remark 10.5. In the N-picture where the open Bruhat cells for the pair of the real flag manifolds $G'/P' \subset G/P$ are represented by $\mathbb{R}^{n-1} \subset \mathbb{R}^n$, we have $\widetilde{\mathbb{C}}^{i,i}_{n-i,i} = \mathrm{Rest}_{x_n=0}$ in Lemma 10.4 (1), $\widetilde{\mathbb{C}}^{i-1,i}_{i-1,i} = \mathrm{Rest}_{x_n=0} \circ d_{\mathbb{R}^n}$ in (2), $\widetilde{\mathbb{C}}^{i,i}_{i,i} = \mathrm{Rest}_{x_n=0}$ in (3), and $\widetilde{\mathbb{C}}^{i-1,i-1}_{i-1,i-1} = \mathrm{Rest}_{x_n=0}$ in (4).

The following lemma is used for an alternative construction (see Proposition 10.13 below) of symmetry breaking operators $\Pi_{i,\delta} \to \pi_{i,\delta}$.

Lemma 10.6. *Suppose* $1 \leq i \leq n$ *and* $\delta \in \{\pm\}$. *Then we have*

$$\mathrm{Hom}_{G'}(I_\delta(i, i)|_{G'}, J_{-\delta}(i - 1, n - i)) = \mathbb{C}\widetilde{\mathrm{A}}^{i,i-1}_{i,n-i,-}.$$

Proof. By Theorem 3.25 (2), $\widetilde{\mathrm{A}}^{i,i-1}_{i,n-i,-} \neq 0$, and therefore the lemma follows from Theorem 3.26. □

10.2.2 Multiplicity-free Property of Symmetry Breaking

In this subsection, we prove the following multiplicity-free property:

Proposition 10.7. *For any* $0 \le i \le n+1$, $0 \le j \le n$, *and* $\delta, \varepsilon \in \{\pm\}$, *we have*

$$\dim_{\mathbb{C}} \mathrm{Hom}_{G'}(\Pi_{i,\delta}|_{G'}, \pi_{j,\varepsilon}) \le 1. \tag{10.5}$$

Proposition 10.7 is a very special case of the multiplicity-free theorem which was proved in Sun–Zhu [55], however, we give a different proof based on Lemmas 10.1 and 10.4 because the following short proof illustrates the idea of this chapter.

Proof of Proposition 10.7. Owing to the vanishing results (Theorem 4.1), it suffices to show (10.5) when $j \in \{i - 1, i\}$ and $\delta\varepsilon = +$. Moreover, the case $j = i - 1$ can be reduced to the case $j = i$ by the duality between the spaces of symmetry breaking operators (Proposition 3.39). Henceforth, we assume $j = i \in \{0, 1, \ldots, n\}$ and $\delta\varepsilon = +$. Then, owing to the embedding results given in Lemma 10.1, the multiplicity-free property (10.5) holds for $1 \le i \le n - 1$ by Lemma 10.4 (2), and for $i = 0$ and n by Lemma 10.4 (1) and (4). Thus Proposition 10.7 is proved. □

10.2.3 Multiplicity-one Property: Proof of Theorem 4.2

In proving Theorem 4.2, we use the following proposition, whose proof is deferred at the next subsection.

Proposition 10.8. $\mathrm{Hom}_{G'}(\Pi_{i,\delta}|_{G'}, \pi_{i,\delta}) \ne \{0\}$ *for all* $0 \le i \le n$ *and* $\delta \in \{\pm\}$.

Remark 10.9. Obviously Proposition 10.8 holds for $i = 0$ because $\Pi_{0,\delta}|_{G'} \simeq \pi_{0,\delta}$ as G'-modules for $\delta \in \{\pm\}$. Indeed, the G-modules $\Pi_{0,+}$ and $\Pi_{0,-}$ are the one-dimensional representations $\mathbf{1}$ and respectively χ_{+-} (Theorem 2.20 (4)), and likewise for the G'-modules $\pi_{0,\pm}$.

Before giving a proof of Proposition 10.8, we show that Proposition 10.8 implies Theorem 4.2.

Proof of Theorem 4.2. By the duality among the spaces of symmetry breaking operators (Proposition 3.39), we may and do assume $j = i$ and $\delta = \varepsilon = +$ because $\widetilde{j} := n - j$ and $\widetilde{i} := n + 1 - i$ satisfy $\widetilde{j} = \widetilde{i} - 1$ if and only if $j = i$. Then Theorem 4.2 follows from Propositions 10.7 (uniqueness) and 10.8 (existence). □

For later purpose, we need a refinement of Proposition 10.8 by providing information of (K, K')-spectrum in Proposition 10.12 below. For this, we fix some terminology:

Definition 10.10 (minimal K-type). We set $m := [\frac{n+1}{2}]$. Suppose $\mu \in \widehat{K}$. To describe an irreducible finite-dimensional representation μ of $K = O(n+1) \times$

$O(1)$, we use the notation in Section 14.1 in Appendix I rather than the previous one in Section 2.2.1, and write

$$\mu = F^{O(n+1)}(\sigma_1, \cdots, \sigma_m)_\varepsilon \boxtimes \delta$$

for $\sigma = (\sigma_1, \cdots, \sigma_m) \in \Lambda^+(m)$ and $\varepsilon, \delta \in \{\pm\}$. We define $\|\mu\| > 0$ by

$$\|\mu\|^2 = \sum_{j=1}^m (\sigma_j + n + 1 - 2j)^2 \quad (= \|\sigma + 2\rho_c\|^2),$$

where $2\rho_c = (n-1, n-3, \cdots, n+1-2m)$ is the sum of positive roots for $\mathfrak{k}_{\mathbb{C}} = \mathfrak{o}(n+1, \mathbb{C})$ in the standard coordinates. For a nonzero admissible representation Π of G, the set of *minimal K-types* of Π is

$$\{\mu \in \widehat{K} : \mu \text{ occurs in } \Pi, \text{ and } \|\mu\| \text{ is minimal with this property}\},$$

see [24, Chap. 2] or [59, Def. 5.4.18].

We then observe:

Remark 10.11 (minimal K-type).   The basic K-type (see Definition 2.17) of the principal series representation $I_\delta(i, \lambda)$ is the unique minimal K-type of the irreducible G-module $\Pi_{i,\delta}$, as stated in Theorem 2.20 (3).

Proposition 10.12. *Let* $(G, G') = (O(n+1, 1), O(n, 1))$, $0 \le i \le n$ *and* $\delta \in \{\pm\}$. *Then there exists a nonzero symmetry breaking operator*

$$A_{i,i} \colon \Pi_{i,\delta} \to \pi_{i,\delta} \tag{10.6}$$

such that its (K, K')-*spectrum for the minimal* K'- *and* K-*types* $\mu^\flat(i, \delta)' (\hookrightarrow \mu^\flat(i, \delta))$ *is nonzero.*

Proposition 10.12 is an *existence* theorem, however, we shall prove it by *constructing* nonzero symmetry breaking operators $\Pi_{i,\delta} \to \pi_{i,\delta}$, see Proposition 10.13 in the next subsection. Alternative constructions are also given in Sections 10.2.5 and 10.2.6, and thus we construct symmetry breaking operators $\Pi_{i,\delta} \to \pi_{i,\delta}$ in the following three ways:

- $\widetilde{\mathbb{A}}^{i,i-1}_{i,n-i,-} \colon I_\delta(i, i) \to J_{-\delta}(i-1, n-i)$, (Proposition 10.13),

- $\widetilde{\mathbb{A}}^{i,i}_{i,i,+} \colon I_\delta(i, i) \to J_\delta(i, i)$,   (Proposition 10.15),

- $\widetilde{\mathbb{A}}^{i,i}_{n-i,i,+} \colon I_\delta(i, n-i) \to J_\delta(i, i)$,   (Proposition 10.16).

10.2.4   First Construction $\Pi_{i,\delta} \to \pi_{i,\delta}$ $(1 \leq i \leq n)$

In this subsection, we construct a nonzero symmetry breaking operator

$$\Pi_{i,\delta} \to \pi_{i,\delta} \qquad \text{for } 1 \leq i \leq n, \ \delta \in \{\pm\},$$

by using Lemma 10.6.

Proposition 10.13. *Suppose $1 \leq i \leq n$ and $\delta \in \{\pm\}$. Then the normalized symmetry breaking operator*

$$\widetilde{\mathbb{A}}^{i,i-1}_{i,n-i,-} : I_\delta(i,i) \to J_{-\delta}(i-1, n-i)$$

satisfies the following:

(1) Image$(\widetilde{\mathbb{A}}^{i,i-1}_{i,n-i,-})_{K'} = (\pi_{i,\delta})_{K'}$ *as $(\mathfrak{g}', K')$-modules;*

(2) $\widetilde{\mathbb{A}}^{i,i-1}_{i,n-i,-}|_{\Pi_{i,\delta}} \neq 0$.

In particular, it induces a symmetry breaking operator $\Pi_{i,\delta} \to \pi_{i,\delta}$ as in the diagram below. Moreover, the (K, K')-spectrum of the resulting operator for the minimal K'- and K-types $\mu^\flat(i,\delta)'$ $(\hookrightarrow \mu^\flat(i,\delta))$ is nonzero.

$$
\begin{array}{ccc}
I_\delta(i,i) & \xrightarrow{\ \widetilde{\mathbb{A}}^{i,i-1}_{i,n-i,-}\ } & J_{-\delta}(i-1, n-i) \\
\cup & & \cup \\
\Pi_{i,\delta} & \dashrightarrow & \pi_{i,\delta}
\end{array}
$$

Convention 10.14. *Hereafter, by abuse of notation, we shall write simply as* Image$(\widetilde{\mathbb{A}}^{i,i-1}_{i,n-i,-}) = \pi_{i,\delta}$ *if their underlying $(\mathfrak{g}', K')$-modules coincide (cf. Proposition 10.13 (1)).*

Proof of Proposition 10.13.

(1) First we observe

$$\text{Image}(\widetilde{\mathbb{A}}^{i,i-1}_{i,n-i,-}) \subset \text{Ker}(\widetilde{\mathbb{T}}^{i-1}_{n-i,i-1})$$

because Theorem 9.24 with $\nu = n - i$ tells the functional equation $\widetilde{\mathbb{T}}^{i-1}_{n-i,i-1} \circ \widetilde{\mathbb{A}}^{i,i-1}_{i,n-i,-} = 0$.

When $n \neq 2i - 1$, we conclude Image$(\widetilde{\mathbb{A}}^{i,i-1}_{i,n-i,-}) = \pi_{i,\delta}$ by Proposition 8.11 because $\pi_{i,\delta}$ is irreducible as a G'-module. When $n = 2i - 1$, the Knapp–Stein operator $\widetilde{\mathbb{T}}^{i-1}_{n-i,i-1}$ vanishes (Proposition 8.12). Instead we use the following renormalized Knapp–Stein operator (see (8.21)):

$$\widetilde{\widetilde{\mathbb{T}}}^{i-1}_{\nu,n-1-\nu} = \frac{1}{\nu-i+1}\widetilde{\mathbb{T}}^{i-1}_{\nu,n-1-\nu}.$$

Then the functional equation given in Theorem 9.24 implies

$$\left(\widetilde{\widetilde{\mathbb{T}}}^{i-1}_{i-1,i-1} + \frac{\pi^{i-1}}{(i-1)!}\,\mathrm{id}\right)\circ\widetilde{\mathbb{A}}^{i,i-1}_{\lambda,i-1,-} = 0.$$

By Lemma 8.15 applied to the subgroup $G' = O(n,1)$ $(= O(2i-1,1))$, we conclude $\mathrm{Image}(\widetilde{\mathbb{A}}^{i,i-1}_{\lambda,n-i,-}) = \mathrm{Image}(\widetilde{\mathbb{A}}^{i,i-1}_{\lambda,i-1,-}) = \pi_{i,\delta}$ in the case $n = 2i-1$, too.

(2) The second statement follows from the fact that the (K,K')-spectrum of $\widetilde{\mathbb{A}}^{i,i-1}_{\lambda,\nu,-}$ (Theorem 9.8) for the basic K-types $(\mu,\mu') = (\mu^\flat(i,\delta),\mu^\sharp(i-1,-\delta)')$ does not vanish. The last assertion is derived from the following observation (see (2.32)): there are isomorphisms of representations of $K' = O(n)\times O(1)$,

$$\mu^\sharp(i-1,-\delta)' \simeq \mu^\flat(i,\delta)'.$$

Hence Proposition 10.13 is proved.   □

Proof of Proposition 10.12. Clear from Proposition 10.13 and Remark 10.9.   □

Thus, the proof of Theorem 4.2 has been completed.

For the rest of this chapter, we give alternative constructions of symmetry breaking operators for later purposes.

10.2.5 Second Construction $\Pi_{i,\delta} \to \pi_{i,\delta}$ $(0 \le i \le n-1)$

In this subsection, we provide another construction of a nonzero symmetry breaking operator

$$\Pi_{i,\delta} \to \pi_{i,\delta} \quad \text{for } 0 \le i \le n-1,\ \delta \in \{\pm\},$$

by using Lemma 10.4 (3).

Proposition 10.15. *Suppose* $0 \le i \le n-1$ *and* $\delta \in \{\pm\}$. *Then the renormalized operator*

$$\widetilde{\mathbb{A}}^{i,i}_{i,i,+} : I_\delta(i,i) \to J_\delta(i,i)$$

satisfies the following:

(1) Image($\tilde{\mathbb{A}}^{i,i}_{i,i,+}$) $= \pi_{i,\delta}$;

(2) $\tilde{\mathbb{A}}^{i,i}_{i,i,+}|_{\Pi_{i,\delta}} \neq 0$.

In particular, it induces a symmetry breaking operator $\Pi_{i,\delta} \to \pi_{i,\delta}$ as in the diagram below. Moreover, the (K, K')-spectrum of the resulting operator for the minimal K'- and K-types $\mu^{\flat}(i,\delta)'$ ($\hookrightarrow \mu^{\flat}(i,\delta)$) is nonzero.

$$
\begin{array}{ccc}
I_\delta(i,i) & \xrightarrow{\ \ \tilde{\mathbb{A}}^{i,i}_{i,i,+}\ \ } & J_\delta(i,i) \\[4pt]
\cup & & \cup \\[4pt]
\Pi_{i,\delta} & \dashrightarrow & \pi_{i,\delta}
\end{array}
$$

Proof of Proposition 10.15.

(1) By the functional equation (9.40), we have

$$
\mathrm{Image}(\tilde{\mathbb{A}}^{i,i}_{\lambda,i,+}) \subset \mathrm{Ker}(\tilde{\mathbb{T}}^{i}_{i,n-1-i}).
$$

When $n \neq 2i + 1$, we conclude $\mathrm{Image}(\tilde{\mathbb{A}}^{i,i}_{\lambda,i,+}) = \pi_{i,\delta}$ by Proposition 8.11.

When $n = 2i + 1$, the Knapp–Stein operator $\tilde{\mathbb{T}}^{i}_{i,n-1-i} \equiv \tilde{\mathbb{T}}^{i}_{i,i}$ vanishes (Proposition 8.12). Instead we use the functional equation (9.49) for the renormalized operators $\tilde{\mathbb{T}}^{i}_{i,i}$ and $\tilde{\mathbb{A}}^{i,i}_{\lambda,i,+}$, which tells that

$$
\mathrm{Image}(\tilde{\mathbb{A}}^{i,i}_{\lambda,i,+}) \subset \mathrm{Ker}\left(\tilde{\mathbb{T}}^{i}_{i,i} - \frac{\pi^i}{\Gamma(i+1)}\,\mathrm{id} \right).
$$

By Lemma 8.15, we conclude $\mathrm{Image}(\tilde{\mathbb{A}}^{i,i}_{\lambda,i,+}) = \pi_{i,\delta}$ because $\tilde{\mathbb{A}}^{i,i}_{\lambda,i,+}$ is nonzero and $\pi_{i,\delta}$ is irreducible as a G'-module.

(2) The assertion follows readily from the (K, K')-spectrum of the renormalized operator $\tilde{\mathbb{A}}^{i,i}_{\lambda,i,+}$ (see (9.42)) for the basic K- and K'-types $(\mu, \mu') = (\mu^{\flat}(i,\delta), \mu^{\flat}(i,\delta)')$. $\square$

10.2.6   Third Construction $\Pi_{i,\delta} \to \pi_{i,\delta}$

We give yet another construction of a nonzero symmetry breaking operator $\Pi_{i,\delta} \to \pi_{i,\delta}$ in the case $n \neq 2i$. In the case $n = 2i$, the normalized operator $\tilde{\mathbb{A}}^{i,i}_{n-i,i,+}$ vanishes. We shall discuss this case separately in Section 10.3.1, see Proposition 10.19.

Proposition 10.16. *If* $2i \neq n$, *then* $\widetilde{\mathbb{A}}^{i,i}_{n-i,i,+} \in \mathrm{Hom}_{G'}(I_\delta(i, n-i)|_{G'}, J_\delta(i,i))$ *satisfies*

$$\widetilde{\mathbb{A}}^{i,i}_{n-i,i,+}|_{\Pi_{i+1,-\delta}} \equiv 0 \quad and \quad \mathrm{Image}(\widetilde{\mathbb{A}}^{i,i}_{n-i,i,+}) = \pi_{i,\delta}.$$

Thus it induces a symmetry breaking operator $\Pi_{i,\delta} \to \pi_{i,\delta}$ *as in the diagram below. Moreover, the* (K, K')-*spectrum of the resulting operator for the minimal* K'- *and* K-*types* $\mu^\flat(i, \delta)'$ $(\hookrightarrow \mu^\flat(i, \delta))$ *is nonzero.*

$$
\begin{array}{ccc}
I_\delta(i, n-i) & \xrightarrow{\;\;\widetilde{\mathbb{A}}^{i,i}_{n-i,i,+}\;\;} & J_\delta(i, i) \\[4pt]
\downarrow & & \cup \\[4pt]
\Pi_{i,\delta} \simeq I_\delta(i, n-i)/\Pi_{i+1,-\delta} & \dashrightarrow & \pi_{i,\delta}
\end{array}
$$

Proof. Since $\widetilde{\mathbb{A}}^{i,i}_{i,i,+} = 0$ by Theorem 3.19 (1), the composition $\widetilde{\mathbb{A}}^{i,i}_{n-i,i,+} \circ \widetilde{\mathbb{T}}^i_{i,n-i}$ vanishes by the functional equation (Theorem 9.25). Thus $\widetilde{\mathbb{A}}^{i,i}_{n-i,i,+}$ is identically zero on $\mathrm{Image}(\widetilde{\mathbb{T}}^i_{i,n-i}) \simeq \Pi_{i+1,-\delta}$ (see Proposition 8.11).

For the second assertion, we use another functional equation (Theorem 9.24) to get $\widetilde{\mathbb{T}}^i_{i,n-1-i} \circ \widetilde{\mathbb{A}}^{i,i}_{n-i,i,+} = 0$. Hence

$$\mathrm{Image}(\widetilde{\mathbb{A}}^{i,i}_{n-i,i,+}) \subset \mathrm{Ker}(\widetilde{\mathbb{T}}^i_{i,n-1-i}) \simeq \pi_{i,\delta}$$

by Proposition 8.11. Since $\widetilde{\mathbb{A}}^{i,i}_{n-i,i,+} \neq 0$ (see Theorem 3.19 (1)) and since $\pi_{i,\delta}$ is irreducible, the underlying $(\mathfrak{g}', K')$-modules of $\mathrm{Image}(\widetilde{\mathbb{A}}^{i,i}_{n-i,i,+})$ and $\pi_{i,\delta}$ coincide. $\square$

10.3 Splitting of $I_\delta(m, m)$ and Its Symmetry Breaking for $(G, G') = (O(2m+1, 1), O(2m, 1))$

Suppose n is even, say $n = 2m$. A distinguished feature in this setting is that the principal series representation $I_\delta(m, \lambda)$ of $G = O(2m+1, 1)$ splits into the direct sum of two irreducible G-modules when $\lambda = m$: for $\delta \in \{\pm\}$,

$$I_\delta(m, m) \simeq \Pi_{m,\delta} \oplus \Pi_{m+1,-\delta}, \tag{10.7}$$

both of which are smooth irreducible tempered representations of G, see Theorem 2.20 (1) and (8). Accordingly, the space of symmetry breaking operators has a direct sum decomposition:

$$\mathrm{Hom}_{G'}(I_\delta(m,m)|_{G'}, J_\varepsilon(m,m))$$

$$\simeq \mathrm{Hom}_{G'}(\Pi_{m,\delta}|_{G'}, J_\varepsilon(m,m)) \oplus \mathrm{Hom}_{G'}(\Pi_{m+1,-\delta}|_{G'}, J_\varepsilon(m,m)), \qquad (10.8)$$

for each $\varepsilon \in \{\pm\}$. The left-hand side of (10.8) has been understood by the classification of symmetry breaking operators given in Theorem 3.26 (see (10.11) as below). On the other hand, the target space $J_\varepsilon(m,m)$ is not irreducible as a G'-module. We recall from Theorem 2.20 (1) that the principal series representation $J_\varepsilon(m,\nu)$ of $G' = O(2m,1)$ at $\nu = m$ has a nonsplitting exact sequence of G'-modules:

$$0 \to \pi_{m,\varepsilon} \to J_\varepsilon(m,m) \to \pi_{m+1,-\varepsilon} \to 0. \qquad (10.9)$$

With this in mind, we shall take a closer look at the right-hand side of (10.8) and determine each summand as follows:

	$\delta\varepsilon = +$	$\delta\varepsilon = -$	
$\mathrm{Hom}_{G'}(\Pi_{m,\delta}|_{G'}, J_\varepsilon(m,m))$	$\mathbb{C}$	$\{0\}$	(10.10)
$\mathrm{Hom}_{G'}(\Pi_{m+1,-\delta}|_{G'}, J_\varepsilon(m,m))$	$\mathbb{C}$	$\mathbb{C}$	

See Section 10.3.1 for the left column of (10.10) in detail, and for Section 10.3.2 for the right column.

10.3.1 $\mathrm{Hom}_{G'}(I_\delta(m,m)|_{G'}, J_\varepsilon(m,m))$ with $\delta\varepsilon = +$

We begin with the case $\delta\varepsilon = +$. Without loss of generality, we may and do assume $\delta = \varepsilon = +$.

Then Lemma 10.4 (3) with Remark 10.5 tells that

$$\mathrm{Hom}_{G'}(I_+(m,m)|_{G'}, J_+(m,m)) = \mathbb{C}\tilde{\mathbb{A}}_{m,m,+}^{m,m} \oplus \mathbb{C}\mathrm{Rest}. \qquad (10.11)$$

The first generator $\tilde{\mathbb{A}}_{m,m,+}^{m,m}$ is defined as the renormalization (Theorem 5.45)

$$\tilde{\mathbb{A}}_{m,m,+}^{m,m} = \lim_{\lambda \to m} \Gamma\left(\frac{\lambda - m}{2}\right) \tilde{\mathbb{A}}_{\lambda,m,+}^{m,m} \qquad (10.12)$$

of the normalized regular symmetry breaking operator $\tilde{\mathbb{A}}_{\lambda,m,+}^{m,m}$ which vanishes at $\lambda = m$ (Theorem 3.19). The second generator, $\mathrm{Rest} \equiv \mathrm{Rest}_{x_n=0}$, is the obvious symmetry breaking operator (cf. Lemma 9.32), given by $\mathrm{Rest}_{x_n=0}$ in the N-picture. By using the second generator, we obtain the following.

Proposition 10.17. *Let* $(G, G') = (O(2m + 1, 1), O(2m, 1))$. *Then we have*

$$\mathrm{Hom}_{G'}(\Pi_{m,+}|_{G'}, J_+(m, m)) = \mathbb{C}\mathrm{Rest}|_{\Pi_{m,+}},$$

$$\mathrm{Hom}_{G'}(\Pi_{m+1,-}|_{G'}, J_+(m, m)) = \mathbb{C}\mathrm{Rest}|_{\Pi_{m+1,-}}.$$

Proof. By the direct sum decompositions (10.11) and (10.7), we have

$$2 = \dim_{\mathbb{C}} \mathrm{Hom}_{G'}(I_+(m, m)|_{G'}, J_+(m, m))$$

$$= \dim_{\mathbb{C}} \mathrm{Hom}_{G'}(\Pi_{m,+}|_{G'}, J_+(m, m)) + \dim_{\mathbb{C}} \mathrm{Hom}_{G'}(\Pi_{m+1,-}|_{G'}, J_+(m, m)).$$

On the other hand, we know from Lemma 9.32 that $\mathrm{Rest}|_{\Pi_{m,+}} \neq 0$ and $\mathrm{Rest}|_{\Pi_{m+1,-}} \neq 0$. Hence we have proved the proposition. □

We have not used the other generator $\tilde{\mathbb{A}}_{m,m,+}^{m,m}$ in (10.11) for the previous proposition. For the sake of completeness, we investigate its restriction to each of the irreducible components in (10.7).

Proposition 10.18. *Retain the notation as in* (10.11).

$$\tilde{\mathbb{A}}_{m,m,+}^{m,m}|_{\Pi_{m+1,-}} \equiv 0.$$

$$\tilde{\mathbb{A}}_{m,m,+}^{m,m}|_{\Pi_{m,+}} = \frac{2\pi^{m-\frac{1}{2}}}{m!}\mathrm{Rest}|_{\Pi_{m,+}}.$$

We also determine the image of the nonzero symmetry breaking operators $\tilde{\mathbb{A}}_{m,m,+}^{m,m}$ and Rest on each irreducible summand in (10.7).

Proposition 10.19. *With Convention 10.14, we have*

$$\mathrm{Image}(\tilde{\mathbb{A}}_{m,m,+}^{m,m}|_{\Pi_{m,+}}) = \mathrm{Image}(\mathrm{Rest}|_{\Pi_{m,+}}) = \pi_{m,+},$$

$$\mathrm{Image}(\mathrm{Rest}|_{\Pi_{m+1,-}}) = J_+(m, m).$$

For the proof of Propositions 10.18 and 10.19, we use Lemma 9.28 about functional equations with appropriate renormalizations. We set

$$c(m) := \frac{\pi^m}{m!}. \tag{10.13}$$

Proof of Proposition 10.18. It follows from the functional equation (9.43) for the renormalized Knapp–Stein operator $\tilde{\mathbb{T}}_{m,m}^m$ that

$$\tilde{\mathbb{A}}_{m,m,+}^{m,m} \circ (c(m)\,\mathrm{id} - \tilde{\mathbb{T}}_{m,m}^m) = 0.$$

On the other hand, Lemma 8.15 implies that the renormalized Knapp–Stein operator satisfies

$$c(m)\,\mathrm{id} - \widetilde{\widetilde{\mathbb{T}}}{}^m_{m,m} = 0\,\mathrm{id}_{\Pi_{m,+}} \oplus 2\,\mathrm{id}_{\Pi_{m+1,-}},$$

which implies $\mathrm{Image}(c(m)\,\mathrm{id} - \widetilde{\widetilde{\mathbb{T}}}{}^m_{m,m}) = \Pi_{m+1,-}$. Therefore, $\widetilde{\widetilde{\mathbb{A}}}{}^{m,m}_{m,m,+}$ is identically zero on the irreducible G-submodule $\Pi_{m+1,-}$.

To see the second statement, we use Proposition 10.17, which shows that $\widetilde{\widetilde{\mathbb{A}}}{}^{m,m}_{m,m,+}|_{\Pi_{m,+}}$ must be proportional to $\mathrm{Rest}|_{\Pi_{m,+}}$. Comparing the (K, K')-spectrum of the two operators $\widetilde{\widetilde{\mathbb{A}}}{}^{m,m}_{m,m,+}$ and Rest with respect to basic K'- and K-types $\mu^\flat(m, +)' \hookrightarrow \mu^\flat(m, +)$ (see the formula (9.42) for $\widetilde{\widetilde{\mathbb{A}}}{}^{m,m}_{m,m,+}$ and Lemma 9.32 for Rest), we get the second statement. $\qquad\square$

Proof of Proposition 10.19. By the functional equation (9.40),

$$\mathrm{Image}(\widetilde{\widetilde{\mathbb{A}}}{}^{m,m}_{m,m,+}|_{\Pi_{m,+}}) \subset \mathrm{Ker}(\widetilde{\mathbb{T}}{}^m_{m,m-1}) = \pi_{m,+}.$$

Since $\widetilde{\widetilde{\mathbb{A}}}{}^{m,m}_{m,m,+}|_{\Pi_{m,+}}$ is nonzero, and since $\pi_{m,+}$ is an irreducible G'-module, we get the first statement. For the second one, we compare the (K, K')-spectrum of $\widetilde{\widetilde{\mathbb{A}}}{}^{m,m}_{m,m,+}$ (see (9.42)) and that of Rest (see (9.51)) in Lemma 9.32. $\qquad\square$

10.3.2 $\mathrm{Hom}_{G'}(I_\delta(m,m)|_{G'}, J_\varepsilon(m,m))$ with $\delta\varepsilon = -$

The case $\delta\varepsilon = -$ is much simpler because the space of symmetry breaking operators is one-dimensional:

$$\mathrm{Hom}_{G'}(I_\delta(m,m)|_{G'}, J_\varepsilon(m,m)) = \mathbb{C}\widetilde{\mathbb{A}}{}^{m,m}_{m,m,-},$$

see Theorem 3.26. Without loss of generality, we may and do assume $(\delta, \varepsilon) = (+, -)$. The restriction of the generator $\widetilde{\mathbb{A}}{}^{m,m}_{m,m,-}$ to each irreducible component in (10.7) is given as follows.

Proposition 10.20. *Let $(G, G') = (O(2m+1, 1), O(2m, 1))$. Then we have*

$$\widetilde{\mathbb{A}}{}^{m,m}_{m,m,-}|_{\Pi_{m,+}} \equiv 0.$$

$$\mathrm{Image}(\widetilde{\mathbb{A}}{}^{m,m}_{m,m,-}|_{\Pi_{m+1,-}}) = \pi_{m,-}.$$

The proof of Proposition 10.20 relies on the functional equations given in Lemma 9.29.

Proof. The functional equation (9.45) implies

$$\widetilde{\mathbb{A}}_{m,m,-}^{m,m} \circ (\widetilde{\widetilde{\mathbb{T}}}_{m,m}^{m} + c(m)\mathrm{id}) = 0.$$

By Lemma 8.15, $\mathrm{Image}(\widetilde{\widetilde{\mathbb{T}}}_{m,m}^{m} + c(m)\mathrm{id}) = \Pi_{m,+}$. Hence the first statement is proved.

The second statement follows from the functional equation (9.46) and from the fact that $\mathrm{Ker}\,(\widetilde{\mathbb{T}}_{m,m-1}^{m} : J_-(m,m) \to J_-(m,m-1)) = \pi_{m,-}$ (see Proposition 8.11).

$\square$

10.4 Splitting of $J_\varepsilon(m,m)$ and Symmetry Breaking Operators for $(G, G') = (O(2m+2,1), O(2m+1,1))$

Suppose n is odd, say $n = 2m+1$. In contrast to the n even case treated in Section 10.3, a distinguished feature in this setting is that the principal series representation $J_\varepsilon(m,v)$ of the subgroup $G' = O(2m+1,1)$ splits into the direct sum of two irreducible tempered representations when $v = m$: for $\varepsilon \in \{\pm\}$,

$$J_\varepsilon(m,m) \simeq \pi_{m,\varepsilon} \oplus \pi_{m+1,-\varepsilon}, \tag{10.14}$$

see Theorem 2.20 (1) and (8). Accordingly, the space of symmetry breaking operators has a direct sum decomposition:

$\mathrm{Hom}_{G'}(I_\delta(i,\lambda)|_{G'}, J_\varepsilon(m,m))$

$\simeq \mathrm{Hom}_{G'}(I_\delta(i,\lambda)|_{G'}, \pi_{m,\varepsilon}) \oplus \mathrm{Hom}_{G'}(I_\delta(i,\lambda)|_{G'}, \pi_{m+1,-\varepsilon}) \tag{10.15}$

for any $\lambda \in \mathbb{C}$. The left-hand side of (10.15) is understood via explicit generators given in Theorem 3.26 (classification). In this section, we examine the following two cases:

$$\mathrm{Hom}_{G'}(I_\delta(m+1,m)|_{G'}, J_\delta(m,m)) = \mathbb{C}\widetilde{\mathbb{A}}_{m,m,+}^{m+1,m} \oplus \mathbb{C}\widetilde{\mathbb{C}}_{m,m}^{m+1,m}, \tag{10.16}$$

$$\mathrm{Hom}_{G'}(I_{-\varepsilon}(m,m)|_{G'}, J_\varepsilon(m,m)) = \mathbb{C}\widetilde{\mathbb{A}}_{m,m,-}^{m,m}, \tag{10.17}$$

in connection with the decomposition in the right-hand side of (10.16).

We retain the notation (10.13) in the previous section, that is,

$$c(m) = \frac{\pi^m}{m!}.$$

Then the irreducible G'-modules $\pi_{m,\varepsilon}$ and $\pi_{m+1,-\varepsilon}$ in (10.14) are the eigenspaces of the renormalized Knapp–Stein operator $\widetilde{\mathbb{T}}^m_{m,m}$ for the subgroup G' with eigenvalues $c(m)$ and $-c(m)$, respectively, by Lemma 8.15.

The case (10.16) will be discussed in Section 10.4.1 and the case (10.17) in Section 10.4.2. In particular, we shall see in Section 10.5, that both $\mathbb{A}' := \frac{1}{2}\widetilde{\mathbb{A}}^{m+1,m}_{m,m,+} + c(m)\widetilde{\mathbb{C}}^{m+1,m}_{m,m}$ in (10.16) and $\frac{1}{2}(-1)^{m+1}\widetilde{\mathbb{A}}^{m,m}_{m,m,-}$ in (10.17) yield the same symmetry breaking operator

$$A_{m+1,m}\colon \Pi_{m+1,\delta} \to \pi_{m,\delta},$$

which will be utilized in the construction of nonzero periods in Chapter 12, see Theorem 12.5.

10.4.1 $\mathrm{Hom}_{G'}(I_\delta(m+1,m)|_{G'}, J_\delta(m,m))$ for $n = 2m+1$

We recall from Theorem 3.19 (2) that the regular symmetry breaking operator $\widetilde{\mathbb{A}}^{i,j}_{\lambda,\nu,+}$ vanishes when $(n,i,j,\lambda,\nu) = (2m+1, m+1, m, m, m)$, and therefore, the left-hand side of (10.15) at $\lambda = m$ is two-dimensional by Theorem 3.25 (2). More precisely, the classification of symmetry breaking operators given in Theorem 3.26 shows (10.16).

On the other hand, we recall from Theorem 2.20 (1) that the principal series representation $I_\delta(m+1,m)$ has a nonsplitting exact sequence of G-modules:

$$0 \to \Pi_{m+2,-\delta} \to I_\delta(m+1,m) \to \Pi_{m+1,\delta} \to 0.$$

The irreducible G-submodule $\Pi_{m+2,-\delta}$ is the image of the Knapp–Stein operator $\widetilde{\mathbb{T}}^{m+1}_{m+1,m}$ for the group G. With this in mind, we shall take a closer look at the right-hand side of (10.15).

We introduce the following element in (10.16):

$$\mathbb{A}' := \frac{1}{2}\widetilde{\mathbb{A}}^{m+1,m}_{m,m,+} + c(m)\widetilde{\mathbb{C}}^{m+1,m}_{m,m}. \tag{10.18}$$

The main result of this subsection is the following.

Proposition 10.21. *Let* $(G, G') = (O(2m+2,1), O(2m+1,1))$. *Then the linear map* $\mathbb{A}'\colon I_\delta(m+1,m) \to J_\delta(m,m)$ *is a symmetry breaking operator satisfying*

$$\mathbb{A}' \circ \widetilde{\mathbb{T}}^{m+1}_{m+1,m} = 0,$$

$$\widetilde{\mathbb{T}}^{m,m}_{m,m} \circ \mathbb{A}' = c(m)\mathbb{A}',$$

$$S(\mathbb{A}') = c(m)\begin{pmatrix} 1 & 0 \\ 0 & 0 \end{pmatrix}.$$

Proposition 10.21 follows from the corresponding results for the renormalized operator $\widetilde{\mathbb{A}}_{m,m,+}^{m+1,m}$ (Lemma 10.22 below) and for the differential operator $\widetilde{\mathbb{C}}_{m,m}^{m+1,m}$ (Lemmas 10.23, 10.25, and 10.26). We begin with the functional equations and the (K, K')-spectrum of the first generator $\widetilde{\mathbb{A}}_{m,m,+}^{m+1,m}$ in (10.16).

Lemma 10.22. *Retain the setting where* $(G, G') = (O(2m + 2, 1), O(2m + 1, 1))$. *Then the renormalized regular symmetry breaking operator* $\widetilde{\mathbb{A}}_{m,m,+}^{m+1,m}$ *satisfies the following:*

$$\widetilde{\mathbb{A}}_{m,m,+}^{m+1,m} \circ \widetilde{\mathbb{T}}_{m+1,m}^{m+1} = -2c(m)\widetilde{\mathbb{A}}_{m+1,m,+}^{m+1,m},$$

$$\widetilde{\mathbb{T}}_{m,m}^{m} \circ \widetilde{\mathbb{A}}_{m,m,+}^{m+1,m} = -c(m)\widetilde{\mathbb{A}}_{m,m,+}^{m+1,m},$$

$$S(\widetilde{\mathbb{A}}_{m,m,+}^{m+1,m}) = c(m)\begin{pmatrix} 0 & 0 \\ 0 & 2 \end{pmatrix}.$$

Proof. See Lemma 9.30 for the first and third equalities, and Lemma 9.31 for the second.   □

For the differential symmetry breaking operator $\widetilde{\mathbb{C}}_{m,m}^{m+1,m}$ in (10.16), we recall from [34, Thm. 1.3] the residue formula of the regular symmetry breaking operators $\widetilde{\mathbb{A}}_{\lambda,\nu,\delta\varepsilon}^{i,j}: I_\delta(i, \lambda) \to J_\varepsilon(j, \nu)$ when $(\lambda, \nu, \delta, \varepsilon) \in \Psi_{\mathrm{sp}}$ for $j = i - 1$ and i, see Fact 9.3. Applying (9.6) to $(n, i, j, \lambda, \nu) = (2m + 1, m + 1, m, \lambda, \lambda)$, we obtain

$$\widetilde{\mathbb{A}}_{\lambda,\lambda,+}^{m+1,m} = \frac{(m - \lambda)\pi^m}{\Gamma(\lambda + 1)}\widetilde{\mathbb{C}}_{\lambda,\lambda}^{m+1,m}, \qquad (10.19)$$

where we recall from (3.17) that the differential symmetry breaking operator $\mathbb{C}_{\lambda,\nu}^{i,i-1}$ vanishes for the parameter that we are dealing with, namely, when $\lambda = \nu = n - i$. So we use the renormalized operator $\widetilde{\mathbb{C}}_{\lambda,\nu}^{i,i-1}$ instead. We note that $\widetilde{\mathbb{C}}_{\lambda,\lambda}^{i,i-1} = \mathrm{Rest}_{x_n=0} \circ \iota_{\frac{\partial}{\partial x_n}}$.

Lemma 10.23. *The* (K, K')-*spectrum of* $\widetilde{\mathbb{C}}_{m,m}^{m+1,m}$ *is given by*

$$S(\widetilde{\mathbb{C}}_{m,m}^{m+1,m}) = \begin{pmatrix} 1 & 0 \\ 0 & -1 \end{pmatrix}.$$

Proof. By the residue formula (10.19), we have

$$\lim_{\lambda \to m} \frac{1}{\lambda - m}\widetilde{\mathbb{A}}_{\lambda,\lambda,+}^{m+1,m} = -c(m)\widetilde{\mathbb{C}}_{m,m}^{m+1,m}.$$

Now the lemma follows from the (K, K')-spectrum of the regular symmetry breaking operator $\widetilde{\mathbb{A}}_{\lambda,\nu,\pm}^{i,j}$ given in Theorem 9.8.   □

The symmetry breaking operator $\widetilde{\mathbb{A}}^{m+1,m}_{\lambda,\nu,+}$ vanishes at $(\lambda,\nu) = (m,m)$. We recall from Lemma 5.43 and Definition 5.44 that

$$(\widetilde{\mathbb{A}}^{m+1,m}_{m,m,+})_{k,l} := \left. \frac{\partial^{k+l}}{\partial\lambda^k \partial\nu^l}\right|_{\substack{\lambda=m \\ \nu=m}} \widetilde{\mathbb{A}}^{m+1,m}_{\lambda,\nu,+} \in \operatorname{Hom}_{G'}(I_\delta(m+1,m)|_{G'}, J_\delta(m,m))$$

for $(k,l) = (1,0)$ and $(0,1)$.

The base change of the vector space $\operatorname{Hom}_{G'}(I_\delta(m+1,m)|_{G'}, J_\delta(m,m))$, see (10.16), is given as follows.

Lemma 10.24.

(1) $2(\widetilde{\mathbb{A}}^{m+1,m}_{m,m,+})_{1,0} = \widetilde{\widetilde{\mathbb{A}}}^{m+1,m}_{m,m,+}$,

(2) $(\widetilde{\mathbb{A}}^{m+1,m}_{m,m,+})_{1,0} + (\widetilde{\mathbb{A}}^{m+1,m}_{m,m,+})_{0,1} = -c(m)\widetilde{\mathbb{C}}^{m+1,m}_{m,m}$.

Proof. The first assertion is immediate from the definition of the renormalized operator $\widetilde{\widetilde{\mathbb{A}}}^{m+1,m}_{m,m,+}$, see (5.57). The second assertion follows from the residue formula (10.19). □

It follows from Lemma 10.24 that

$$\lim_{\lambda\to m}\frac{1}{\lambda - m}\widetilde{\mathbb{A}}^{m+1,m}_{\lambda,2m-\lambda,+} = \widetilde{\widetilde{\mathbb{A}}}^{m+1,m}_{m,m,+} + c(m)\widetilde{\mathbb{C}}^{m+1,m}_{m,m}. \tag{10.20}$$

Now we give functional equations of the differential symmetry breaking operators $\widetilde{\mathbb{C}}^{m+1,m}_{m,m}$ and the (renormalized) Knapp–Stein operators for G' and G as follows.

Lemma 10.25. $\widetilde{\mathbb{T}}^m_{m,m}\circ\widetilde{\mathbb{C}}^{m+1,m}_{m,m} = \widetilde{\widetilde{\mathbb{A}}}^{m+1,m}_{m,m,+} + c(m)\widetilde{\mathbb{C}}^{m+1,m}_{m,m}$.

Proof. By the functional equation in Theorem 9.24, we have

$$\widetilde{\mathbb{T}}^m_{\lambda,2m-\lambda}\circ\widetilde{\mathbb{A}}^{m+1,m}_{\lambda,\lambda,+} = \frac{(m-\lambda)\pi^m}{\Gamma(\lambda+1)}\widetilde{\mathbb{A}}^{m+1,m}_{\lambda,2m-\lambda,+}.$$

Hence we get from the residue formula (10.19)

$$\widetilde{\mathbb{T}}^m_{\lambda,2m-\lambda}\circ\widetilde{\mathbb{C}}^{m+1,m}_{\lambda,\lambda} = \widetilde{\mathbb{A}}^{m+1,m}_{\lambda,2m-\lambda,+}.$$

Now Lemma 10.25 follows from (10.20). □

Lemma 10.26. $\widetilde{\mathbb{C}}^{m+1,m}_{m,m}\circ\widetilde{\mathbb{T}}^{m+1}_{m+1,m} = \widetilde{\widetilde{\mathbb{A}}}^{m+1,m}_{m+1,m,+}$.

Proof. By the functional equation in Theorem 9.25, we have

$$\widetilde{\mathbb{A}}^{m+1,m}_{\lambda,\lambda,+}\circ\widetilde{\mathbb{T}}^{m+1}_{2m+1-\lambda,\lambda} = \frac{\pi^{m+\frac{1}{2}}(m-\lambda)}{\Gamma(\lambda+1)}\widetilde{\mathbb{A}}^{m+1,m}_{2m+1-\lambda,\lambda,+}.$$

By the residue formula (10.19) and by analytic continuation, we get

$$\widetilde{\mathbb{C}}^{m+1,m}_{\lambda,\lambda} \circ \widetilde{\mathbb{T}}^{m+1}_{2m+1-\lambda,\lambda} = \pi^{\frac{1}{2}} \widetilde{\mathbb{A}}^{m+1,m}_{2m+1-\lambda,\lambda,+}.$$

Since $\widetilde{\mathbb{A}}^{m+1,m}_{m+1,m,+} = \pi^{\frac{1}{2}} \widetilde{\mathbb{A}}^{m+1,m}_{m+1,m,+}$ by the definition (9.38) of the renormalized operator $\widetilde{\mathbb{A}}^{i,j}_{\lambda,\nu,\pm}$, the lemma is proved. $\square$

10.4.2 $\mathrm{Hom}_{G'}(I_{-\varepsilon}(m,m)|_{G'}, J_{\varepsilon}(m,m))$ for $n = 2m+1$

In this subsection, we examine

$$\mathrm{Hom}_{G'}(I_{-\varepsilon}(m,m)|_{G'}, J_{\varepsilon}(m,m)) = \mathbb{C}\widetilde{\mathbb{A}}^{m,m}_{m,m,-},$$

as stated in (10.17), which is derived from Theorems 3.25 and 3.26. We recall from
Theorem 2.20 (1) that there is a nonsplitting exact sequence of G-modules:

$$0 \to \Pi_{m,-\delta} \to I_{-\delta}(m,m) \to \Pi_{m+1,\delta} \to 0.$$

Concerning the regular symmetry breaking operator $\widetilde{\mathbb{A}}^{m,m}_{m,m,-}$, we have the following.

Lemma 10.27. *Let* $(G, G') = (O(2m+2, 1), O(2m+1, 1))$. *Then we have*

$$\widetilde{\widetilde{\mathbb{T}}}^m_{m,m} \circ \widetilde{\mathbb{A}}^{m,m}_{m,m,-} = c(m)\widetilde{\mathbb{A}}^{m,m}_{m,m,-},$$

$$\widetilde{\mathbb{A}}^{m,m}_{m,m,-} \circ \widetilde{\mathbb{T}}^m_{m+1,m} = 0,$$

$$S(\widetilde{\mathbb{A}}^{m,m}_{m,m,-}) = 2(-1)^{m+1} c(m) \begin{pmatrix} 0 & 0 \\ 1 & 0 \end{pmatrix}.$$

Proof. The proof of first formula parallels to that of Lemma 9.31, and the second
formula is a special case of Theorem 9.25. The third formula follows from
Theorem 9.8. $\square$

10.5   Symmetry Breaking Operators from $\Pi_{i,\delta}$ to $\pi_{i-1,\delta}$

In Sections 10.2 and 10.3, we constructed nontrivial symmetry breaking operators
from the irreducible representation $\Pi_{i,\delta}$ of $G = O(n+1, 1)$ to the irreducible one
$\pi_{i,\delta}$ of $G' = O(n, 1)$. This is sufficient for the proof of Theorem 4.2 by the duality
theorem (Proposition 3.39) between symmetry breaking operators for the indices:

$$(i, j) \text{ and } (\widetilde{i}, \widetilde{j}) := (n+1-i, n-j).$$

Nevertheless, we give in this section an explicit construction of the normalized symmetry breaking operators $\Pi_{i,\delta} \to \pi_{j,\varepsilon}$ also for $j = i - 1$, and determine their (K, K')-spectrum of symmetry breaking operators from $\Pi_{i,\delta}$ to $\pi_{i-1,\delta}$. The results will be used in the computation of *periods* of admissible smooth representations in Chapter 12.

We begin with some basic properties of the regular symmetry breaking operator

$$\widetilde{\mathbb{A}}^{i,i-1}_{\lambda,\nu,\delta}: I_\delta(i, \lambda) \to J_\delta(i - 1, \nu)$$

for $(\lambda, \nu) = (n - i, i - 1)$.

Proposition 10.28. *Suppose $1 \leq i \leq n$ and $\delta \in \{\pm\}$.*

(1) $\Pi_{i+1,-\delta} \subset \mathrm{Ker}\,(\widetilde{\mathbb{A}}^{i,i-1}_{n-i,i-1,+})$.

(2) Image $(\widetilde{\mathbb{A}}^{i,i-1}_{n-i,i-1,+}) \simeq \pi_{i-1,\delta}$ *if* $n \neq 2i - 1$; $\widetilde{\mathbb{A}}^{i,i-1}_{n-i,i-1,+} = 0$ *if* $n = 2i - 1$.

Proof.

(1) Applying the functional equation given in Theorem 9.25 with $\lambda = i$, we see that the symmetry breaking operator $\widetilde{\mathbb{A}}^{i,i-1}_{n-i,\nu,\gamma}$ vanishes on the image of the Knapp–Stein intertwining operator $\widetilde{\mathbb{T}}^i_{i,n-i}: I_\delta(i, i) \to I_\delta(i, n - i)$, namely, on the irreducible submodule $\Pi_{i+1,-\delta}$ (see Theorem 2.20 (1)).

(2) By Theorem 3.19, $\widetilde{\mathbb{A}}^{i,i-1}_{n-i,i-1,+} = 0$ if and only if $n = 2i - 1$.

Suppose from now that $n \neq 2i - 1$. Applying the functional equation given in Theorem 9.24 with $(\lambda, \nu, \gamma) = (n - i, i - 1, +)$, we see that the composition $\widetilde{\mathbb{T}}^{i-1}_{\nu,n-1-\nu} \circ \widetilde{\mathbb{A}}^{i,i-1}_{\lambda,\nu,+}$ is a scalar multiple of the symmetry breaking operator $\widetilde{\mathbb{A}}^{i,i-1}_{n-i,n-i,+}$, which vanishes by Theorem 3.19. In turn, applying Proposition 8.11 to $G' = O(n, 1)$, we get

$$\mathrm{Ker}(\widetilde{\mathbb{T}}^{i-1}_{\nu,n-1-\nu}: J_\delta(i - 1, \nu) \to J_\delta(i - 1, n - 1 - \nu)) \simeq \pi_{i-1,\delta}$$

because $i - 1 \neq \frac{1}{2}(n - 1)$. Hence the second statement is also proved. $\qquad\square$

Since $\widetilde{\mathbb{A}}^{i,i-1}_{n-i,i-1,+} = 0$ for $n = 2i - 1$, we treat this case separately as follows. Suppose $n = 2m + 1$. We recall that there are a nonsplitting exact sequence of G-modules

$$0 \to \Pi_{m,-\delta} \to I_{-\delta}(m, m) \to \Pi_{m+1,\delta} \to 0$$

and a direct sum decomposition of irreducible G'-modules

$$J_\delta(m, m) \simeq \pi_{m,\delta} \oplus \pi_{m+1,-\delta}.$$

We use the following regular symmetry breaking operator

$$\widetilde{\mathbb{A}}^{m,m}_{m,m,-}: I_{-\delta}(m, m) \to J_\delta(m, m).$$

Proposition 10.29. *Suppose* $(G, G') = (O(2m+2, 1), O(2m+1, 1))$ *and* $\delta \in \{\pm\}$.

(1) $\mathrm{Ker}(\widetilde{\mathbb{A}}_{m,m,-}^{m,m}) \supset \Pi_{m,-\delta}$.

(2) $\mathrm{Image}(\widetilde{\mathbb{A}}_{m,m,-}^{m,m}) = \pi_{m,\delta}$.

Proof. The assertions follow from Lemma 10.27. □

It follows from Proposition 10.28 that if $n \neq 2i - 1$ then the normalized symmetry breaking operator $\widetilde{\mathbb{A}}_{n-i,n-i,+}^{i,i-1}$ yields a surjective G'-homomorphism

$$A_{i,i-1} \colon \Pi_{i,\delta} \to \pi_{i-1,\delta} \tag{10.21}$$

by the following diagram.

$$
I_\delta(i, n-i) \xrightarrow{\widetilde{\mathbb{A}}_{n-i,n-i,+}^{i,i-1}} \pi_{i-1,\delta} \lhook\joinrel\longrightarrow J_\delta(i-1, i-1)
$$

$$
\Pi_{i,\delta} \xleftarrow{\;\sim\;} I_\delta(i, n-i)/\Pi_{i+1,-\delta}
$$

If $n = 2i - 1$, we set $(n, i) = (2m+1, m+1)$. Then, similarly to the aforementioned case $n \neq 2i - 1$, Proposition 10.21 shows that the symmetry breaking operator $\mathbb{A}' \colon I_\delta(m+1, m) \to J_\delta(m, m)$ defined in (10.18) yields a surjective G'-homomorphism

$$A_{m+1,m} \colon \Pi_{m+1,\delta} \to \pi_{m,\delta} \tag{10.22}$$

by the following diagram.

$$
I_\delta(m+1, m) \xrightarrow{\;\;\mathbb{A}'\;\;} \pi_{m,\delta} \subset J_\delta(m, m)
$$

$$
\Pi_{m+1,\delta} \xleftarrow{\;\sim\;} I_\delta(m+1, m)/\Pi_{m+2,-\delta}
$$

In order to define the (K, K')-spectrum, we need to fix an inclusive map from the K'-type into the K-type, see Definition 9.7. In our setting, we use the natural embedding of the minimal K- and K'-types

$$\mu^\flat(i, \delta) \hookleftarrow \mu^\flat(i-1, \delta)' \tag{10.23}$$

of the irreducible representations $\Pi_{i,\delta}$ and $\pi_{i-1,\delta}$ of G and the subgroup G', respectively, as in Section 9.6. Then we get the following formula for the (K, K')-spectrum.

Proposition 10.30. *Let* $(G, G') = (O(n+1, 1), O(n, 1))$ *and* $1 \leq i \leq n+1$. *Then the symmetry breaking operator*

$$A_{i,i-1} \colon \Pi_{i,\delta} \to \pi_{i-1,\delta}$$

acts on $\mu^{\flat}(i-1, \delta)'$ $(\hookrightarrow \mu^{\flat}(i, \delta))$ *as the following scalar:*

$$\begin{cases} \dfrac{\pi^{\frac{n-1}{2}}(n-2i+1)}{(n-i)!} & \text{if } n \neq 2i - 1, \\[2mm] \dfrac{\pi^{\frac{n-1}{2}}}{(n-i)!} & \text{if } n = 2i - 1. \end{cases}$$

Proof. For $n \neq 2i - 1$, the assertion follows directly from the $(1, 1)$-component of the matrix $S(\widetilde{\mathbb{A}}_{\lambda,\nu,+}^{i,i-1})$ in Theorem 9.8 with $(\lambda, \nu) = (n - i, i - 1)$. For $n = 2i - 1$, the $(1,1)$-component of $S(\mathbb{A}')$ in Proposition 10.21 with $(n, i) = (2m + 1, m + 1)$ shows the desired formula. $\qquad\square$

Remark 10.31. When $n = 2i - 1$, we set $(n, i) = (2m + 1, m + 1)$ as above. In this case we may use $\widetilde{\mathbb{A}}_{m,m,-}^{m,m}$ in Lemma 10.27 for an alternative construction of $A_{m+1,m} \in \mathrm{Hom}_{G'}(\Pi_{m+1,\delta}|_{G'}, \pi_{m,\delta})$. To see this, we recall from Section 10.4.2 the following natural inclusion

$$\mathrm{Hom}_{G'}(\Pi_{m+1,\delta}|_{G'}, \pi_{m,\delta}) \subset \mathrm{Hom}_{G'}(I_{-\delta}(m, m)|_{G'}, J_{\delta}(m, m)) = \mathbb{C}\widetilde{\mathbb{A}}_{m,m,-}^{m,m},$$

and therefore any element in $\mathrm{Hom}_{G'}(\Pi_{m+1,\delta}|_{G'}, \pi_{m,\delta})$ is proportional to the one which is induced from $\widetilde{\mathbb{A}}_{m,m,-}^{m,m}$. On the other hand, Proposition 10.29 tells that the symmetry breaking operator $\widetilde{\mathbb{A}}_{m,m,-}^{m,m}$ yields a surjective G'-homomorphism $\Pi_{m+1,\delta} \to \pi_{m,\delta}$ by the following diagram.

$$\begin{array}{ccc} I_{-\delta}(m, m) & \xrightarrow{\widetilde{\mathbb{A}}_{m,m,-}^{m,m}} & \pi_{m,\delta} \hookrightarrow J_{\delta}(m, m) \\ \downarrow & \circlearrowleft & \\ \Pi_{m+1,\delta} \xleftarrow{\;\sim\;} I_{-\delta}(m, m)/\Pi_{m,-\delta} & & \end{array}$$

By Lemma 10.27, $\frac{1}{2}(-1)^{m+1}\widetilde{\mathbb{A}}_{m,m,-}^{m,m}$ has the (K, K')-spectrum for the basic K- and K'-types

$$S\left(\frac{1}{2}(-1)^{m+1}\widetilde{\mathbb{A}}_{m,m,-}^{m,m}\right) = c(m)\begin{pmatrix} 0 & 0 \\ 1 & 0 \end{pmatrix}.$$

In view of the $(2,1)$-component, the resulting symmetry breaking operator from $\Pi_{m+1,\delta}$ to $\pi_{m,\delta}$ has the (K, K')-spectrum $c(m)$ for the embedding of the K- and K'-types $\mu^{\flat}(m + 1, \delta) \hookleftarrow \mu^{\flat}(m, \delta)'$. This is the same with the (K, K')-spectrum of $A_{m+1,m}$ which is induced from $\mathbb{A}' \in \mathrm{Hom}_{G'}(I_{\delta}(m + 1, m)|_{G'}, J_{\delta}(m, m))$. Hence $\frac{1}{2}(-1)^{m+1}\widetilde{\mathbb{A}}_{m,m,-}^{m,m}$ induces the same symmetry breaking operator with $A_{m+1,m}$.

Chapter 11
Application I: Some Conjectures by B. Gross and D. Prasad: Restrictions of Tempered Representations of $SO(n+1,1)$ to $SO(n,1)$

Inspired by automorphic forms and L-functions, B. Gross and D. Prasad published in 1992 a conjecture about the restriction of irreducible tempered representations of special orthogonal groups $SO(p+1,q)$ to a special orthogonal subgroup $SO(p,q)$, see [14]. B. Sun and C.-B. Zhu [55] proved that in this case the multiplicities are at most one, and B. Gross and D. Prasad conjectured that given a Vogan packet of tempered representations of $SO_{n+2} \times SO_{n+1}$ there exist exactly one group $SO(p+1,q) \times SO(p,q)$ with $p+q = n+1$ and one (tempered) representation $\overline{U}_1 \boxtimes \overline{U}_2$ of this group with $m(\overline{U}_1 \boxtimes \overline{U}_2, \mathbb{C}) = 1$. They also stated a conjectured algorithm to determine the group and the representation $\overline{U}_1 \boxtimes \overline{U}_2$ in the Vogan packet with $m(\overline{U}_1 \boxtimes \overline{U}_2, \mathbb{C}) = 1$.

In this chapter we prove that the algorithm of B. Gross and D. Prasad predicts the multiplicity correctly for representations in Vogan packets of tempered principal series representations of $SO(n+1,1) \times SO(n,1)$ as well as for the 3 irreducible representations $\overline{\Pi}, \overline{\pi}, \overline{\varpi}$ of $SO(2m+2,1), SO(2m+1,1), SO(2m,1)$ with trivial infinitesimal character ρ.

The Gross–Prasad conjectures are stated only for representations of special orthogonal groups in [14]. Thus we are considering in this chapter symmetry breaking for tempered representations of $\overline{G} \times \overline{G'} = SO(n+1,1) \times SO(n,1)$ and not as in the previous chapters for $G \times G' = O(n+1,1) \times O(n,1)$. We refer to Appendix II (Chapter 15) for notation and for results about the restriction of representations from orthogonal groups to special orthogonal groups.

11.1 Vogan Packets of Tempered Induced Representations

We use a bar over representations to distinguish between representations of the special orthogonal group and those of the orthogonal group.

© Springer Nature Singapore Pte Ltd. 2018
T. Kobayashi, B. Speh, *Symmetry Breaking for Representations of Rank One Orthogonal Groups II*, Lecture Notes in Mathematics 2234, https://doi.org/10.1007/978-981-13-2901-2_11

Every tempered principal series representation of $SO(n+1,1)$ is of the form

$$\overline{I}_\delta(\overline{V}, \lambda) \equiv \operatorname{Ind}_{\overline{P}}^{\overline{G}}(\overline{V} \boxtimes \delta, \lambda) \quad \text{for } (\overline{\sigma}, \overline{V}) \in \widehat{SO(n)}, \ \delta \in \{\pm\}, \ \lambda \in \frac{n}{2} + \sqrt{-1}\mathbb{R},$$

which is the smooth representation of a unitarily induced principal series representation from a finite-dimensional representation of the minimal parabolic subgroup $\overline{P}$ of $\overline{G} = SO(n+1,1)$.

For n even, we assume that the central element $-I_{n+2}$ of the special orthogonal group $\overline{G} = SO(n+1,1)$ acts nontrivially on the principal series representation $\overline{I}_\delta(\overline{V}, \lambda)$, and thus $\overline{I}_\delta(\overline{V}, \lambda)$ is a genuine representation of $\overline{G}$, i.e., that $-I_{n+2}$ is not in the kernel of $\overline{V} \boxtimes \delta$. For n odd, $\overline{G} = SO(n+1,1)$ does not have a nontrivial center, and we do not need an assumption on the pair $(\overline{V}, \delta)$.

We observe if n is odd, the Langlands parameter of the representations of $SO(n,0)$ factors through the identity component of its L-group, and it defines a representation of $SO(n-2p, 2p)$ and not of $O(n-2p, 2p)$, see [3].

The Langlands parameter of the induced representations $\overline{I}_\delta(\overline{V}, \lambda)$ factors through the Levi subgroup of a maximal parabolic subgroup of the Langlands dual group $^L G$ [49]. This parabolic subgroup corresponds to a maximal parabolic subgroup of $SO(n+1,1)$ whose Levi subgroup L is a real form of $SO(n, \mathbb{C}) \times SO(2, \mathbb{C})$ and thus is isomorphic to $SO(n,0) \times SO(1,1) \simeq SO(n) \times GL(1, \mathbb{R})$. Note that $SO(1,1) \simeq GL(1, \mathbb{R})$ is a disconnected group and so determines the character δ.

The pure inner real forms of $SO(n, \mathbb{C})$ with a compact Cartan subgroup are $SO(n-2p, 2p)$, $0 \le p \le \frac{n}{2}$. For n even, we assume that the center of the group $SO(n-2p, 2p)$ is not contained in the kernel of the discrete series representation, see Proposition 15.11 (6).

By [12, p. 35], if G is $SO(2m+2,1)$ or $SO(2m+1,1)$, then there are 2^m representations in the Vogan packet containing a tempered representation $\overline{I}_\delta(\overline{V}, \lambda)$ and they are parametrized by characters of a finite group $\mathcal{A}_1 \simeq (\mathbb{Z}/2\mathbb{Z})^m$. We write $VP(\overline{I}_\delta(\overline{V}, \lambda))$ for this Vogan packet.

The representations in the Vogan packet $VP(\overline{I}_\delta(\overline{V}, \lambda))$ can be described as follows: we call a real form $SO(\ell, k)$ of $SO(\ell+k, \mathbb{C})$ pure if ℓ is even and thus admits discrete series representations. We consider parabolic subgroups of $SO(n-2p+1, 1+2p)$ with Levi subgroups L, which are pure inner forms of $SO(n) \times GL(1, \mathbb{R})$. Hence they are isomorphic to

$$L \simeq SO(n-2p, 2p) \times GL(1, \mathbb{R}).$$

The Vogan packet $VP(\overline{I}_\delta(\overline{V}, \lambda))$ contains the **tempered principal series representations** of $SO(n-2p+1, 1+2p)$ which have the same infinitesimal character as $\overline{I}_\delta(\overline{V}, \lambda)$, and which are induced from the outer tensor product of a discrete series representation of $SO(n-2p, 2p)$, with the same infinitesimal character as $\overline{V}$ and a one-dimensional representation χ_λ of $GL(1, \mathbb{R})$, [60]. See also [1].

We use the same conventions for a Vogan packet $VP(\overline{J}_\varepsilon(\overline{W}, \nu))$ of the tempered principal series representation $\overline{J}_\varepsilon(\overline{W}, \nu)$ of $\overline{G}'$.

11.2 Vogan Packets of Discrete Series Representations with Integral Infinitesimal Character of $SO(2m, 1)$

We begin with the case $n = 2m - 1$. In this case $SO(n + 1, 1) = SO(2m, 1)$ has discrete series representations. We fix a set of positive roots $\Delta^+ \subset \mathfrak{t}_{\mathbb{C}}^*$ for the root system $\Delta(\mathfrak{so}(2m + 1, \mathbb{C}), \mathfrak{t}_{\mathbb{C}})$ and denote by ρ half the sum of positive roots as before. Let η be an integral infinitesimal character, which is dominant with respect to Δ^+. For $\ell + k = 2m + 1$, we call a real form $SO(\ell, k)$ *pure* if ℓ is even. The Vogan packet containing the discrete series representation with infinitesimal character η is the disjoint union of discrete series representations with infinitesimal character η of the pure inner forms. The cardinality of this packet is

$$2^m = \sum_{\substack{0 \le \ell \le 2m \\ \ell : \text{even}}} \binom{m}{\frac{\ell}{2}}.$$

There exists a finite group $\mathcal{A}_2 \simeq (\mathbb{Z}/2\mathbb{Z})^m$ whose characters parametrize the representations in the Vogan packet. For the discrete series representation with parameter $\chi \in \widehat{\mathcal{A}_2}$ we write $\overline{\pi}(\chi)$. For more details see [14] or [60]. If $\overline{\pi}$ is a discrete series representation of $SO(2m, 1)$ we write $VP(\overline{\pi})$ for the Vogan packet containing $\overline{\pi}$.

Example 11.1. Suppose that $\overline{\pi}$ is a discrete series representation of $SO(2m, 1)$ with trivial infinitesimal character ρ.

(1) The trivial one-dimensional representation **1** of the inner form $SO(0, 2m + 1)$ is in $VP(\overline{\pi})$.
(2) We can define similarly a Vogan packet $VP(\overline{\pi})$ containing $(SO(1, 2m), \overline{\pi})$.

11.3 Embedding the Group $\overline{G'} = SO(n - 2p, 2p + 1)$ into the Group $\overline{G} = SO(n - 2p + 1, 2p + 1)$

To formulate the Gross–Prasad conjecture we have to fix an embedding of $\overline{G'}$ into $\overline{G}$. We observe:

(1) The quasisplit forms of the odd special orthogonal group are $SO(m, m + 1)$ and $SO(m + 1, m)$. The pure inner forms in the same class as $SO(m, m + 1)$ are $SO(m - 2p, m + 2p + 1)$ and those in the same class as $SO(m + 1, m)$ are $SO(m + 1 - 2p, m + 2p)$.
(2) The quasisplit forms of the even special orthogonal group are $SO(m, m)$, $SO(m - 1, m + 1)$, and $SO(m + 1, m - 1)$. Then the pure inner forms are $SO(n - 2p, n + 2p)$ and $SO(m + 1 - 2p, m - 1 - 2p)$, respectively, with $p \le \frac{m}{2}$.

So

1. if $n = 2m$, then the orthogonal group $SO(2m + 1, 1)$ is a pure inner form of $SO(m + 1, m + 1)$ if m is even and of $SO(m + 2, m)$ if m is odd;
2. if $n = 2m - 1$, then the orthogonal group $SO(2m, 1)$ is a pure inner form of $SO(m + 1, m)$ if m is odd and of $SO(m, m + 1)$ if m is even.

We consider an indefinite quadric form

$$Q_{n-2p+1,2p+1}(x) = x_1^2 + \cdots + x_{n-2p+1}^2 - x_{n-2p+2}^2 - \cdots - x_{n+2}^2$$

of signature $(n - 2p + 1, 2p + 1)$. We assume that $n - 2p + 1 > 0$ and identify $SO(n - 2p, 2p + 1)$ with the subgroup of $SO(n - 2p + 1, 2p + 1)$ which stabilizes the basis vector e_{n-2p+1}. This allows us to identify the Levi subgroup of the maximal parabolic subgroup of $SO(n - 2p, 2p + 1)$ with the intersection of the corresponding maximal parabolic subgroup of $\overline{G}$. This embedding of $SO(n, 1)$ into $SO(n + 1, 1)$ is conjugate to the one we consider in Section 2.1. We use this embedding in the formulation of the Gross–Prasad conjectures.

For tempered principal series representations we consider symmetry breaking operators, namely, $SO(n - 2p, 2p + 1)$-homomorphisms from representations in $VP(\overline{I}_\delta(\overline{V}, \lambda))$ to representations in $VP(\overline{J}_\varepsilon(\overline{W}, \nu))$, see Section 11.4.

If the tempered representation of $\overline{G}$ or of $\overline{G}'$ is a discrete series representation, we consider symmetry breaking from a Vogan packet of discrete series representations to a Vogan packet of tempered principal series representations, respectively from a Vogan packet of tempered principal series representations to a Vogan packet of discrete series representations (Section 11.5).

11.4   The Gross–Prasad Conjecture I: Tempered Principal Series Representations

By Theorem 3.30, there is a nontrivial symmetry breaking operator between the **tempered** principal representations $I_\delta(V, \lambda)$ of $G = O(n + 1, 1)$ and $J_\varepsilon(W, \nu)$ of $G' = O(n, 1)$ if and only if $(\sigma, V) \in \widehat{O(n)}$ and $(\tau, W) \in \widehat{O(n - 1)}$ satisfy

$$[V : W] = \dim_{\mathbb{C}} \operatorname{Hom}_{O(n-1)}(V|_{O(n-1)}, W) \neq 0.$$

An analogous result holds for a pair of the *special* orthogonal groups $(\overline{G}, \overline{G}') = (SO(n + 1, 1), SO(n, 1))$. We set

$$[\overline{V} : \overline{W}] \equiv [\overline{V}|_{SO(n-1)} : \overline{W}] := \dim_{\mathbb{C}} \operatorname{Hom}_{SO(n-1)}(\overline{V}|_{SO(n-1)}, \overline{W}).$$

In Theorem 15.14 in Appendix II we prove:

Theorem 11.2. *There is a nontrivial symmetry breaking operator between the* **tempered** *principal series representations* $\overline{I}_\delta(\overline{V}, \lambda)$ *of* $\overline{G} = SO(n + 1, 1)$ *and* $\overline{J}_\varepsilon(\overline{W}, \nu)$

of $\overline{G'} = O(n, 1)$ if and only if $(\overline{\sigma}, \overline{V}) \in \widehat{SO(n)}$ and $(\overline{\tau}, \overline{W}) \in \widehat{SO(n-1)}$ satisfy

$$[\overline{V}|_{SO(n-1)} : \overline{W}] \neq 0.$$

In their article B. Gross and D. Prasad presented a conjectured algorithm to determine the pair of representations in the Vogan packets $VP(\overline{I}_\delta(\overline{V}, \lambda))$ and $VP(\overline{J}_\varepsilon(\overline{W}, \nu))$ with a nontrivial $SO(n, 1)$-symmetry breaking operator. We prove next that the algorithm in fact predicts :

$$[\overline{V}|_{SO(n-1)} : \overline{W}] \neq 0 \quad \text{if and only if} \quad \operatorname{Hom}_{\overline{G'}}(\overline{I}_\delta(\overline{V}, \lambda)|_{\overline{G'}}, \overline{J}_\varepsilon(\overline{W}, \nu)) \neq \{0\}.$$

Observation 11.3. A Levi subgroup L with $[L, L] = SO(r, s)$ of the maximal parabolic subgroup determines the class of pure inner forms of $SO(r + 1, s + 1)$. So for any algorithm to determine the pair $(SO(r + 1, s), SO(r, s))$ of the groups in the Gross–Prasad conjectures it is enough to determine the pair of the Levi subgroups and their corresponding discrete series representations.

First Case. *Suppose that* $(\overline{G}, \overline{G'}) = (SO(2m + 1, 1), SO(2m, 1))$.
     Let $T_{\mathbb{C}}$ be a torus in $SO(2m + 2, \mathbb{C}) \times SO(2m + 1, \mathbb{C})$, and $X^*(T_{\mathbb{C}})$ the character group. Fix a basis

$$X^*(T_{\mathbb{C}}) = \mathbb{Z}e_1 \oplus \mathbb{Z}e_2 \oplus \cdots \oplus \mathbb{Z}e_{m+1} \oplus \mathbb{Z}f_1 \oplus \mathbb{Z}f_2 \oplus \cdots \oplus \mathbb{Z}f_m$$

such that the standard root basis Δ_0 is given by

$$e_1 - e_2, e_2 - e_3, \ldots, e_m - e_{m+1}, e_m + e_{m+1}, f_1 - f_2, f_2 - f_3, \ldots, f_{m-1} - f_m, f_m$$

if $m \geq 1$.
     We fix $\delta, \varepsilon \in \{\pm\}$ as in Section 11.1.
     Recall that all representations in a Vogan packet have the same Langlands parameter. We identify the Langlands parameter of the representations in the same Vogan packet as

$$(SO(2m + 1, 1) \times SO(2m, 1), \overline{I}_\delta(\overline{V}, \lambda) \boxtimes \overline{J}_\varepsilon(\overline{W}, \nu))$$

for a pair $(\overline{V}, \overline{W})$ of irreducible finite-dimensional representations with infinitesimal character

$$(v_1 + m - 1)e_1 + (v_2 + m - 2)e_2 + \cdots + (v_m)e_m - (\lambda - m)e_{m+1}$$
$$+ (u_1 + m - \frac{3}{2})f_1 + (u_2 + m - \frac{5}{2})f_2 + \cdots + (u_{m-1} + \frac{1}{2})f_{m-1}$$
$$- (v - m + \frac{1}{2})f_m,$$

see (2.26). Here $(v_1, v_2, \ldots, v_m)$ is the highest weight of the $SO(2m)$-module $\overline{V}$, $(u_1, u_2, \ldots, u_{m-1})$ is the highest weight of the $SO(2m-1)$-module $\overline{W}$ and the continuous parameter $\lambda - m$ and $\nu - m + \frac{1}{2}$ are purely imaginary, and thus $\overline{I}_\delta(\overline{V}, \lambda)$ and $\overline{J}_\varepsilon(\overline{W}, \nu)$ are (smooth) tempered principal series representations of $\overline{G}$ and $\overline{G'}$, respectively.

As discussed before, to determine the pair

$$(SO(n - 2p + 1, 2p + 1), SO(n - 2p, 2p - 1))$$

it suffices to solve this problem for the Levi subgroups. Hence it suffices to consider the Langlands parameter

$$(v_1 + m - 2)e_1 + (v_2 + m - 3)e_2 + \cdots + (v_m)e_m$$
$$+ (u_1 + m - \frac{5}{2})f_1 + (u_2 + m - \frac{7}{2})f_2 + \cdots + (u_{m-1} + \frac{1}{2})f_{m-1}.$$

Let δ_i be the element which is -1 in the i-th factor of $\mathcal{A}_1$ and equal to 1 everywhere else, and ε_j the element which is -1 in the j-th factor of $\mathcal{A}_2$ and 1 everywhere else. Then the algorithm [14, p. 993] determines $\chi_1 \in \widehat{\mathcal{A}_1}$ and $\chi_2 \in \widehat{\mathcal{A}_2}$ by

$$\chi_1(\delta_i) = (-1)^{\#m-i+1>} \quad \text{and} \quad \chi_2(\varepsilon_j) = (-1)^{\#m-j+\frac{1}{2}<},$$

where $\#m - i + 1 >$ is the cardinality of the set

$$\{j : v_j + m - i > \text{the coefficients of } f_j\},$$

and $\#m - j + \frac{1}{2} <$ is the cardinality of the set

$$\{i : v_i + m - j - 1 + \frac{1}{2} < \text{the coefficients of } e_i\}.$$

If $\text{Hom}_{SO(n-1)}(\overline{V}|_{SO(n-1)}, \overline{W}) \neq \{0\}$, then $v_1 \leq u_1 \leq v_2 \leq \cdots \leq u_{m-1} \leq |v_m|$. Hence we deduce that both characters are alternating characters if and only if $\text{Hom}_{SO(n-1)}(\overline{V}|_{SO(n-1)}, \overline{W}) \neq \{0\}$.

Second Case. *Suppose that* $(\overline{G}, \overline{G'}) = (SO(2m, 1), SO(2m - 1, 1))$.

We use the same arguments for the pair

$$(\overline{G}, \overline{G'}) = (SO(2m, 1), SO(2m - 1, 1)).$$

We normalize the quasisplit forms by

$$SO(m+1,m) \times SO(m,m) \qquad \text{if } m \text{ is even,}$$

$$SO(m,m+1) \times SO(m-1,m+1) \quad \text{if } m \text{ is odd.}$$

Applying the formulæ in [14, (12.21)], we define the integers p and q with $0 \le p \le m$ and $0 \le q \le m$ by

$$p = \#\{i : \chi_1(\delta_i) = (-1)^i\} \quad \text{and} \quad q = \#\{j : \chi_2(\varepsilon_j) = (-1)^{m+j}\},$$

and we get the pure forms

$$SO(2m-2p+1,2p) \times SO(2q,2m-2q) \qquad \text{if } m \text{ is even,}$$

$$SO(2p+1,2m-2p+1) \times SO(2m-2q,2q+1) \quad \text{if } m \text{ is odd.}$$

In our setting, we get the pair of integers $(p,q) = (0,m)$ for m even; $(p,q) = (m,0)$ for m odd. Applying [14, (12.22)] with correction by changing n by m *loc. cit.*, we deduce that the alternating character χ defines the pure inner form

$$SO(2m+1,0) \times SO(2m,0) \qquad \text{for } m \text{ is even and odd.}$$

Hence

$$\overline{G} = SO(2m,1) \text{ and } \overline{G'} = SO(2m-1,1).$$

The only representation in $VP(\overline{I}_\delta(\overline{V},\lambda)) \times VP(\overline{J}_\varepsilon(\overline{W},\nu))$ for this pair of pure inner forms is

$$\overline{I}_\delta(\overline{V},\lambda) \boxtimes \overline{J}_\varepsilon(\overline{W},\nu).$$

If χ is not the alternating character, the calculation shows that we obtain a different pair of groups. Thus we can rephrase the conjecture by B. Gross and D. Prasad as follows:

Conjecture 11.4 (Gross–Prasad conjecture I). *Suppose that* $\overline{I}_\delta(\overline{V},\lambda) \boxtimes \overline{J}_\varepsilon(\overline{W},\nu)$ *are tempered principal series representations of* $SO(n+1,1) \times SO(n,1)$. *Then*

$$\mathrm{Hom}_{SO(n,1)}(\overline{I}_\delta(\overline{V},\lambda) \boxtimes \overline{J}_\varepsilon(\overline{W},\nu), \mathbb{C}) = \mathbb{C}$$

if and only if $\overline{V} \in \widehat{SO(n)}$ *and* $\overline{W} \in \widehat{SO(n-1)}$ *satisfies*

$$[\overline{V}|_{SO(n-1)} : \overline{W}] \ne 0.$$

Theorem 11.5 (see Theorem 15.14). *The Gross–Prasad conjecture I holds.*

We can deduce Theorem 11.5 from the corresponding results (Theorem 3.30) for the orthogonal groups $O(n + 1, 1) \times O(n, 1)$ by using results about the reduction from $O(N, 1)$ to the special orthogonal group $SO(N, 1)$. See the proof of Theorem 15.14 in Section 15.6 of Appendix II for details.

11.5 The Gross–Prasad Conjecture II: Tempered Representations with Trivial Infinitesimal Character ρ

For completeness, we include the discussion of the Gross–Prasad conjectures for tempered representations with trivial infinitesimal character ρ which we also discussed in detail in [43].

We modify here the notation from [43] by denoting the restriction of a representation Π of $O(n + 1, 1)$ to the subgroup $SO(n + 1, 1)$ by $\overline{\Pi}$.

The Gross–Prasad conjecture I in the previous section treated the case where both $\overline{\Pi}$ and $\overline{\pi}$ are tempered *principal series representations* of the group $G = SO(n + 1, 1)$ and $G' = SO(n, 1)$, respectively.

Thus the remaining cases are when $\overline{\Pi}$ or $\overline{\pi}$ are *discrete series representations*. We note that both $\overline{\Pi}$ and $\overline{\pi}$ cannot be discrete series representations in our setting because $\overline{G}$ admit discrete series representations if and only if n is odd and $\overline{G'}$ admit those if and only if n is even. Thus we discuss the Gross–Prasad conjecture in this case separately depending on the parity of n, with the following notation.

Consider symmetry breaking operators for tempered representations with trivial infinitesimal character ρ of the group $SO(n + 1, 1)$ for $n = 2m, 2m - 1$, and $2m - 2$. We denote the corresponding representations by Π, π, and ϖ, respectively, using the subscripts defined in Section 15.5 in Appendix II. We thus consider symmetry breaking from $SO(2m + 1, 1)$ to $SO(2m, 1)$ and further to $SO(2m - 1, 1)$:

$$\overline{\Pi}_{m,(-1)^{m+1}} \to \overline{\pi}_m \to \overline{\varpi}_{m-1,(-1)^m}.$$

Here $\overline{\Pi}_{m,(-1)^{m+1}}$ and $\overline{\varpi}_{m-1,(-1)^m}$ are tempered principal series representations which are nontrivial on the center of $SO(2m + 1, 1)$, respectively $SO(2m - 1, 1)$, and thus are genuine representations of the special orthogonal groups, see Proposition 15.11 (6). Since $\overline{\pi}_{m,+} \simeq \overline{\pi}_{m,-}$ as $SO(2m, 1)$-modules, we simply write $\overline{\pi}_m$ for $\overline{\pi}_{m,\pm}$, which is a discrete series representation of $SO(2m, 1)$. All representations have the trivial infinitesimal character ρ.

11.5.1   The Gross–Prasad Conjecture II: Symmetry Breaking from $\overline{\Pi}_{m,(-1)^{m+1}}$ to the Discrete Series Representation $\overline{\pi}_m$

We consider first the Vogan packet of tempered representations which contains the pair $(SO(2m+1,1) \times SO(2m,1), \overline{\Pi}_{m,\delta} \boxtimes \overline{\pi}_m)$ or the Vogan packet which contains the pair $(SO(1,1+2m) \times SO(1,2m), \overline{\Pi}_{m,\delta} \boxtimes \overline{\pi}_m)$. The representations in these packets are parametrized by characters of

$$\mathcal{A}_1 \times \mathcal{A}_2 \simeq (\mathbb{Z}/2\mathbb{Z})^m \times (\mathbb{Z}/2\mathbb{Z})^m \simeq (\mathbb{Z}/2\mathbb{Z})^{2m}.$$

We recall the algorithm proposed by B. Gross and D. Prasad which determines a pair $(\chi_1, \chi_2) \in \widehat{\mathcal{A}_1} \times \widehat{\mathcal{A}_2}$, hence representations

$$(\overline{\Pi}(\chi_1), \overline{\pi}(\chi_2)) \in VP(\overline{\Pi}_{m,\delta}) \times VP(\overline{\pi}_m)$$

so that

$$\mathrm{Hom}_{\overline{G}(\chi_2)}(\overline{\Pi}(\chi_1)|_{\overline{G}(\chi_2)}, \overline{\pi}(\chi_2)) \neq \{0\},$$

where $\overline{G}(\chi_2)$ is the pure inner form determined by χ_2.

Let $T_{\mathbb{C}}$ be a torus in $SO(2m+2,\mathbb{C}) \times SO(2m+1,\mathbb{C})$, and $X^*(T_{\mathbb{C}})$ the character group. As before the standard root basis Δ_0 is given by

$$e_1 - e_2, e_2 - e_3, \ldots, e_m - e_{m+1}, e_m + e_{m+1}, f_1 - f_2, f_2 - f_3, \ldots, f_{m-1} - f_m, f_m$$

if $m \geq 1$.

We fix $\delta = (-1)^{m+1}$ so that $\overline{\Pi}_{m,\delta}$ is a genuine representation of $SO(2m+1,1)$. We can identify the Langlands parameter of the Vogan packet containing

$$(SO(2m+1,1) \times SO(2m,1), \overline{\Pi}_{m,\delta} \boxtimes \overline{\pi}_m)$$

with

$$me_1 + (m-1)e_2 + \cdots + e_m + 0e_{m+1} + \left(m - \frac{1}{2}\right)f_1 + \left(m - \frac{3}{2}\right)f_2 + \cdots + \frac{1}{2}f_m.$$

Let δ_i be the character in $\widehat{\mathcal{A}_1}$ which is -1 in the i-th factor of $\mathcal{A}_1$ and equal to 1 everywhere else, and ε_j be the character which is -1 in the j-th factor of $\mathcal{A}_2$ and 1 everywhere else.

Then the algorithm by B. Gross and D. Prasad [14, p. 993] determines characters $\chi_1 \in \widehat{\mathcal{A}_1}$ and $\chi_2 \in \widehat{\mathcal{A}_2}$ by

$$\chi_1(\delta_i) = (-1)^{\#m-i+1>} \quad \text{and} \quad \chi_2(\varepsilon_j) = (-1)^{\#m-j+\frac{1}{2}<},$$

where $\#m - i + 1 >$ is the cardinality of the set

$$\{j : m - i + 1 > \text{the coefficients of } f_j\},$$

and $\#m - j + \frac{1}{2} <$ is the cardinality of the set

$$\{i : m - j + \frac{1}{2} < \text{the coefficients of } e_i\}.$$

As discussed before we normalize the quasisplit form by

$$SO(m + 1, m + 1) \times SO(m, m + 1) \quad \text{if } m \text{ is even},$$
$$SO(m + 2, m) \times SO(m + 1, m) \quad \text{if } m \text{ is odd}.$$

Applying the formulæ in [14, (12.21)] we define the integers p and q with $0 \le p \le m$ and $0 \le q \le m$ by

$$p = \#\{i : \chi_1(\delta_i) = (-1)^i\} \quad \text{and} \quad q = \#\{j : \chi_2(\varepsilon_j) = (-1)^{m+j}\}$$

and we get the pure forms

$$SO(2m - 2p + 1, 2p + 1) \times SO(2q, 2m - 2q + 1) \quad \text{if } m \text{ is even}, \tag{11.1}$$
$$SO(2p + 1, 2m - 2p + 1) \times SO(2m - 2q, 2q + 1) \quad \text{if } m \text{ is odd}. \tag{11.2}$$

In our setting, we get the pair of integers $(p, q) = (0, m)$ for m even; $(p, q) = (m, 0)$ for m odd. Applying [14, (12.22)] with correction by changing n by m *loc.cit.*, we deduce that this character defines the pure inner form

$$SO(2m + 1, 1) \times SO(2m, 1) \text{ for } m \text{ even and odd}.$$

The only representation in $VP(\overline{\Pi}_{m,\delta}) \times VP(\overline{\pi}_m)$ for this pair of pure inner forms is $\overline{\Pi}_{m,\delta} \boxtimes \overline{\pi}_m$. Hence Theorem 15.19 implies the Gross–Prasad conjecture in that case.

11.5.2 The Gross–Prasad Conjecture II: Symmetry Breaking from the Discrete Series Representation π_m to $\varpi_{m-1,(-1)^m}$

We now consider the Vogan packet of tempered representations containing the pair $(SO(2m, 1) \times SO(2m - 1, 1), \overline{\pi}_m \boxtimes \overline{\varpi}_{m-1,(-1)^m})$, i.e., the Vogan packet

$$VP(\overline{\pi}_m \boxtimes \overline{\varpi}_{m-1,(-1)^m}) \subset VP(\overline{\pi}_m) \times VP(\overline{\varpi}_{m-1,(-1)^m}).$$

The packet $VP(\overline{\pi}_m) \times VP(\overline{\varpi}_{m,(-1)^m})$ is parametrized by characters of the finite group

$$\mathcal{A}_2 \times \mathcal{A}_3 \simeq (\mathbb{Z}/2\mathbb{Z})^m \times (\mathbb{Z}/2\mathbb{Z})^{m-1} \simeq (\mathbb{Z}/2\mathbb{Z})^{2m-1}.$$

Again the algorithm by B. Gross and D. Prasad determines a pair $(\chi_2, \chi_3) \in \widehat{\mathcal{A}}_2 \times \widehat{\mathcal{A}}_3$ and hence representations

$$(\overline{\pi}(\chi_2), \overline{\varpi}(\chi_3)) \in VP(\overline{\pi}_m) \times VP(\overline{\varpi}_{m-1,(-1)^m})$$

so that

$$\mathrm{Hom}_{\overline{G}(\chi_3)}(\overline{\pi}(\chi_2)|_{\overline{G}(\chi_3)}, \overline{\varpi}(\chi_3)) \neq \{0\},$$

where $\overline{G}(\chi_3)$ is the pure inner form determined by χ_3.

Let $T_{\mathbb{C}}$ be a torus in $SO(2m+1, \mathbb{C}) \times SO(2m, \mathbb{C})$ and $X^*(T_{\mathbb{C}})$ the character group. Fix a basis

$$X^*(T_{\mathbb{C}}) = \mathbb{Z}f_1 \oplus \mathbb{Z}f_2 \oplus \cdots \oplus \mathbb{Z}f_m \oplus \mathbb{Z}g_1 \oplus \mathbb{Z}g_2 \oplus \cdots \oplus \mathbb{Z}g_m$$

such that the standard root basis Δ_0 is given by

$$f_1 - f_2, f_2 - f_3, \cdots, f_{m-1} - f_m, f_m, g_1 - g_2, g_2 - g_3, \cdots, g_{m-1} - g_m, g_{m-1} + g_m$$

for $m \geq 2$. Take $\varepsilon = (-1)^m$ as before.

We identify the Langlands parameter of the Vogan packet

$$VP(\overline{\pi}_m) \times VP(\overline{\varpi}_{m,(-1)^m})$$

with

$$(m - \frac{1}{2})f_1 + (m - \frac{3}{2})f_2 + \ldots + \frac{1}{2}f_m + (m-1)g_1 + (m-2)g_2 + \cdots + g_{m-1} + 0g_m.$$

Again applying [14, Prop. 12.18] we define characters $\chi_2 \in \widehat{\mathcal{A}}_2$, $\chi_3 \in \widehat{\mathcal{A}}_3$ as follows: Let $\varepsilon_j \in \mathcal{A}_2 \simeq (\mathbb{Z}/2\mathbb{Z})^m$ be the element which is -1 in the j-th factor and equal to 1 everywhere else as in Section 11.4; $\gamma_k \in \mathcal{A}_3 \simeq (\mathbb{Z}/2\mathbb{Z})^{m-1}$ the element which is -1 in the k-th factor and 1 everywhere else. Then $\chi_2 \in \widehat{\mathcal{A}}_2$ and $\chi_3 \in \widehat{\mathcal{A}}_3$ are determined by

$$\chi_2(\varepsilon_j) = (-1)^{\#m-j+1/2<} \quad \text{and} \quad \chi_3(\gamma_k) = (-1)^{\#m-k>},$$

where $\#m - j + \frac{1}{2} <$ is the cardinality of the set

$$\{k : m - j + \frac{1}{2} < \text{the coefficients of } g_k\},$$

and $\#m - k >$ is the cardinality of the set

$$\{j : m - k > \text{ the coefficients of } f_j\}.$$

As discussed we normalize the quasisplit form by

$$SO(m+1, m) \times SO(m+1, m-1) \quad \text{if } m \text{ is even,}$$
$$SO(m, m+1) \times SO(m, m) \qquad\qquad \text{if } m \text{ is odd.}$$

We define the integers p and q with $0 \le p \le m$ and $0 \le q \le m - 1$ by

$$p = \#\{j : \chi_2(\varepsilon_j) = (-1)^j\} \quad \text{and} \quad q = \#\{k : \chi_3(\gamma_k) = (-1)^{m+k}\},$$

and we get

$$SO(2m - 2p + 1, 2p) \times SO(2q + 1, 2m - 2q - 1) \quad \text{if } m \text{ is even,}$$
$$SO(2p + 1, 2m - 2p) \times SO(2m - 2q - 1, 2q + 1) \quad \text{if } m \text{ is odd.}$$

In our setting, the pair of integers (p, q) is given by $(p, q) = (m, 0)$ for m even; $(p, q) = (0, m - 1)$ for m odd. We deduce that this character defines the pure inner form

$$SO(1, 2m) \times SO(1, 2m - 1) \text{ for } m \text{ even and odd.}$$

The only representation in $VP(\overline{\pi}_m) \times VP(\overline{\omega}_{m-1, (-1)^m})$ with this pair of pure inner forms is $(\overline{\pi}_m, \overline{\omega}_{m-1, (-1)^m})$.

In Chapter 4, we have determined

$$\text{Hom}_{G'}(\Pi \boxtimes \pi, \mathbb{C}) \quad \text{for all } \Pi \in \text{Irr}(G)_\rho \text{ and } \pi \in \text{Irr}(G')_\rho,$$

see Theorems 4.1 and 4.2 and also Theorem 5.4 for orthogonal groups

$$G \times G' = O(n+1, 1) \times O(n, 1),$$

from which we deduce analogous results about

$$\text{Hom}_{\overline{G'}}(\overline{\Pi} \boxtimes \overline{\pi}, \mathbb{C}) \quad \text{for all } \overline{\Pi} \in \text{Irr}(\overline{G})_\rho \text{ and } \overline{\pi} \in \text{Irr}(\overline{G'})_\rho,$$

for the special orthogonal groups

$$\overline{G} \times \overline{G'} = SO(n+1, 1) \times SO(n, 1),$$

in Theorem 15.19. By the aforementioned argument, Theorem 15.19 implies the following.

Theorem 11.6. *The conjectures by B. Gross and D. Prasad [14] for tempered representations of special orthogonal groups $SO(n+1,1) \times SO(n,1)$ with trivial infinitesimal character ρ hold.*

Remark 11.7. The Gross–Prasad conjectures concern tempered representations with trivial infinitesimal character ρ, but one may expect similar results for unitary representations of orthogonal groups with integral infinitesimal character. Considering "Arthur–Vogan packets" instead of the Vogan packets will include other unitary representations which are of interest to number theory for example to the representation $A_\mathfrak{q}(\lambda)$. Low dimensional examples and our results suggest that there exists pairs of groups $\overline{G} \times \overline{G'} = SO(p+1,q) \times SO(p,q)$ and of representations $U_1 \boxtimes U_2$ in this "Arthur–Vogan packet" so that $\mathrm{Hom}_{\overline{G'}}(U_1|_{\overline{G'}} \boxtimes U_2, \mathbb{C}) \neq \{0\}$. The examples also suggest an algorithm to determine pairs of groups and the pairs of representations with nontrivial multiplicity.

Chapter 12
Application II: Periods, Distinguished Representations and $(\mathfrak{g}, K)$-cohomologies

Let H be a subgroup of G. Following the terminology used in automorphic forms and the relative trace formula, we say that a smooth representation U of G is H-distinguished if there exists a nontrivial H-invariant linear functional

$$F^H : U \to \mathbb{C},$$

i.e., if U has a nontrivial H-period F^H. We consider first irreducible representations of G with infinitesimal character ρ which are H-distinguished for the pair $(G, H) = (O(n+1, 1), O(m+1, 1))$ or for the pair $(G, H) = (O(n, 1) \times O(m, 1), O(m, 1))$ with $m \leq n$. We then discuss a bilinear form on the $(\mathfrak{g}, K)$-cohomology of the representations of $(O(n+1, 1) \times O(n, 1))$ with infinitesimal character ρ which is induced by a symmetry breaking operator.

12.1 Periods and $O(n, 1)$-distinguished Representations

12.1.1 Periods

Let $\mathbb{K}$ be a number field, $\mathbb{A}$ its adels and let $G_1 \times G_2$ be a direct product of semisimple groups over a number field $\mathbb{K}$. We assume that $G_2 \subset G_1$. If the outer tensor product representation $\Pi_\mathbb{A} \boxtimes \pi_\mathbb{A}$ is an automorphic representation of the direct product group $G_1(\mathbb{A}) \times G_2(\mathbb{A})$, then the G_2-period integral is defined as

$$\int_{G_2(\mathbb{K})\backslash G_2(\mathbb{A})} \Phi_1(h)\phi_2(h)dh.$$

© Springer Nature Singapore Pte Ltd. 2018
T. Kobayashi, B. Speh, *Symmetry Breaking for Representations of Rank One Orthogonal Groups II*, Lecture Notes in Mathematics 2234,
https://doi.org/10.1007/978-981-13-2901-2_12

Here Φ_1 and ϕ_2 are smooth vectors for the representation $\Pi_{\mathbb{A}} \boxtimes \pi_{\mathbb{A}}$. If $\Pi_{\mathbb{A}} \boxtimes \pi_{\mathbb{A}}$ is cuspidal, then the integral converges and it defines a $G_2(\mathbb{A})$-invariant linear functional on the smooth vectors of $\Pi_{\mathbb{A}} \boxtimes \pi_{\mathbb{A}}$. If this linear functional is not zero, then $\Pi_{\mathbb{A}} \boxtimes \pi_{\mathbb{A}}$ is called G_2-*distinguished*. Conjecturally for certain pairs of groups the value of this integral is a multiple of the central value of an L-function, see [12, 18, 19].

Often this period integral factors into a product of local integrals. Following the global terminology we say that an admissible smooth representation $\Pi \boxtimes \pi$ of the direct product group $G_1(\mathbb{R}) \times G_2(\mathbb{R})$ is $G_2(\mathbb{R})$-*distinguished* if there is a nontrivial continuous linear functional

$$F^{G_2(\mathbb{R})} : \Pi \boxtimes \pi \to \mathbb{C}$$

which is invariant by $G_2(\mathbb{R})$ under the diagonal action. Here we recall Section 5.1.2 for the topology on the tensor product. If $\Pi \boxtimes \pi$ is $G_2(\mathbb{R})$-distinguished, we say that $F^{G_2(\mathbb{R})}$ is a *period* of $\Pi \boxtimes \pi$. We say that the period is nontrivial on a vector $\Phi \otimes \phi \in \Pi \boxtimes \pi$ if $\Phi \otimes \phi$ is not in the kernel of $F^{G_2(\mathbb{R})}$. If the period is nontrivial on a unit function $\Phi \otimes \phi$, we refer to its image as the value of the period on $\Phi \otimes \phi$.

Remark 12.1. The integral

$$\int_{G_2(\mathbb{R})} \Phi(h)\phi(h)dh$$

converges for some smooth vectors of discrete series representations $\Pi \boxtimes \pi$ for some symmetric pairs $(G_1(\mathbb{R}), G_2(\mathbb{R}))$. This was used by J. Vargas [58] to determine some subrepresentations in the restriction of some discrete series representations Π of $G_1(\mathbb{R})$ to the subgroup $G_2(\mathbb{R})$.

We recall from Theorem 5.4 that the space of symmetry breaking operators

$$\mathrm{Hom}_{G_2(\mathbb{R})}(\Pi|_{G_2(\mathbb{R})}, \pi^{\vee})$$

and the space of $G_2(\mathbb{R})$-invariant continuous linear functionals

$$\mathrm{Hom}_{G_2(\mathbb{R})}(\Pi \boxtimes \pi, \mathbb{C})$$

are naturally isomorphic to each other. Thus, instead of considering a $G_2(\mathbb{R})$-equivariant continuous linear functional defined by an integral, we may use symmetry breaking operators to construct $G_2(\mathbb{R})$-invariant continuous linear functionals. This technique allows us to obtain $G_2(\mathbb{R})$-invariant continuous linear functionals not only for discrete series representations but also for nontempered representations. Thus we can determine for the pair $(G, G') = (O(n + 1, 1), O(n, 1))$ the dimension of the space $\mathrm{Hom}_{G'}(\Pi \boxtimes \pi, \mathbb{C})$ for all $\Pi \in \mathrm{Irr}(G)_\rho$ and $\pi \in \mathrm{Irr}(G')_\rho$ as follows.

Corollary 12.2. *Suppose* $0 \le i \le n+1$, $0 \le j \le n$, *and* δ, $\varepsilon \in \{\pm\}$. *Let* $\Pi_{i,\delta}$ *and* $\pi_{j,\varepsilon}$ *be irreducible admissible smooth representations of* $G = O(n+1, 1)$ *and* $G' = O(n, 1)$, *respectively, that have the trivial infinitesimal character* ρ *as in* (2.35). *Then the following three conditions on* $(i, j, \delta, \varepsilon)$ *are equivalent:*

(i) $\mathrm{Hom}_{G'}(\Pi_{i,\delta} \boxtimes \pi_{j,\varepsilon}, \mathbb{C}) \ne \{0\}$;

(ii) $\dim_{\mathbb{C}} \mathrm{Hom}_{G'}(\Pi_{i,\delta} \boxtimes \pi_{j,\varepsilon}, \mathbb{C}) = 1$;

(iii) $j \in \{i, i-1\}$ *and* $\delta = \varepsilon$.

Proof. Owing to Theorem 5.4, this is a restatement of Theorems 4.1 and 4.2. □

12.1.2 Distinguished Representations

Let G be a reductive group, and H a reductive subgroup. We regard H as a subgroup of the direct product group $G \times H$ via the diagonal embedding $H \hookrightarrow G \times H$.

Definition 12.3. Let ψ be a one-dimensional representation of H. We say an admissible smooth representation Π of G is (H, ψ)-*distinguished* if

$$\mathrm{Hom}_H(\Pi \boxtimes \psi^\vee, \mathbb{C}) \simeq \mathrm{Hom}_H(\Pi|_H, \psi) \ne \{0\}.$$

If the character ψ is trivial, we say Π is H-*distinguished*.

In what follows, we deal mainly with the pair

$$(G, H) = (O(n+1, 1), O(m+1, 1)) \quad \text{for } m \le n.$$

Theorem 12.4. *Let* $0 \le i \le n+1$. *Then the representations* $\Pi_{i,\delta}$ ($\delta \in \{\pm\}$) *of* $G = O(n+1, 1)$ *are* $O(n+1-i, 1)$-*distinguished.*

The period is given by the composition of the symmetry breaking operators that we constructed in Chapter 10 with respect to the chain of subgroups

$$G = O(n+1, 1) \supset O(n, 1) \supset O(n-1, 1) \supset \cdots \supset O(m+1, 1) = H, \qquad (12.1)$$

as we shall see in the proof in Section 12.2. Without loss of generality, we consider the case $\delta = +$, and write simply Π_i for $\Pi_{i,+}$. We recall from Theorem 2.20 (3) that $\Pi_i \equiv \Pi_{i,+}$ has a minimal K-type $\mu^\flat(i, +) = \bigwedge^i(\mathbb{C}^{n+1}) \boxtimes \mathbf{1}$.

Let $v \in \bigwedge^i(\mathbb{C}^{n+1})$ be the image of $1 \in \mathbb{C}$ via the following successive inclusions:

$$\textstyle\bigwedge^i(\mathbb{C}^{n+1}) \supset \bigwedge^{i-1}(\mathbb{C}^n) \supset \cdots \supset \bigwedge^{i-l}(\mathbb{C}^{n+1-l}) \supset \cdots \supset \bigwedge^0(\mathbb{C}^{n+1-i}) \simeq \mathbb{C} \ni 1,$$

and we regard v as an element of the minimal K-type $\mu^\flat(i, +)$ of Π_i.

Theorem 12.5. *Let* Π_i *be the irreducible representation of* $G = O(n+1, 1)$, *and* v *be the normalized element of its minimal K-type as above. For* $0 \le i \le n$, *the value*

$F(v)$ *of the* $O(n+1-i, 1)$*-period* F *on* $v \in \Pi_i$ *is*

$$\frac{\pi^{\frac{1}{4}i(2n-i-1)}}{((n-i)!)^{i-1}} \times \begin{cases} \frac{1}{(n-2i)!} & \text{if } 2i < n+1, \\ (-1)^{n+1}(2i-n-1)! & \text{if } 2i \geq n+1. \end{cases}$$

12.1.3   Symmetry Breaking Operators from $\Pi_{i,\delta}$ to $\pi_{j,\delta}$ $(j \in \{i-1, i\})$

Let $(G, G') = (O(n+1, 1), O(n, 1))$. We recall from Theorem 2.20 (2) that

$$\mathrm{Irr}(G)_\rho = \{\Pi_{i,\delta} : 0 \leq i \leq n+1, \delta = \pm\},$$
$$\mathrm{Irr}(G')_\rho = \{\pi_{j,\varepsilon} : 0 \leq j \leq n, \varepsilon = \pm\}.$$

In Chapter 10, we constructed nontrivial symmetry breaking operators

$$A_{i,j} \colon \Pi_{i,\delta} \to \pi_{j,\varepsilon}$$

for $j \in \{i-1, i\}$ and $\delta = \varepsilon$, and investigated their (K, K')-spectrum for minimal K- and K'-types,

$$(\mu, \mu') = (\mu^\flat(i, \delta), \mu^\flat(j, \delta)'),$$

see Proposition 10.30 in the case $j = i-1$ and Proposition 10.12 in the case $j = i$.

For the proofs of Theorems 12.4 and 12.5, we use these operators $A_{i,j}$ in the case $j = i-1$. For the study of the bilinear forms on $(\mathfrak{g}, K)$-cohomologies (see Section 12.4 below), we shall use them in the case $j = i$.

12.2   Proofs of Theorems 12.4 and 12.5

We are ready to prove Theorems 12.4 and 12.5 by using Proposition 10.30 successively.

Proof of Theorem 12.4. Consider the chain (12.1) of orthogonal subgroups with $m = n - i$. For $1 \leq \ell \leq i$, we denote by

$$A_{\ell, \ell-1} \colon \Pi_{i-\ell+1}^{O(n-\ell+2, 1)} \to \Pi_{i-\ell}^{O(n-\ell+1, 1)}$$

the symmetry breaking operator given in Proposition 10.30 for the pair $(O(n-\ell+2, 1), O(n-\ell+1, 1))$ of groups. Here "$\Pi_{i-\ell}^{O(n-\ell+1, 1)}$" stands for the irreducible representation "$\Pi_{i-\ell,+}$" of the group $O(n-\ell+1,1)$ as given in Theorem 2.20, by a little abuse of notation. Then the composition

$$F := A_{1,0} \circ \cdots \circ A_{i-1,i-2} \circ A_{i,i-1} \tag{12.2}$$

defines a nonzero $O(n+1-i,1)$-invariant functional on the irreducible representation $\Pi_i \equiv \Pi_{i,+}$ of $G = O(n+1,1)$. $\square$

Proof of Theorem 12.5. The irreducible representation $\Pi_{i-\ell}^{O(n+1-\ell,1)}$, namely, "$\Pi_{i-\ell,+}$" of the group $O(n+1-\ell,1)$ has a minimal K-type

$$\mu^\flat(i-\ell,+)^{(\ell)} := \bigwedge^{i-\ell}(\mathbb{C}^{n+1-\ell}) \boxtimes \mathbf{1} \in \widehat{O(n+1-\ell)} \times \widehat{O(1)}.$$

The (K,K')-spectrum of the symmetry breaking operator $A_{\ell,\ell-1}: \Pi_{i-\ell+1}^{O(n-\ell+2,1)} \to \Pi_{i-\ell}^{O(n-\ell+1,1)}$ for the minimal K-types $\mu^\flat(i-\ell+1,+)^{(\ell-1)} \leftrightarrow \mu^\flat(i-\ell,+)^{(\ell)}$ is given by

$$\frac{\pi^{\frac{n-\ell}{2}}}{(n-i)!} \times \begin{cases} n-2i+\ell & \text{if } n \neq 2i-\ell, \\ 1 & \text{if } n = 2i-\ell, \end{cases}$$

by Proposition 10.30. Applying this formula successively to the sequence of minimal K-types:

$$\mu^\flat(i,+) \equiv \mu^\flat(i,+)^{(0)} \leftrightarrow \cdots \leftrightarrow \mu^\flat(i-\ell,+)^{(\ell)} \leftrightarrow \cdots \leftrightarrow \mu^\flat(0,+)^{(i)} = \mathbb{C},$$

we get

$$F(v) = \prod_{\ell=1}^{i} \frac{\pi^{\frac{n-\ell}{2}}(n-2i+\ell)}{(n-i)!} = \frac{\pi^{\frac{1}{4}i(2n-i-1)}}{((n-i)!)^{i-1}(n-2i)!}$$

if $n > 2i-1$.

On the other hand, if $n < 2i-1 < 2n-1$, then

$$F(v) = \left(\prod_{\ell=1}^{2i-n-1} \frac{\pi^{\frac{n-\ell}{2}}(n-2i+\ell)}{(n-i)!} \right) \cdot \frac{\pi^{n-i}}{(n-i)!} \cdot \left(\prod_{\ell=2i-n+1}^{i} \frac{\pi^{\frac{n-\ell}{2}}(n-2i+\ell)}{(n-i)!} \right)$$

$$= \frac{\pi^{\frac{1}{4}i(2n-i-1)}}{((n-i)!)^i} (((-1)^{2i-n-1}(2i-n-1)!) \cdot 1 \cdot ((n-i)!))$$

$$= \frac{(-1)^{n+1}\pi^{\frac{1}{4}i(2n-i-1)}(2i-n-1)!}{((n-i)!)^{i-1}}.$$

The cases $i = \frac{n+1}{2}$ (n: odd) or $i = n$ are treated separately, and it turns out that the formula of $F(v)$ coincides with the one for $i < 2i - 1 < 2n - 1$. Thus we have completed the proof of Theorem 12.5.   □

In the next theorem, we consider the pair

$$(G, H) = (O(n+1, 1), O(m+1, 1)) \text{ with } m \le n.$$

We write Π_i^G ($0 \le i \le n+1$) for the irreducible representation $\Pi_{i,+}$ of G (see (2.35)), and write π_j^H for the irreducible representation "$\Pi_{j,+}$" of the subgroup H for $0 \le j \le m+1$. Theorem 12.6 below generalizes Theorem 12.4, which corresponds to the case $j = 0$.

Theorem 12.6. *Let $0 \le i \le n+1$ and $0 \le j \le m+1$.*

(1) *The outer tensor product representation $\Pi_i^G \boxtimes \pi_j^H$ of the direct product group $G \times H$ has an H-period if $0 \le i - j \le n - m$.*
(2) *The period constructed by the composition of the symmetry breaking operators via the sequence (12.1) is nontrivial on the minimal K-type.*

Proof of Theorem 12.6. The proof is essentially the same with the one for Theorem 12.4 except that we use not only the surjective symmetry breaking operator $A_{i,i-1} \colon \Pi_{i,+} \to \pi_{i-1,+}$ for the pair $(G, G') = (O(n+1, 1), O(n, 1))$ but also the one

$$A_{i,i} \colon \Pi_{i,+} \to \pi_{i,+}$$

for which the (K, K')-spectrum on minimal K-types $\mu^\flat(i, +) \hookleftarrow \mu^\flat(i, +)'$ is nonzero by Proposition 10.12.

Composing the symmetry breaking operators $A_{k,k-1}$ or $A_{k,k}$ successively to the sequence (12.1) of orthogonal groups, we get a nonzero continuous H-homomorphism $\Pi_i^G \to \pi_j^H$ if $0 \le i - j \le n - m$. Then the first statement follows because π_j^H is self-dual. The second statement is clear by the construction and by the (K, K')-spectrum.   □

12.3   Bilinear Forms on $(\mathfrak{g}, K)$-cohomologies via Symmetry Breaking: General Theory for Nonvanishing

For the rest of this chapter, we discuss $(\mathfrak{g}, K)$-cohomologies via symmetry breaking. In this section, we deal with a general setting where $G \supset G'$ is a pair of real reductive Lie groups. We shall define natural bilinear forms on $(\mathfrak{g}, K)$-cohomologies and $(\mathfrak{g}', K')$-cohomologies via symmetry breaking $G \downarrow G'$, and prove a nonvanishing result (Theorem 12.11) in the general setting generalizing a theorem of B. Sun [54].

12.3.1   Pull-back of $(\mathfrak{g}, K)$-cohomologies via Symmetry Breaking

Let G be a real reductive Lie group, and K a maximal compact subgroup. We recall that the $(\mathfrak{g}, K)$-cohomology groups are the right derived functor of

$$\mathrm{Hom}_{\mathfrak{g}, K}(\mathbb{C}, *)$$

from the category of $(\mathfrak{g}, K)$-modules. Suppose further that G' is a real reductive subgroup such that $K' := K \cap G'$ is a maximal compact subgroup of G'. We write $\mathfrak{g}_{\mathbb{C}} = \mathfrak{k}_{\mathbb{C}} + \mathfrak{p}_{\mathbb{C}}$ and $\mathfrak{g}'_{\mathbb{C}} = \mathfrak{k}'_{\mathbb{C}} + \mathfrak{p}'_{\mathbb{C}}$ for the complexifications of the corresponding Cartan decompositions. In what follows, we set

$$d := \dim G'/K' = \dim_{\mathbb{C}} \mathfrak{p}'_{\mathbb{C}}.$$

We shall use the Poincaré duality for the subgroup G', which may be disconnected. In order to deal with disconnected groups, we consider the natural one-dimensional representation of K' defined by

$$\chi : K' \to GL_{\mathbb{C}}(\textstyle\bigwedge^d \mathfrak{p}'_{\mathbb{C}}) \simeq \mathbb{C}^{\times}. \tag{12.3}$$

The differential $d\chi$ is trivial on the Lie algebra $\mathfrak{k}'$. We extend χ to a $(\mathfrak{g}', K')$-module by letting $\mathfrak{g}'$ act trivially. Then we have

$$H^d(\mathfrak{g}', K'; \chi) \simeq \mathbb{C}. \tag{12.4}$$

Example 12.7.   For $G' = O(n, 1)$, the adjoint action of $K' \simeq O(n) \times O(1)$ on $\mathfrak{p}'_{\mathbb{C}} \simeq \mathbb{C}^n$ gives rise to the one-dimensional representation

$$\textstyle\bigwedge^n (\mathfrak{p}'_{\mathbb{C}}) \simeq \bigwedge^n (\mathbb{C}^n) \boxtimes (-1)^n.$$

Hence, in terms of the one-dimensional character χ_{ab} of $O(n, 1)$ defined in (2.13), the $(\mathfrak{g}', K')$-module χ defined in (12.3) is isomorphic to $\chi_{-, (-1)^n}$. See also Example 12.16 below.

Now we recall the Poincaré duality for $(\mathfrak{g}, K)$-cohomologies of $(\mathfrak{g}, K)$-modules when G is not necessarily connected:

Lemma 12.8 (Poincaré duality).   *Let χ be the one-dimensional $(\mathfrak{g}', K')$-module as in (12.3). Then for any irreducible $(\mathfrak{g}', K')$-module Y, there is a canonical perfect pairing*

$$H^j(\mathfrak{g}', K'; Y) \times H^{d-j}(\mathfrak{g}', K'; Y^{\vee} \otimes \chi) \to H^d(\mathfrak{g}', K'; \chi) \simeq \mathbb{C} \tag{12.5}$$

for all $j \in \mathbb{N}$.

Proof. See [24, Cor. 3.6] (see also [9, Chap. I, Sect. 1] when K is connected).   □

We use the terminology "symmetry breaking operator" also in the category of $(\mathfrak{g}, K)$-modules, when we are given a pair $(\mathfrak{g}, K)$ and $(\mathfrak{g}', K')$ such that $\mathfrak{g} \supset \mathfrak{g}'$ and $K \supset K'$. We prove the following.

Proposition 12.9. *Let X be a $(\mathfrak{g}, K)$-module, Y a $(\mathfrak{g}', K')$-module, and $Y^\vee$ the contragredient $(\mathfrak{g}', K')$-module of Y. Suppose $T: X \to Y$ is a $(\mathfrak{g}', K')$-homomorphism, where we regard the $(\mathfrak{g}, K)$-module X as a $(\mathfrak{g}', K')$ by restriction. Then the symmetry breaking operator T induces a canonical homomorphism*

$$T_*: H^j(\mathfrak{g}, K; X) \to H^j(\mathfrak{g}', K'; Y) \tag{12.6}$$

and a canonical bilinear form

$$B_T: H^j(\mathfrak{g}, K; X) \times H^{d-j}(\mathfrak{g}', K'; Y^\vee \otimes \chi) \to \mathbb{C} \tag{12.7}$$

for all $j \in \mathbb{N}$.

Proof. The $(\mathfrak{g}, K)$-module X is viewed as a $(\mathfrak{g}', K')$-module by restriction. Then the map of pairs $(\mathfrak{g}', K') \hookrightarrow (\mathfrak{g}, K)$ induces natural homomorphisms

$$H^j(\mathfrak{g}, K; X) \to H^j(\mathfrak{g}', K'; X) \quad \text{for all } j \in \mathbb{N}.$$

On the other hand, since $T: X \to Y$ is a $(\mathfrak{g}', K')$-homomorphism, it induces natural homomorphisms

$$H^j(\mathfrak{g}', K'; X) \to H^j(\mathfrak{g}', K'; Y) \quad \text{for all } j \in \mathbb{N}.$$

Composing these two maps, we get the homomorphisms (12.6).

In turn, combining the morphism (12.6) with the Poincaré duality in (12.5) in Lemma 12.8, we get the bilinear map B_T as desired.   □

12.3.2   Nonvanishing of Pull-back of $(\mathfrak{g}, K)$-cohomologies of $A_{\mathfrak{q}}$ via Symmetry Breaking

Retain the setting where (G, G') is a pair of real reductive Lie groups. In this subsection, we discuss a nonvanishing result for morphisms between $(\mathfrak{g}, K)$-cohomologies and $(\mathfrak{g}', K')$-cohomologies under certain assumption on the (K, K')-spectrum of the symmetry breaking operator, see Theorem 12.11 and Remark 12.12 below.

In order to formulate a nonvanishing theorem, we begin with a setup for finite-dimensional representations of compact Lie groups. Let U be a K-module, U' a K'-module, and $\varphi: U \to U'$ a K'-homomorphism. Via the inclusion map $\mathfrak{p}' \hookrightarrow \mathfrak{p}$, the composition of the following two morphisms

$$\mathrm{Hom}_K(\textstyle\bigwedge^j \mathfrak{p}_{\mathbb{C}}, U) \to \mathrm{Hom}_{K'}(\textstyle\bigwedge^j \mathfrak{p}_{\mathbb{C}}, U') \to \mathrm{Hom}_{K'}(\textstyle\bigwedge^j \mathfrak{p}'_{\mathbb{C}}, U')$$

induces natural homomorphisms

$$\varphi_* : \mathrm{Hom}_K(\textstyle\bigwedge^j \mathfrak{p}_{\mathbb{C}}, U) \to \mathrm{Hom}_{K'}(\textstyle\bigwedge^j \mathfrak{p}'_{\mathbb{C}}, U') \tag{12.8}$$

for all $j \in \mathbb{N}$.

Definition 12.10. A K'-homomorphism φ is said to be $\mathfrak{p}$-*nonvanishing at degree j* if the induced morphism φ_* in (12.8) is nonzero.

By a theorem of Vogan–Zuckerman [61] every irreducible representation of G with nontrivial $(\mathfrak{g}, K)$-cohomology is equivalent to the representation, to be denoted usually by $A_{\mathfrak{q}}$ for some θ-stable parabolic subalgebra $\mathfrak{q}$. Here $A_{\mathfrak{q}}$ is a $(\mathfrak{g}, K)$-module cohomologically induced from the trivial one-dimensional representation of the Levi subgroup $L = N_G(\mathfrak{q}) := \{g \in G : \mathrm{Ad}(g)\mathfrak{q} = \mathfrak{q}\}$. Suppose $\mathfrak{q} = \mathfrak{l}_{\mathbb{C}} + \mathfrak{u}$ and $\mathfrak{q}' = \mathfrak{l}'_{\mathbb{C}} + \mathfrak{u}'$ be θ-stable parabolic subalgebras of $\mathfrak{g}_{\mathbb{C}}$ and $\mathfrak{g}'_{\mathbb{C}}$, respectively. In general, we do not assume an inclusive relation of $\mathfrak{q}$ and $\mathfrak{q}'$. We shall work with a symmetry breaking operator $T : X \to Y$, where X is a $(\mathfrak{g}, K)$-module $A_{\mathfrak{q}}$ and Y is a $(\mathfrak{g}', K')$-module $A_{\mathfrak{q}'}$. We note that Y contains a unique minimal K'-type, say μ'. Let Y' be the K'-submodule containing all the remaining K'-types in Y, and

$$\mathrm{pr} : Y \to \mu'$$

be the first projection of the direct sum decomposition $Y = \mu' \oplus Y'$.

Theorem 12.11. *Let $T : X \to Y$ be a $(\mathfrak{g}', K')$-homomorphism, where X is a $(\mathfrak{g}, K)$-module $A_{\mathfrak{q}}$ and Y is a $(\mathfrak{g}', K')$-module $A_{\mathfrak{q}'}$. Let U be the representation space of the minimal K-type μ in X, and U' that of the minimal K'-type μ' in Y. We define a K'-homomorphism by*

$$\varphi_T := \mathrm{pr} \circ T|_U : U \to U'. \tag{12.9}$$

(1) *If φ_T is zero, then the homomorphisms $T_* : H^j(\mathfrak{g}, K; X) \to H^j(\mathfrak{g}', K'; Y)$ (see (12.6)) and the bilinear form B_T (see (12.7)) vanish for all degrees $j \in \mathbb{N}$.*
(2) *If φ_T is $\mathfrak{p}$-nonvanishing at degree j, then T_* and the bilinear forms B_T are nonzero for this degree j.*

Proof of Theorem 12.11. By Vogan–Zuckerman [61, Cor. 3.7 and Prop. 3.2], we have natural isomorphisms:

$$\mathrm{Hom}_K(\textstyle\bigwedge^j \mathfrak{p}_{\mathbb{C}}, U) \overset{\sim}{\to} \mathrm{Hom}_K(\textstyle\bigwedge^j \mathfrak{p}_{\mathbb{C}}, X) \overset{\sim}{\to} H^j(\mathfrak{g}, K; A_{\mathfrak{q}}). \tag{12.10}$$

By the definition (12.6) of T_* in Proposition 12.9 and φ_* (see (12.8)), the following diagram commutes:

$$
\begin{array}{ccc}
\operatorname{Hom}_K(\bigwedge^j \mathfrak{p}_{\mathbb{C}}, U) & \overset{\sim}{\to} \operatorname{Hom}_K(\bigwedge^j \mathfrak{p}_{\mathbb{C}}, X) \overset{\sim}{\to} H^j(\mathfrak{g}, K; X) \\
(T|_U)_* \downarrow & \circlearrowleft & \downarrow T_* \\
\operatorname{Hom}_{K'}(\bigwedge^j \mathfrak{p}'_{\mathbb{C}}, T(U)) \subset \operatorname{Hom}_{K'}(\bigwedge^j \mathfrak{p}'_{\mathbb{C}}, Y) \overset{\sim}{\to} H^j(\mathfrak{g}', K'; Y).
\end{array}
$$

Since $\operatorname{Hom}_{K'}(\bigwedge^j \mathfrak{p}'_{\mathbb{C}}, Y') = \{0\}$ for all j where $Y = \mu' \oplus Y'$ is the decomposition as a K'-module as before, we obtain the following commutative diagram by replacing $(T|_U)_*$ with $(\varphi_T)_*$:

$$
\begin{array}{ccc}
\operatorname{Hom}_K(\bigwedge^j \mathfrak{p}_{\mathbb{C}}, U) & \overset{\sim}{\to} \operatorname{Hom}_K(\bigwedge^j \mathfrak{p}_{\mathbb{C}}, X) \overset{\sim}{\to} H^j(\mathfrak{g}, K; X) \\
(\varphi_T)_* \downarrow & \circlearrowleft & \downarrow T_* \\
\operatorname{Hom}_{K'}(\bigwedge^j \mathfrak{p}'_{\mathbb{C}}, U') \overset{\sim}{\to} \operatorname{Hom}_{K'}(\bigwedge^j \mathfrak{p}'_{\mathbb{C}}, Y) \overset{\sim}{\to} H^j(\mathfrak{g}', K'; Y).
\end{array}
$$

Hence T_* is a nonzero map if and only if $(\varphi_T)_*$ is nonzero. Since the bilinear map (12.5) is a perfect pairing, we conclude Theorem 12.11.   □

Remark 12.12.

(1) The nonvanishing assumption of φ_T in the first statement of Theorem 12.11 can be reformulated as the nonvanishing of the (K, K')-spectrum (see Section 9.3) of the symmetry breaking operator T at (μ, μ').
(2) The verification of the $\mathfrak{p}$-vanishing assumption of φ_T in the second statement of Theorem 12.11 reduces to a computation of finite-dimensional representations of compact Lie groups K and K'.
(3) If we set $R := \dim_{\mathbb{C}}(\mathfrak{u} \cap \mathfrak{p}_{\mathbb{C}})$ and $R' := \dim_{\mathbb{C}}(\mathfrak{u}' \cap \mathfrak{p}'_{\mathbb{C}})$, then the isomorphisms [61, Cor. 3.7] show

$$
\operatorname{Hom}_K(\bigwedge^j \mathfrak{p}_{\mathbb{C}}, \mu) \simeq \operatorname{Hom}_{L \cap K}(\bigwedge^{j-R}(\mathfrak{l}_{\mathbb{C}} \cap \mathfrak{p}_{\mathbb{C}}), \mathbb{C}),
$$

$$
\operatorname{Hom}_{K'}(\bigwedge^j \mathfrak{p}'_{\mathbb{C}}, \mu') \simeq \operatorname{Hom}_{L' \cap K'}(\bigwedge^{j-R'}(\mathfrak{l}'_{\mathbb{C}} \cap \mathfrak{p}'_{\mathbb{C}}), \mathbb{C}).
$$

12.4 Nonvanishing Bilinear Forms on $(\mathfrak{g}, K)$-cohomologies via Symmetry Breaking for $(G, G') = (O(n+1, 1), O(n, 1))$

12.4.1 Nonvanishing Theorem for $O(n+1, 1) \downarrow O(n, 1)$

In this section, we apply the general result (Theorem 12.11) to the pair $(G, G') = (O(n+1, 1), O(n, 1))$.

In Proposition 14.45 in Appendix I, we shall see that if Π is an irreducible unitary representation of $G = O(n+1, 1)$ with $H^*(\mathfrak{g}, K; \Pi_K) \neq \{0\}$, then the smooth representation Π^∞ must be isomorphic to $\Pi_{\ell,\delta}$ defined in (2.35) for some $0 \leq \ell \leq n + 1$ and $\delta \in \{\pm\}$. Thus, we shall apply Theorem 12.11 to the representations $\Pi_{\ell,\delta}$ of G and similar representations $\pi_{m,\varepsilon}$ of the subgroup $G' = O(n, 1)$.

In what follows, by abuse of notation, we denote an admissible smooth representation and its underlying $(\mathfrak{g}, K)$-module by the same letter when we discuss their $(\mathfrak{g}, K)$-cohomologies.

Theorem 12.13. *Let* $(G, G') = (O(n+1, 1), O(n, 1))$, $0 \leq i \leq n$, *and* $\delta \in \{\pm\}$. *Let* $T := A_{i,i}$ *be the symmetry breaking operator* $\Pi_{i,\delta} \to \pi_{i,\delta}$ *given in Proposition 10.12.*

(1) *T induces bilinear forms*

$$B_T : H^j(\mathfrak{g}, K; \Pi_{i,\delta}) \times H^{n-j}(\mathfrak{g}', K'; \pi_{n-i,(-1)^n\delta}) \to \mathbb{C} \quad \text{for all } j.$$

(2) *The bilinear form B_T is nonzero if and only if $j = i$ and $\delta = (-1)^i$.*

Remark 12.14. A similar theorem was proved by B. Sun [54] for the $(\mathfrak{g}, K)$-cohomology with nontrivial coefficients of a tempered representation of the pair $(GL(n, \mathbb{R}), GL(n-1, \mathbb{R}))$.

We begin with the computation of the $(\mathfrak{g}, K)$-cohomologies of the irreducible representation $\Pi_{\ell,\delta}$ of $G = O(n+1, 1)$.

Lemma 12.15. *Suppose $0 \leq \ell \leq n+1$, $j \in \mathbb{N}$, and $\delta \in \{\pm\}$. Then*

$$H^j(\mathfrak{g}, K; \Pi_{\ell,\delta}) = \begin{cases} \mathbb{C} & \text{if } j = \ell \text{ and } \delta = (-1)^\ell, \\ \{0\} & \text{otherwise.} \end{cases}$$

In view of Theorem 2.20 (4), we have:

Example 12.16. For $G' = O(n, 1)$, we have $\pi_{n,(-1)^n} \simeq \chi_{-,(-1)^n}$ from Theorem 2.20 (4). In turn, the assertion $H^n(\mathfrak{g}', K'; \chi_{-,(-1)^n}) \simeq \mathbb{C}$ from Lemma 12.15 corresponds to the equation (12.4) by Example 12.7.

By Proposition 14.44 in Appendix I, Lemma 12.15 may be reformulated in terms of the cohomologically induced representations

$$(A_{\mathfrak{q}_i})_{ab} = A_{\mathfrak{q}_i} \otimes \chi_{ab} \simeq \mathcal{R}_{\mathfrak{q}_i}^{S_i}(\chi_{ab} \otimes \mathbb{C}_{\rho(\mathfrak{u}_i)})$$

(see Section 14.9.1 for notation) as follows:

Lemma 12.17. *Suppose $0 \leq i \leq [\frac{n+1}{2}]$ and $j \in \mathbb{N}$. Then we have*

$$H^j(\mathfrak{g}, K; (A_{\mathfrak{q}_i})_{++}) = \mathbb{C} \quad \text{if } j = i \quad \in 2\mathbb{N}; \quad = \{0\} \quad \text{otherwise,}$$

$$H^j(\mathfrak{g}, K; (A_{\mathfrak{q}_i})_{+-}) = \mathbb{C} \quad \text{if } j = i \quad \in 2\mathbb{N}+1; = \{0\} \quad \text{otherwise,}$$

$$H^j(\mathfrak{g}, K; (A_{\mathfrak{q}_i})_{-+}) = \mathbb{C} \quad \textit{if } j = n+1-i \in 2\mathbb{N}; \qquad = \{0\} \quad \textit{otherwise,}$$

$$H^j(\mathfrak{g}, K; (A_{\mathfrak{q}_i})_{--}) = \mathbb{C} \quad \textit{if } j = n+1-i \in 2\mathbb{N}+1; \ = \{0\} \quad \textit{otherwise.}$$

Proof of Lemma 12.17. We recall from Theorem 2.20 (3) (see also Proposition 14.44 in Appendix I) that the irreducible G-module $\Pi_{i,\delta}$ contains $\mu^\flat(i,\delta) \simeq \bigwedge^i(\mathbb{C}^{n+1}) \boxtimes \delta$ as its minimal K-type. By [61], we have then a natural isomorphism

$$\mathrm{Hom}_K(\textstyle\bigwedge^j \mathfrak{p}_\mathbb{C}, \mu^\flat(i,\delta)) \simeq H^j(\mathfrak{g}, K; \Pi_{i,\delta}).$$

On the other hand, the adjoint action of $K = O(n+1) \times O(1)$ on $\mathfrak{p}_\mathbb{C} \simeq \mathbb{C}^{n+1}$ gives rise to the j-th exterior tensor representation

$$\textstyle\bigwedge^j(\mathfrak{p}_\mathbb{C}) \simeq \bigwedge^j(\mathbb{C}^{n+1}) \boxtimes (-1)^j.$$

Now the lemma follows. □

Lemma 12.18. *Let φ_T be the K'-homomorphism defined in* (12.9) *for the symmetry breaking operator $T : \Pi_{i,\delta} \to \pi_{i,\delta}$ in Theorem 12.13. Then φ_T is $\mathfrak{p}$-nonvanishing at degree j (Definition 12.10) if and only if $j = i$ and $\delta = (-1)^i$.*

Proof. Similarly to the G-module $\Pi_{i,\delta}$, the G'-module $\pi_{i,\delta}$ contains $\mu^\flat(i,\delta)' \simeq \bigwedge^i(\mathbb{C}^n) \boxtimes \delta$ as its minimal K-type. Then φ_T in Theorem 12.11 amounts to a nonzero multiple of the projection (see (7.2)),

$$\mathrm{pr}_{i\to i} : \textstyle\bigwedge^i(\mathbb{C}^{n+1}) \boxtimes \delta \to \bigwedge^i(\mathbb{C}^n) \boxtimes \delta.$$

Then $(\varphi_T)_*$ is a nonzero multiple of the natural map from

$$\mathrm{Hom}_{O(n+1)\times O(1)}(\textstyle\bigwedge^j(\mathbb{C}^{n+1}) \boxtimes (-1)^j, \bigwedge^i(\mathbb{C}^{n+1}) \boxtimes \delta)$$

to

$$\mathrm{Hom}_{O(n)\times O(1)}(\textstyle\bigwedge^j(\mathbb{C}^n) \boxtimes (-1)^j, \bigwedge^i(\mathbb{C}^n) \boxtimes \delta)$$

induced by the projection $\mathrm{pr}_{i\to i}$. Now the lemma is clear. □

We are ready to apply the general result (Theorem 12.11) to prove Theorem 12.13.

Proof of Theorem 12.13. By Example 12.7, we have an isomorphism $\chi \simeq \chi_{-,(-1)^n}$ as $(\mathfrak{g}', K')$-modules. Then it follows from Theorem 2.20 (5) and (6) that there are natural G'-isomorphisms:

$$\pi_{i,\delta}^\vee \otimes \chi_{-,(-1)^n} \simeq \pi_{i,\delta} \otimes \chi_{-,(-1)^n} \simeq \pi_{n-i,(-1)^n\delta}.$$

Thus Theorem 12.13 (1) follows from Proposition 12.9. It then follows from Lemma 12.18 that Theorem 12.13 (2) holds as a special case of Theorem 12.11. □

In Proposition 14.44, we shall see that the underlying $(\mathfrak{g}', K')$-module of $\pi_{n-i,(-1)^n\delta}$ is isomorphic to $(A_{\mathfrak{q}'_i})_{-,(-1)^n\delta}$ if $0 \leq i \leq [\frac{n}{2}]$. The symmetry breaking operator $A_{i,i} : \Pi_{i,\delta} \to \pi_{i,\delta}$ given in Proposition 10.12 induces a $(\mathfrak{g}', K')$-homomorphism $(A_{\mathfrak{q}_i})_{+,\delta} \to (A_{\mathfrak{q}'_i})_{+,\delta}$.

Corollary 12.19. *If* $0 \leq 2i \leq n$, *then the symmetry breaking operator* $A_{i,i} : \Pi_{i,\delta} \to \pi_{i,\delta}$ *induces bilinear forms*

$$H^j(\mathfrak{g}, K; (A_{\mathfrak{q}_i})_{+,\delta}) \times H^{n-j}(\mathfrak{g}', K'; (A_{\mathfrak{q}'_i})_{-,(-1)^n\delta}) \to \mathbb{C}$$

and linear maps

$$H^j(\mathfrak{g}, K; (A_{\mathfrak{q}_i})_{+,\delta}) \to H^j(\mathfrak{g}', K'; (A_{\mathfrak{q}'_i})_{+,\delta})$$

for all j. *They are nontrivial if and only if* $j = i$ *and* $\delta = (-1)^i$.

Composing the symmetry breaking operators we deduce the following.

Corollary 12.20. *If* $0 \leq 2i \leq n$ *and* $H = O(n+1-i, 1)$, *then the composition of the symmetry breaking operators induces a linear map*

$$H^j(\mathfrak{g}, K; (A_{\mathfrak{q}_i})_{+,\delta}) \to H^j(\mathfrak{h}, K \cap H'; (A_{\mathfrak{q}_i\cap\mathfrak{h}})_{+,\delta}) \quad \text{for all } j.$$

It is nontrivial if and only if $j = i$ *and* $\delta = (-1)^{n+1-i}$.

Remark 12.21. Y. Tong and S. P. Wang [56, 63] considered representations of $SO(n+1, 1)$ with nontrivial $(\mathfrak{g}, K)$-cohomology which are $SO(n-i) \times SO(n+1-i, 1)$-distinguished. Independently S. Kudla and J. Millson [46] considered representations of $O(n+1, 1)$ with nontrivial $(\mathfrak{g}, K)$-cohomology which are $O(n-i) \times O(n+1-i, 1)$-distinguished. Since $O(n-i)$ commutes with $O(n-i+1, 1)$, we have an action of $O(n-i)$ on $\mathrm{Hom}_{O(n-i+1,1)}(\Pi_{i,\delta}, \mathbb{C})$ and $\mathrm{Hom}_{O(n-i+1,1)}(\Pi_{i,\delta}, \mathbb{C})^{O(n-i)}$ is isomorphic to $\mathrm{Hom}_{O(n-i)\times O(n-i+1,1)}(\Pi_{i,\delta}, \mathbb{C})$. By results in [46] this induces a nontrivial linear map on the $(\mathfrak{g}, K)$-cohomology.

12.4.2 Special Cycles

Geometric, topological and arithmetic properties of hyperbolic symmetric spaces $X_\Gamma = \Gamma\backslash O(n+1, 1)/K$ for a discrete subgroup Γ have been studied extensively using representation theoretic and geometric techniques. See for example [5, 6] and references therein. If X_Γ is compact, then the Matsushima–Murakami formula [9, Chap. VII, Thm. 3.2] shows

$$H^*(X_\Gamma, \mathbb{C}) \simeq \bigoplus_{\Pi\in\widehat{G}} m(\Gamma, \Pi) H^*(\mathfrak{g}, K; \Pi_K),$$

where $\widehat{G}$ is the set of equivalence classes of irreducible unitary representations of G (*i.e.*, the *unitary dual* of G), and we set for $\Pi \in \widehat{G}$

$$m(\Gamma, \Pi) := \dim_{\mathbb{C}} \mathrm{Hom}_G(\Pi, L^2(\Gamma \backslash G)).$$

By abuse of notation, we shall omit the subscript K in the underlying $(\mathfrak{g}, K)$-module Π_K of Π when we discuss its $(\mathfrak{g}, K)$-cohomologies.

In Proposition 14.45 in Appendix I, we shall show that every irreducible unitary representations with nontrivial $(\mathfrak{g}, K)$-cohomology is isomorphic to a representations $\Pi_{i,\delta}$ for some i and $\delta \in \{\pm\}$, see also Theorem 2.20 (9). Thus

$$H^*(X_\Gamma, \mathbb{C}) = \bigoplus_{i,\delta} m(\Gamma, \Pi_{i,\delta}) H^*(\mathfrak{g}, K; \Pi_{i,\delta}).$$

To obtain arithmetic information about the cohomology and the homology of X_Γ, special cycles, *i.e.,* orbits of subgroups $H \subset G$ on X_Γ, and their homology classes are frequently used. Suppose $0 \le 2i \le n + 1$. We let

$$G_i = O(n + 1 - i, 1), \quad K_i := K \cap G_i \simeq O(n + 1 - i) \times O(1),$$

and X_i be the Riemannian symmetric space G_i/K_i. Let $b_i := n + 1 - i$, the dimension of X_i. We set $\delta = (-1)^{n+1-i}$. By Corollary 12.20 there exists a nontrivial linear map $A^{n+1-i,n+1-i}$:

$$H^{n+1-i}(\mathfrak{g}, K; (A_{\mathfrak{q}_{n+1-i}})_{+,\delta}) \to H^{n+1-i}(\mathfrak{g}_i, K_i; (A_{\mathfrak{q}_{n+1-i} \cap (\mathfrak{g}_i)_{\mathbb{C}}})_{+,\delta}).$$

Note that $(A_{\mathfrak{q}_{n+1-i} \cap (\mathfrak{g}_i)_{\mathbb{C}}})_{+,\delta}$ is one-dimensional and the image of $A^{n+1-i,n+1-i}$ is isomorphic to

$$\mathrm{Hom}_{K_i}(\textstyle\bigwedge^{n+1-i}(\mathfrak{p}_{\mathbb{C}} \cap (\mathfrak{g}_i)_{\mathbb{C}}), \chi_{+,\delta}) \simeq \mathrm{Hom}_{K_i}(\textstyle\bigwedge^{n+1-i}(\mathbb{C}^{n+1-i}) \boxtimes \mathbf{1}, \mathbf{1}).$$

Since the nonzero element of

$$\mathrm{Hom}_{K_i}(\textstyle\bigwedge^{n+1-i}(\mathbb{C}^{n+1-i}) \boxtimes \mathbf{1}, \mathbf{1})$$

gives a volume form on the symmetric space $X_i = G_i/K_i$, this suggests that the homology classes defined by the orbits of $O(n + 1 - i, 1)$ for $0 \le 2i \le n + 1$ on X_Γ are related to the contribution of $H^{n+1-i}(\mathfrak{g}, K; \Pi_{i,\delta})$ to the cohomology of X_Γ. The work of S. Kudla and J. Millson confirms this. We sketch their results following the exposition in [46–48].

We have an embedding

$$\iota_{X_i} : X_i \hookrightarrow X = G/K.$$

We fix an orientation of X and X_i which is invariant under the connected component of G respectively G_i. Let $\mathbb{A}$ be the adels of the real number field $\mathbb{K}$. Then

$$X_{\mathbb{A}} = X \otimes G(\mathbb{A}_f)$$

is the adelic symmetric space. We set $G^+ := \prod SO_0(p, q)$ where we take the product over all real places of $\mathbb{K}$ and $G^+(\mathbb{K}) := G(\mathbb{K}) \cap G^+ G(\mathbb{A}_f)$. Then

$$H^*(G(\mathbb{K})\backslash G(\mathbb{A}_f); \mathbb{C}) = H^*(\mathfrak{g}, K; C^\infty(G(\mathbb{Q})\backslash G(\mathbb{A})))$$

and

$$H^*(G^+(\mathbb{K})\backslash G(\mathbb{A})/K K_f; \mathbb{C}) = H^*(X_{\mathbb{A}}; \mathbb{C})^{K_f}.$$

The cohomology here is the de Rham cohomology if $X_{\mathbb{A}}$ is compact, otherwise the cohomology with compact support.

Following the exposition and notation in [48, Sect. 2] we have an inclusion

$$\iota_{X_i} : X_i \times G_i(\mathbb{A}_f) \to X \times G(\mathbb{A}_f)$$

which is equivariant under the right action of $G(\mathbb{A}_f)$. For $g \in G(\mathbb{A}_f)$ we obtain a special cycle

$$X_{i,g} = X_i G_i(\mathbb{A}_f)/(g K_f g^{-1} \cap G_i(\mathbb{A}_f)).$$

Consider the subspace $SX_i(X_{\mathbb{A}})$ spanned by special cycles in the homology group $H_i(X_{\mathbb{A}})$.

We now assume that all but one factor of G_∞ is compact and thus that $X_{\mathbb{A}}/K_f$ is compact. Using the theta correspondence, S. Kudla and J. Millson show that there exist a subgroup K_f and nontrivial homomorphisms

$$\Psi : H^{b_i}(\mathfrak{g}, K; \Pi) \to H^{b_i}(X_{\mathbb{A}}/K_f; \mathbb{C}) \subset H^{b_i}(X_{\mathbb{A}})$$

for some irreducible representation Π of G.

Using integration, S. Kudla and J. Millson [46], [48, Thm. 7.1] prove the following:

Theorem 12.22. *There exists a nontrivial pairing*

$$\Psi(H^{n+1-i}(\mathfrak{g}, K; \Pi)) \times SX_i(X_{\mathbb{A}}) \to \mathbb{C}.$$

Remark 12.23.

(1) As we see in Theorem 2.20 (9), Lemma 12.17 and Proposition 14.44, the irreducible representation Π of G with $H^{n+1-i}(\mathfrak{g}, K; \Pi) \neq \{0\}$ must be of the

form

$$\Pi \simeq \Pi_{n+1-i,(-1)^{n+1-i}},$$

namely, $\Pi_K \simeq (A_{\mathfrak{q}_i})_{-,(-1)^{n+1-i}}$.

(2) The nontrivial pairing in Theorem 12.22 defines an $O(n+1-i, 1)$-invariant linear functional on the irreducible G-module $\Pi_{n+1-i,(-1)^{n+1-i}}$ which is non-trivial on the minimal K-type.

Chapter 13
A Conjecture: Symmetry Breaking for Irreducible Representations with Regular Integral Infinitesimal Character

We conjecture that Theorems 4.1 and 4.2 hold in more generality. We will formalize and explain this conjecture in this chapter more precisely and provide some supporting evidence.

As before we assume that $G = O(n + 1, 1)$ and $G' = O(n, 1)$.

13.1 Hasse Sequences and Standard Sequences of Irreducible Representations with Regular Integral Infinitesimal Character and Their Langlands Parameters

Before stating the conjecture we define Hasse sequences and standard sequences of irreducible representations, and collect more information about the representations which occur in the Hasse and standard sequences. In Chapter 14 (Appendix I) we determine their θ-stable parameters.

13.1.1 Definition of Hasse Sequence and Standard Sequence

Definition-Theorem 13.1 (Hasse sequence). Let $n = 2m$ or $2m - 1$. For every irreducible finite-dimensional representation F of the group $G = O(n + 1, 1)$, there exists uniquely a sequence

$$U_0 , \dots , U_{m-1} , U_m$$

© Springer Nature Singapore Pte Ltd. 2018
T. Kobayashi, B. Speh, *Symmetry Breaking for Representations of Rank One Orthogonal Groups II*, Lecture Notes in Mathematics 2234,
https://doi.org/10.1007/978-981-13-2901-2_13

of irreducible admissible smooth representations $U_i \equiv U_i(F)$ of G such that

1. $U_0 \simeq F$;
2. consecutive representations are composition factors of a principal series representation;
3. U_i $(0 \leq i \leq m)$ are pairwise inequivalent as G-modules.

We refer to the sequence

$$U_0 , \ldots , U_{m-1} , U_m$$

as the *Hasse sequence* of irreducible representations starting with the finite-dimensional representation $U_0 = F$. We shall write $U_j(F)$ for U_j if we emphasize the sequence $\{U_j(F)\}$ starts with $U_0 = F$.

Sketch of the Proof. D. Collingwood [11, Chap. 6] computed embeddings of irreducible Harish-Chandra modules into principal series representations for all connected simple groups of real rank one, which allowed him to define a diagrammatic description of irreducible representations with regular integral infinitesimal character of the connected group $G_0 = SO_0(n+1, 1)$. For the disconnected group $G = O(n+1, 1)$, we can determine similarly the composition factors of principal series representations, as in Theorems 13.7 and 13.9 below (see Sections 15.1–15.5 in Appendix II for the relationship between irreducible representations of the disconnected group $G = O(n+1, 1)$ and those of a normal subgroup of finite index). To show the existence and the uniqueness of the Hasse sequence, we note that there exists uniquely a principal series representation that contains a given irreducible finite-dimensional representation F as a subrepresentation. Then there exists only one irreducible composition factor other than F, which is defined to be U_1. Repeating this procedure, we can find irreducible representations U_2, U_3, $\cdots$, whence the existence and the uniqueness of the Hasse sequence is shown for the disconnected group $G = O(n+1, 1)$. □

As we have seen in Theorem 2.20 (1) when F is the trivial one-dimensional representation **1**, the representations U_i and U_{i+1} in this sequence have different signatures. The standard sequence (Definition 2.21) starting with **1** has an adjustment for the different signatures. Extending this definition for the sequence starting with an *arbitrary* irreducible finite-dimensional representation F, we define the standard sequence of irreducible representations starting with F as follows:

Definition 13.2 (standard sequence). If

$$U_0 , \ldots , U_{m-1} , U_m$$

is the Hasse sequence starting with an irreducible finite-dimensional representation F of G, then we refer to

$$\Pi_0 := U_0 , \ldots , \Pi_{m-1} := U_{m-1} \otimes (\chi_{+-})^{m-1} , \Pi_m := U_m \otimes (\chi_{+-})^m$$

as the *standard sequence* of irreducible representations $\Pi_i = \Pi_i(F)$ starting with $\Pi_0 = U_0 = F$, where χ_{+-} is the one-dimensional representation of G defined in (2.13).

Remark 13.3. Clearly, any $U_j(F)$ in the Hasse sequence (or any $\Pi_j(F)$ in the standard sequence) starting with an irreducible finite-dimensional representation F of G has a regular integral $\mathfrak{Z}_G(\mathfrak{g})$-infinitesimal character (Definition 2.1).

The next proposition follows readily from the definition.

Proposition 13.4 (tensor product with characters). *Let F be an irreducible finite-dimensional representation of G, and χ a one-dimensional representation of G. Then the representations in the Hasse sequences (and in the standard sequence) starting with F and $F \otimes \chi$ have the following relations:*

$$\text{(Hasse sequence)} \qquad U_i(F) \otimes \chi \simeq U_i(F \otimes \chi),$$

$$\text{(standard sequence)} \qquad \Pi_i(F) \otimes \chi \simeq \Pi_i(F \otimes \chi).$$

The Hasse sequences and the standard sequences starting with one-dimensional representations of G are described as follows.

Example 13.5. We recall from Theorem 2.20 that $\Pi_{\ell,\delta}$ $(0 \le \ell \le n+1, \delta \in \{\pm\})$ are irreducible representations of $G = O(n+1, 1)$ with $\mathfrak{Z}_G(\mathfrak{g})$-infinitesimal character ρ_G. Then for each one-dimensional representation $F \simeq \chi_{\pm\pm}$ of G (see (2.13)), the Hasse sequence $U_i(F)$ $(0 \le i \le [\frac{n+1}{2}])$ that starts with $U_0(F) \simeq F$, and the standard sequence $\Pi_i(F) := U_i(F) \otimes (\chi_{+-})^i$ are given as follows.

$$
\begin{aligned}
U_i(\mathbf{1}) &= \Pi_{i,(-1)^i}, & \Pi_i(\mathbf{1}) &= \Pi_{i,+}, \\
U_i(\chi_{+-}) &= \Pi_{i,(-1)^{i+1}}, & \Pi_i(\chi_{+-}) &= \Pi_{i,-}, \\
U_i(\chi_{-+}) &= \Pi_{n+1-i,(-1)^i}, & \Pi_i(\chi_{-+}) &= \Pi_{n+1-i,+}, \\
U_i(\chi_{--}) &= \Pi_{n+1-i,(-1)^{i+1}}, & \Pi_i(\chi_{--}) &= \Pi_{n+1-i,-}.
\end{aligned}
$$

13.1.2 Existence of Hasse Sequence

In Section 13.2, we formalize a conjecture about when

$$\mathrm{Hom}_{G'}(\Pi|_{G'}, \pi) \ne \{0\}$$

for $\Pi \in \mathrm{Irr}(G)$ and $\pi \in \mathrm{Irr}(G')$ that have regular integral infinitesimal characters by using the standard sequence (Definition 13.2). The formulation is based on the following theorem which asserts that the converse statement to Remark 13.3 is also true.

Theorem 13.6. *Any irreducible admissible representation of G of moderate growth with regular integral $\mathfrak{Z}_G(\mathfrak{g})$-infinitesimal character is of the form $U_j(F)$ in the Hasse sequence for some j $(0 \le j \le [\frac{n+1}{2}])$ and for some irreducible finite-dimensional representation F of G.*

Similarly, any irreducible admissible representation of G of moderate growth with regular $\mathfrak{Z}_G(\mathfrak{g})$-infinitesimal character is of the form $\Pi_j(F')$ in the standard sequence for some j $(0 \le j \le [\frac{n+1}{2}])$ and for some irreducible finite-dimensional representation F' of G.

The proof of Theorem 13.6 follows from the classification of $\mathrm{Irr}(G)$ (Theorem 14.36 in Appendix I) and the Langlands parameter of the representations in the Hasse sequence below (see also Theorem 14.35).

13.1.3   Langlands Parameter of the Representations in the Hasse Sequence

Let F be an irreducible finite-dimensional representation of $G = O(n+1,1)$. We now determine the Langlands parameter of the representations in the Hasse sequence $\{U_i(F)\}$ (and the standard sequence $\{\Pi_i(F)\}$) for $0 \le i \le [\frac{n+1}{2}]$ and their K-types.

We use the parametrization of the finite-dimensional representation of $O(n,1)$ introduced in Section 14.1 in Appendix I.

We begin with the case where F is obtained from an irreducible representation of $O(n+2)$ of type I (Definition 2.4) via the unitary trick. The description of $U_i(F)$ and $\Pi_i(F)$ for more general F can be derived from this case by taking the tensor product with one-dimensional representations $\chi_{\pm\pm}$ of G, see Theorem 13.11 below.

Case 1. $n = 2m$ and $G = O(2m+1,1)$.

For $F \in \widehat{O(n+2)}$ of type I, we define $\sigma^{(i)} \equiv \sigma^{(i)}(F) \in \widehat{O(n)}$ of type I for $0 \le i \le m = \frac{n}{2}$ as follows. We write $F = F^{O(n+2,\mathbb{C})}(s)$ with

$$s = (s_0, \cdots, s_m, 0^{m+1}) \in \Lambda^+(n+2) \equiv \Lambda^+(2m+2)$$

as in (2.20), and regard it as an irreducible finite-dimensional representation of $G = O(n+1,1)$. We set

$$\sigma^{(i)} := F^{O(n)}(s^{(i)}) \in \widehat{O(n)} \quad \text{for } 0 \le i \le m,$$

where $s^{(i)} \in \Lambda^+(n) \equiv \Lambda^+(2m)$ is given for $0 \le i \le m$ as follows:

$$s^{(i)} := (s_0 + 1, \cdots, s_{i-1} + 1, \widehat{s_i}, s_{i+1}, \cdots, s_m, 0^m). \tag{13.1}$$

It is convenient to introduce the *extended Hasse sequence* $\{U_i \equiv U_i(F)\}$ $(0 \leq i \leq 2m + 1)$ by defining

$$U_i(F) := U_{n+1-i}(F) \otimes \chi_{--} \quad \text{for } m + 1 \leq i \leq n + 1 = 2m + 1. \tag{13.2}$$

Theorem 13.7 ($n = 2m$). *Given an irreducible finite-dimensional representation $F^{O(n+2,\mathbb{C})}(s)$ of $G = O(n + 1, 1)$ with*

$$s = (s_0, s_1, \cdots, s_m, 0, \cdots, 0) \in \Lambda^+(n + 2)(= \Lambda^+(2m + 2)),$$

there exists uniquely an extended Hasse sequence U_0, U_1, $\cdots$, U_{2m+1} starting with the irreducible finite-dimensional representation $U_0 = F^{O(n+2,\mathbb{C})}(s)$. Moreover, the extended Hasse sequence U_0, $\cdots$, U_{2m+1} satisfies the following properties.

(1) *There exist exact sequences of G-modules:*

$$0 \to U_i \to I_{(-1)^{i-s_i}}(\sigma^{(i)}, i - s_i) \to U_{i+1} \to 0 \qquad (0 \leq i \leq m),$$

$$0 \to U_i \to I_{(-1)^{n-i-s_{n-i}}}(\sigma^{(n-i)} \otimes \det, i + s_{n-i}) \to U_{i+1} \to 0 \ (m \leq i \leq 2m).$$

(2) *The K-type formula of the irreducible G-module U_i $(0 \leq i \leq m)$ is given by*

$$\bigoplus_b F^{O(n+1)}(b) \boxtimes (-1)^{\sum_{k=0}^{m}(b_k - s_k)},$$

where $b = (b_0, b_1, \cdots, b_m, 0, \cdots, 0)$ runs over $\Lambda^+(n + 1) \equiv \Lambda^+(2m + 1)$ subject to

$$b_0 \geq s_0 + 1 \geq b_1 \geq s_1 + 1 \geq \cdots \geq b_{i-1} \geq s_{i-1} + 1,$$

$$s_i \geq b_i \geq s_{i+1} \geq b_{i+1} \geq \cdots \geq s_m \geq b_m \geq 0,$$

$$b_m \in \{0, 1\}.$$

In particular, the minimal K-type(s) of the G-module U_i $(0 \leq i \leq m)$ are given as follows:

for $s_m = 0$,

$$F^{O(n+1)}(s^{(i)}, 0) \boxtimes (-1)^{i-s_i}$$

$$= F^{O(n+1)}(s_0 + 1, \cdots, s_{i-1} + 1, \widehat{s_i}, s_{i+1}, \cdots, s_m, 0^{m+1}) \boxtimes (-1)^{i-s_i};$$

for $s_m > 0$,

$$F^{O(n+1)}(s^{(i)}, 0) \boxtimes (-1)^{i-s_i} \quad and \quad (F^{O(n+1)}(s^{(i)}, 0) \otimes \det) \boxtimes (-1)^{i-s_i+1}.$$

Sketch of the Proof.

(1) By the translation principle, the first exact sequence follows from Theorem 2.20 (1) which corresponds to the case $F \simeq \mathbf{1}$. Taking its dual, we obtain another exact sequence

$$0 \to U_{i+1} \to I_{(-1)^{i-s_i}}(\sigma^{(i)}, n-i+s_i) \to U_i \to 0 \quad \text{for } 0 \le i \le m,$$

because U_i is self-dual. Taking the tensor product with the one-dimensional representation χ_{--} of G, we obtain by (13.2) and by Lemma 2.14 another exact sequence of G-modules:

$$0 \to U_{n-i} \to I_{(-1)^{i-s_i}}(\sigma^{(i)} \otimes \det, n-i+s_i) \to U_{n+1-i} \to 0.$$

Replacing i $(0 \le i \le m)$ by $n-i$ $(m \le n-i \le 2m)$, we have shown the second exact sequence.

(2) The K-type formula of the irreducible finite-dimensional representation $U_0 = F^{O(n+1,1)}(s)$ of G is known by the classical branching law (see Fact 2.12). Since the K-type formula of the principal series representation is given by the Frobenius reciprocity which we can compute by using Fact 2.12 again, the K-type formula of U_{i+1} follows inductively from that of U_i by the exact sequence in the first statement.

$\square$

See also Theorem 14.50 in Appendix I for another description of the irreducible representation $U_i(F)$ in terms of θ-*stable parameters.*

Remark 13.8. When $i = m$ and $n = 2m$, $s^{(i)}$ is of the form

$$s^{(m)} = (s_0 + 1, \cdots, s_{m-1} + 1, 0^m) \in \Lambda^+(2m),$$

and therefore the irreducible $O(n)$-module $\sigma^{(m)} = F^{O(2m)}(s^{(m)})$ is of type Y (Definition 2.6). Hence we have an isomorphism

$$\sigma^{(m)} \simeq \sigma^{(m)} \otimes \det \qquad\qquad (13.3)$$

as $O(2m)$-modules by Lemma 2.9. We recall from Theorem 13.7 (1) that there is an exact sequence of G-modules as follows:

$$0 \to U_m \to I_{(-1)^{m-s_m}}(\sigma^{(m)}, m-s_m) \to U_{m+1} \to 0.$$

Taking the tensor product with the character $\chi_{--} \simeq \det$, we obtain from (13.2) and Lemma 2.14 another exact sequence of G-modules:

$$0 \to U_{m+1} \to I_{(-1)^{m-s_m}}(\sigma^{(m)} \otimes \det, m-s_m) \to U_m \to 0.$$

By (13.3), the principal series representations

$$I_{(-1)^{m-s_m}}(F^{O(2m)}(s^{(m)}), m - s_m) \simeq I_{(-1)^{m-s_m}}(F^{O(2m)}(s^{(m)})) \otimes \det, m - s_m)$$

split into a direct sum of two irreducible G-modules U_m and U_{m+1} (see also Theorem 14.46 (3) in Appendix I).

Case 2. $n = 2m - 1$ and $G = O(2m, 1)$.

For $F \in \widehat{O(n+2)}$ of type I, we define $\sigma^{(i)} \equiv \sigma^{(i)}(F) \in \widehat{O(n)}$ for $0 \le i \le m - 1 = \frac{1}{2}(n-1)$ as follows. We write $F = F^{O(n+2)}(s)$ with

$$s = (s_0, s_1, \cdots, s_{m-1}, 0^{m+1}) \in \Lambda^+(n+2) \equiv \Lambda^+(2m+1),$$

as in (2.20). Then we define $s^{(i)} \in \Lambda^+(n) \equiv \Lambda^+(2m-1)$ $(0 \le i \le m-1)$ by

$$s^{(i)} := (s_0 + 1, \cdots, s_{i-1} + 1, \widehat{s_i}, s_{i+1}, \cdots, s_{m-1}, 0^m), \tag{13.4}$$

and define irreducible finite-dimensional representations by

$$\sigma^{(i)} := F^{O(n)}(s^{(i)}) \in \widehat{O(n)} \quad \text{for } 0 \le i \le m - 1.$$

For later purpose, we set

$$s^{(m)} := (s_0 + 1, \cdots, s_{m-2} + 1, 1, 0^{m-1}) \in \Lambda^+(n).$$

Then there is an isomorphism as $O(n)$-modules:

$$\sigma^{(m-1)} \otimes \det \simeq F^{O(n)}(s^{(m)}).$$

It is convenient to introduce the *extended Hasse sequence* $\{U_i \equiv U_i(F)\}$ $(0 \le i \le 2m)$ by defining for $m + 1 \le i \le 2m$

$$U_i(F) := U_{n+1-i}(F) \otimes \chi_{-+}. \tag{13.5}$$

Implicitly, the definition (13.5) includes a claim that there is an isomorphism of discrete series representations (*cf.* Remark 13.10 below):

$$U_m(F) \simeq U_m(F) \otimes \chi_{-+} \tag{13.6}$$

when $G = O(n+1, 1)$ with $n = 2m - 1$.

We note that the one-dimensional representations χ_{--} and χ_{-+} in (13.2) and (13.5) are chosen differently according to the parity of n.

The proof of the following theorem goes similarly to that of Theorem 13.7.

Theorem 13.9 ($n = 2m - 1$). *Given an irreducible finite-dimensional representation* $F^{O(n+2,\mathbb{C})}(s)$ *of* $G = O(n+1, 1)$ *with*

$$s = (s_0, s_1, \cdots, s_{m-1}, 0^{m+1}) \in \Lambda^+(n+2)(= \Lambda^+(2m+1)),$$

there exists uniquely an extended Hasse sequence $U_0, U_1, \cdots, U_{2m}$ *of* $G = O(2m, 1)$ *starting with the irreducible finite-dimensional representation* $U_0 = F^{O(n+2,\mathbb{C})}(s)$. *Moreover, the extended Hasse sequence* $U_0, U_1, \cdots, U_{2m}$ *satisfies the following properties.*

(1) *There exist exact sequences of* G-modules:

$$0 \to U_i \to I_{(-1)^{i-s_i}}(\sigma^{(i)}, i - s_i) \to U_{i+1} \to 0 \quad (0 \le i \le m - 1),$$

$$0 \to U_i \to I_{(-1)^{n-i-s_{n-i}}}(\sigma^{(n-i)} \otimes \det, i + s_{n-i})$$

$$\to U_{i+1} \to 0 \qquad\qquad (m \le i \le 2m - 1).$$

(2) *The* K-*type formula of the irreducible* G-*module* U_i ($0 \le i \le m$) *is given by*

$$\bigoplus_b F^{O(n+1)}(b) \boxtimes (-1)^{\sum_{k=0}^m b_k - \sum_{k=0}^{m-1} s_k},$$

where $b = (b_0, b_1, \cdots, b_{m-1}, 0, \cdots, 0)$ *runs over* $\Lambda^+(n+1) \equiv \Lambda^+(2m)$ *subject to the following conditions:*

$$b_0 \ge s_0 + 1 \ge b_1 \ge s_1 + 1 \ge \cdots \ge b_{i-1} \ge s_{i-1} + 1,$$

$$s_i \ge b_i \ge s_{i+1} \ge b_{i+1} \ge \cdots \ge s_{m-1} \ge b_{m-1} \ge 0.$$

In particular, the minimal K-*type of the* G-*module* U_i ($0 \le i \le m$) *is given by*

$$F^{O(n+1)}(s^{(i)}, 0) \boxtimes (-1)^i$$

$$= F^{O(n+1)}(s_0 + 1, \cdots, s_{i-1} + 1, \widehat{s_i}, s_{i+1}, \cdots, s_{m-1}, 0^{m+1}) \boxtimes (-1)^{i-s_i}.$$

Remark 13.10. U_m is a discrete series representation of $G = O(2m, 1)$.

See also Theorem 14.51 in Appendix I for another description of the irreducible representation $U_i(F)$ in terms of θ-stable parameters.

By applying Proposition 13.4 and Lemma 2.14, we may unify the first statement of Theorems 13.7 and 13.9 as follows.

Theorem 13.11. *Let* F *be an irreducible finite-dimensional representation of* $G = O(n+1, 1)$ *of type I (see Definition 14.2 in Appendix I), and* $a, b \in \{\pm\}$. *Then for* $F_{a,b} := F \otimes \chi_{ab}$, *there exists uniquely a Hasse sequence* $U_i(F_{a,b})$ ($0 \le i \le [\frac{n+1}{2}]$) *starting with* $U_0(F_{a,b}) = F_{a,b}$. *Moreover, the irreducible* G-*modules* $U_i(F_{a,b})$ *occur* *in the following exact sequence of G-modules*

$$0 \to U_i(F_{a,b}) \to I_{ab(-1)^{i-s_i}}(\sigma_a^{(i)}, i - s_i) \to U_{i+1}(F_{a,b}) \to 0$$

for $0 \le i \le [\frac{n-1}{2}]$. Here $\sigma_a^{(i)} = \sigma^{(i)}$ if $a = +$; $\sigma^{(i)} \otimes$ det if $a = -$.

Remark 13.12. By (3.22), we have linear bijections for all i, j:

$$\mathrm{Hom}_{G'}(U_i(F)|_{G'}, U'_j(F')) \simeq \mathrm{Hom}_{G'}(U_{n+1-i}(F)|_{G'}, U'_{n-j}(F') \otimes \chi_{+-}).$$

Remark 13.13. Using the definition of the extended Hasse sequence we also define an extended standard sequence.

By abuse of notation we will from now on not distinguish between Hasse sequences and extended Hasse sequences and refer to both as Hasse sequences. A similar convention applies to standard sequences.

The following observation will be used in Section 13.3.4 for the proof of Evidence E.4 of Conjecture 13.15 below.

Proposition 13.14. *Suppose F and F' are irreducible finite-dimensional representations of $G = O(n+1, 1)$ and $G' = O(n, 1)$, respectively, such that $\mathrm{Hom}_{G'}(F|_{G'}, F') \ne \{0\}$. Suppose the principal series representations $I_\delta(V, \lambda)$ of G and $J_\varepsilon(W, \nu)$ of G' contain F and F', respectively, as subrepresentations. Then the following hold.*

(1) $[V : W] = 1$;
(2) $(\lambda, \nu, \delta, \varepsilon) \in \Psi_{\mathrm{sp}}$ *(see (1.3)), namely, the quadruple $(\lambda, \nu, \delta, \varepsilon)$ does not satisfy the generic parameter condition (3.2).*

Proof. For the proof, we use a description of irreducible finite-dimensional representations of the disconnected group $G = O(n+1, 1)$ in Section 14.1 of Appendix I. In particular, using Lemma 14.3, we may write

$$F = F^{O(n+1,1)}(\lambda_0, \cdots, \lambda_{[\frac{n}{2}]})_{a,b}$$

for some $(\lambda_0, \cdots, \lambda_{[\frac{n}{2}]}) \in \Lambda^+([\frac{n}{2}] + 1)$ and $a, b \in \{\pm\}$. By the branching rule for $O(n+1, 1) \downarrow O(n, 1)$ (see Theorem 14.7), an irreducible summand F' of $F|_{O(n,1)}$ is of the form

$$F' = F^{O(n,1)}(\nu_0, \cdots, \nu_{[\frac{n-1}{2}]})_{a,b}$$

for some $(\nu_0, \cdots, \nu_{[\frac{n-1}{2}]}) \in \Lambda^+([\frac{n+1}{2}])$ such that

$$\lambda_0 \ge \nu_0 \ge \lambda_1 \ge \cdots \ge \nu_{[\frac{n-1}{2}]} \ge 0 \qquad \text{for } n \text{ odd,}$$

$$\lambda_0 \ge \nu_0 \ge \lambda_1 \ge \cdots \ge \nu_{[\frac{n-1}{2}]} \ge \lambda_{[\frac{n}{2}]} \quad \text{for } n \text{ even.}$$

We recall that for every irreducible finite-dimensional representation F of a real reductive Lie group there exists only one principal series representation that contains F as a subrepresentation. By Theorem 13.11 with $i = 0$, the unique parameter (V, δ, λ) is given by

$$V = F^{O(n)}(\lambda_1, \cdots, \lambda_{[\frac{n}{2}]}) \ (\otimes \det \text{ if } a = -), \ \lambda = -\lambda_0 \text{ and } \delta = ab(-1)^{-\lambda_0}.$$

Likewise, the unique parameter (W, ε, ν) for F' is given by

$$W = F^{O(n-1)}(\nu_1, \cdots, \nu_{[\frac{n-1}{2}]}) \ (\otimes \det \text{ if } a = -), \ \nu = -\nu_0, \text{ and } \varepsilon = ab(-1)^{-\nu_0}.$$

Hence $[V : W] \neq 0$, or equivalently, $[V : W] = 1$ by the branching rule for $O(n) \downarrow O(n-1)$. Moreover, $\delta\varepsilon = (-1)^{\nu_0 + \lambda_0}$ and $\nu - \lambda = \lambda_0 - \nu_0 \in \mathbb{N}$. Hence the generic parameter condition (3.2) fails, or equivalently, $(\lambda, \nu, \delta, \varepsilon) \in \Psi_{\text{sp}}$.   □

13.2   The Conjecture

We propose a conjecture about when

$$\text{Hom}_{G'}(\Pi|_{G'}, \pi) = \mathbb{C}$$

where $\Pi \in \text{Irr}(G)$ and $\pi \in \text{Irr}(G')$ have regular integral infinitesimal characters (Definition 2.1). We give two formulations of the conjecture, see Conjectures 13.15 and 13.17 below. Supporting evidence is given in Section 13.3.

13.2.1   Conjecture: Version 1

We begin with a formulation of the conjecture in terms of a standard sequence (Definition-Theorem 13.1) of irreducible representations Π_i of $G = O(n + 1, 1)$ and that of irreducible representations π_j of the subgroup $G' = O(n, 1)$. We note that both Π_i and π_j have regular integral infinitesimal characters because both $F := \Pi_0$ and $F' := \pi_0$ are irreducible *finite-dimensional* representations of G and G', respectively.

Conjecture 13.15. *Let F be an irreducible finite-dimensional representations of $G = O(n + 1, 1)$, and $\{\Pi_i(F)\}$ be the standard sequence starting at $\Pi_0(F) = F$. Let F' be an irreducible finite-dimensional representation of the subgroup $G' = O(n, 1)$, and $\{\pi_j(F')\}$ the standard sequence starting at $\pi_0(F') = F'$. Assume that*

$$\text{Hom}_{G'}(F|_{G'}, F') \neq \{0\}.$$

Then the symmetry breaking for representations $\Pi_i(F)$, $\pi_j(F')$ in the standard sequences is represented graphically in Diagrams 13.1 and 13.2. In the first row are representations of G, in the second row are representations of G'. Symmetry breaking operators are represented by arrows, namely, there exist nonzero symmetry breaking operators if and only if there are arrows in the diagram.

$$
\begin{array}{ccccccc}
\Pi_0(F) & \Pi_1(F) & \cdots & \Pi_{m-1}(F) & & \Pi_m(F) \\
\downarrow \swarrow & \downarrow \swarrow & \swarrow & \downarrow & \swarrow & \downarrow \\
\pi_0(F') & \pi_1(F') & \cdots & \pi_{m-1}(F') & & \pi_m(F')
\end{array}
$$

Diagram. 13.1 Symmetry breaking for $O(2m+1,1) \downarrow O(2m,1)$

$$
\begin{array}{cccccccc}
\Pi_0(F) & \Pi_1(F) & \cdots & \Pi_{m-1}(F) & & \Pi_m(F) & & \Pi_{m+1}(F) \\
\downarrow \swarrow & \downarrow \swarrow & \swarrow & \downarrow & \swarrow & \downarrow & \swarrow & \\
\pi_0(F') & \pi_1(F') & \cdots & \pi_{m-1}(F') & & \pi_m(F') & &
\end{array}
$$

Diagram. 13.2 Symmetry breaking for $O(2m+2,1) \downarrow O(2m+1,1)$

Remark 13.16. Instead of using standard sequences to state the conjecture it may be also useful to rephrase it using extended Hasse sequences.

13.2.2 Conjecture: Version 2

We rephrase the conjecture using θ-stable parameters, which will be introduced in Section 14.9 of Appendix I, and restate Conjecture 13.15 as an algorithm in this notation.

In Theorems 14.50 and 14.51 of Appendix I, we shall give the θ-stable parameters of the representations of the standard sequence starting with an irreducible finite-dimensional representation F summarized as follows.

1. Suppose that $n = 2m$. Let

$$
F = F^{O(2m+1,1)}(\mu)_{a,b} = F^{O(2m+1,1)}(\mu) \otimes \chi_{ab}
$$

for $\mu \in \Lambda^+(m+1)$ and $a, b \in \{\pm\}$ be an irreducible finite-dimensional representation of $O(2m+1,1)$, see Section 14.1 in Appendix I. Its θ-stable parameter is

$$
(\parallel \mu_1, \mu_2, \ldots, \mu_m, \mu_{m+1})_{a,b}
$$

and we have the θ-stable parameters of the representations in the standard sequence (written in column).

$$\Pi_0(F) = (\ \|\ \mu_1, \mu_2, \ldots, \mu_m, \mu_{m+1})_{a,b}$$

$$\Pi_1(F) = (\mu_1\ \|\ \mu_2, \ldots, \mu_m, \mu_{m+1})_{a,b}$$

$$\vdots \qquad\qquad \vdots$$

$$\Pi_m(F) = (\mu_1, \mu_2, \ldots, \mu_m\ \|\ \mu_{m+1})_{a,b}.$$

2. Suppose that $n = 2m + 1$. Let

$$F = F^{O(2m+2,1)}(\mu)_{a,b} = F^{O(n+1,1)}(\mu) \otimes \chi_{ab}$$

for $\mu \in \Lambda^+(m+1)$ and $a, b \in \{\pm\}$ be an irreducible finite-dimensional representation of $O(2m+2,1)$. Its θ-stable parameter is

$$(\ \|\ \mu_1, \mu_2, \ldots, \mu_m, \mu_{m+1})_{a,b}$$

and we have the θ-stable parameters of the representations in standard sequence (written in column).

$$\Pi_0(F) = (\|\ \mu_1, \mu_2, \ldots, \mu_{m+1})_{a,b}$$

$$\Pi_1(F) = (\mu_1\ \|\ \mu_2, \ldots, \mu_{m+1})_{a,b}$$

$$\vdots \qquad\qquad \vdots$$

$$\Pi_{m+1}(F) = (\mu_1, \mu_2, \ldots, \mu_{m+1}\ \|)_{a,b}.$$

We refer to the finite-dimensional representation $\Pi_0(F) = F$ as the *starting representation* of the standard sequence and to the tempered representation $\Pi_m(F)$ (when $n = 2m$) or the discrete series representation $\Pi_{m+1}(F)$ (when $n = 2m + 1$) as the *last representation* of the standard sequence (see Remarks 13.8 and 13.10).

Conjecture 13.17. *Let $F^G(\mu)_{a,b}$ be an irreducible finite-dimensional representation of $G = O(n+1,1)$, and $F^{G'}(\nu)_{a,b}$ be an irreducible finite-dimensional representation of the subgroup $G' = O(n,1)$, where $\mu \in \Lambda^+([\frac{n+2}{2}])$, $\nu \in \Lambda^+([\frac{n+1}{2}])$, and $a, b, c, d \in \{\pm\}$, see (14.5) and (14.8) in Appendix I. Assume that*

$$\mathrm{Hom}_{G'}(F^G(\mu)_{a,b}|_{G'}, F^{G'}(\nu)_{c,d}) \neq \{0\}. \tag{13.7}$$

In (1) and (2) below, nontrivial symmetry breaking operators are represented by arrows connecting the θ-stable parameters of the representations.

(1) *Suppose that* $n = 2m$. *Then* $\mu = (\mu_1, \cdots, \mu_{m+1}) \in \Lambda^+(m+1)$ *and* $\nu = (\nu_1, \cdots, \nu_m) \in \Lambda^+(m)$. *Then two representations in the standard sequences have a nontrivial symmetry breaking operator if and only if the θ-stable parameters of the representations satisfy one of the following conditions.*

$$(\mu_1, \ldots, \mu_i \parallel \mu_{i+1}, \ldots, \mu_{m+1})_{a,b}$$

$$\Downarrow$$

$$(\nu_1, \ldots, \nu_i \parallel \nu_{i+1}, \ldots, \nu_m)_{c,d}$$

or

$$(\mu_1, \ldots, \mu_i \parallel \mu_{i+1}, \ldots, \mu_{m+1})_{a,b}$$

$$\Downarrow$$

$$(\nu_1, \ldots, \nu_{i-1} \parallel \nu_i, \nu_{i+1}, \ldots, \nu_m)_{c,d}$$

(2) *Suppose that $n = 2m + 1$. Then two infinite-dimensional representations in the standard sequences have a nontrivial symmetry breaking operator if and only if the θ-stable parameters of the representations satisfy one of the following conditions:*

$$(\mu_1, \ldots, \mu_i \parallel \mu_{i+1}, \ldots, \mu_{m+1})_{a,b}$$

$$\Downarrow$$

$$(\nu_1, \ldots, \nu_i \parallel \nu_{i+1}, \ldots, \nu_{m+1})_{c,d}$$

or

$$(\mu_1, \ldots, \mu_i \parallel \mu_{i+1}, \ldots, \mu_{m+1})_{a,b}$$

$$\Downarrow$$

$$(\nu_1, \ldots, \nu_{i-1} \parallel \nu_i, \ldots, \nu_{m+1})_{c,d}$$

Remark 13.18. See Theorem 14.7 in Appendix I for the condition on the parameters μ, ν, and a, b, c, d such that (13.7) holds. In particular, (13.7) implies either $(a, b) = (c, d)$ or $(a, b) = (-c, -d)$. See also Lemma 14.4 (2) for the description of overlaps in the expressions of irreducible finite-dimensional representations of $O(N - 1, 1)$ when N is even.

13.3 Supporting Evidence

In this section, we provide some evidence supporting our conjecture.

E.1 If $F \in \mathrm{Irr}(G)_\rho$ and $F' \in \mathrm{Irr}(G')_\rho$, the Conjecture 13.15 is true. (Equivalently, if $F^{O(n+1,1)}(\mu)_{+,+}$ and $F^{O(n,1)}(\nu)_{+,+}$ are both the trivial one-dimensional representations, Conjecture 13.17 is true.)

E.2   Some vanishing results for symmetry breaking operators.
E.3   Our conjecture is consistent with the Gross–Prasad conjecture for *tempered* representations of the special orthogonal group.
E.4   There exists a nontrivial symmetry breaking operator $\Pi_1 \to \pi_1$.

13.3.1 Evidence E.1

This was proved in Theorems 4.1 and 4.2.

13.3.2 Evidence E.2

Detailed proofs of the following propositions will be published in a sequel to this monograph.

Recall from Definition-Theorem 13.1 that $U_i(F_{a,b})$ refers to the i-th term in the Hasse sequence starting with the finite-dimensional representation $F_{a,b} = F \otimes \chi_{ab}$ of G and $U_j(F'_{c,d})$ to the j-th term in the Hasse sequence starting with the finite-dimensional representation $F'_{c,d} = F' \otimes \chi_{cd}$ of G'.

Proposition 13.19. *Let* $a, b, c, d \in \{\pm\}$, $0 \le i \le [\frac{n+1}{2}]$ *and* $0 \le j \le [\frac{n}{2}]$. *Then*

$$\mathrm{Hom}_{G'}(U_i(F_{a,b})|_{G'}, U_j(F'_{c,d})) = \{0\} \quad \text{if } j \ne i - 1, i.$$

If one of the representations of $G = O(n+1, 1)$ respectively of $G' = O(n, 1)$ is tempered then the following vanishing theorems hold.

- Assume first $(G, G') = (O(2m, 1), O(2m - 1, 1))$.

Let $s = (s_0, \cdots, s_{m-1}, 0^{m+1}) \in \Lambda^+(2m + 1)$ and $t = (t_0, \cdots, t_{m-1}, 0^m) \in \Lambda^+(2m)$ satisfy $t \prec s$ (see Definition 2.11 for the notation).

Proposition 13.20. *Let* $U_0, \cdots, U_m, U_{m+1}$ *be the Hasse sequence of* $G = O(2m, 1)$ *with* $U_0 = F^{O(2m+1, \mathbb{C})}(s)$, *and* $U'_0, \cdots, U'_{m-1}$ *be that of* $G' = O(2m - 1, 1)$ *with* $U'_0 = F^{O(2m, \mathbb{C})}(t)$. *Then*

$$\mathrm{Hom}_{G'}(U_m|_{G'}, U'_j) = \{0\} \quad \text{if } 0 \le j \le m - 2.$$

- Assume now $(G, G') = (O(2m + 1, 1), O(2m, 1))$.

Let $s = (s_0, \cdots, s_m, 0^{m+1}) \in \Lambda^+(2m + 2)$ and $t = (t_0, \cdots, t_{m-1}, 0^{m+1}) \in \Lambda^+(2m + 1)$ satisfy $t \prec s$.

Proposition 13.21. *Let* $U_0, \cdots, U_m$ *be the Hasse sequence of* $G = O(2m + 1, 1)$ *with* $U_0 = F^{O(2m+2, \mathbb{C})}(s)$, *and* $U'_0, \cdots, U'_m$ *be that of* $G' = O(2m, 1)$ *with* $U'_0 =$

$F^{O(2m+1,\mathbb{C})}(t)$. *Then*

$$\mathrm{Hom}_{G'}(U_i|_{G'}, U_m') = \{0\} \quad if \quad 0 \leq i \leq m-1.$$

Remark 13.22. These propositions prove only part of the vanishing statement of symmetry breaking operators formulated in Conjecture 13.17.

13.3.3 Evidence E.3

We use the notations and assumptions of the previous section, and show that our conjecture is consistent with the original Gross–Prasad conjecture for *tempered representations* [14]. For simplicity, we treat here only the case $(G, G') = (O(n + 1, 1), O(n, 1))$ with $n = 2m$. We shall see that a special case of Conjecture 13.17 (*i.e.*, the conjecture for the last representation of the standard sequence) implies some results (see (13.9) below) that were predicted by the original conjecture of Gross and Prasad for tempered representations of special orthogonal groups.

Assume that the irreducible finite-dimensional representations Π_0 of G and π_0 of G' are of type I (Definition 14.2) and that $(\mu_1, \ldots, \mu_m, \mu_{m+1})$ and $(\nu_1, \ldots, \nu_m)$ are their highest weights.

By the branching law for *finite-dimensional* representations with respect to $G \supset G'$ (see Theorem 14.7 in Appendix I), the condition

$$\mathrm{Hom}_{O(n,1)}(\Pi_0|_{G'}, \pi_0) \neq \{0\}$$

is equivalent to

$$\mu_1 \geq \nu_1 \geq \mu_2 \geq \cdots \geq \nu_m \geq \mu_{m+1} \geq 0. \tag{13.8}$$

Let U_m (resp. $\Pi_m = U_m \otimes (\chi_{+-})^m$) be the m-th term of the Hasse sequence (resp. the standard sequence) starting with the irreducible finite-dimensional representation $\Pi_0 = U_0$ (see Definitions 13.1 and 13.2). Then we have a direct sum decomposition of the principal series representation

$$I_{(-1)^{m-\mu_{m+1}}}(F^{O(2m)}(\mu_1 + 1, \cdots, \mu_m + 1, 0^m), m - \mu_{m+1}) \simeq U_m \oplus (U_m \otimes \det)$$

by Theorem 13.7 (1) and Remark 13.8. Assume that Π_m is tempered. Then U_m is also tempered, and the continuous parameter of the principal series representation must lie on the unitary axis, that is, $m - \mu_{m+1} \in m + \sqrt{-1}\mathbb{R}$. Hence $\mu_{m+1} = 0$.

Since $\mu_{m+1} = 0$, the θ-stable parameters of the tempered representations Π_m, $\Pi_m \otimes \det$ are given by

$$(\mu_1, \ldots, \mu_m \| 0)_{+,+}, \quad (\mu_1, \cdots, \mu_m \| 0)_{-,-},$$

whereas the θ-stable parameter of the discrete series representation of $G' = O(2m, 1)$ is given by

$$(\nu_1, \ldots, \nu_m \, \|)_{+,+}.$$

In view of the K-type formula in Theorem 13.7 (2), we see

$$U_m \not\simeq U_m \otimes \det$$

as G-modules, and thus $\Pi_m \not\simeq \Pi_m \otimes \det$. Therefore, the restriction of the principal series representation Π_m of $G = O(2m + 1, 1)$ to the subgroup $\overline{G} = SO(2m + 1, 1)$ is irreducible by Lemma 15.2 (1) in Appendix II. We set

$$\overline{\Pi}_m := \Pi_m |_{\overline{G}},$$

which is an irreducible tempered representation of $\overline{G}$.

We now consider representations of the subgroups $G' = O(2m, 1)$ and $\overline{G'} = SO(2m, 1)$. We observe that there is at most one discrete series representation of $\overline{G'} = SO(n, 1)$ for each infinitesimal character (see Proposition 14.41 in Appendix I). Therefore the restriction of the discrete series representation π_m of $G' = O(2m, 1)$ to the subgroup $\overline{G'} = SO(2m, 1)$ is irreducible, which is denoted by $\overline{\pi}_m$.

With these notations, Proposition 15.13 in Appendix II yields a natural linear isomorphism:

$$\mathrm{Hom}_{G'}(\Pi_m |_{G'}, \pi_m) \oplus \mathrm{Hom}_{G'}((\Pi_m \otimes \det)|_{G'}, \pi_m) \simeq \mathrm{Hom}_{\overline{G'}}(\overline{\Pi}_m |_{\overline{G'}}, \overline{\pi}_m).$$

Conjecture 13.17 for the pair $(G, G') = (O(n + 1, 1), O(n, 1))$ is applied to this specific situation; the first term in the left-hand side equals $\mathbb{C}$ and the second term vanishes. Thus Conjecture 13.17 in this case implies the following statement for the pair $(\overline{G}, \overline{G'}) = (SO(n + 1, 1), SO(n, 1))$ of special orthogonal groups:

$$\mathrm{Hom}_{\overline{G'}}(\overline{\Pi}_m |_{\overline{G'}}, \overline{\pi}_m) = \mathbb{C} \quad \text{if } \mu_{m+1} = 0 \text{ and } (13.8) \text{ is satisfied.} \tag{13.9}$$

We now assume that the representation Π_m is nontrivial on the center. This determines the Langlands parameters of the Vogan packets $VP(\overline{\Pi}_m)$ and $VP(\overline{\pi}_m)$ of $\overline{G}$ respectively $\overline{G'}$, and we follow exactly the steps of the algorithm by Gross and Prasad outlined in Chapter 11. We conclude again that the Gross–Prasad conjecture predicts that $\{\overline{\Pi}_m, \overline{\pi}_m\}$ is the unique pair of representation in $VP(\overline{\Pi}_m) \times VP(\overline{\pi}_m)$ with a nontrivial symmetry breaking operator.

13.3.4 Evidence E.4

We will prove the existence of a nontrivial symmetry breaking operator

$$\Pi_1 \to \pi_1.$$

We first introduce graphs to encode information about the images and kernels of symmetry breaking operators between reducible principal series representations as well as information about the images of the subrepresentation under the symmetry breaking operators. This will be helpful to visualize the composition of an symmetry breaking operator with a Knapp–Stein operator.

Admissible Graphs

Consider the vertices of a square. We call the following set of directed graphs *admissible*:

$$
\begin{array}{cccc}
O \to O & O \quad O & O \to O & O \quad O \\
\quad \nearrow & \quad \nearrow & & \quad \searrow \\
O \to O & O \to O & O \to O & O \to O \\[2mm]
O \to O & O \quad O & O \quad O & \\
\quad \searrow & \quad \searrow & & \\
O \quad O & O \quad O & O \to O &
\end{array}
$$

and the zero graph without arrows:

$$
\begin{array}{cc}
O & O \\[2mm]
O & O
\end{array}
$$

Admissible graphs will encode information about the images and kernels of symmetry breaking operators. In the setting we shall use later, it is convenient to define the following equivalence relation among graphs, see Lemma 13.28.

Convention 13.23. *We identify two graphs $\mathcal{G}_1$ and $\mathcal{G}_2$ if*

$$\mathcal{G}_1 = \mathcal{G}_2 \cup \{\ell\}$$

where ℓ is an arrow ending at the lower right vertex and $\mathcal{G}_2$ already contains an arrow which starts from the same vertex as ℓ and which ends at the upper right vertex.

Example 13.24. The following graphs are pairwise equivalent.

$$
\begin{array}{ccc}
\begin{matrix} \mathrm{O} \to \mathrm{O} \\ \searrow \\ \mathrm{O} \quad \mathrm{O} \end{matrix}
& \equiv &
\begin{matrix} \mathrm{O} \to \mathrm{O} \\ \\ \mathrm{O} \quad \mathrm{O} \end{matrix} \,,
\end{array}
$$

$$
\begin{array}{ccc}
\begin{matrix} \mathrm{O} \to \mathrm{O} \\ \\ \mathrm{O} \to \mathrm{O} \end{matrix}
& \equiv &
\begin{matrix} \mathrm{O} \to \mathrm{O} \\ \searrow \\ \mathrm{O} \to \mathrm{O} \end{matrix} \,,
\end{array}
$$

$$
\begin{array}{ccccc}
\begin{matrix} \mathrm{O} \to \mathrm{O} \\ \nearrow \\ \mathrm{O} \to \mathrm{O} \end{matrix}
& \equiv &
\begin{matrix} \mathrm{O} \to \mathrm{O} \\ \diagdown\!\!\!\!\diagup \\ \mathrm{O} \to \mathrm{O} \end{matrix}
& \equiv &
\begin{matrix} \mathrm{O} \to \mathrm{O} \\ \nearrow \\ \mathrm{O} \quad \mathrm{O} \end{matrix} \,,
\end{array}
$$

$$
\begin{array}{ccc}
\begin{matrix} \mathrm{O} \quad \mathrm{O} \\ \nearrow \\ \mathrm{O} \to \mathrm{O} \end{matrix}
& \equiv &
\begin{matrix} \mathrm{O} \quad \mathrm{O} \\ \nearrow \\ \mathrm{O} \quad \mathrm{O} \end{matrix} \,.
\end{array}
$$

We obtain a colored graph by coloring the vertices of the graph by 4 different colors, each with a different color. We typically use the colors blue and red for the vertices in the left column and green and orange for the vertices in the right column.

Mutation of Admissible Graphs

We obtain a new colored graph $\mathcal{G}_2$ from a graph $\mathcal{G}_1$ by "mutation". The rules of the mutation are given as follows.

Rule 1. Consider the colored vertices on the right. Remove any arrow which ends at the lower right vertex. Interchange the two colored vertices on the right. The arrows which used to end at the upper right vertex now end at the lower right vertex.

Rule 2. Consider the colored vertices on the left. Remove any arrow which starts at the upper left corner. Interchange the two colored vertices on the left. The arrows which used to start at the lower left vertex now start at the upper left vertex.

Rule 3. If the mutated graph $\mathcal{G}_2$ has no arrows, *i.e.,* $\mathcal{G}_2$ is the zero graph, the mutation is not allowed.

We write **R** for the mutation on the right column and **L** for the mutation on the left column. We sometimes refer to **R** and **L** as *mutation rules*.

It is easy to see the following.

Lemma 13.25.

(1) *The mutated graph is again admissible.*
(2) *Mutation is well-defined for the equivalence relations given in Convention 13.23.*
(3) *Admissible graphs for which no mutation is allowed do not have an arrow except for the one from the upper left vertex to the lower right vertex.*
(4) **R** ∘ **R** *and* **L** ∘ **L** *are not allowed mutations.*
(5) **R** ∘ **L** = **L** ∘ **R**.

Definition 13.26 (source and sink). We call an admissible graph $\mathcal{G}$ a *source* of a set of graphs if all other graphs of the set are obtained through mutations of $\mathcal{G}$. We call a graph $\mathcal{G}$ a *sink* in a set of admissible graphs if neither **R** nor **L** is an allowed mutation of $\mathcal{G}$.

Applying these rules, we obtain the following families of mutated graphs with one source. The source for the first, second, and third types is at the top right corner, applying **R** changes the right column and applying **L** changes the left column.

First Type

Second Type

Third Type

$$
\begin{array}{cc}
\mathrm{O} \rightarrow \mathrm{O} \\
\quad\searrow \\
\mathrm{O} \qquad \mathrm{O}
\end{array}
\quad\overset{\mathbf{L}}{\Longleftarrow}\quad
\begin{array}{cc}
\mathrm{O} \qquad \mathrm{O} \\
\qquad\nearrow \\
\mathrm{O} \rightarrow \mathrm{O}
\end{array}
$$

$$\mathbf{R}\big\Downarrow \qquad\qquad\qquad \big\Downarrow\mathbf{R}$$

$$
\begin{array}{cc}
\mathrm{O} \qquad \mathrm{O} \\
\quad\searrow \\
\mathrm{O} \qquad \mathrm{O}
\end{array}
\quad\overset{\mathbf{L}}{\Longleftarrow}\quad
\begin{array}{cc}
\mathrm{O} \qquad \mathrm{O} \\
\\
\mathrm{O} \rightarrow \mathrm{O}
\end{array}
$$

Type A

$$
\begin{array}{cc}
\mathrm{O} \rightarrow \mathrm{O} \\
\quad\searrow \\
\mathrm{O} \qquad \mathrm{O}
\end{array}
$$

$$\big\Downarrow\mathbf{R}$$

$$
\begin{array}{cc}
\mathrm{O} \qquad \mathrm{O} \\
\quad\searrow \\
\mathrm{O} \qquad \mathrm{O}
\end{array}
$$

Type B

$$
\begin{array}{cc}
\mathrm{O} \qquad \mathrm{O} \\
\quad\searrow \\
\mathrm{O} \qquad \mathrm{O}
\end{array}
\quad\overset{\mathbf{L}}{\Longleftarrow}\quad
\begin{array}{cc}
\mathrm{O} \qquad \mathrm{O} \\
\quad\searrow \\
\mathrm{O} \rightarrow \mathrm{O}
\end{array}
$$

Type C

$$
\begin{array}{cc}
\mathrm{O} \qquad \mathrm{O} \\
\quad\searrow \\
\mathrm{O} \qquad \mathrm{O}
\end{array}
\quad\overset{\mathbf{L}}{\Longleftarrow}\quad
\begin{array}{cc}
\mathrm{O} \qquad \mathrm{O} \\
\\
\mathrm{O} \rightarrow \mathrm{O}
\end{array}
$$

This proves the following.

Lemma 13.27. *Let $\mathcal{F}$ be the family of admissible graphs that are obtained through mutations of a nonzero admissible graph.*

(1) *If $\mathcal{F}$ is not a singleton, it is one of the above six types.*
(2) *If $\mathcal{F}$ is a singleton, it is a coloring of the following graph.*

$$
\begin{array}{cc}
\mathrm{O} \qquad \mathrm{O} \\
\quad\searrow \\
\mathrm{O} \qquad \mathrm{O}
\end{array}
$$

From Symmetry Breaking Operators to Admissible Graphs

Assume that a principal series representation $I_\delta(V, \lambda)$ of G has exactly two composition factors Π^1 and Π^2, which are not equivalent to each other. (The assumption is indeed satisfied for $G = O(n+1, 1)$ whenever $I_\delta(V, \lambda)$ is reducible.) Thus there is an exact sequence of G-modules:

$$0 \to \Pi^1 \to I_\delta(V, \lambda) \to \Pi^2 \to 0. \tag{13.10}$$

Graphically, the irreducible inequivalent composition factors are represented by circles with different colors. The bottom circle represents the socle as follows.

$$O$$

$$O$$

Later we shall assume in addition that the exact sequence (13.10) does not split. (The assumption is satisfied if one of Π^1 or Π^2 is finite-dimensional. More generally, the assumption is indeed satisfied for most of the pairs of the composition factors of the principal series representations of $G = O(n+1, 1)$ with regular integral infinitesimal characters, see Theorem 2.20 for example.)

An analogous notation will be applied to principal series representations $J_\varepsilon(W, \nu)$ of the subgroup $G' = O(n, 1)$ with two composition factors. Thus we represent the two composition factors of the reducible principal series representations $I_\delta(V, \lambda)$ and of $J_\varepsilon(W, \nu)$ by four differently colored circles in a square; both the composition factors of a principal series representation are represented by circles vertically.

We have the convention that the composition factors of the representation $I_\delta(V, \lambda)$ of G are represented by the circles on the left, those of $J_\varepsilon(W, \nu)$ of the subgroup on the right. Using this convention we get four squares with colored circles which are obtained by changing the colors in each vertical column.

To a symmetry breaking operator

$$\mathbb{B}_{\lambda, \nu}^{V, W} : I_\delta(V, \lambda) \to J_\varepsilon(W, \nu)$$

we associate a graph which encodes information about the image and kernel of the symmetry breaking operator $\mathbb{B}_{\lambda, \nu}^{V, W}$ as well as information about the image of the irreducible subrepresentation of the principal series representation $I_\delta(V, \lambda)$ of G under the symmetry breaking operator. We proceed as follows: we obtain the arrows of the graph by considering the action of symmetry breaking operator $\mathbb{B}_{\lambda, \nu}^{V, W}$ on the composition factors. If no arrow starts at a circle, then this means that the corresponding composition factor is in the kernel of the symmetry breaking operator. If no arrow ends at a circle, then this means that the G'-submodule of $J_\varepsilon(W, \nu)$ corresponding to the circle is not in the image of the symmetry breaking operator. Then we have:

Lemma 13.28. *Assume that both principal series representations $I_\delta(V, \lambda)$ and $J_\varepsilon(W, \nu)$ have exactly two inequivalent composition factors with nontrivial extensions. Then with Convention 13.23 the graph associated to our symmetry breaking operator $\mathbb{B}_{\lambda,\nu}^{V,W} \in \mathrm{Hom}_{G'}(I_\delta(V, \lambda)|_{G'}, J_\varepsilon(W, \nu))$ is an admissible graph.*

The proof of Lemma 13.28 is straightforward. We illustrate it by examples as below.

Example 13.29 (Graph of symmetry breaking operators).

(1) Suppose that the symmetry breaking operator is surjective and its restriction to the socle O is also surjective. Then the associated graph is given by

$$
\begin{array}{ccc}
\mathrm{O} & \to & \mathrm{O} \\
& \diagdown\!\!\!\!\diagup & \\
\mathrm{O} & \to & \mathrm{O}
\end{array}
$$

by definition. With Convention 13.23, we have

$$
\begin{array}{ccc}
\mathrm{O} & \to & \mathrm{O} \\
& \nearrow & \\
\mathrm{O} & \to & \mathrm{O}
\end{array}
\quad \equiv \quad
\begin{array}{ccc}
\mathrm{O} & \to & \mathrm{O} \\
& \diagdown\!\!\!\!\diagup & \\
\mathrm{O} & \to & \mathrm{O}
\end{array} \quad,
$$

see Example 13.24. Then the graph in the left-hand side is admissible.

(2) Suppose that the symmetry breaking operator is zero. Then it is depicted by the zero graph.

$$
\begin{array}{ccc}
\mathrm{O} & & \mathrm{O} \\
& & \\
\mathrm{O} & & \mathrm{O}
\end{array}
$$

To reduce the clutter in a diagram representing a set of mutated graphs we often omit the zero graph, *i.e.,* the zero symmetry breaking operator.

We would like to encode information about a symmetry breaking operator and all its compositions with the Knapp–Stein operators at the same time. Composing symmetry breaking operators $\mathbb{B}_{\lambda,\nu}^{V,W}$ with a Knapp–Stein intertwining operator

$$
\widetilde{\mathbb{T}}_{\lambda,n-\lambda}^{V} : I_\delta(V, \lambda) \to I_\delta(V, n - \lambda)
$$

for the group G (see (8.12)), respectively

$$
\widetilde{\mathbb{T}}_{\nu,n-1-\nu}^{W} : J_\varepsilon(W, \nu) \to J_\varepsilon(W, n - 1 - \nu)
$$

for the subgroup G', we obtain another symmetry breaking operator. If this new operator is not zero then it can be represented again by an admissible graph. The graphs of these operators are arranged compatible with our previous article [42, Figs. 2.1–2.5] where we draw ν-value on the x-axis and the λ-value on the y-axis. We place the corresponding symmetry breaking operator in the corresponding quadrant. For example, if $\lambda \geq \frac{n}{2}$ and $\nu \geq \frac{n-1}{2}$, then the parameters are arranged as

$$(n-1-\nu, \lambda) \qquad\qquad (\nu, \lambda)$$
$$(n-1-\nu, n-\lambda) \qquad\qquad (\nu, n-\lambda)$$

in the (ν, λ)-plane, and accordingly these symmetry breaking operators are arranged as follows.

$$\widetilde{\mathbb{T}}^W_{\nu, n-1-\nu} \circ \mathbb{B}^{V,W}_{\lambda, \nu} \qquad\qquad\qquad\qquad \mathbb{B}^{V,W}_{\lambda, \nu}$$

$$\widetilde{\mathbb{T}}^W_{\nu, n-1-\nu} \circ \mathbb{B}^{V,W}_{\lambda, \nu} \circ \widetilde{\mathbb{T}}^V_{n-\lambda, \lambda} \qquad\qquad \mathbb{B}^{V,W}_{\lambda, \nu} \circ \widetilde{\mathbb{T}}^V_{n-\lambda, \lambda}$$

Accordingly, we shall consider four graphs of these four symmetry breaking operators.

By the definition of the mutation rule, we obtain:

Lemma 13.30. *Assume that a principal series representation $I_\delta(V, \lambda)$ has two irreducible composition factors Π^1 and Π^2 with nonsplitting exact sequence (13.10) and that the Knapp–Stein operator $\widetilde{\mathbb{T}}^V_{\lambda, n-\lambda} : I_\delta(V, \lambda) \to I_\delta(V, n-\lambda)$ is nonzero but vanishes on the subrepresentation Π^1. Then the graph associated to a symmetry breaking operator composed with $\widetilde{\mathbb{T}}^V_{n-\lambda, \lambda}$ for the group G is obtained by using the mutation rule $\mathbf{L}$ for graphs. Similarly, the graph associated to a symmetry breaking operator composed with a nonzero Knapp–Stein operator $\widetilde{\mathbb{T}}^W_{\nu, n-1-\nu} : J_\varepsilon(W, \nu) \to J_\varepsilon(W, n-1-\nu)$ for the subgroup G' (with an analogous assumption on $J_\varepsilon(W, \nu)$) is obtained by using the mutation rule $\mathbf{R}$ for graphs.*

Example 13.31. In the Memoirs article [42] we considered the case of two spherical principal series representations $I(\lambda)$ and $J(\nu)$ for integral parameters i, j. If $(-i, -j) \in L_{even}$, namely, if $i \geq j \geq 0$ and $i \equiv j \mod 2$, then the normalized regular symmetry breaking operator $I(-i) \to J(-j)$ is zero [42, Thm. 8.1]. The other symmetry breaking operators for spherical principal series representations with the same infinitesimal character are nonzero and we have functional equations with nonvanishing coefficients [42, Thm. 8.5]. Thus the family of mutated graphs associated to the regular symmetry breaking operators is given as follows.

We recall from [42, Chap. 1] (or from Theorem 2.20 in a more general setting) that both the G-module $I(-i)$ and the G'-module $J(-j)$ contain irreducible finite-dimensional representations as their subrepresentations (red and orange circles) and irreducible infinite-dimensional representations $T(i)$ and $T(j)$ (blue and green) as their quotients, respectively. The corresponding socle filtrations are given graphically as follows.

$$I(-i) = \quad \begin{matrix} O \\ \\ O \end{matrix} \qquad\qquad J(-j) = \quad \begin{matrix} O \\ \\ O \end{matrix}$$

Note that, under the assumption $i \geq j \geq 0$ and $i \equiv j \mod 2$, we have a nontrivial symmetry breaking operator between the two finite-dimensional representations (red and orange circles) and as well as between the nontrivial composition factors $T(i) \to T(j)$ (blue and green circles), see [42, Thm 1.2 (1-a)].

Example 13.32. More generally in Corollary 3.18 we proved that

$$\widetilde{\mathbb{A}}^{V,W}_{\lambda_0,\nu_0,\gamma} = 0$$

for negative integers λ_0, ν_0 implies that

$$\widetilde{\mathbb{A}}^{V,W}_{n-\lambda_0,n-1-\nu_0,\gamma} \neq 0.$$

Since $(n-1-\nu_0, n-\lambda_0) \in \mathbb{N}^2$, we may place the graph associated to the regular symmetry breaking operator $\widetilde{\mathbb{A}}^{V,W}_{n-\lambda_0,n-1-\nu_0,\gamma}$ in the NE corner according to the position in the (ν, λ)-plane as in [42, Fig. 2.1, III.A or III.B].

On the other hand, since $(\nu_0, \lambda_0) \in (-\mathbb{N})^2$, we may place a zero graph associated to the zero operator $\widetilde{\mathbb{A}}^{V,W}_{\lambda_0,\nu_0,\gamma}$ in the SW corner according to the position in the (ν, λ)-plane as in [42, Fig. 2.1, I.A. or I.B.].

Example 13.33. In the Memoirs article [42, Thm. 11.1] we prove that there is a differential symmetry breaking operator in the SW corner if the regular symmetry breaking operator is zero. To this operator and its composition with the Knapp–Stein operators the assigned graph is given as follows.

Note that the differential operator gives a source in the mutation graphs in the SW corner in this setting.

Existence of a Nontrivial Symmetry Breaking Operators $\Pi_1 \to \pi_1$
Recall that we assume that

$$m(\Pi_0, \pi_0) = 1$$

for the irreducible finite-dimensional representations Π_0 of G and π_0 of the subgroup G'. We consider now a pair of reducible principal series representations $I_\delta(V, \lambda)$ of G and $J_\varepsilon(W, \nu)$ of G' with finite-dimensional composition factors Π_0, π_0, respectively.

Lemma 13.34. *Suppose that both O and O are representing irreducible finite-dimensional representations of G and G'. We assume that O and O respectively O and O are representing the composition factors of a principal series representation of G, respectively G'. Then the following graphs are not associated to a symmetry breaking operator.*

$$
\begin{array}{cccccccc}
O & O & \quad & O & O & \quad O & O & \quad O \to O \\
& \searrow & & & \searrow & \quad \nearrow & & \quad \nearrow \\
O & O & \quad & O & \to O & \quad O \to O & \quad O \to O
\end{array}
$$

Proof. The representations O and O are finite-dimensional. The image of a finite-dimensional representation by a symmetry breaking operator is finite-dimensional. $\qquad\square$

Lemma 13.35. *We keep Convention 13.23 and the assumptions of Lemma 13.34.*

(1) *Suppose that O and O stand for both irreducible subrepresentations of the principal series representations of G and G', respectively. The graph associated to a nontrivial symmetry breaking operator is one of the following.*

$$
\begin{array}{cccccc}
O & O & \quad O \to O & \quad O \to O \\
& \nearrow & \quad \nearrow & \\
O \to O & & \quad O \to O & \quad O \to O
\end{array}
$$

(2) *Suppose that O and O stand for both irreducible finite-dimensional subrepresentations of the principal series representations. The graph associated to a nontrivial symmetry breaking operator is one of the following.*

$$
\begin{array}{cccccccc}
O \to O & \quad O \to O & \quad O & O & \quad O & O \\
& & \quad \searrow & & \quad \searrow & \\
O \to O & \quad O & O & \quad O \to O & \quad O & O
\end{array}
$$

Using the composition with the Knapp–Stein operators we obtain an action of the (little) Weyl group of $O(n+1,1) \times O(n,1)$ on the continuous parameters of the symmetry breaking operators, hence on the symmetry breaking operators and also on their associated admissible graphs through the mutation rules.

Example 13.36. Let $\mathcal{F}$ be a family of mutated graphs such that the graph associated to the symmetry breaking operator $\widetilde{\mathbb{A}}^{V,W}_{n-\lambda_0,n-1-\nu_0,\gamma}$ is a source. If $\mathcal{F}$ is of first type, then the graph in the SE corner shows that there is a nontrivial symmetry breaking operator $\Pi_1 \to \pi_1$.

Using functional equations and the information about (K, K')-spectrum of regular symmetry breaking operators it is in some cases possible (see for example [42]) to show that the associated graph is of first type, but in general we do not have such explicit information about the regular symmetry breaking operators and so we have to proceed differently.

Suppose that Π_0 and π_0 are irreducible finite-dimensional subrepresentations of $I_\delta(V, \lambda)$ and $J_\varepsilon(W, \nu)$ with $\mathrm{Hom}_{G'}(\Pi_0|_{G'}, \pi_0) \neq \{0\}$. By Proposition 13.14, $[V : W] \neq 0$ and $(\lambda, \nu, \delta, \varepsilon) \in \Psi_{\text{sing}}$, namely, the quadruple $(\lambda, \nu, \delta, \varepsilon)$ does not satisfy the generic parameter condition (3.2). By Theorem 3.5 (see also Theorem 6.1 (1)), there exists a nonzero differential symmetry breaking operator

$$\mathbf{D}: I_\delta(V, \lambda) \to J_\varepsilon(W, \nu),$$

which we denote by $\mathbf{D}$. The image of $\mathbf{D}$ is infinite-dimensional by Theorem 6.8. Thus by Lemma 13.35 (2), we obtain the following.

Lemma 13.37. *The graph associated to $\mathbf{D}$ is one of the following.*

$$
\begin{array}{ccccccc}
O & \to & O & \quad & O & \to & O \\
  &   &   & \quad &   & \searrow & \\
O & \to & O & \quad & O &   & O
\end{array}
$$

Mutating the graph of $\mathbf{D}$ by $\mathbf{R}$ we get the following.

$$
\begin{array}{ccc}
O & & O \\
 & \searrow & \\
O & & O
\end{array}
$$

Thus composing the differential symmetry breaking operator with a Knapp–Stein operator on the right we obtain a nontrivial symmetry breaking operator with this diagram and thus a symmetry breaking operator $U_1(F) \to U_1(F')$. We are ready to prove the following theorem, which gives evidence of our conjecture.

Theorem 13.38. *Suppose that F and F' are irreducible finite-dimensional representations of G and G', respectively. Let Π_i, π_j be the standard sequences starting at F, F', respectively. Then there exists a nontrivial symmetry breaking operator*

$$\Pi_1 \to \pi_1$$

if $\mathrm{Hom}_{G'}(F|_{G'}, F') \neq \{0\}$.

Proof. Recall from Definition 13.2 that $\Pi_0 = F$, $\pi_0 = F'$ and $\Pi_1 = U_1(F) \otimes \chi_{+-}$, $\pi_1 = U_1(F') \otimes \chi_{+-}$ and so

$$\text{Hom}_{G'}(\Pi_1|_{G'}, \pi_1) \simeq \text{Hom}_{G'}(U_1(F)|_{G'}, U_1(F')).$$

$\square$

Chapter 14
Appendix I: Irreducible Representations of $G = O(n+1, 1)$, θ-stable Parameters, and Cohomological Induction

In Appendix I, we give a classification of irreducible admissible representations of $G = O(n+1, 1)$ in Theorem 14.36. In particular, we give a number of equivalent descriptions of irreducible representations with integral infinitesimal character (Definition 2.1) by means of Langlands quotients (or subrepresentations), coherent continuation starting at $\Pi_{i,\delta}$, and cohomologically induced representations from finite-dimensional representations of θ-stable parabolic subalgebras, see Theorem 14.35. Our results include a description of the following irreducible representations:

- "Hasse sequence" starting with arbitrary finite-dimensional irreducible representations (Theorems 14.50 and 14.51);
- complementary series representations with singular integral infinitesimal character (Theorem 14.53).

Since the Lorentz group $G = O(n+1, 1)$ has four connected components, we need a careful treatment even in dealing with finite-dimensional representations because not all of them extend holomorphically to $O(n+2, \mathbb{C})$. Thus Appendix I starts with irreducible finite-dimensional representations (Section 14.1), and then discuss infinite-dimensional admissible representations for the rest of the chapter.

14.1 Finite-dimensional Representations of $O(N-1, 1)$

In this section we give a parametrization of irreducible finite-dimensional representations of the disconnected groups $O(N-1, 1)$ and $O(N)$. The description here fits well with the θ-stable parameters (Definition 14.42) for the Hasse sequence, see Theorem 14.50. We note that the parametrization here for irreducible finite-dimensional representations of $O(N)$ is different from what was defined in Section 2.2.1, although the "dictionary" is fairly simple, see Remark 14.1.

© Springer Nature Singapore Pte Ltd. 2018   251
T. Kobayashi, B. Speh, *Symmetry Breaking for Representations*
of Rank One Orthogonal Groups II, Lecture Notes in Mathematics 2234,
https://doi.org/10.1007/978-981-13-2901-2_14

There are two connected components in the compact Lie group $O(N)$. We recall from Definition 2.4 that the set of equivalence classes of irreducible finite-dimensional representations of the orthogonal group $O(N)$ can be divided into two types, namely, type I and II. On the other hand, there are four connected components in the noncompact Lie group $O(N-1,1)$, and the division into two types is not sufficient for the classification of irreducible finite-dimensional representations of $O(N-1,1)$. We observe that some of the irreducible finite-dimensional representations of $O(N-1,1)$ cannot be extended to holomorphic representations of $O(N,\mathbb{C})$. For example, neither the one-dimensional representation χ_{+-} nor χ_{-+} of $O(N-1,1)$ (see (2.13)) comes from a holomorphic character of $O(N,\mathbb{C})$. We shall use only representations of "type I" and tensoring them with four characters χ_{ab} $(a,b \in \{\pm\})$ to describe all irreducible finite-dimensional representations of $O(N-1,1)$.

First of all, we recall from (2.17) that $\Lambda^+(k)$ is the set of $\lambda \in \mathbb{Z}^k$ with $\lambda_1 \geq \lambda_2 \geq \cdots \geq \lambda_k \geq 0$.

Let $N \geq 2$. For $\lambda \in \Lambda^+([\frac{N}{2}])$, we extend it to

$$\widetilde{\lambda} := (\lambda_1, \cdots, \lambda_{[\frac{N}{2}]}, \underbrace{0, \cdots, 0}_{[\frac{N+1}{2}]}) \in \mathbb{Z}^N, \tag{14.1}$$

and define

$$F^{O(N,\mathbb{C})}(\lambda)_+ \equiv F^{O(N,\mathbb{C})}(\widetilde{\lambda}), \tag{14.2}$$

to be the unique irreducible summand of $O(N,\mathbb{C})$ in the irreducible finite-dimensional representation $F^{GL(N,\mathbb{C})}(\widetilde{\lambda})$ of $GL(N,\mathbb{C})$ that contains a highest weight vector corresponding to $\widetilde{\lambda}$, see (2.20). Its restriction to the real forms $O(N)$ and $O(N-1,1)$ will be denoted by $F^{O(N)}(\lambda)_+$ and $F^{O(N-1,1)}(\lambda)_{+,+}$, respectively. Then the irreducible $O(N)$-module $F^{O(N)}(\lambda)_+$ is a representation of type I. We may summarize these notations as follows.

$$F^{O(N)}(\lambda)_+ \underset{\mathrm{rest}_{O(N)}}{\xleftarrow{\sim}} F^{O(N,\mathbb{C})}(\widetilde{\lambda}) \underset{\mathrm{rest}_{O(N-1,1)}}{\xrightarrow{\sim}} F^{O(N-1,1)}(\lambda)_{+,+}. \tag{14.3}$$

Remark 14.1. With the notation as in (2.20), we have

$$F^{O(N)}(\lambda)_+ \simeq F^{O(N)}(\widetilde{\lambda})$$

for $\lambda \in \Lambda^+([\frac{N}{2}])$. This is a general form of representations of $O(N)$ of type I (Definition 2.4). Then other representations of $O(N)$, *i.e.*, representations of type II are obtained from the tensor product of those of type I with the one-dimensional representation, det, as we recall now.

Suppose $0 \leq 2\ell \leq N$. If $\lambda \in \Lambda^+([\frac{N}{2}])$ is of the form

$$\lambda = (\lambda_1, \cdots, \lambda_\ell, \underbrace{0, \cdots, 0}_{[\frac{N}{2}]-\ell})$$

with $\lambda_\ell > 0$, then by (2.23), we have an isomorphism as representations of $O(N)$:

$$F^{O(N)}(\lambda)_+ \otimes \det \simeq F^{O(N)}(\lambda_1, \cdots, \lambda_\ell, \underbrace{1, \cdots, 1}_{\ell}, \underbrace{0, \cdots, 0}_{N-2\ell}, \underbrace{}_{\ell}),$$

which is of type II if $N \neq 2\ell$. We shall denote this representation by $F^{O(N)}(\lambda)_-$ as (14.4) below.

Analogously, $F^{O(N-1,1)}(\lambda)_{+,+}$ is a general form of representations of the Lorentz group $O(N-1,1)$ of type I in the following sense.

Definition 14.2 (representation of type I for $O(N-1,1)$). An irreducible finite-dimensional representation of $O(N-1,1)$ is said to be of *type I* if it is obtained as the holomorphic continuation of an irreducible representation of $O(N)$ of type I (see Definition 2.4).

We define for $\lambda \in \Lambda^+([\frac{N}{2}])$

$$F^{O(N)}(\lambda)_- := F^{O(N)}(\lambda)_+ \otimes \det, \tag{14.4}$$

$$F^{O(N-1,1)}(\lambda)_{a,b} := F^{O(N-1,1)}(\lambda)_{+,+} \otimes \chi_{ab} \quad (a, b \in \{\pm\}). \tag{14.5}$$

These are irreducible representations of $O(N)$ and $O(N-1,1)$, respectively.

With the notation (14.4) and (14.5), irreducible finite-dimensional representations of $O(N)$ and of $O(N-1,1)$, respectively, are described as follows:

Lemma 14.3.

(1) *Any irreducible finite-dimensional representation of $O(N)$ is of the form $F^{O(N)}(\lambda)_+$ or $F^{O(N)}(\lambda)_-$ for some $\lambda \in \Lambda^+([\frac{N}{2}])$.*

(2) *Suppose $N \geq 3$. Any irreducible finite-dimensional representation of the group $O(N-1,1)$ is of the form $F^{O(N-1,1)}(\lambda)_{a,b}$ for some $\lambda \in \Lambda^+([\frac{N}{2}])$ and $a, b \in \{\pm\}$.*

The point of Lemma 14.3 (2) is that an analogous statement of Weyl's unitary trick may fail for the disconnected group $O(N-1,1)$, that is, not all irreducible finite-dimensional representations of $O(N-1,1)$ cannot extend to holomorphic representations of $O(N, \mathbb{C})$.

Proof of Lemma 14.3.

(1) This is a restatement of Weyl's description (2.20) of $\widehat{O(N)}$.

(2) Take any irreducible finite-dimensional representation σ of $O(N-1,1)$. By the Frobenius reciprocity, σ occurs as an irreducible summand of the induced representation $\mathrm{Ind}_{SO_0(N-1,1)}^{O(N-1,1)}(\sigma|_{SO_0(N-1,1)})$. Since $N \geq 3$, the fundamental group of $SO(N,\mathbb{C})/SO_0(N-1,1)$ is trivial because it is homotopic to $SO(N)/SO(N-1) \simeq S^{N-1}$, see [26, Lem. 6.1]. Hence the irreducible finite-dimensional representation τ of $SO_0(N-1,1)$ extends to a holomorphic representation of $SO(N,\mathbb{C})$, which we shall denote by $\tau_\mathbb{C}$.

Let $\lambda \in \Lambda^+([\frac{N}{2}])$ be the highest weight of the irreducible $SO(N,\mathbb{C})$-module $\tau_\mathbb{C}$. Then $\tau_\mathbb{C}$ occurs in the restriction $F^{O(N,\mathbb{C})}(\widetilde{\lambda})|_{SO(N,\mathbb{C})}$, and therefore the $SO_0(N-1,1)$-module τ occurs in the restriction $F^{O(N-1,1)}(\lambda)_{+,+}|_{SO(N-1,1)}$. Hence σ occurs as an irreducible summand of the induced representation

$$\mathrm{Ind}_{SO_0(N-1,1)}^{O(N-1,1)}(F^{O(N-1,1)}(\lambda)_{+,+}|_{SO_0(N-1,1)}). \tag{14.6}$$

In light that $F^{O(N-1,1)}(\lambda)_{+,+}$ is a representation of $O(N-1,1)$, we can compute the induced representation (14.6) as follows.

$$(14.6) \simeq F^{O(N-1,1)}(\lambda)_{+,+} \otimes \mathrm{Ind}_{SO_0(N-1,1)}^{O(N-1,1)}(\mathbf{1})$$

$$\simeq F^{O(N-1,1)}(\lambda)_{+,+} \otimes \left(\bigoplus_{a,b \in \{\pm\}} \chi_{ab} \right)$$

$$\simeq \bigoplus_{a,b \in \{\pm\}} F^{O(N-1,1)}(\lambda)_{a,b}.$$

Thus Lemma 14.3 is proved.   □

There are a few overlaps in the expressions (14.4) for $O(N)$-modules and (14.5) for $O(N-1,1)$-modules. We give a necessary and sufficient condition for two expressions, which give the same irreducible representation as follows.

Lemma 14.4.

(1) *The following two conditions on* $\lambda, \mu \in \Lambda^+([\frac{N}{2}])$ *and* $a,b \in \{\pm\}$ *are equivalent:*

    (i) $F^{O(N)}(\lambda)_a \simeq F^{O(N)}(\mu)_b$ *as* $O(N)$-*modules;*
    (ii) "$\lambda = \mu$ *and* $a = b$" *or the following condition holds:*

$$\lambda = \mu, \ N \text{ is even}, \ \lambda_{\frac{N}{2}} > 0, \text{ and } a = -b. \tag{14.7}$$

(2) *Suppose* $N \geq 2$. *Then the following two conditions on* $\lambda, \mu \in \Lambda^+([\frac{N}{2}])$ *and* $a,b,c,d \in \{\pm\}$ *are equivalent:*

    (i) $F^{O(N-1,1)}(\lambda)_{a,b} \simeq F^{O(N-1,1)}(\mu)_{c,d}$ *as* $O(N-1,1)$-*modules;*
    (ii) "$\lambda = \mu$ *and* $(a,b) = (c,d)$" *or the following condition holds:*

$$\lambda = \mu, \ N \text{ is even, } \lambda_{\frac{N}{2}} > 0, \text{ and } (a,b) = -(c,d). \tag{14.8}$$

Proof.

(1) The $O(N)$-isomorphism $F^{O(N)}(\lambda)_a \simeq F^{O(N)}(\mu)_b$ implies an obvious isomorphism $F^{O(N)}(\lambda)_a|_{SO(N)} \simeq F^{O(N)}(\mu)_b|_{SO(N)}$ as $SO(N)$-modules, whence $\lambda = \mu$ by the classical branching law (Lemma 2.7) for the restriction $O(N) \downarrow SO(N)$. Then the equivalence (i) $\Leftrightarrow$ (ii) follows from the equivalence (i) $\Leftrightarrow$ (iii) in Lemma 2.13.

(2) Similarly to the proof for the first statement, we may and do assume $\lambda = \mu$ by considering of the restriction $O(N-1,1) \downarrow SO(N-1,1)$. Then the proof of the equivalence (i) $\Leftrightarrow$ (ii) for $O(N-1,1)$ reduces to the case for $O(N,1)$ and the following lemma. $\qquad\square$

Lemma 14.5. *Suppose σ is an irreducible finite-dimensional representation of $O(N-1,1)$.*

(1) *Suppose $N \geq 2$. If σ is extended to a holomorphic representation of $O(N,\mathbb{C})$, then neither $\sigma \otimes \chi_{+-}$ nor $\sigma \otimes \chi_{-+}$ can be extended to a holomorphic representation of $O(N,\mathbb{C})$.*

(2) *Suppose $N \geq 3$. If σ cannot be extended to a holomorphic representation of $O(N,\mathbb{C})$, then both $\sigma \otimes \chi_{+-}$ and $\sigma \otimes \chi_{-+}$ can be extended to a holomorphic representation of $O(N,\mathbb{C})$.*

Proof.

(1) If $\sigma \otimes \chi_{ab}$ extends to a holomorphic representation of $O(N,\mathbb{C})$, then so does the subrepresentation χ_{ab} in the tensor product $(\sigma \otimes \chi_{ab}) \otimes \sigma^\vee$, where $\sigma^\vee$ stands for the contragredient representation of σ. Since χ_{ab} is the restriction of some holomorphic character of $O(N,\mathbb{C})$ if and only if $(a,b) = (+,+)$ or $(-,-)$, the first statement is proved.

(2) As in the proof of Lemma 14.3 (2), we see that at least one element in $\{\sigma \otimes \chi_{ab} : a,b \in \{\pm\}\}$ can be extended to a holomorphic representation of $O(N,\mathbb{C})$. Then the second statement follows from the first one. $\qquad\square$

Example 14.6. The natural action of $O(N)$ on i-th exterior algebra $\bigwedge^i(\mathbb{C}^N)$ is given as

$$\bigwedge^i(\mathbb{C}^N) \simeq \begin{cases} F^{O(N)}(1^i, 0^{[\frac{N}{2}]-i})_+ & \text{if } i \leq \frac{N}{2}, \\ F^{O(N)}(1^{N-i}, 0^{i-[\frac{N+1}{2}]})_- & \text{if } i \geq \frac{N}{2}, \end{cases}$$

with the notation in this section, whereas the same representation was described as

$$\bigwedge^i(\mathbb{C}^N) \simeq F^{O(N)}(\underbrace{1,\cdots,1}_{i}, \underbrace{0,\cdots,0}_{N-i})$$

with the notation (2.20) in Section 2.2.

As in the classical branching rule for $O(N) \downarrow O(N-1)$ given in Fact 2.12, we give the irreducible decomposition of finite-dimensional representations of $O(N, 1)$ when restricted to the subgroup $O(N-1, 1)$ as follows:

Theorem 14.7 (branching rule for $O(N, 1) \downarrow O(N-1, 1)$). *Let $N \geq 2$. Suppose that $(\lambda_1, \cdots, \lambda_{[\frac{N+1}{2}]}) \in \Lambda^+([\frac{N+1}{2}])$ and $a, b \in \{\pm\}$. Then the irreducible finite-dimensional representation $F^{O(N,1)}(\lambda_1, \cdots, \lambda_{[\frac{N+1}{2}]})_{a,b}$ of $O(N, 1)$ decomposes into a multiplicity-free sum of irreducible representations of $O(N-1, 1)$ as follows:*

$$F^{O(N,1)}(\lambda_1, \cdots, \lambda_{[\frac{N+1}{2}]})_{a,b}|_{O(N-1,1)} \simeq \bigoplus F^{O(N-1,1)}(\nu_1, \cdots, \nu_{[\frac{N}{2}]})_{a,b},$$

where the summation is taken over $(\nu_1, \cdots, \nu_{[\frac{N}{2}]}) \in \mathbb{Z}^{[\frac{N}{2}]}$ subject to

$$\lambda_1 \geq \nu_1 \geq \lambda_2 \geq \cdots \geq \nu_{\frac{N}{2}} \geq 0 \qquad \text{for } N \text{ even},$$

$$\lambda_1 \geq \nu_1 \geq \lambda_2 \geq \cdots \geq \nu_{\frac{N-1}{2}} \geq \lambda_{\frac{N+1}{2}} \quad \text{for } N \text{ odd}.$$

Proof. The assertion follows in the case $(a, b) = (+, +)$ from Fact 2.12. The general case follows from the definition (14.5) and from the observation that the restriction $\chi_{a,b}|_{G'}$ of the G-character $\chi_{a,b}$ gives the same type of a character for $G' = O(N-1, 1)$, see (3.23). □

14.2   Singular Parameters for $V \in \widehat{O(n)}$: $S(V)$ and $S_Y(V)$

In this section we prepare some notation that describes the parameters of *reducible* principal series representations $I_\delta(V, \lambda)$ of $G = O(n+1, 1)$.

We recall from Lemma 2.14 that both of the following subsets

$$\{(\delta, V, \lambda) : I_\delta(V, \lambda) \text{ has regular integral } \mathfrak{Z}_G(\mathfrak{g})\text{-infinitesimal character}\},$$

$$\{(\delta, V, \lambda) : I_\delta(V, \lambda) \text{ is reducible}\}$$

of $\{\pm\} \times \widehat{O(n)} \times \mathbb{C}$ are preserved under the following transforms:

$$(\delta, V, \lambda) \mapsto (-\delta, V, \lambda),$$

$$(\delta, V, \lambda) \mapsto (\delta, V \otimes \det, \lambda).$$

Thus we omit the signature δ in our notation, and focus on the second and third components.

Definition 14.8. We define two subsets of $\widehat{O(n)} \times \mathbb{C}$ (actually, of $\widehat{O(n)} \times \mathbb{Z}$) by

$$\mathcal{RInt} := \{(V, \lambda) : I_\delta(V, \lambda) \text{ has regular integral } \mathfrak{Z}_G(\mathfrak{g})\text{-infinitesimal character}\},$$
$$\mathcal{Red} := \{(V, \lambda) : I_\delta(V, \lambda) \text{ is reducible}\}. \qquad (14.9)$$

Both the sets $\mathcal{RInt}$ and $\mathcal{Red}$ are preserved by the transformations

$$(V, \lambda) \mapsto (V \otimes \det, \lambda),$$
$$(V, \lambda) \mapsto (V, n - \lambda).$$

This is clear for $\mathcal{RInt}$, whereas the assertions for $\mathcal{Red}$ follows from the G-isomorphism $I_\delta(V, \lambda) \otimes \chi_{--} \simeq I_\delta(V \otimes \det, \lambda)$ by Lemma 2.14 and from the fact that $I_\delta(V, n - \lambda)$ is isomorphic to the contragredient representation of $I_\delta(V, \lambda)$. We shall introduce two discrete sets $S(V)$ and $S_Y(V)$ for $V \in \widehat{O(n)}$ in Definition 14.10 below, and prove in Lemma 14.12 and in Theorem 14.15

$$\mathcal{RInt} = \{(V, \lambda) \in \widehat{O(n)} \times \mathbb{Z} : \lambda \notin S(V)\}$$

$$\cup \qquad\qquad\qquad \cup$$

$$\mathcal{Red} = \{(V, \lambda) \in \widehat{O(n)} \times \mathbb{Z} : \lambda \notin S(V) \cup S_Y(V)\},$$

see also Convention 14.11.

14.2.1 Infinitesimal Character $r(V, \lambda)$ of $I_\delta(V, \lambda)$

Suppose that $V \in \widehat{O(n)}$ is given as

$$V = F^{O(n)}(\sigma)_\varepsilon \quad \text{for some } \sigma \in \Lambda^+\left(\left[\frac{n}{2}\right]\right) \text{ and } \varepsilon \in \{\pm\}$$

with the notation as in Section 14.1. We define an element of $\mathfrak{h}_{\mathbb{C}}^* \simeq \mathbb{C}^{\left[\frac{n}{2}\right]+1}$ by

$$r(V, \lambda) := (\sigma_1 + \frac{n}{2} - 1, \sigma_2 + \frac{n}{2} - 2, \cdots, \sigma_{\left[\frac{n}{2}\right]} + \frac{n}{2} - \left[\frac{n}{2}\right], \lambda - \frac{n}{2}). \qquad (14.10)$$

The ordering in (14.10) will play a crucial role in a combinatorial argument in later sections, whereas, up to the action of the Weyl group W_G, $r(V, \lambda)$ gives the $\mathfrak{Z}_G(\mathfrak{g})$-infinitesimal character of the unnormalized induced representation $I_\delta(V, \lambda)$ of $G = O(n + 1, 1)$, see (2.26).

Example 14.9. For $0 \leq i \leq n$, we set $\ell := \min(i, n - i)$ and

$$\rho^{(i)} := r(\textstyle\bigwedge^i(\mathbb{C}^n), i)$$

$$= (\underbrace{\frac{n}{2}, \frac{n}{2} - 1, \cdots, \frac{n}{2} - \ell + 1}_{\ell}, \underbrace{\frac{n}{2} - \ell - 1, \cdots, \frac{n}{2} - [\frac{n}{2}]}_{[\frac{n}{2}] - \ell}, i - \frac{n}{2})$$

$$= \begin{cases} (\underbrace{\frac{n}{2}, \cdots, \frac{n}{2} - i + 1}_{i}, \underbrace{\frac{n}{2} - i - 1, \cdots, \frac{n}{2} - [\frac{n}{2}]}_{[\frac{n}{2}] - i}, i - \frac{n}{2}) & \text{for } i \leq [\frac{n}{2}], \\[4mm] (\underbrace{\frac{n}{2}, \cdots, -\frac{n}{2} + i + 1}_{n - i}, \underbrace{-\frac{n}{2} + i - 1, \cdots, \frac{n}{2} - [\frac{n}{2}]}_{i - [\frac{n+1}{2}]}, i - \frac{n}{2}) & \text{for } [\frac{n+1}{2}] \leq i. \end{cases}$$

Here are some elementary properties.

(1) The following equations hold:

$$\rho^{(i)} - \rho^{(0)} = (\underbrace{1, \cdots, 1}_{\ell}, \underbrace{0, \cdots, 0}_{[\frac{n}{2}] - \ell}, i) \tag{14.11}$$

$$= \begin{cases} (\underbrace{1, \cdots, 1}_{i}, \underbrace{0, \cdots, 0}_{[\frac{n}{2}] - i}, i) & \text{for } 0 \leq i \leq [\frac{n}{2}], \\[4mm] (\underbrace{1, \cdots, 1}_{n - i}, \underbrace{0, \cdots, 0}_{i - [\frac{n+1}{2}]}, i) & \text{for } [\frac{n+1}{2}] \leq i \leq n. \end{cases}$$

(2) Let $r(V, \lambda)$ be defined as in (14.10). Then for any i $(0 \leq i \leq n)$, we have

$$r(V, \lambda) = (\sigma_1, \cdots, \sigma_{[\frac{n}{2}]}, \lambda) + \rho^{(0)}$$

$$= (\sigma_1 - 1, \cdots, \sigma_\ell - 1, \sigma_{\ell+1}, \cdots, \sigma_{[\frac{n}{2}]}, \lambda - i) + \rho^{(i)},$$

where we retain the notation $\ell = \min(i, n - i)$.

(3) For all i $(0 \leq i \leq n)$,

$$\rho_G \equiv \rho^{(i)} \mod W_G. \tag{14.12}$$

14.2.2 Singular Integral Parameter: $S(V)$ and $S_Y(V)$

Retain the setting as in Section 14.2.1. Let $G = O(n+1,1)$ and $m = [\frac{n}{2}]$. Suppose $V \in \widehat{O(n)}$ is given as $V = F^{O(n)}(\sigma)_\varepsilon$ with $\sigma = (\sigma_1, \cdots, \sigma_m)$ and $\varepsilon \in \{\pm\}$. Since σ_1, $\cdots, \sigma_m \in \mathbb{Z}$, the following three conditions on $\lambda \in \mathbb{C}$ are equivalent:

(i) The $\mathfrak{z}_G(\mathfrak{g})$-infinitesimal character of $I_\delta(V, \lambda)$ is integral in the sense of Definition 2.1;

(ii) $\langle r(V, \lambda), \alpha^\vee \rangle \in \mathbb{Z}$ for any $\alpha \in \Delta(\mathfrak{g}_\mathbb{C}, \mathfrak{h}_\mathbb{C})$;

(iii) $\lambda \in \mathbb{Z}$.

For each $V \in \widehat{O(n)}$, we introduce a subset $S(V)$ in $\mathbb{Z}$ (and a subset $S_Y(V)$ in $\mathbb{Z}$ for V of type Y) as follows.

Definition 14.10 ($S(V)$ and $S_Y(V)$). Let $m = [\frac{n}{2}]$. For $V = F^{O(n)}(\sigma)_\varepsilon$ with $\sigma = (\sigma_1, \cdots, \sigma_m) \in \Lambda^+(m)$ and $\varepsilon \in \{\pm\}$, we define a finite subset of $\mathbb{Z}$ by

$$S(V) := \{j - \sigma_j, n + \sigma_j - j : 1 \le j \le m\}. \tag{14.13}$$

When the irreducible $O(n)$-module V is of type Y (see Definition 2.6), namely, when n is even ($= 2m$) and $\sigma_m > 0$, we define also the following finite set

$$S_Y(V) := \{\lambda \in \mathbb{Z} : 0 < |\lambda - m| < \sigma_m\}. \tag{14.14}$$

We note that

$$S(V) \cap S_Y(V) = \emptyset$$

by definition. We shall sometimes adopt the following convention:

Convention 14.11. *When V is of type X (see Definition 2.6), we set*

$$S_Y(V) = \emptyset.$$

The definitions imply the following lemma.

Lemma 14.12. *The $\mathfrak{z}_G(\mathfrak{g})$-infinitesimal character of $I_\delta(V, \lambda)$ is regular integral (see Definition 2.1) if and only if $\lambda \in \mathbb{Z} - S(V)$. Thus, we have*

$$\mathcal{RInt} = \{(V, \lambda) \in \widehat{O(n)} \times \mathbb{Z} : \lambda \notin S(V)\}.$$

We refer to $S(V)$ as the set of *singular integral parameters*. It should be noted that $I_\delta(V, \lambda)$ has regular integral infinitesimal character if $\lambda \in S_Y(V)$, since $S_Y(V) \subset \mathbb{Z} - S(V)$.

We shall see in Theorem 14.15 below that the principal series representation $I_\delta(V, \lambda)$ is irreducible if and only if $\lambda \in (\mathbb{C} - \mathbb{Z}) \cup S(V) \cup S_Y(V)$.

We end this section with a lemma that will be used in Appendix III (Chapter 16) when we discuss translation functors.

Lemma 14.13. *Let $V \in \widehat{O(n)}$ and $\lambda \in \mathbb{Z} - S(V)$.*

(1) *Suppose V is of type X (Definition 2.6). Then the $W_{\mathfrak{g}}$- and W_G-orbits through $r(V, \lambda) \in \mathfrak{h}_{\mathbb{C}}^* \simeq \mathbb{C}^{[\frac{n}{2}]+1}$ coincide:*

$$W_{\mathfrak{g}} r(V, \lambda) = W_G r(V, \lambda). \tag{14.15}$$

(2) *Suppose V is of type Y. Then (14.15) holds if and only if $\lambda = \frac{n}{2}$.*

Proof.

(1) The assertion is obvious when n is odd because $W_{\mathfrak{g}} = W_G$ in this case. Suppose n is even, say, $n = 2m$. It is sufficient to show that $r(V, \lambda)$ contains zero in its entries. Since V is of type X, we have $\sigma_m = 0$, and therefore, the m-th entry of $r(V, \lambda)$ amounts to $\sigma_m + m - m = 0$ by the definition (14.10). Thus the lemma is proved.

(2) Since V is of type Y, n is even ($= 2m$) and $W_G \supsetneq W_{\mathfrak{g}}$. Since $\lambda \notin S(V)$, $r(V, \lambda)$ is $W_{\mathfrak{g}}$-regular. Hence (14.15) holds if and only if at least one of the entries in $r(V, \lambda)$ equals zero. Since $\sigma_1 \geq \sigma_2 \geq \cdots \geq \sigma_m > 0$, this happens only when the $(m+1)$-th entry of $r(V, \lambda)$ vanishes, *i.e.*, $\lambda = \frac{n}{2}(= m)$. Hence Lemma 14.13 is proved. □

Remark 14.14. For $n = 2m$ (even), if V is of type X or if $\lambda = m$, then the $\mathfrak{z}_G(\mathfrak{g})$-infinitesimal character (14.10) is regular for $W_{\mathfrak{g}}$ in the sense of Definition 2.1, but is "singular" with respect to the Weyl group W_G for the *disconnected* group $G = O(n+1, 1)$ which is not in the Harish-Chandra class.

14.3   Irreducibility Condition of $I_\delta(V, \lambda)$

We are ready to state a necessary and sufficient condition for the principal series representation $I_\delta(V, \lambda)$ of $G = O(n+1, 1)$ to be irreducible.

We recall from (14.13) and (14.14) the definitions of $S(V)$ and $S_Y(V)$, respectively.

Theorem 14.15 (irreducibility criterion of $I_\delta(V, \lambda)$). *Let $G = O(n+1, 1)$, $\delta \in \{\pm\}$, $V \in \widehat{O(n)}$, and $\lambda \in \mathbb{C}$.*

(1) *If $\lambda \in \mathbb{C} - \mathbb{Z}$, then the principal series representation $I_\delta(V, \lambda)$ of G is irreducible.*

(2) *Suppose $\lambda \in \mathbb{Z}$. Then $I_\delta(V, \lambda)$ is irreducible if and only if*

$$\lambda \in S(V) \qquad\qquad \text{when } V \text{ is of type X,}$$

$$\lambda \in S(V) \cup S_Y(V) \quad \text{when } V \text{ is of type Y.}$$

Thus $\mathcal{R}ed$ *(see (14.9)) is given by*

$$\mathcal{R}ed = \{(V, \lambda) \in \widehat{O(n)} \times \mathbb{Z} : \lambda \notin S(V) \cup S_Y(V)\} \tag{14.16}$$

with Convention 14.11.

The proof of Theorem 14.15 will be given in Section 15.3 in Appendix II by inspecting the restriction of $I_\delta(V, \lambda)$ of $G = O(n + 1, 1)$ to its subgroups $\overline{G} = SO(n + 1, 1)$ and $G_0 = SO_0(n + 1, 1)$. See also [4, 16] for representations of the connected group.

Example 14.16. Let $0 \leq i \leq n$. The exterior tensor representation on $\bigwedge^i(\mathbb{C}^n)$ is of type X if and only if $n \neq 2i$ (see Example 2.8). A simple computation shows

$$S(\bigwedge^{(i)}(\mathbb{C}^n)) = \mathbb{Z} - (\{i, n - i\} \cup (-\mathbb{N}_+) \cup (n + \mathbb{N}_+)) \qquad \text{for } 0 \leq i \leq n,$$

$$S_Y(\bigwedge^{(m)}(\mathbb{C}^n)) = \emptyset \qquad \text{for } n = 2m,$$

see also Example 14.25. Hence $I_\delta(i, \lambda)$ is reducible if and only if

$$\lambda \in \{i, n - i\} \cup (-\mathbb{N}_+) \cup (n + \mathbb{N}_+)$$

by Theorem 14.15. See Theorem 2.20 for the socle filtration of $I_\delta(i, \lambda)$ for $\lambda = i$ or $n - i$.

For later purpose, we decompose $\mathcal{R}ed$ into two disjoint subsets as follows:

Definition 14.17. We recall from Definition 2.6 that any $V \in \widehat{O(n)}$ is either of type X or of type Y for $\widehat{O(n)}$. We set

$$\mathcal{R}ed_I := \{(V, \lambda) \in \mathcal{R}ed : V \text{ is of type X or } \lambda = \frac{n}{2}\},$$

$$\mathcal{R}ed_{II} := \{(V, \lambda) \in \mathcal{R}ed : V \text{ is of type Y and } \lambda \neq \frac{n}{2}\}.$$

Then we have a disjoint union

$$\mathcal{R}ed = \mathcal{R}ed_I \amalg \mathcal{R}ed_{II}.$$

Remark 14.18. If n is odd, then

$$\mathcal{R}ed_{II} = \emptyset \quad \text{and} \quad \mathcal{R}ed = \mathcal{R}ed_I.$$

14.4   Subquotients of $I_\delta(V, \lambda)$

By Theorem 14.15, the principal series representation $I_\delta(V, \lambda)$ of $G = O(n+1, 1)$ is reducible *i.e.*, $(V, \lambda) \in \mathcal{R}ed$ if and only if

$$\lambda \in \mathbb{Z} - (S(V) \cup S_Y(V))$$

with Convention 14.11. In this section, we explain the socle filtration of $I_\delta(V, \lambda)$. A number of different characterizations of the subquotients will be given in later sections, see Theorem 14.35 for summary. We divide the arguments into the following two cases:

Case 1. $\lambda \neq \frac{n}{2}$, see Section 14.4.1;
Case 2. $\lambda = \frac{n}{2}$, see Section 14.4.2.

14.4.1   Subquotients of $I_\delta(V, \lambda)$ for V of Type X

We begin with the case where $\lambda \neq \frac{n}{2}$. This means that we treat the following cases:

- V is of type X, and $\lambda \in \mathbb{Z} - S(V)$;
- V is of type Y, and $\lambda \in \mathbb{Z} - (S(V) \cup S_Y(V) \cup \{\frac{n}{2}\})$.

Proposition 14.19. *Let* $G = O(n+1, 1)$, $V \in \widehat{O(n)}$, $\delta \in \{\pm\}$, *and* $\lambda \in \mathbb{Z} - (S(V) \cup S_Y(V))$. *Assume further that* $\lambda \neq \frac{n}{2}$. *Then there exists a unique proper submodule of the principal series representation* $I_\delta(V, \lambda)$, *to be denoted by* $I_\delta(V, \lambda)^\flat$. *In particular, the quotient G-module*

$$I_\delta(V, \lambda)^\sharp := I_\delta(V, \lambda)/I_\delta(V, \lambda)^\flat$$

is irreducible.

The proof of Proposition 14.19 will be given in Section 15.4 of Appendix II.

Remark 14.20. The K-type formulæ and the minimal K-types of the irreducible G-modules $I_\delta(V, \lambda)^\flat$ and $I_\delta(V, \lambda)^\sharp$ will be given in Propositions 14.30 and 14.34, respectively.

14.4.2   Subrepresentations of $I_\delta(V, \frac{n}{2})$ for V of Type Y

Next we discuss the case:

- V is of type Y and $\lambda = \frac{n}{2}$.

In this case $I_\delta(V, \lambda)$ is the smooth representation of a tempered unitary representation.

Proposition 14.21 (reducible tempered principal series representation). *Let $G = O(n+1, 1)$ with $n = 2m$, $V \in \widehat{O(n)}$ be of type Y, and $\delta \in \{\pm\}$. Then the principal series representation $I_\delta(V, m)$ of G is decomposed into the direct sum of two irreducible representations of G, to be written as:*

$$I_\delta(V, m) \simeq I_\delta(V, m)^\flat \oplus I_\delta(V, m)^\sharp.$$

If we express $V = F^{O(n)}(\sigma)_\varepsilon$ by $\sigma = (\sigma_1, \cdots, \sigma_m) \in \Lambda^+(m)$ with $\sigma_m > 0$ and $\varepsilon \in \{\pm\}$, then the irreducible G-modules $I_\delta(V, m)^\flat$ and $I_\delta(V, m)^\sharp$ are characterized by their minimal K-types given respectively by the following:

$$F^{O(n+1)}(\sigma_1, \cdots, \sigma_m)_\varepsilon \boxtimes \delta,$$

$$F^{O(n+1)}(\sigma_1, \cdots, \sigma_m)_{-\varepsilon} \boxtimes (-\delta).$$

Proof. This is proved in Proposition 15.5 (2) except for the assertion on the K-types. The last assertion on the minimal K-types follow from the K-type formula of $I_\delta(V, m)^\flat$ and $I_\delta(V, m)^\sharp$ in Proposition 14.30 (2).   □

14.4.3   Socle Filtration of $I_\delta(V, \lambda)$

By Theorem 14.15 together with Propositions 14.19 and 14.21, we obtain the following:

Corollary 14.22. *Let $G = O(n+1, 1)$ for $n \geq 2$. Then the principal series representation $I_\delta(V, \lambda)$ ($\delta \in \{\pm\}$, $V \in \widehat{O(n)}$, $\lambda \in \mathbb{C}$) of G is either irreducible or of composition series of length two.*

14.5   Definition of the Height $i(V, \lambda)$

In this section we introduce the "height"

$$i: \mathcal{RInt} \to \{0, 1, \ldots, n\}, \quad (V, \lambda) \mapsto i(V, \lambda)$$

which plays an important role in the study of the principal series representation $I_\delta(V, \lambda)$ of G. For instance, we shall see in Section 14.7 that the K-type formula for subquotients of $I_\delta(V, \lambda)$ is described by using the height $i(V, \lambda)$ when $(V, \lambda) \in \mathcal{Red}$ (Definition 14.8). Moreover, we shall prove in Theorem 16.6 that the G-module $I_\delta(V, \lambda)$ is obtained by the translation functor applied to the principal series representation $I_\pm(i, i)$ with the trivial infinitesimal character ρ_G without "crossing the wall" if we take i to be the height $i(V, \lambda)$, see Theorem 16.6. We note that the group $G = O(n+1, 1)$ is not in the Harish-Chandra class when n is even, and will discuss carefully a *translation functor* in Appendix III (Chapter 16).

We recall from (14.10) that

$$r(V, \lambda) = (\sigma_1 + \frac{n}{2} - 1, \sigma_2 + \frac{n}{2} - 2, \cdots, \sigma_m + \frac{n}{2} - m, \lambda - \frac{n}{2}),$$

where $m := [\frac{n}{2}]$. To specify the Weyl chamber for $W_{\mathfrak{g}}$ that $r(V, \lambda) \in (\frac{1}{2}\mathbb{Z})^{m+1}$ belongs to, we label the places where $\lambda - \frac{n}{2}$ is located with respect to the following inequalities.

Case 1. $n = 2m$:

$$-\sigma_1 - m + 1 < -\sigma_2 - m + 2 < \cdots < -\sigma_m \leq \sigma_m < \cdots < \sigma_2 + m - 2 < \sigma_1 + m - 1;$$

Case 2. $n = 2m + 1$:

$$-\sigma_1 - m + \frac{1}{2} < -\sigma_2 - m + \frac{3}{2} < \cdots < -\sigma_m - \frac{1}{2} < 0 < \sigma_m + \frac{1}{2} < \cdots < \sigma_1 + m - \frac{1}{2}.$$

Unifying these inequalities by adding $\frac{n}{2}$ to each term, we may write as

$$1 - \sigma_1 < 2 - \sigma_2 < \cdots < m - \sigma_m \leq \frac{n}{2} \leq \sigma_m + n - m < \cdots < \sigma_2 + n - 2 < \sigma_1 + n - 1.$$

Definition 14.23. For $0 \leq i \leq n$, we define the following subsets $R(V; i)$ of $\mathbb{Z}$:

$$\{\lambda \in \mathbb{Z} : i - \sigma_i < \lambda < i + 1 - \sigma_{i+1}\} \qquad \text{for } 0 \leq i < \frac{n-1}{2},$$

$$\{\lambda \in \mathbb{Z} : \frac{n-1}{2} - \sigma_{\frac{n-1}{2}} < \lambda < \frac{n}{2}\} \qquad \text{for } i = \frac{n-1}{2} \quad (n \text{ odd}),$$

$$\{\lambda \in \mathbb{Z} : \frac{n}{2} - \sigma_{\frac{n}{2}} < \lambda < \sigma_{\frac{n}{2}} + \frac{n}{2}\} \qquad \text{for } i = \frac{n}{2} \quad (n \text{ even}),$$

$$\{\lambda \in \mathbb{Z} : \frac{n}{2} < \lambda < \sigma_{\frac{n-1}{2}} + \frac{n+1}{2}\} \qquad \text{for } i = \frac{n+1}{2} \quad (n \text{ odd}),$$

$$\{\lambda \in \mathbb{Z} : \sigma_{n-i+1} + i - 1 < \lambda < \sigma_{n-i} + i\} \quad \text{for } \frac{n+1}{2} < i \leq n.$$

Here we regard $\sigma_0 = \infty$.

Lemma 14.24. *Let $V \in \widehat{O(n)}$. We recall from (14.13) that $S(V)$ is the set of singular integral parameters.*

(1) *The set of regular integral parameters has the following disjoint decomposition:*

$$\mathbb{Z} - S(V) = \coprod_{i=0}^{n} R(V; i).$$

In particular, there exists a map

$$i(V, \cdot) : \mathbb{Z} - S(V) \to \{0, 1, \ldots, n\} \tag{14.17}$$

such that $\lambda \in R(V; i(V, \lambda))$.

(2) *The set $S(V)$ is preserved by the transformations $\lambda \mapsto n - \lambda$ and $V \mapsto V \otimes \det$, and we have*

$$i(V, n - \lambda) = n - i(V, \lambda)$$

$$i(V \otimes \det, \lambda) = i(V, \lambda)$$

for any $\lambda \in \mathbb{Z} - S(V)$.

(3) *$R(V; \frac{n}{2}) \neq \emptyset$ if and only if n is even and the irreducible $O(n)$-module V is of type Y. In this case, we have*

$$R(V; \frac{n}{2}) = \{\frac{n}{2}\} \cup S_Y(V) \quad \text{(disjoint union).} \tag{14.18}$$

Example 14.25. Let $0 \leq i \leq n$. For the i-th exterior tensor representation $V = \bigwedge^i (\mathbb{C}^n)$ of $O(n)$, we have

$$S(\bigwedge^{(i)} (\mathbb{C}^n)) = \mathbb{Z} - (\{i, n - i\} \cup (-\mathbb{N}_+) \cup (n + \mathbb{N}_+)).$$

Furthermore, we see from Example 14.6 that the set $R(V; j)$ is given as follows.

(1) For $1 \leq i \leq n - 1$,

$$R(\bigwedge^i (\mathbb{C}^n); j) = \begin{cases} -\mathbb{N}_+ & \text{if } j = 0, \\ \{j\} & \text{if } j = i \text{ or } n - i, \\ n + \mathbb{N}_+ & \text{if } j = n, \\ \emptyset & \text{otherwise.} \end{cases}$$

(2) For $i = 0$ or n,

$$R(\bigwedge^i (\mathbb{C}^n); j) = \begin{cases} -\mathbb{N} & \text{if } j = 0, \\ n + \mathbb{N} & \text{if } j = n, \\ \emptyset & \text{otherwise.} \end{cases}$$

We recall from Definition 14.8 that $\mathcal{R}\mathcal{I}nt$ is a subset of $\widehat{O(n)} \times \mathbb{Z}$.

Definition 14.26 (height $i(V, \lambda)$). By (14.17) in Lemma 14.24, we define a map

$$i \colon \mathcal{R}\mathcal{I}nt \to \{0, 1, \ldots, n\},$$

see Lemma 14.12. We refer to $i(V, \lambda)$ as the *height* of (V, λ). We also refer it to as the height of the principal series representation $I_\delta(V, \lambda)$.

Example 14.27. We illustrate the definition of the height $i(V, \lambda) \in \{0, 1, \ldots, n\}$ for $(V, \lambda) \in \mathcal{R}\mathcal{I}nt$ by a graphic description when $m (= [\frac{n}{2}]) = 1$, namely, when $n = 2$ or 3. In this case $G = O(n+1, 1)$ is either $O(3, 1)$ or $O(4, 1)$, and $V \in \widehat{O(n)}$ is given by $V = F^{O(n)}(\sigma_1)_\varepsilon$ with $\sigma_1 \in \mathbb{N}$ and $\varepsilon \in \{\pm\}$. Then

$$\mathcal{R}\mathcal{I}nt \simeq \begin{cases} \{(\sigma_1, \varepsilon, \lambda) \in \mathbb{N} \times \{\pm\} \times \mathbb{Z} : \lambda - 1 \neq \pm\sigma_1\} & \text{if } n = 2, \\ \{(\sigma_1, \varepsilon, \lambda) \in \mathbb{N} \times \{\pm\} \times \mathbb{Z} : \lambda - 2 \neq \pm\sigma_1, \lambda \neq 2\} & \text{if } n = 3. \end{cases}$$

In the (σ_1, λ)-plane, the height $i(V, \lambda)$ is given as in Figure 14.1.

The red dots stand for $(V, \lambda) = (\bigwedge^j(\mathbb{C}^n), j)$ when $j = 0, 1, \ldots, n$.

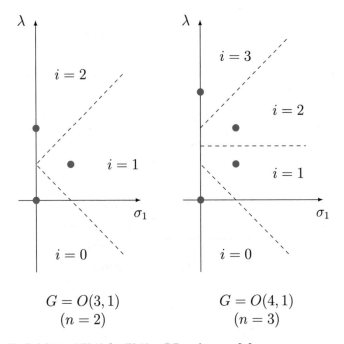

Fig. 14.1   The height $i = i(V, \lambda)$ for $(V, \lambda) \in \mathcal{R}\mathcal{I}nt$ when $n = 2, 3$

The case where the height $i(V, \lambda)$ is equal to $\frac{n}{2}$ requires a special attention.

Lemma 14.28. *Let* $m := [\frac{n}{2}]$. *Suppose that* $V = F^{O(n)}(\sigma)_\varepsilon$ *with* $\sigma \in \Lambda^+(m)$ *and* $\varepsilon \in \{\pm\}$, *and* $\lambda \in \mathbb{Z} - S(V)$.

(1) *The height* $i(V, \lambda)$ *is equal to* $\frac{n}{2}$ *if and only if* n *is even* $(= 2m)$ *and* $\sigma_m > |\lambda - m|$.

(2) *If* $\lambda \in S_Y(V)$ *(see (14.14)), then* n *is even* $(= 2m)$ *and* $i(V, \lambda) = m$.

(3) *Suppose that* V *is of type* Y *(Definition 2.6). Then, for* $(V, \lambda) \in \mathcal{R}ed$, *the following two conditions are equivalent:*

   (i) $i(V, \lambda) = \frac{n}{2}$;
   (ii) n *is even and* $\lambda = \frac{n}{2}$.

14.6   *K*-type Formulæ of Irreducible *G*-modules

In this section we provide explicit K-type formulæ of irreducible representations of $G = O(n + 1, 1)$. The height $i(V, \lambda)$ plays a crucial role in describing the K-type formulæ of irreducible subquotients of $I_\delta(V, \lambda)$, see Proposition 14.30 (1).

14.6.1   K-type Formula of $I_\delta(V, \lambda)$

We begin with the K-type formula of the principal series representation $I_\delta(V, \lambda)$ which generalizes Lemma 2.16 for $I_\delta(i, i)$ in the setting that $V = \bigwedge^i(\mathbb{C}^n)$.

Proposition 14.29 (K-type formula of $I_\delta(V, \lambda)$). *Let* $G = O(n + 1, 1)$ *and* $m = [\frac{n}{2}]$. *Suppose that* $V = F^{O(n)}(\sigma)_\varepsilon$ *with* $\sigma = (\sigma_1, \cdots, \sigma_m) \in \Lambda^+(m)$ *and* $\varepsilon \in \{\pm\}$.

(1) *For* $n = 2m + 1$, *the* K-*type formula of the principal series representation* $I_\delta(V, \lambda)$ *is given by*

$$\bigoplus_\mu F^{O(n+1)}(\mu_1, \cdots, \mu_{m+1})_\varepsilon \boxtimes \delta(-1)^{\sum_{j=1}^{m+1} \mu_j - \sum_{j=1}^m \sigma_j},$$

where $\mu = (\mu_1, \cdots, \mu_{m+1})$ *runs over* $\Lambda^+(m + 1)$ *subject to*

$$\mu_1 \geq \sigma_1 \geq \mu_2 \geq \sigma_2 \geq \cdots \geq \mu_m \geq \sigma_m \geq \mu_{m+1} \geq 0. \tag{14.19}$$

(2) *For* $n = 2m$ *and* $V \in \widehat{O(n)}$ *of type* X *(Definition 2.6), the* K-*type formula of* $I_\delta(V, \lambda)$ *is given by*

$$\bigoplus_\mu F^{O(n+1)}(\mu_1, \cdots, \mu_m)_\varepsilon \boxtimes \delta(-1)^{\sum_{j=1}^m \mu_j - \sum_{j=1}^m \sigma_j},$$

where $\mu = (\mu_1, \cdots, \mu_m)$ *runs over* $\Lambda^+(m+1)$ *subject to*

$$\mu_1 \geq \sigma_1 \geq \cdots \geq \mu_m \geq \sigma_m (= 0). \tag{14.20}$$

(3) *For* $n = 2m$ *and* $V \in \widehat{O(n)}$ *of type Y, the K-type formula of* $I_\delta(V, \lambda)$ *is given by*

$$\bigoplus_{\kappa = \pm} \bigoplus_{\mu} F^{O(n+1)}(\mu_1, \cdots, \mu_m)_{\kappa\varepsilon} \boxtimes \kappa\delta(-1)^{\sum_{j=1}^m \mu_j - \sum_{j=1}^m \sigma_j},$$

where $\mu = (\mu_1, \cdots, \mu_m)$ *runs over* $\Lambda^+(m)$ *subject to*

$$\mu_1 \geq \sigma_1 \geq \cdots \geq \mu_m \geq \sigma_m (> 0). \tag{14.21}$$

Proof. By the Frobenius reciprocity, Proposition 14.29 follows from the classical branching rule for the restriction $O(n+1) \downarrow O(n)$, see Fact 2.12. □

Since the principal series representation $I_\delta(V, \lambda)$ of G is multiplicity-free as a K-module, any subquotient of $I_\delta(V, \lambda)$ can be characterized by its K-types. In the next subsection, we provide K-type formulæ of subquotients of $I_\delta(V, \lambda)$ based on Proposition 14.29.

14.6.2   K-types of Subquotients $I_\delta(V, \lambda)^\flat$ and $I_\delta(V, \lambda)^\sharp$

We recall from (14.9) and Theorem 14.15 that the following two conditions on $(V, \lambda) \in \widehat{O(n)} \times \mathbb{C}$ are equivalent.

(i) $(V, \lambda) \in \mathcal{R}ed$, *i.e.*, the G-module $I_\delta(V, \lambda)$ is reducible;
(ii) $\lambda \in \mathbb{Z} - (S(V) \cup S_Y(V))$.

We note that $\lambda = \frac{n}{2}$ belongs to $\mathbb{Z} - (S(V) \cup S_Y(V))$ when n is even.

In this section, we describe the K-types of the subquotients $I_\delta(V, \lambda)^\flat$ and $I_\delta(V, \lambda)^\sharp$ when the principal series representation $I_\delta(V, \lambda)$ is reducible, *i.e.*, when $(V, \lambda) \in \mathcal{R}ed$, see (14.16).

We shall see that the description depends on the height $i(V, \lambda)$ (Definition 14.26) when $\lambda = \frac{n}{2}$. To be more precise, let $m = [\frac{n}{2}]$ and $V \in \widehat{O(n)}$. Suppose $\lambda \in \mathbb{Z} - (S(V) \cup S_Y(V))$ and we define i to be the height $i(V, \lambda) \in \{0, 1, \ldots, n\}$. We write $V = F^{O(n)}(\sigma)_\varepsilon$ with $\sigma = (\sigma_1, \cdots, \sigma_m) \in \Lambda^+(m)$ and $\varepsilon \in \{\pm\}$ as before. We observe the following:

- if $i < \frac{n-1}{2}$, then $1 \leq i+1 \leq m$ and the condition $i - \sigma_i < \lambda < i+1 - \sigma_{i+1}$ (Definition 14.23) amounts to

$$\sigma_{i+1} \leq i - \lambda \quad \text{and} \quad i - \lambda + 1 \leq \sigma_i; \tag{14.22}$$

- if $i = \frac{n-1}{2}$, then n is odd ($= 2m + 1$) and we have

$$0 \leq m - \lambda \quad \text{and} \quad m - \lambda + 1 \leq \sigma_m; \tag{14.23}$$

- if $i = \frac{n+1}{2}$, then n is odd ($= 2m + 1$) and we have

$$0 \leq \lambda - m - 1 \quad \text{and} \quad \lambda - m \leq \sigma_m; \tag{14.24}$$

- if $\frac{n+1}{2} < i$, then $1 \leq n - i + 1 \leq m$ and the condition $\sigma_{n-i+1} + i - 1 < \lambda < \sigma_{n-i} + i$ amounts to

$$\sigma_{n-i+1} \leq \lambda - i \quad \text{and} \quad \lambda - i + 1 \leq \sigma_{n-i}. \tag{14.25}$$

We recall that the principal series representation $I_\delta(V, \lambda)$ of $G = O(n + 1, 1)$ is K-multiplicity-free, and its K-type formula is given explicitly in Proposition 14.29. To describe the K-type formulæ of subquotients of $I_\delta(V, \lambda)$, we use the inequalities (14.22)–(14.25) in Proposition 14.30 (1) below.

Proposition 14.30 (*K*-type formulæ of subquotients). *Suppose that* $(V, \lambda) \in \mathcal{R}ed$, *or equivalently,* $V \in \widehat{O(n)}$ *and* $\lambda \in \mathbb{Z} - (S(V) \cup S_Y(V))$, *see Theorem 14.15. Let* $i := i(V, \lambda) \in \{0, 1, \ldots, n\}$ *be the height of* (V, λ) *as in Definition 14.26.*

(1) *Suppose* $\lambda \neq \frac{n}{2}$. *In this case* $i \neq \frac{n}{2}$. *Then the K-types of the submodule* $I_\delta(V, \lambda)^\flat$ *and the quotient* $I_\delta(V, \lambda)^\sharp$ *of* $I_\delta(V, \lambda)$, *see Proposition 14.19, are subsets of the K-types of* $I_\delta(V, \lambda)$ (*Proposition 14.29) characterized by the following additional inequalities:*

- *for* $i \leq \frac{n-1}{2}$, *the condition* $\sigma_{i+1} \leq \mu_{i+1} \leq \sigma_i$ *in (14.19)–(14.21) is divided as follows:*

$$(\sigma_{i+1} \leq) \, \mu_{i+1} \quad \leq i - \lambda \qquad \text{for } I_\delta(V, \lambda)^\flat,$$

$$i - \lambda + 1 \leq \, \mu_{i+1} \quad (\leq \sigma_i) \qquad \text{for } I_\delta(V, \lambda)^\sharp;$$

- *for* $\frac{n+1}{2} \leq i$, *the condition* $\sigma_{n-i+1} \leq \mu_{n-i+1} \leq \sigma_{n-i}$ *in (14.19)–(14.21) is divided as follows:*

$$\lambda - i + 1 \leq \, \mu_{n-i+1} (\leq \sigma_{n-i}) \qquad \text{for } I_\delta(V, \lambda)^\flat,$$

$$(\sigma_{n-i+1} \leq) \, \mu_{n-i+1} \quad \leq \lambda - i \qquad \text{for } I_\delta(V, \lambda)^\sharp.$$

Here we regard $\sigma_{m+1} = 0$ (*this happens when* $i = \frac{n \pm 1}{2}$).

(2) *Suppose $\lambda = \frac{n}{2}$. In this case n is even ($= 2m$) and $i = m$. Then the K-types of the submodules $I_\delta(V, \lambda)^\flat$ and $I_\delta(V, \lambda)^\sharp$ of the (tempered) principal series representation $I_\delta(V, \lambda)$, see Proposition 14.21, are given by*

$$\bigoplus_\mu F^{O(n+1)}(\mu_1, \cdots, \mu_m)_\varepsilon \boxtimes \delta(-1)^{\sum_{j=1}^m \mu_j - \sum_{j=1}^m \sigma_j} \qquad \text{for } I_\delta(V, \lambda)^\flat,$$

$$\bigoplus_\mu F^{O(n+1)}(\mu_1, \cdots, \mu_m)_{-\varepsilon} \boxtimes \delta(-1)^{\sum_{j=1}^m \mu_j - \sum_{j=1}^m \sigma_j - 1} \quad \text{for } I_\delta(V, \lambda)^\sharp$$

where $\mu = (\mu_1, \cdots, \mu_m)$ runs over $\Lambda^+(m)$ subject to (14.21).

Proof. The K_0-types for all irreducible subquotients of principal series representations of the connected Lie group $G_0 = SO_0(n+1, 1)$ were obtained in Hirai [16], from which analogous results for the group $\overline{G} = SO(n+1, 1)$ are easily shown. Our concern is with the group $G = O(n+1, 1)$. Then the first assertion follows from Proposition 14.29 on the K-type formula of $I_\delta(V, \lambda)$ and from the branching rule for the restriction $G \downarrow \overline{G}$ in Propositions 15.7 and 15.8 in Appendix II. The second assertion follows from the branching rule of $I_\delta(V, \frac{n}{2})$ for the restriction $G \downarrow \overline{G}$ in Proposition 15.5. $\qquad\square$

14.7 $(\delta, V, \lambda) \rightsquigarrow (\delta^\uparrow, V^\uparrow, \lambda^\uparrow)$ and $(\delta^\downarrow, V^\downarrow, \lambda^\downarrow)$

In this section, we introduce a correspondence

$$\delta \in \{\pm\}, \quad V \in \widehat{O(n)}, \quad \text{and } \lambda \in \mathbb{Z} - (S(V) \cup S_Y(V))$$

$$\updownarrow$$

$$\delta^\uparrow \in \{\pm\}, \ V^\uparrow \in \widehat{O(n)}, \text{ and } \lambda^\uparrow \in \mathbb{Z} - (S(V^\uparrow) \cup S_Y(V^\uparrow))$$

satisfying the following two properties (Proposition 14.33):

$$i(V^\uparrow, \lambda^\uparrow) = i(V, \lambda) + 1$$
$$I_{\delta^\uparrow}(V^\uparrow, \lambda^\uparrow)^\flat \simeq I_\delta(V, \lambda)^\sharp.$$

We retain the notation that $G = O(n+1, 1)$ and $m = [\frac{n}{2}]$.

Definition 14.31. Suppose that $V = F^{O(n)}(\sigma)_\varepsilon$ with $\sigma \in \Lambda^+(m)$ and $\varepsilon \in \{\pm\}$, and $\lambda \in \mathbb{Z} - S(V)$. Let $i := i(V, \lambda) \in \{0, 1, \ldots, n\}$ be the height of (V, λ) as in Lemma 14.24.

(1) We assume $0 \le i \le n - 1$, or equivalently, $\lambda \le \sigma_1 + n - 1$. We define

$$(\delta, V, \lambda)^\uparrow \equiv (\delta^\uparrow, V^\uparrow, \lambda^\uparrow) \in \mathbb{Z}/2\mathbb{Z} \times \widehat{O(n)} \times \mathbb{Z} \qquad (14.26)$$

with $V^\uparrow := F^{O(n)}(\sigma^\uparrow)_\varepsilon$ as follows:

- For $\lambda < \frac{n}{2}$, we have $0 \le i < \frac{n}{2}$ and set

$$\delta^\uparrow := \delta(-1)^{i+1-\sigma_{i+1}-\lambda},$$
$$\sigma^\uparrow := (\sigma_1, \cdots, \sigma_i, i + 1 - \lambda, \sigma_{i+2}, \cdots, \sigma_m),$$
$$\lambda^\uparrow := i + 1 - \sigma_{i+1}.$$

- For $\frac{n}{2} \le \lambda \le \sigma_1 + n - 1$, we have $\frac{n}{2} \le i \le n - 1$ and set

$$\delta^\uparrow := \delta(-1)^{\lambda - \sigma_{n-i} - i},$$
$$\sigma^\uparrow := (\sigma_1, \cdots, \sigma_{n-i-1}, \lambda - i, \sigma_{n-i+1}, \cdots, \sigma_m),$$
$$\lambda^\uparrow := \sigma_{n-i} + i.$$

(2) Conversely, for $1 \le i \le n$, namely, for $1 - \sigma_1 \le \lambda$, we define

$$(\delta, V, \lambda)^\downarrow \equiv (\delta^\downarrow, V^\downarrow, \lambda^\downarrow) \qquad (14.27)$$

as the inverse of the correspondence

$$(\delta, V, \lambda) \mapsto (\delta, V, \lambda)^\uparrow.$$

A prototype for Definition 14.31 appeared implicitly in Theorem 2.20 for the principal series representations $I_\delta(i, i)$ having the trivial $\mathfrak{z}_G(\mathfrak{g})$-infinitesimal character ρ_G. We now explain this explicitly as an example for $(V, \lambda) = (\bigwedge^i(\mathbb{C}^n), i)$ $(1 \le i \le n)$:

Example 14.32. For the exterior representations $\bigwedge^i(\mathbb{C}^n)$ of $O(n)$, we have

$$(\delta, \bigwedge^i(\mathbb{C}^n), i)^\uparrow = (-\delta, \bigwedge^{i+1}(\mathbb{C}^n), i + 1) \quad \text{for } 0 \le i \le n - 1,$$
$$(\delta, \bigwedge^i(\mathbb{C}^n), i)^\downarrow = (-\delta, \bigwedge^{i-1}(\mathbb{C}^n), i - 1) \quad \text{for } 1 \le i \le n.$$

The proof follows directly from the definition, see Example 14.6.

Here are basic properties of the correspondence

$$(\delta, V, \lambda) \mapsto (\delta^\uparrow, V^\uparrow, \lambda^\uparrow) \quad \text{or} \quad (\delta^\downarrow, V^\downarrow, \lambda^\downarrow).$$

Proposition 14.33. *Suppose that $(V,\lambda) \in \mathcal{RInt}$, i.e., $V \in \widehat{O(n)}$ and $\lambda \in \mathbb{Z} - S(V)$. In what follows, we assume the height $i(V,\lambda)$ is not equal to n when we consider $(V^\uparrow, \lambda^\uparrow)$, and is nonzero when we consider $(V^\downarrow, \lambda^\downarrow)$.*

(1) $r(V^\uparrow, \lambda^\uparrow), r(V^\downarrow, \lambda^\downarrow) \in W_G r(V,\lambda)$, *see* (14.10). *In particular,* $(V^\uparrow, \lambda^\uparrow)$, $(V^\downarrow, \lambda^\downarrow) \in \mathcal{RInt}$.

(2) $i(V^\uparrow, \lambda^\uparrow) - 1 = i(V,\lambda) = i(V^\downarrow, \lambda^\downarrow) + 1$.

(3) $\delta^\uparrow(-1)^{\lambda^\uparrow} = \delta(-1)^\lambda = \delta^\downarrow(-1)^{\lambda^\downarrow}$.

(4) $(V^\uparrow, \lambda^\uparrow), (V^\downarrow, \lambda^\downarrow) \in \mathcal{Red}$, *if $(V,\lambda) \in \mathcal{Red}$, see* (14.9).

(5) *Suppose that $(V,\lambda) \in \mathcal{Red}$ and $\lambda \neq \frac{n}{2}$. Then the unique submodule of $I_{-\delta}(V^\uparrow, \lambda^\uparrow)$ is isomorphic to the unique quotient of $I_\delta(V,\lambda)$, that is, we have the following G-isomorphisms with the notation as in Proposition 14.19:*

$$I_{\delta\uparrow}(V^\uparrow, \lambda^\uparrow)^\flat \simeq I_\delta(V,\lambda)^\sharp,$$

$$I_{\delta\downarrow}(V^\downarrow, \lambda^\downarrow)^\sharp \simeq I_\delta(V,\lambda)^\flat.$$

With these notations, we give the formulæ for the minimal K-types of the irreducible subquotients $I_\delta(V,\lambda)^\flat$ and $I_\delta(V,\lambda)^\sharp$ in $I_\delta(V,\lambda)$ in the setting of Proposition 14.19.

Proposition 14.34. *Let $G = O(n+1,1)$ and $m = [\frac{n}{2}]$. Suppose $V = F^{O(n)}(\sigma)_\varepsilon$ with $\sigma = (\sigma_1, \cdots, \sigma_m) \in \Lambda^+(m)$ and $\varepsilon \in \{\pm\}$. Let $\delta \in \{\pm\}$ and $\lambda \in \mathbb{Z} - (S(V) \cup S_Y(V) \cup \{\frac{n}{2}\})$.*

(1) *The minimal K-types of $I_\delta(V,\lambda)^\flat$ for $\lambda < \frac{n}{2}$ and of $I_\delta(V,\lambda)^\sharp$ for $\lambda > \frac{n}{2}$ are given by*

$$F^{O(n+1)}(\sigma)_\varepsilon \boxtimes \delta \qquad\qquad\qquad \text{for } n = 2m \text{ and } \sigma_m = 0,$$

$$F^{O(n+1)}(\sigma)_\varepsilon \boxtimes \delta, \quad F^{O(n+1)}(\sigma)_{-\varepsilon} \boxtimes (-\delta) \qquad \text{for } n = 2m \text{ and } \sigma_m > 0,$$

$$F^{O(n+1)}(\sigma, 0)_\varepsilon \boxtimes \delta \qquad\qquad\qquad \text{for } n = 2m + 1.$$

(2) *The minimal K-types of $I_\delta(V,\lambda)^\sharp$ for $\lambda < \frac{n}{2}$ and of $I_\delta(V,\lambda)^\flat$ for $\lambda > \frac{n}{2}$ are given by*

$$F^{O(n+1)}(\sigma^\uparrow)_\varepsilon \boxtimes \delta^\uparrow \qquad\qquad\qquad \text{for } n = 2m \text{ and } \sigma_m = 0,$$

$$F^{O(n+1)}(\sigma^\uparrow)_\varepsilon \boxtimes \delta^\uparrow, \quad F^{O(n+1)}(\sigma^\uparrow)_{-\varepsilon} \boxtimes (-\delta^\uparrow) \qquad \text{for } n = 2m \text{ and } \sigma_m > 0,$$

$$F^{O(n+1)}(\sigma^\uparrow, 0)_\varepsilon \boxtimes \delta^\uparrow \qquad\qquad\qquad \text{for } n = 2m + 1.$$

14.8 Classification of Irreducible Admissible Representations of $G = O(n+1, 1)$

Irreducible admissible representations of the connected group $G_0 = SO_0(n+1, 1)$ were classified infinitesimally (*i.e.*, on the level of $(\mathfrak{g}, K_0)$-modules) by Hirai [16], see also Borel–Wallach [9] and Collingwood [11, Chap. 5]. However, we could not find in the literature a classification of irreducible admissible representations of the indefinite orthogonal group $G = O(n+1, 1)$, which is not in the Harish-Chandra class when n is even. For the sake of completeness, we give an infinitesimal classification of irreducible admissible representations of G, or equivalently, give a classification of irreducible $(\mathfrak{g}, K)$-modules in this section. Moreover we give three characterizations of the irreducible representations of G when they are neither principal series representations nor tempered representations, see Theorem 14.35.

14.8.1 Characterizations of the Irreducible Subquotients $\Pi_\delta(V, \lambda)$

We recall from Section 2.4.5 the irreducible representations $\Pi_{\ell, \delta}$ of G that have the trivial $\mathfrak{Z}_G(\mathfrak{g})$-infinitesimal character ρ_G. Analogously to the notation $\Pi_{\ell, \delta}$ in (2.35) for $\mathrm{Irr}(G)_\rho$, we set

$$\Pi_\delta(V, \lambda) := I_\delta(V, \lambda)^\flat \tag{14.28}$$

for $\delta \in \{\pm\}$ and $(V, \lambda) \in \mathcal{R}ed$. If $i(V, \lambda) \neq 0$, then we have a G-isomorphism

$$\Pi_\delta(V, \lambda) \simeq I_{\delta\downarrow}(V^\downarrow, \lambda^\downarrow)^\sharp, \tag{14.29}$$

where $(\delta^\downarrow, V^\downarrow, \lambda^\downarrow)$ is given in Definition 14.31. We also have a G-isomorphism

$$\Pi_\delta(V, \lambda) \simeq I_\delta(V, n - \lambda)^\sharp. \tag{14.30}$$

We have already discussed in Proposition 14.21 irreducible subquotients of reducible tempered principal series representations $I_\delta(V, \lambda)$ under the assumption that $(V, \lambda) \in \mathcal{R}ed$ with $\lambda = \frac{n}{2}$. This assumption implies that n is even, V is of type Y and $\lambda = \frac{n}{2}$. The next theorem discusses the remaining (and the important) case when the principal series representation $I_\delta(V, \lambda)$ is reducible, namely, $(V, \lambda) \in \mathcal{R}ed$ with an additional condition $\lambda \neq \frac{n}{2}$.

Theorem 14.35 (characterizations of $\Pi_\delta(V,\lambda)$). *Let $G = O(n+1,1)$, and we set $m := [\frac{n}{2}]$. Assume that $(V,\lambda) \in \mathcal{R}ed$. This means that $V \in \widehat{O(n)}$ and*

$$\lambda \in \begin{cases} \mathbb{Z} - S(V) & \text{if } n = 2m+1, \\ \mathbb{Z} - (S(V) \cup S_Y(V)) & \text{if } n = 2m, \end{cases}$$

see Theorem 14.15. We further assume that $\lambda \neq \frac{n}{2}$.

(1) (Langlands subrepresentation of principal series) *For $\delta \in \{\pm\}$, $\Pi_\delta(V,\lambda)$ is the unique proper G-submodule of $I_\delta(V,\lambda)$.*

(2) (θ-stable parameter) *Let $i := i(V,\lambda) \in \{0,1,\ldots,n\}$ be the height of (V,λ) as in (14.17). We write $V = F^{O(n)}(\sigma)_\varepsilon$ with $\sigma = (\sigma_1,\cdots,\sigma_m) \in \Lambda^+(m)$ and $\varepsilon \in \{\pm\}$. Then the underlying $(\mathfrak{g},K)$-module of $\Pi_\delta(V,\lambda)$ is given by means of θ-stable parameter (see Section 14.9) as*

$$\Pi_\delta(V,\lambda)_K \simeq \begin{cases} (\sigma_1 - 1, \cdots, \sigma_i - 1 \,||\, i - \lambda, \sigma_{i+1}, \cdots, \sigma_m)_{\varepsilon,\delta\varepsilon} \\ \quad \text{if } \lambda < \frac{n}{2}, \\ (\sigma_1 - 1, \cdots, \sigma_{n-i} - 1, \lambda - i \,||\, \sigma_{n-i+1}, \cdots, \sigma_m)_{\varepsilon,-\delta\varepsilon} \\ \quad \text{if } \frac{n}{2} < \lambda. \end{cases}$$

(3) (coherent family starting at $\Pi_{i,\delta} \in \text{Irr}(G)_\rho$) *We set*

$$r(V,\lambda) \in \mathbb{C}^{m+1} \; (\simeq \mathfrak{h}_{\mathbb{C}}^*) \text{ as in (14.10).}$$

Denote by P_μ the projection to the primary component with the generalized $\mathfrak{Z}_G(\mathfrak{g})$-infinitesimal character $\mu \in \mathfrak{h}_{\mathbb{C}}^*$ mod W_G (see Section 16.2.1 in Appendix III). Let $F(V,\lambda)$ be the irreducible finite-dimensional representation of $G = O(n+1,1)$, which will be defined in Definition 16.17 in Appendix III. Then there is a natural G-isomorphism:

$$\Pi_\delta(V,\lambda) \simeq P_{r(V,\lambda)}(\Pi_{i,\delta} \otimes F(V,\lambda)).$$

(4) (Hasse sequence and standard sequence starting at $F(V,\lambda)$) *Let $\Pi_j(F)$ $(j = 0,1,\cdots,n)$ be the standard sequence starting with an irreducible finite-dimensional representation F of G (Definition 13.2), and $i = i(V,\lambda)$ the height of (V,λ). Then there is a natural G-isomorphism:*

$$\Pi_\delta(V,\lambda) \simeq \Pi_i(F(V,\lambda)) \otimes \chi_{+\delta}.$$

See Proposition 14.19 for (1), Theorem 14.46 for (2), Theorem 16.6 for (3) in Chapter 16 (Appendix III), and Theorems 14.50 and 14.51 for (4).

14.8.2 Classification of $\mathrm{Irr}(G)$

We give an infinitesimal classification of irreducible admissible representations of $G = O(n+1,1)$. One may reduce the proof to the case of connected groups, by inspecting the restriction to the subgroup $\overline{G} = SO(n+1,1)$ or the identity component group $G_0 = SO_0(n+1,1)$, see Chapter 15 (Appendix II).

Theorem 14.36 (classification of $\mathrm{Irr}(G)$). *Irreducible admissible representations of moderate growth of $G = O(n+1,1)$ are listed as follows:*

- $I_\delta(V,\lambda)$ $\lambda \in (\mathbb{C} - \mathbb{Z}) \cup S(V) \cup S_Y(V)$,

- $\Pi_\delta(V,\lambda)$ $\lambda \in \mathbb{Z} - (S(V) \cup S_Y(V))$ *and* $\lambda \le \dfrac{n}{2}$,

where $V \in \widehat{O(n)}$ and $\delta \in \{\pm\}$.

We note that there is an isomorphism of irreducible G-modules:

$$I_\delta(V,\lambda) \simeq I_\delta(V, n-\lambda)$$

when $\lambda \in (\mathbb{C} - \mathbb{Z}) \cup S(V) \cup S_Y(V)$.

14.9 θ-stable Parameters and Cohomological Parabolic Induction

In this section we give a parametrization of irreducible subquotients of the principal series representations

$$I_\delta(V,\lambda) = \mathrm{Ind}_P^G(V \otimes \delta \otimes \mathbb{C}_\lambda)$$

of the group $G = O(n+1,1)$ in terms of cohomological parabolic induction.

14.9.1 Cohomological Parabolic Induction
$$A_{\mathfrak{q}}(\lambda) = \mathcal{R}_{\mathfrak{q}}^S(\mathbb{C}_{\lambda+\rho(\mathfrak{u})})$$

We fix some notation of cohomological parabolic induction. A basic reference is Vogan [59] and Knapp–Vogan [24].

We begin with a *connected* real reductive Lie group G. Let K be a maximal compact subgroup, and θ the corresponding Cartan involution. Given an element $X \in \mathfrak{k}$, the complexified Lie algebra $\mathfrak{g}_{\mathbb{C}} = \mathrm{Lie}(G) \otimes_{\mathbb{R}} \mathbb{C}$ is decomposed into the eigenspaces of $\sqrt{-1}\mathrm{ad}(X)$, and we write

$$\mathfrak{g}_{\mathbb{C}} = \mathfrak{u}_- + \mathfrak{l}_{\mathbb{C}} + \mathfrak{u}$$

for the sum of the eigenspaces with negative, zero, and positive eigenvalues. Then $\mathfrak{q} := \mathfrak{l}_{\mathbb{C}} + \mathfrak{u}$ is a θ-stable parabolic subalgebra with Levi subgroup

$$L = \{g \in G : \mathrm{Ad}(g)\mathfrak{q} = \mathfrak{q}\}. \tag{14.31}$$

The homogeneous space G/L is endowed with a G-invariant complex manifold structure with holomorphic cotangent bundle $G \times_L \mathfrak{u}$. As an algebraic analogue of Dolbeault cohomology groups for G-equivariant holomorphic vector bundle over G/L, Zuckerman introduced a cohomological parabolic induction functor $\mathcal{R}_{\mathfrak{q}}^j (\cdot \otimes \mathbb{C}_{\rho(\mathfrak{u})})$ ($j \in \mathbb{N}$) from the category of $(\mathfrak{l}, L \cap K)$-modules to the category of $(\mathfrak{g}, K)$-modules. We adopt here the normalization of the cohomological parabolic induction $\mathcal{R}_{\mathfrak{q}}^j$ from a θ-stable parabolic subalgebra $\mathfrak{q} = \mathfrak{l}_{\mathbb{C}} + \mathfrak{u}$ so that the $\mathfrak{z}(\mathfrak{g})$-infinitesimal character of the $(\mathfrak{g}, K)$-module $\mathcal{R}_{\mathfrak{q}}^j(F)$ equals

the $\mathfrak{z}(\mathfrak{l})$-infinitesimal character of the $\mathfrak{l}$-module F

modulo the Weyl group via the Harish-Chandra isomorphism.

We note that if F' is an $(\mathfrak{l}, L \cap K)$-module then $F := F' \otimes \mathbb{C}_{\rho(\mathfrak{u})}$ may not be defined as an $(\mathfrak{l}, L \cap K)$-module, but can be defined as a module of the metaplectic covering group of L. When F satisfies a positivity condition called "good range of parameters", the cohomology $\mathcal{R}_{\mathfrak{q}}^j(F)$ concentrates on the degree

$$S := \dim_{\mathbb{C}}(\mathfrak{u} \cap \mathfrak{k}_{\mathbb{C}}).$$

For a one-dimensional representation F, we also use another convention "$A_{\mathfrak{q}}(\lambda)$". Following the normalization of Vogan–Zuckerman [61], we set

$$A_{\mathfrak{q}}(\lambda) := \mathcal{R}_{\mathfrak{q}}^S(\mathbb{C}_{\lambda + \rho(\mathfrak{u})})$$

for a one-dimensional representation $\mathbb{C}_{\lambda}$ of L. In particular, we set

$$A_{\mathfrak{q}} := A_{\mathfrak{q}}(0) = \mathcal{R}_{\mathfrak{q}}^S(\mathbb{C}_{\rho(\mathfrak{u})}),$$

which is an irreducible $(\mathfrak{g}, K)$-module with the same $\mathfrak{z}(\mathfrak{g})$-infinitesimal character ρ as that of the trivial one-dimensional representation $\mathbf{1}$ of G.

Similar notation will be used for disconnected groups G. For a character χ of the component group G/G_0, we have an isomorphism of $(\mathfrak{g}, K)$-modules:

$$(A_{\mathfrak{q}})_{\chi} := A_{\mathfrak{q}} \otimes \chi \simeq \mathcal{R}_{\mathfrak{q}}^S(\chi \otimes \mathbb{C}_{\rho(\mathfrak{u})}).$$

14.9.2 θ-stable Parabolic Subalgebra $\mathfrak{q}_i$ for $G = O(n+1, 1)$

We apply the general theory reviewed in Section 14.9.1 to the indefinite orthogonal group $G = O(n+1, 1)$. For this, we set up some notation for θ-stable parabolic subalgebra $\mathfrak{q}_i$ and $\mathfrak{q}_{\frac{n+1}{2}}^{\pm}$ of $\mathfrak{g}_{\mathbb{C}} = \mathrm{Lie}(G) \otimes_{\mathbb{R}} \mathbb{C} \simeq \mathfrak{o}(n+2, \mathbb{C})$ as follows.

We take a Cartan subalgebra $\mathfrak{t}^c$ of $\mathfrak{k}$, and extend it to a fundamental Cartan subalgebra $\mathfrak{h} = \mathfrak{t}^c + \mathfrak{a}^c$. If n is odd then $\mathfrak{a}^c = \{0\}$. Choose the standard coordinates $\{f_k : 1 \leq k \leq [\frac{n}{2}] + 1\}$ on $\mathfrak{h}_{\mathbb{C}}^*$ such that the root system of $\mathfrak{g}$ and $\mathfrak{k}$ are given by

$$\Delta(\mathfrak{g}_{\mathbb{C}}, \mathfrak{h}_{\mathbb{C}}) = \{\pm(f_i \pm f_j) : 1 \leq i < j \leq [\frac{n}{2}] + 1\}$$
$$\left(\cup \{\pm f_\ell : 1 \leq \ell \leq [\frac{n}{2}] + 1\} \quad (n: \text{odd}) \right),$$
$$\Delta(\mathfrak{k}_{\mathbb{C}}, \mathfrak{t}_{\mathbb{C}}) = \{\pm(f_i \pm f_j) : 1 \leq i < j \leq [\frac{n+1}{2}]\}$$
$$\left(\cup \{\pm f_\ell : 1 \leq \ell \leq [\frac{n+1}{2}]\} \quad (n: \text{even}) \right).$$

For $1 \leq i \leq [\frac{n+1}{2}]$, we define elements of $\mathfrak{t}_{\mathbb{C}}^*$ by

$$\mu_i := \sum_{k=1}^{i} (\frac{n}{2} + 1 - k) f_k,$$
$$\mu_i^- := \mu_i - (n + 2 - 2i) f_i.$$

It is convenient to set $\mu_0 = \mu_0^- = 0$. (We shall use μ_i^- only when we consider the identity component group $G_0 = SO_0(n+1, 1)$ with n odd and when $n + 1 = 2i$ for later arguments.) Let $\langle , \rangle$ be the standard bilinear form on $\mathfrak{h}_{\mathbb{C}}^* \simeq \mathbb{C}^{[\frac{n}{2}]+1}$.

Definition 14.37 (θ-stable parabolic subalgebra $\mathfrak{q}_i$). For $0 \leq i \leq [\frac{n+1}{2}]$, we define θ-stable parabolic subalgebras $\mathfrak{q}_i \equiv \mathfrak{q}_i^+ = (\mathfrak{l}_i)_{\mathbb{C}} + \mathfrak{u}_i$ and $\mathfrak{q}_i^- = (\mathfrak{l}_i)_{\mathbb{C}} + \mathfrak{u}_i^-$ in $\mathfrak{g}_{\mathbb{C}} = \mathrm{Lie}(G) \otimes_{\mathbb{R}} \mathbb{C}$ by the condition that $\mathfrak{q}_i$ and $\mathfrak{q}_i^-$ contain the fundamental Cartan subalgebra $\mathfrak{h}$ and that their nilradicals $\mathfrak{u}_i$ and $\mathfrak{u}_i^-$ are given respectively by

$$\Delta(\mathfrak{u}_i, \mathfrak{h}_{\mathbb{C}}) = \{\alpha \in \Delta(\mathfrak{g}_{\mathbb{C}}, \mathfrak{h}_{\mathbb{C}}) : \langle \alpha, \mu_i \rangle > 0\},$$
$$\Delta(\mathfrak{u}_i^-, \mathfrak{h}_{\mathbb{C}}) = \{\alpha \in \Delta(\mathfrak{g}_{\mathbb{C}}, \mathfrak{h}_{\mathbb{C}}) : \langle \alpha, \mu_i^- \rangle > 0\}.$$

Then the Levi subgroup of $\mathfrak{q} = \mathfrak{q}_i$ and $\mathfrak{q}_i^-$ is given by

$$L_i := N_G(\mathfrak{q}) \equiv \{g \in G : \mathrm{Ad}(g)\mathfrak{q} = \mathfrak{q}\} \simeq SO(2)^i \times O(n - 2i + 1, 1). \qquad (14.32)$$

We note that L_i is not in the Harish-Chandra class if n is even, as is the case $G = O(n+1, 1)$.

If we write $\rho(u_i)$ and $\rho(u_i^-)$ for half the sum of roots in u_i and u_i^-, respectively, then

$$\rho(u_i) = \mu_i \quad \text{and} \quad \rho(u_i^-) = \mu_i^-.$$

We suppress the superscript $+$ for q_i^+ except for the case $n+1 = 2i$. For later purpose, we compare the following three groups with the same Lie algebras:

$$G_0 = SO_0(n+1, 1) \hookrightarrow \overline{G} = SO(n+1, 1) \hookrightarrow G = O(n+1, 1) \qquad (14.33)$$

with maximal compact subgroups

$$K_0 = SO(n+1) \quad \hookrightarrow \overline{K} = O(n+1) \quad \hookrightarrow K = O(n+1) \times O(1).$$

Lemma 14.38.

(1) *A complete system of the K_0-conjugacy classes of θ-stable parabolic subalgebras of $\mathfrak{g}_{\mathbb{C}}$ with Levi subgroup L_i (14.32) is given by*

$$\{q_i\} \qquad \text{for } 0 \le i < [\frac{n+1}{2}],$$

$$\{q_{\frac{n+1}{2}}^+, q_{\frac{n+1}{2}}^-\} \quad \text{for } i = \frac{n+1}{2} \quad (n\text{:odd}).$$

(2) *The θ-stable parabolic subalgebra q_i with the property (14.32) is unique up to conjugation by the disconnected group $\overline{K}$ (and therefore, also by K) for all i $(0 \le i \le [\frac{n+1}{2}])$.*

We also make the following two observations:

Lemma 14.39. *L_i is compact if and only if n is odd and $2i = n+1$. In this case, $L_i \simeq SO(2)^{\frac{n+1}{2}} \times O(1)$.*

Lemma 14.40. *The inclusion maps (14.33) induce the following inclusion and bijection:*

$$G_0/N_{G_0}(q_i) \hookrightarrow \overline{G}/N_{\overline{G}}(q_i) \xrightarrow{\sim} G/N_G(q_i) = G/L_i$$

for all i $(0 \le i \le [\frac{n+1}{2}])$. The first inclusion is bijective if $n+1 \ne 2i$.

The second bijection is reflected by the irreducibility of the G-module $\Pi_{\ell,\delta}$ when restricted to the subgroup $\overline{G} = SO(n+1, 1)$, see Proposition 15.11 (1) in Appendix II.

Lemmas 14.38 and 14.39 yield the following (well-known) representation theoretic results:

Proposition 14.41.

(1) G (or $\overline{G}$, G_0) admits a discrete series representation if and only if n is odd.
(2) Suppose n is odd. Then there exists only one discrete series representation of $\overline{G}$ for each regular integral infinitesimal character; there exist exactly two discrete series representations of G (also of G_0) for each regular integral infinitesimal character.

For $n = 2m - 1$ in the second statement of Proposition 14.41, we note the following properties for the three groups $G \supset \overline{G} \supset G_0$:

- $L_m \simeq SO(2)^m \times O(1)$ has two connected components;
- $L_m \cap \overline{G} = L_m \cap G_0$ are connected;
- $\mathfrak{q}_m^+$ and $\mathfrak{q}_m^-$ are not conjugate by G_0; they are conjugate by $\overline{G}$ or G.

See [27, Thm. 3 (0)] for results in a more general setting of the indefinite orthogonal group $O(p, q)$.

For $\nu = (\nu_1, \cdots, \nu_i) \in \mathbb{Z}^i$, $\mu \in \Lambda^+([\frac{n}{2}] - i + 1)$, and $a, b \in \{\pm\}$, we consider an irreducible finite-dimensional L_i-module

$$F^{O(n-2i+1,1)}(\mu)_{a,b} \otimes \mathbb{C}_\nu$$

and define an admissible smooth representation of G of moderate growth, to be denoted by

$$(\nu_1, \cdots, \nu_i \,||\, \mu_1, \cdots, \mu_{[\frac{n}{2}]-i+1})_{a,b},$$

whose underlying $(\mathfrak{g}, K)$-module is given by the cohomological parabolic induction

$$\mathcal{R}_{\mathfrak{q}_i}^{S_i}(F^{O(n-2i+1,1)}(\mu)_{a,b} \otimes \mathbb{C}_{\nu+\rho(\mathfrak{u}_i)}) \tag{14.34}$$

of degree S_i, where we set

$$S_i := \dim_{\mathbb{C}}(\mathfrak{u}_i \cap \mathfrak{k}_{\mathbb{C}}) = i(n - i). \tag{14.35}$$

We note that if $i = 0$ then $(\,||\, \mu_1, \cdots, \mu_{[\frac{n}{2}]+1})_{a,b}$ is finite-dimensional.

Definition 14.42 (θ-stable parameter). We call $(\nu_1, \cdots, \nu_i \,||\, \mu_1, \cdots, \mu_{[\frac{n}{2}]-i+1})_{a,b}$ the θ-stable parameter of the representation (14.34).

If the θ-stable parameter of a representation Π of G is given by

$$(\nu_1, \cdots, \nu_i \,||\, \mu_1, \cdots, \mu_{[\frac{n}{2}]-i+1})_{a,b},$$

then that of $\Pi \otimes \chi_{cd}$ for $c, d \in \{\pm\}$ is given by

$$(\nu_1, \cdots, \nu_i \,||\, \mu_1, \cdots, \mu_{[\frac{n}{2}]-i+1})_{ac,bd}. \tag{14.36}$$

The $\mathfrak{Z}_G(\mathfrak{g})$-infinitesimal character of $(\nu_1, \cdots, \nu_i \,\|\, \mu_1, \cdots, \mu_{[\frac{n}{2}]-i+1})_{a,b}$ is given by

$$(\nu_1, \cdots, \nu_i, \mu_1, \cdots, \mu_{[\frac{n}{2}]-i+1}) + (\frac{n}{2}, \frac{n}{2} - 1, \cdots, \frac{n}{2} - [\frac{n}{2}]).$$

In particular, the G-module

$$(\underbrace{0, \cdots, 0}_{i} \,\|\, \underbrace{0, \cdots, 0}_{[\frac{n}{2}]-i+1})_{a,b}$$

has the trivial infinitesimal character ρ_G. In this case we shall write

$$(A_{\mathfrak{q}_i})_{a,b} := \mathcal{R}_{\mathfrak{q}_i}^{S_i}(\chi_{ab} \otimes \mathbb{C}_{\rho(\mathfrak{u}_i)}) \tag{14.37}$$

for its underlying $(\mathfrak{g}, K)$-module, see Proposition 14.44 below.

Sometimes we suppress the subscript $+, +$ and write simply $A_{\mathfrak{q}_i}$ to denote the $(\mathfrak{g}, K)$-module $(A_{\mathfrak{q}_i})_{+,+}$.

Remark 14.43.

(1) (good range) The irreducible finite-dimensional representation $F^{O(n-2i+1,1)}(\mu)_{a,b} \otimes \mathbb{C}_{\nu+\rho(\mathfrak{u})}$ of the metaplectic cover of L_i is in the *good range* with respect to the θ-stable parabolic subalgebra $\mathfrak{q}_i$ (see [24, Def. 0.49] for the definition) if and only if

$$\nu_1 \geq \nu_2 \geq \cdots \geq \nu_i \geq \mu_1.$$

In this case, the $(\mathfrak{g}, K)$-module (14.34) is nonzero and irreducible, and therefore $(\nu_1, \cdots, \nu_i \,\|\, \mu_1, \cdots, \mu_{[\frac{n}{2}]-i+1})_{a,b}$ is a nonzero irreducible G-module. For the description of the Hasse sequence (Theorem 14.46 below), we need only the parameter in the good range.

(2) (weakly fair range) If $\mu = (0, \cdots, 0)$, then the $(\mathfrak{g}, K)$-module (14.34) reduces to

$$A_{\mathfrak{q}_i}(\nu)_{a,b} := \mathcal{R}_{\mathfrak{q}_i}^{S_i}(\chi_{ab} \otimes \mathbb{C}_{\nu+\rho(\mathfrak{u}_i)})$$

cohomologically induced from the one-dimensional representation $\chi_{ab} \otimes \mathbb{C}_{\nu+\rho(\mathfrak{u})}$. We note that $\chi_{ab} \otimes \mathbb{C}_{\nu+\rho(\mathfrak{u})}$ is in the *weakly fair range* with respect to $\mathfrak{q}_i$ (see [24, Def. 0.52] for the definition) if and only if

$$\nu_1 + \frac{n}{2} \geq \nu_2 + \frac{n}{2} - 1 \geq \cdots \geq \nu_i + \frac{n}{2} - i + 1 \geq 0. \tag{14.38}$$

In this case the $(\mathfrak{g}, K)$-module $A_{\mathfrak{q}_i}(\nu)_{a,b}$ may or may not vanish. See [27, Thm. 3] for the conditions on $\nu \in \mathbb{Z}^i$ in the weakly fair range that assure the nonvanishing and the irreducibility of $A_{\mathfrak{q}_i}^{S_i}(\mathbb{C}_\nu)_{a,b}$. We shall see in Section 14.11

that the underlying $(\mathfrak{g}, K)$-modules of singular complementary series representations are isomorphic to these modules.

14.9.3 Irreducible Representations $\Pi_{\ell,\delta}$ and $(A_{\mathfrak{q}_i})_{\pm,\pm}$

In this subsection, we give a description of the underlying $(\mathfrak{g}, K)$-modules of the subquotients $\Pi_{\ell,\delta}$ of the principal series representation of the disconnected group $G = O(n+1, 1)$ in terms of the cohomologically parabolic induced modules $(A_{\mathfrak{q}_i})_{\pm,\pm}$.

We recall from (2.35) the definition of the irreducible representations $\Pi_{\ell,\delta}$ ($0 \leq \ell \leq n+1, \delta = \pm$) of $G = O(n+1, 1)$. The set

$$\{\Pi_{\ell,\delta} : 0 \leq \ell \leq n+1, \delta = \pm\}$$

exhausts irreducible admissible representations of moderate growth having $\mathfrak{Z}_G(\mathfrak{g})$-infinitesimal character ρ_G, see Theorem 2.20 (2). Their underlying $(\mathfrak{g}, K)$-modules $(\Pi_{\ell,\delta})_K$ can be given by cohomologically parabolic induced modules as follows.

Proposition 14.44. *For $0 \leq i \leq [\frac{n+1}{2}]$, let $\mathfrak{q}_i$ be the θ-stable parabolic subalgebras with the Levi subgroup $L_i \simeq SO(2)^i \times O(n-2i+1, 1)$ as in Definition 14.37.*

(1) *The underlying $(\mathfrak{g}, K)$-modules of the irreducible G-modules $\Pi_{\ell,\delta}$ ($0 \leq \ell \leq n+1, \delta \in \{\pm\}$) are given by the cohomological parabolic induction as follows:*

$$(\Pi_{i,+})_K \simeq (A_{\mathfrak{q}_i})_{+,+} \supset \bigwedge\nolimits^i(\mathbb{C}^{n+1}) \boxtimes \mathbf{1},$$

$$(\Pi_{i,-})_K \simeq (A_{\mathfrak{q}_i})_{+,-} \supset \bigwedge\nolimits^i(\mathbb{C}^{n+1}) \boxtimes \mathrm{sgn},$$

$$(\Pi_{n+1-i,+})_K \simeq (A_{\mathfrak{q}_i})_{-,+} \supset \bigwedge\nolimits^{n+1-i}(\mathbb{C}^{n+1}) \boxtimes \mathbf{1},$$

$$(\Pi_{n+1-i,-})_K \simeq (A_{\mathfrak{q}_i})_{-,-} \supset \bigwedge\nolimits^{n+1-i}(\mathbb{C}^{n+1}) \boxtimes \mathrm{sgn}.$$

For later purpose, we also indicated their minimal K-types in the right column (see Theorem 2.20 (3)).

(2) *If n is even or if $2i \neq n+1$, then the four $(\mathfrak{g}, K)$-modules $(A_{\mathfrak{q}_i})_{a,b}$ ($a, b \in \{\pm\}$) are not isomorphic to each other.*
If $2i = n+1$, then there are isomorphisms

$$(A_{\mathfrak{q}_{\frac{n+1}{2}}})_{+,+} \simeq (A_{\mathfrak{q}_{\frac{n+1}{2}}})_{-,+} \quad and \quad (A_{\mathfrak{q}_{\frac{n+1}{2}}})_{+,-} \simeq (A_{\mathfrak{q}_{\frac{n+1}{2}}})_{-,-}$$

as $(\mathfrak{g}, K)$-modules for the disconnected group $O(n+1, 1)$.

Thus the left-hand sides of the formulæ in Proposition 14.44 (1) have overlaps when n is odd and $i = \frac{n+1}{2}$. In fact, the Levi part in this case is of the form $L_{\frac{n+1}{2}} \simeq SO(2)^{\frac{n+1}{2}} \times O(0, 1)$, and $\chi_{-+} \simeq \mathbf{1}$ and $\chi_{+-} \simeq \chi_{--}$ as $O(0, 1)$-modules.

14.9.4   Irreducible Representations with Nonzero $(\mathfrak{g}, K)$-cohomologies

In this section, we prove Theorem 2.20 (9) on the classification of irreducible unitary representations of $G = O(n+1, 1)$ with nonzero $(\mathfrak{g}, K)$-cohomologies. We have already seen in Lemma 12.15 that $H^*(\mathfrak{g}, K; (\Pi_{\ell,\delta})_K) \neq \{0\}$ for all $0 \leq \ell \leq n+1$ and $\delta \in \{\pm\}$. Hence the proof of Theorem 2.20 (9) will be completed by showing the following.

Proposition 14.45. *Let Π be an irreducible unitary representation of the group $G = O(n+1, 1)$ such that $H^*(\mathfrak{g}, K; \Pi_K) \neq \{0\}$. Then the smooth representation Π^∞ is isomorphic to $\Pi_{\ell,\delta}$ (see (2.35)) for some $0 \leq \ell \leq n+1$ and $\delta \in \{\pm\}$.*

Proof. We begin with representations of the identity component group $G_0 = SO_0(n+1, 1)$. In this case, we write $A_\mathfrak{q}^0$ by putting superscript 0 to denote the $(\mathfrak{g}, K_0)$-module which is cohomologically induced from the trivial one-dimensional representation of a θ-stable parabolic subalgabra $\mathfrak{q}$.

By a theorem of Vogan and Zuckerman [61], any irreducible unitary representation Π^0 of G_0 with $H^*(\mathfrak{g}, K_0; (\Pi^0)_{K_0}) \neq \{0\}$ is of the form $(\Pi^0)_{K_0} \simeq A_\mathfrak{q}^0$ for some θ-stable parabolic subalgebra $\mathfrak{q}$ in $\mathfrak{g}_\mathbb{C}$. We recall from Definition 14.37 that $\mathfrak{q}_i$ $(0 \leq i < \frac{n+1}{2})$ and $\mathfrak{q}_i^\pm$ $(i = \frac{n+1}{2})$ are θ-stable parabolic subalgebras such that the Levi subgroup $N_{G_0}(\mathfrak{q}_i)$ (or $N_{G_0}(\mathfrak{q}_i^\pm)$) are isomorphic to $SO(2)^i \times SO_0(n-2i+1, 1)$. They exhaust all θ-stable parabolic subalgebras up to inner automorphisms and up to cocompact Levi factors, namely, there exists $0 \leq i \leq [\frac{n+1}{2}]$ such that

$$\mathfrak{q}_i \subset \mathfrak{q} \quad \text{and} \quad N_{G_0}(\mathfrak{q})/N_{G_0}(\mathfrak{q}_i) \text{ is compact}$$

if we take a conjugation of $\mathfrak{q}$ by an element of G_0. (For $i = \frac{n+1}{2}$, $\mathfrak{q}_i$ is considered as either $\mathfrak{q}_i^+$ or $\mathfrak{q}_i^-$.) Then we have a $(\mathfrak{g}, K_0)$-isomorphism

$$(\Pi^0)_{K_0} \simeq A_\mathfrak{q}^0 \simeq \begin{cases} A_{\mathfrak{q}_i}^0 & \text{if } 2i < n+1, \\ A_{\mathfrak{q}_i^+}^0 \text{ or } A_{\mathfrak{q}_i^-}^0 & \text{if } 2i = n+1. \end{cases} \tag{14.39}$$

Now we consider an irreducible unitary representation Π of the *disconnected* group $G = O(n+1, 1)$ such that $H^*(\mathfrak{g}, K; \Pi_K) \neq \{0\}$. The assumption implies $H^*(\mathfrak{g}, K_0; \Pi_K) \neq \{0\}$, and therefore there exists a G_0-irreducible submodule Π^0 of the restriction $\Pi|_{G_0}$ such that $H^*(\mathfrak{g}, K_0; (\Pi^0)_{K_0}) \neq \{0\}$. By the reciprocity, the underlying $(\mathfrak{g}, K)$-module Π_K must be an irreducible summand in the induced representation

$$\mathrm{ind}_{\mathfrak{g}, K_0}^{\mathfrak{g}, K}((\Pi^0)_{K_0}).$$

It follows from (14.39) and from Proposition 14.44 (2) that

$$\mathrm{ind}_{\mathfrak{g},K_0}^{\mathfrak{g},K}((\Pi^0)_{K_0}) \simeq \begin{cases} \displaystyle\bigoplus_{a,b\in\{\pm\}} (A_{\mathfrak{q}_i})_{a,b} & \text{if } 2i < n+1, \\[2mm] (A_{\mathfrak{q}_{\frac{n+1}{2}}})_{+,+} \oplus (A_{\mathfrak{q}_{\frac{n+1}{2}}})_{-,-} & \text{if } 2i = n+1. \end{cases}$$

Thus Proposition 14.45 follows from Proposition 14.44 (1). □

14.9.5 Description of Subquotients in $I_\delta(V,\lambda)$

We use the θ-stable parameter for the description of irreducible subquotients of the principal series representations $I_\delta(V,\lambda)$ of $G = O(n+1,1)$ with regular integral infinitesimal character.

Theorem 14.46. *Suppose* $V \in \widehat{O(n)}$ *and* $\lambda \in \mathbb{Z} - S(V)$. *Let* $i := i(\lambda, V)$ *be the height as in Lemma 14.24. We write* $V = F^{O(n)}(\sigma)_\varepsilon$ *with* $\sigma = (\sigma_1, \cdots, \sigma_{[\frac{n}{2}]}) \in \Lambda^+([\frac{n}{2}])$ *and* $\varepsilon \in \{\pm\}$. *Let* $\delta \in \{\pm\}$.

(1) *Suppose* $\lambda \geq \frac{n}{2}$. *Then* $\frac{n}{2} \leq i \leq n$.
    If $i \neq \frac{n}{2}$, *then we have the following nonsplit exact sequence of G-modules of moderate growth:*

$$0 \to (\sigma_1 - 1, \cdots, \sigma_{n-i} - 1, \lambda - i \,||\, \sigma_{n-i+1}, \cdots, \sigma_{[\frac{n}{2}]})_{\varepsilon, -\delta\varepsilon}$$

$$\to I_\delta(V,\lambda)$$

$$\to (\sigma_1 - 1, \cdots, \sigma_{n-i} - 1 \,||\, \lambda - i, \sigma_{n-i+1}, \cdots, \sigma_{[\frac{n}{2}]})_{\varepsilon, \delta\varepsilon} \to 0. \qquad (14.40)$$

(2) *Suppose* $\lambda \leq \frac{n}{2}$. *Then* $0 \leq i \leq \frac{n}{2}$.
    If $i \neq \frac{n}{2}$, *then we have the following nonsplit exact sequence of G-modules of moderate growth:*

$$0 \to (\sigma_1 - 1, \cdots, \sigma_i - 1 \,||\, i - \lambda, \sigma_{i+1}, \cdots, \sigma_{[\frac{n}{2}]})_{\varepsilon, \delta\varepsilon}$$

$$\to I_\delta(V,\lambda)$$

$$\to (\sigma_1 - 1, \cdots, \sigma_i - 1, i - \lambda \,||\, \sigma_{i+1}, \cdots, \sigma_{[\frac{n}{2}]})_{\varepsilon, -\delta\varepsilon} \to 0. \qquad (14.41)$$

(3) *Suppose* $i = \frac{n}{2}$, *or equivalently, suppose that n is even and* $\sigma_{\frac{n}{2}} > |\lambda - \frac{n}{2}|$.
    If $\lambda \neq \frac{n}{2}$, *then* $\lambda \in S_Y(V)$ *(see (14.18)). In this case, $I_\delta(V,\lambda)$ is irreducible and we have a G-isomorphism:*

$$I_\delta(V,\lambda) \simeq (\sigma_1 - 1, \cdots, \sigma_{\frac{n}{2}} - 1 \,||\, |\lambda - \frac{n}{2}|)_{a,b}$$

whenever $a, b \in \{\pm\}$ *satisfies* $ab = \delta$.

If $\lambda = \frac{n}{2}$, then $I_\delta(V, \lambda)$ splits into the direct sum of two irreducible representations of G:

$$I_\delta(V, \lambda) \simeq \bigoplus_{a,b \in \{\pm\}, ab = \delta} (\sigma_1 - 1, \cdots, \sigma_{\frac{n}{2}} - 1 \| 0)_{a,b}. \tag{14.42}$$

Remark 14.47. In Theorem 14.46 (3), we have a G-isomorphism

$$I_\delta(F^{O(n)}(\sigma)_+, \lambda) \simeq I_\delta(F^{O(n)}(\sigma)_-, \lambda) \quad \text{for each } \delta = \pm.$$

In fact, by Lemma 14.28, $i(\lambda, V) = \frac{n}{2}$ implies that V is of type Y, hence there is an $O(N)$-isomorphism $F^{O(N)}(\sigma)_+ \simeq F^{O(N)}(\sigma)_-$ by Lemma 2.9.

Moreover, the restriction of each irreducible summand in (14.42) to the special orthogonal group $SO(n+1, 1)$ is irreducible (see Lemma 15.2 (1) in Appendix II).

14.9.6 Proof of Theorem 14.46

Sketch of the Proof of Theorem 14.46. If the $\mathfrak{Z}_G(\mathfrak{g})$-infinitesimal character of the principal series representation $I_\delta(F^{O(n)}(\sigma)_\varepsilon, \lambda)$ is ρ_G, then Theorem 14.46 is a reformulation of Theorem 2.20 in terms of θ-stable parameters. This is done in Proposition 14.49 below.

The general case is derived from the above case by the translation principle, see Theorems 16.22 and 16.24, and also the argument there (*e.g.*, Lemma 16.12) in Appendix III. □

Suppose $V = \bigwedge^i(\mathbb{C}^n)$. By Example 14.6, the principal series representation $I_\delta(i, \lambda) = \operatorname{Ind}_P^G(\bigwedge^i(\mathbb{C}^n) \otimes \delta \otimes \mathbb{C}_\lambda)$ is expressed as follows.

Lemma 14.48. *There are natural G-isomorphisms:*

$$I_\delta(\ell, \ell) \simeq I_\delta(F^{O(n)}(1^\ell, 0^{[\frac{n}{2}] - \ell})_+, \ell) \qquad \text{if } \ell \leq \frac{n}{2},$$

$$I_\delta(\ell, \ell) \simeq I_\delta(F^{O(n)}(1^{n-\ell}, 0^{\ell - [\frac{n+1}{2}]})_-, \ell) \qquad \text{if } \ell \geq \frac{n}{2}.$$

Proposition 14.49. *Suppose $0 \leq \ell \leq \frac{n}{2}$. Then Theorem 14.46 holds for $\lambda = \ell$ and $\sigma = (1^\ell, 0^{[\frac{n}{2}] - \ell}) \in \Lambda^+([\frac{n}{2}])$.*

Proof. By Theorem 2.20 (1), we have an exact sequence of G-modules

$$0 \to \Pi_{\ell, \delta} \to I_\delta(\ell, \ell) \to \Pi_{\ell+1, -\delta} \to 0,$$

which does not split as far as $\ell \neq \frac{n}{2}$. By Proposition 14.44, this yields an exact sequence of $(\mathfrak{g}, K)$-modules:

$$0 \to (A_{\mathfrak{q}_\ell})_{+,\delta} \to I_\delta(\ell, \ell)_K \to (A_{\mathfrak{q}_{\ell+1}})_{+,-\delta} \to 0.$$

By Lemma 14.48 and the definition of θ-stable parameters, this exact sequence can be written as

$$0 \to (0^\ell \,||\, 0^{[\frac{n}{2}]-\ell+1})_{+,\delta} \to I_\delta(F^{O(n)}(\sigma)_+, \ell) \to (0^{\ell+1} \,||\, 0^{[\frac{n}{2}]-\ell})_{+,-\delta} \to 0.$$

Since the height of $(F^{O(n)}(\sigma)_+, \ell) = (\bigwedge^\ell(\mathbb{C}^n), \ell)$ is given by $i(\bigwedge^\ell(\mathbb{C}^n), \ell) = \ell$. $i(\ell, \sigma) = \ell$ by Example 14.25, we get Proposition 14.49 from Lemma 2.14 and (14.36). $\qquad\square$

14.10 Hasse Sequence in Terms of θ-stable Parameters

This section gives a description of the Hasse sequence (Definition-Theorem 13.1) and the standard sequence (Definition 13.2) in terms of θ-stable parameters.

We set $m := [\frac{n+1}{2}]$, namely $n = 2m - 1$ or $2m$. Let F be an irreducible finite-dimensional representation of $G = O(n+1, 1)$, and $U_i \equiv U_i(F)$ $(0 \leq i \leq [\frac{n+1}{2}])$ be the Hasse sequence with $U_0 \simeq F$. We write

$$F = F^{O(n+1,1)}(s_0, \cdots, s_{[\frac{n}{2}]})_{a,b}$$

as in Lemma 14.3 (2).

Theorem 14.50. *Let* $n = 2m$ *and* $0 \leq i \leq m$.

(1) (Hasse sequence) $U_i(F) \simeq (s_0, \cdots, s_{i-1} \,||\, s_i, \cdots, s_m)_{a,(-1)^{i-s_i}b}$.

(2) (standard sequence) $U_i(F) \otimes \chi_{+-}^i \simeq (s_0, \cdots, s_{i-1} \,||\, s_i, \cdots, s_m)_{a,(-1)^{s_i}b}$.

Proof.

(1) We begin with the case $a = b = +$. Let $s := (s_0, \cdots, s_m, 0^{m+1}) \in \Lambda^+(2m+2)$. As in (13.1) of Section 13.1, we define $s^{(\ell)} \in \Lambda^+(2m)$ for $0 \leq \ell \leq m$. Then by Theorem 13.7, there is an injective G-homomorphism

$$U_\ell(F) \hookrightarrow I_{(-1)^{\ell-s_\ell}}(F^{O(n)}(s^{(\ell)}), \ell - s_\ell).$$

The $O(n)$-module $F^{O(n)}(s^{(\ell)})$ is of type I (Definition 2.4), and we have

$$i(F^{O(n)}(s^{(\ell)}), \ell - s_\ell) = \ell$$

with the notation of Lemma 14.24.

By Theorem 14.46, we get the theorem for $a = b = +$ case. The general case follows from the case $(a, b) = (+, +)$ by the tensoring argument given in Proposition 13.4.

(2) The second statement follows from Definition 13.2 and (14.36). $\square$

The case n odd is given similarly as follows.

Theorem 14.51. *Let $n = 2m - 1$, and $0 \leq i \leq m - 1$.*

(1) (Hasse sequence) $U_i(F) \simeq (s_0, \cdots, s_{i-1} \,||\, s_i, \cdots, s_{m-1})_{a,(-1)^{i-s_i} b}.$

(2) (standard sequence) $U_i(F) \otimes \chi^i_{+-} \simeq (s_0, \cdots, s_{i-1} \,||\, s_i, \cdots, s_{m-1})_{a,(-1)^{s_i} b}.$

Proof.

(1) We begin with the case $a = b = +$. Let $s := (s_0, \cdots, s_{m-1}, 0^{m+1}) \in \Lambda^+(2m + 1)$. As in (13.4), we define $s^{(\ell)} \in \Lambda^+(2m - 1)$ for $0 \leq \ell \leq m - 1$. Then by Theorem 13.9,

$$U_\ell(F) \subset I_{(-1)^{\ell - s_\ell}}(F^{O(n)}(s^{(\ell)}), \ell - s_\ell).$$

The $O(n)$-module $F^{O(n)}(s^{(\ell)})$ is of type I, and we obtain

$$i(F^{O(n)}(s^{(\ell)}), \ell - s_\ell) = \ell$$

with the notation of Lemma 14.24.

By Theorem 14.46, we get the theorem for $a = b = +$ case. The general case follows from the case $(a, b) = (+, +)$ by the tensoring argument given in Proposition 13.4.

(2) The second statement follows from Definition 13.2 and (14.36). $\square$

14.11   Singular Integral Case

We end this chapter with cohomologically induced representations with *singular* parameter, and give a description of complementary series representations with integral parameter (see Section 3.6.3) in terms of θ-stable parameters.

For $0 \leq r \leq [\frac{n+1}{2}]$, we define $\mathfrak{q}_r = (\mathfrak{l}_r)_{\mathbb{C}} + \mathfrak{u}_r$ to be the θ-stable parabolic subalgebra with Levi subgroups $L_r \simeq SO(2)^r \times O(n + 1 - 2r, 1)$ in $G = O(n+1, 1)$ as in Definition 14.37. We set $S_r = r(n - r)$.

For $v = (v_1, \cdots, v_r) \in \mathbb{Z}^r \simeq (SO(2)^r)\hat{}$ and $a, b \in \{\pm\}$, we consider the underlying $(\mathfrak{g}, K)$-modules of the admissible smooth representations of G:

$$(v_1, \cdots, v_r \,||\, \underbrace{0, \cdots, 0}_{[\frac{n}{2}] - r + 1})_{a,b},$$

namely, the following $(\mathfrak{g}, K)$-modules

$$A_{\mathfrak{q}_r}(\nu)_{a,b} = \mathcal{R}^{S_r}_{\mathfrak{q}_r}(\chi_{ab} \otimes \mathbb{C}_{\nu+\rho(\mathfrak{u})}) \simeq \mathcal{R}^{S_r}_{\mathfrak{q}_r}(\mathbb{C}_{\nu+\rho(\mathfrak{u}_r)}) \otimes \chi_{ab},$$

which are cohomologically induced from the one-dimensional representations $\mathbb{C}_\nu \boxtimes \chi_{ab}$ of the Levi subgroup L_r, see Remark 14.43 for our normalization about "ρ-shift".

Sometimes we suppress the subscript $+,+$ and write simply $A_{\mathfrak{q}_r}(\nu)$ for the $(\mathfrak{g}, K)$-module $A_{\mathfrak{q}_r}(\nu)_{+,+}$.

For a description of singular integral complementary series representations $I_\delta(i, s)$ in terms of θ-stable parameters, we need to treat the parameter ν outside the good range [24, Def. 0.49] relative to the θ-stable parabolic subalgebra $\mathfrak{q}_r$ with $r = i + 1$ (see Theorem 14.53 below), for which the general theory about the nonvanishing and irreducibility (e.g.. [24, Thm. 0.50]) does not apply. For instance, the condition on the parameter ν for which $A_{\mathfrak{q}_r}(\nu) \neq 0$ is usually very complicated when ν wanders outside the good range. In our setting, we use the following results from [27]:

Fact 14.52. *Let* $0 \leq r \leq [\frac{n+1}{2}]$*, and* $\mathfrak{q}_r$ *be the* θ*-stable parabolic subalgebra as defined in Definition 14.37. Suppose that* $\nu = (\nu_1, \cdots, \nu_r) \in \mathbb{Z}^r$ *satisfies the weakly fair condition* (14.38) *relative to* $\mathfrak{q}_r$*. Let* $a, b \in \{\pm\}$*.*

(1) *The G-module* $(\nu_1, \cdots, \nu_r \,||\, 0, \cdots, 0)_{a,b}$ *is nonzero if and only if* $r = 1$ *or* $\nu_{r-1} \geq -1$.

(2) *If the condition (1) is fulfilled, then* $(\nu_1, \cdots, \nu_r \,||\, 0, \cdots, 0)_{a,b}$ *is irreducible and unitarizable.*

Proof. This is a special case of [27, Thm. 3] for the indefinite orthogonal group $O(p, q)$ with $(p, q) = (n + 1, 1)$ with the notation there. □

Assume now $\nu_1 = \cdots = \nu_{r-1} = 0$. Then the necessary and sufficient condition for the parameter $\nu = (0, \cdots, 0, \nu_r) \in \mathbb{Z}^r$ to be in the weakly fair range but outside the good range is given by

$$\nu_r \in \{-1, -2, \cdots, r - 1 - [\frac{n}{2}]\}.$$

In this case, the G-module $(0, \cdots, 0, \nu_r \,||\, 0, \cdots, 0)_{a,b}$ is nonzero, irreducible, and unitarizable for $a, b \in \{\pm\}$ as is seen in Fact 14.52. It turns out that these very parameters give rise to the complementary series representations with integral parameter stated in Section 3.6.3 as follows:

Theorem 14.53. *Let $0 \le i \le [\frac{n}{2}] - 1$. Then the underlying $(\mathfrak{g}, K)$-modules of the complementary series representations $I_{\pm}(i, s)$ and $I_{\pm}(n - i, s)$ with integral parameter $s \in \{i + 1, i + 2, \cdots, [\frac{n}{2}]\}$ are given by*

$$I_+(i, s)_K \simeq A_{\mathfrak{q}_{i+1}}(0, \cdots, 0, s - i)_{+,+};$$

$$I_-(i, s)_K \simeq A_{\mathfrak{q}_{i+1}}(0, \cdots, 0, s - i)_{+,-};$$

$$I_+(n - i, s)_K \simeq A_{\mathfrak{q}_{i+1}}(0, \cdots, 0, s - i)_{-,-};$$

$$I_-(n - i, s)_K \simeq A_{\mathfrak{q}_{i+1}}(0, \cdots, 0, s - i)_{-,+}.$$

Hence, their smooth globalizations are described by θ-stable parameters as follows:

$$I_+(i, s) \quad \simeq \underbrace{(0, \cdots, 0, s - i}_{i+1} || \underbrace{0, \cdots, 0)}_{[\frac{n}{2}]-i}{}_{+,+};$$

$$I_-(i, s) \quad \simeq (0, \cdots, 0, s - i || 0, \cdots, 0)_{+,-};$$

$$I_+(n - i, s) \simeq (0, \cdots, 0, s - i || 0, \cdots, 0)_{-,-};$$

$$I_-(n - i, s) \simeq (0, \cdots, 0, s - i || 0, \cdots, 0)_{-,+}.$$

Chapter 15
Appendix II: Restriction
to $\overline{G} = SO(n+1, 1)$

So far we have been working with symmetry breaking for a pair of the orthogonal groups $(O(n+1,1), O(n,1))$. On the other hand, the Gross–Prasad conjectures (Chapters 11 and 13) are formulated for special orthogonal groups rather than orthogonal groups. In this chapter, we explain how to translate the results for $(G, G') = (O(n+1,1), O(n,1))$ to those for the pair of special orthogonal groups $(\overline{G}, \overline{G'}) = (SO(n+1,1), SO(n,1))$. A part of the results here (e.g., Theorem 15.16) was announced in [43].

In what follows, we use a bar over representations of special orthogonal groups to distinguish them from those of orthogonal groups.

15.1 Restriction of Representations of $G = O(n+1, 1)$
to $\overline{G} = SO(n+1, 1)$

It is well-known that any irreducible admissible representation Π of a real reductive group G is decomposed into the direct sum of finitely many irreducible admissible representations of $\overline{G}$ if $\overline{G}$ is an open normal subgroup of G (see [9, Chap. II, Lem. 5.5]). In order to understand how the restriction $\Pi|_{\overline{G}}$ decomposes, we use the action of the quotient group $G/\overline{G}$ on the ring $\mathrm{End}_{\overline{G}}(\Pi|_{\overline{G}}) = \mathrm{Hom}_{\overline{G}}(\Pi|_{\overline{G}}, \Pi|_{\overline{G}})$.

We apply this general observation to our setting where

$$(G, \overline{G}) = (O(n+1,1), SO(n+1,1)).$$

In this case, the quotient group $G/\overline{G} \simeq \mathbb{Z}/2\mathbb{Z}$. With the notation (2.13) of the characters χ_{ab} of G,

$$\{\chi_{++}, \chi_{--}\} = \{\mathbf{1}, \det\}$$

© Springer Nature Singapore Pte Ltd. 2018
T. Kobayashi, B. Speh, *Symmetry Breaking for Representations of Rank One Orthogonal Groups II*, Lecture Notes in Mathematics 2234,
https://doi.org/10.1007/978-981-13-2901-2_15

is the set of irreducible representations of $G = O(n+1,1)$ which are trivial on $\overline{G} = SO(n+1,1)$. In other words, we have a direct sum decomposition as G-modules:

$$\text{Ind}_{\overline{G}}^{G}\mathbf{1} \simeq \mathbf{1} \oplus \det.$$

Then we have the following:

Lemma 15.1. *Let Π be a continuous representation of $G = O(n+1,1)$. Then there is a natural linear bijection:*

$$\text{End}_{\overline{G}}(\Pi|_{\overline{G}}) \simeq \text{Hom}_{G}(\Pi, \Pi) \oplus \text{Hom}_{G}(\Pi, \Pi \otimes \det).$$

Proof. Clear from the following linear bijections:

$$\text{End}_{\overline{G}}(\Pi|_{\overline{G}}) \simeq \text{Hom}_{G}(\Pi, \text{Ind}_{\overline{G}}^{G}(\Pi|_{\overline{G}})) \simeq \text{Hom}_{G}(\Pi, \Pi \otimes \text{Ind}_{\overline{G}}^{G}\mathbf{1}).$$

$\square$

We examine the restriction of irreducible representations of G to the subgroup $\overline{G}$:

Lemma 15.2. *Suppose that Π is an irreducible admissible representation of $G = O(n+1,1)$.*

(1) *If $\Pi \not\simeq \Pi \otimes \det$ as G-modules, then the restriction $\Pi|_{\overline{G}}$ is irreducible.*
(2) *If $\Pi \simeq \Pi \otimes \det$ as G-modules, then the restriction $\Pi|_{\overline{G}}$ is the direct sum of two irreducible admissible representations of $\overline{G}$ that are not isomorphic to each other.*

Proof. By Lemma 15.1, we have

$$\dim_{\mathbb{C}} \text{Hom}_{\overline{G}}(\Pi|_{\overline{G}}, \Pi|_{\overline{G}}) = \dim_{\mathbb{C}} \text{Hom}_{G}(\Pi, \Pi) + \dim_{\mathbb{C}} \text{Hom}_{G}(\Pi, \Pi \otimes \det)$$

$$= \begin{cases} 1 & \text{if } \Pi \not\simeq \Pi \otimes \det, \\ 2 & \text{if } \Pi \simeq \Pi \otimes \det. \end{cases}$$

Since the restriction $\Pi|_{\overline{G}}$ is the direct sum of irreducible admissible representations of $\overline{G}$, we may write the decomposition as

$$\Pi|_{\overline{G}} \simeq \bigoplus_{j=1}^{N} m_{j} \overline{\Pi}_{j},$$

where $\overline{\Pi}_j$ are (mutually inequivalent) irreducible admissible representations of $\overline{G}$ and $m_j \in \mathbb{N}_+$ denote the multiplicity of $\overline{\Pi}_j$ in $\Pi|_{\overline{G}}$ for $1 \leq j \leq N$. By Schur's lemma,

$$\dim_{\mathbb{C}} \operatorname{End}_{\overline{G}}(\Pi|_{\overline{G}}) = \sum_{j=1}^{N} m_j^2.$$

This is equal to 1 or 2 if and only if $N = m_1 = 1$ or $N = 2$ and $m_1 = m_2 = 1$, respectively. Hence we get the conclusion. $\qquad\square$

15.2 Restriction of Principal Series Representation of $G = O(n+1,1)$ to $\overline{G} = SO(n+1,1)$

This section discusses the restriction of the principal series representation $I_\delta(V, \lambda)$ of $G = O(n+1, 1)$ to the normal subgroup $\overline{G} = SO(n+1, 1)$ of index two. First of all, we fix some notation for principal series representations of $\overline{G}$. We set $\overline{P} := P \cap \overline{G}$. Then $\overline{P}$ is a minimal parabolic subgroup of $\overline{G}$, and its Langlands decomposition is given by $\overline{P} = \overline{M} A N_+$, where

$$\overline{M} := M \cap \overline{G} = \left\{ \begin{pmatrix} \varepsilon & & \\ & B & \\ & & \varepsilon \end{pmatrix} : B \in SO(n), \varepsilon = \pm 1 \right\} \simeq SO(n) \times O(1)$$

is a subgroup of M of index two. For an irreducible representation $(\overline{\sigma}, \overline{V})$ of $SO(n)$, $\delta \in \{\pm\}$, and $\lambda \in \mathbb{C}$, we denote by $\overline{I}_\delta(\overline{V}, \lambda)$ the (unnormalized) induced representation $\operatorname{Ind}_{\overline{P}}^{\overline{G}}(\overline{V} \otimes \delta \otimes \mathbb{C}_\lambda)$ of $\overline{G} = SO(n+1, 1)$.

Let us compare principal series representations of G regarded as $\overline{G}$-modules by restriction with principal series representations of $\overline{G}$. For this, we suppose V is an irreducible representation of $O(n)$, $\delta \in \{\pm\}$, and $\lambda \in \mathbb{C}$, and form a principal series representation $I_\delta(V, \lambda)$ of $G = O(n+1, 1)$. Then its restriction to the subgroup $\overline{G} = SO(n+1, 1)$ is isomorphic to $\operatorname{Ind}_{\overline{P}}^{\overline{G}}(V|_{SO(n)} \otimes \delta \otimes \mathbb{C}_\lambda)$ as a $\overline{G}$-module, because the inclusion $\overline{G} \hookrightarrow G$ induces an isomorphism $\overline{G}/\overline{P} \xrightarrow{\sim} G/P$.

Concerning the $SO(n)$-module $V|_{SO(n)}$, we recall from Definition 2.6 that $V \in \widehat{O(n)}$ is said to be of type X or of type Y according to whether V is irreducible or reducible when restricted to $SO(n)$. In the latter case, n is even (see Lemma 2.7) and V is decomposed into the direct sum of two irreducible representations of $SO(n)$:

$$V = V^{(+)} \oplus V^{(-)}, \tag{15.1}$$

where $V^{(-)}$ is isomorphic to the contragredient representation of $V^{(+)}$. Accordingly, we have an isomorphism as $\overline{G}$-modules:

$$I_\delta(V, \lambda)|_{\overline{G}} \simeq \begin{cases} \overline{I}_\delta(V, \lambda) & \text{if } V \text{ is of type X,} \\ \overline{I}_\delta(V^{(+)}, \lambda) \oplus \overline{I}_\delta(V^{(-)}, \lambda) & \text{if } V \text{ is of type Y.} \end{cases} \tag{15.2}$$

By using (15.2), we obtain the structural results of the restriction of the principal series representation $I_\delta(V, \lambda)$ of $G = O(n+1, 1)$ to the subgroup $\overline{G} = SO(n+1, 1)$ and further to the identity component group $G_0 = SO_0(n+1, 1)$.

15.2.1   Restriction $I_\delta(V, \lambda)|_{\overline{G}}$ When $I_\delta(V, \lambda)$ Is Irreducible

We begin with the case where $I_\delta(V, \lambda)$ is irreducible as a G-module.

Lemma 15.3. *Let* $(\sigma, V) \in \widehat{O(n)}$, $\delta \in \{\pm\}$ *and* $\lambda \in \mathbb{C}$. *Suppose* $I_\delta(V, \lambda)$ *is irreducible as a module of* $G = O(n+1, 1)$.

(1) *Suppose* V *is of type X. Then the following three conditions on* (δ, V, λ) *are equivalent:*

    (i) $I_\delta(V, \lambda)$ *is irreducible as a* G-module;
    (ii) *The restriction* $I_\delta(V, \lambda)|_{\overline{G}}$ *is irreducible as a* $\overline{G}$-module;
    (iii) *The restriction* $I_\delta(V, \lambda)|_{G_0}$ *is irreducible as a* G_0-module.

(2) *Suppose* V *is of type Y. If* $I_\delta(V, \lambda)$ *is irreducible as a* G-module, then $I_\delta(V, \lambda)|_{\overline{G}}$ *splits into the direct sum of two irreducible* $\overline{G}$-modules that are not isomorphic *to each other. In this case,* n *is even and we may write the irreducible decomposition of* $V|_{SO(n)}$ *as in* (15.1). *Then there is a natural isomorphism*

$$I_\delta(V, \lambda)|_{\overline{G}} \simeq \overline{I}_\delta(V^{(+)}, \lambda) \oplus \overline{I}_\delta(V^{(-)}, \lambda)$$

as $\overline{G}$-modules. Moreover, both $\overline{I}_\delta(V^{(+)}, \lambda)$ and $\overline{I}_\delta(V^{(-)}, \lambda)$ *stay irreducible when restricted to* G_0, *and they are not isomorphic to each other also as* G_0-modules.

Proof. We observe that the first factor of M is isomorphic to $O(n)$, whereas that of $M \cap \overline{G}$ $(= \overline{M})$ and of $M \cap G_0$ is isomorphic to $SO(n)$. Since the crucial part is the restriction from the Levi subgroup MA of G to that of $\overline{G}$ or of G_0, we focus on the restriction $G \downarrow \overline{G}$, which involves the restriction of V with respect to the inclusion $O(n) \supset SO(n)$. The restriction $G \downarrow G_0$ can be analyzed similarly by using the four characters χ_{ab} $(a, b \in \{\pm\})$ instead of $\chi_{--} = \det$ as in [35, Chap. 2, Sect. 5].

From now on, we consider the restriction $G \downarrow \overline{G}$. We recall from Lemma 2.14 the following isomorphism of G-modules:

$$I_\delta(V, \lambda) \otimes \chi_{--} \simeq I_\delta(V \otimes \det, \lambda).$$

(1) If V is of type X, then $V \not\simeq V \otimes \det$ as $O(n)$-modules. In turn, the G-modules $I_\delta(V, \lambda)$ and $I_\delta(V \otimes \det, \lambda)$ are not isomorphic to each other, because their K-structures are different by the Frobenius reciprocity and the branching rule for $O(n) \downarrow O(n-1)$ (Fact 2.12). Therefore, $I_\delta(V, \lambda)|_{\overline{G}}$ is irreducible by Lemma 15.2 (1).

(2) If V is of type Y, then $V \otimes \det \simeq V$ by Lemma 2.9, and therefore Lemma 15.2 (2) concludes the first assertion. The remaining assertions are now clear.

$\square$

15.2.2 Restriction $I_\delta(V, \lambda)|_{\overline{G}}$ When V Is of Type Y

We take a closer look at the case where $V \in \widehat{O(n)}$ is of type Y (Definition 2.6). This means that n is even, say $n = 2m$, and the representation V is of the form

$$V = F^{O(2m)}(\sigma_1, \cdots, \sigma_m)_\varepsilon$$

with $\sigma_1 \geq \cdots \geq \sigma_m \geq 1$ and $\varepsilon \in \{\pm\}$, see Section 14.1 for the notation. Then the restriction $V|_{SO(n)}$ decomposes as

$$V|_{SO(n)} = V^{(+)} \oplus V^{(-)}$$

as in (15.1), where the highest weights of the irreducible $SO(2m)$-modules $V^{(\pm)}$ are given by $(\sigma_1, \cdots, \sigma_{m-1}, \pm\sigma_m)$. We recall from Definition 14.10 for the subsets $S(V)$ and $S_Y(V)$ of $\mathbb{Z}$.

Proposition 15.4. *Suppose $G = O(n+1, 1)$ with $n = 2m$ and $(\sigma, V) \in \widehat{O(n)}$ is of type Y. Let $\delta \in \{\pm\}$.*

(1) *The following four conditions on $\lambda \in \mathbb{C}$ are equivalent.*

   (i) *$\overline{I}_\delta(V^{(+)}, \lambda)$ is reducible as a representation of $\overline{G} = SO(n+1, 1)$;*
   (ii) *$\overline{I}_\delta(V^{(-)}, \lambda)$ is reducible as a $\overline{G}$-module;*
   (iii) *$\pm(\lambda - m) \in \mathbb{Z} - (\{\sigma_j + m - j : j = 1, \cdots, m\} \cup \{0, 1, 2, \cdots, \sigma_m - 1\})$;*
   (iv) *$\lambda \in \mathbb{Z} - (S(V) \cup S_Y(V) \cup \{m\})$.*

(2) *Suppose that λ satisfies one of (therefore any of) the above equivalent conditions. Then, for $\varepsilon \in \{\pm\}$, the principal series representation $\overline{I}_\delta(V^{(\varepsilon)}, \lambda)$ of $\overline{G}$ has a unique $\overline{G}$-submodule, to be denoted by $\overline{I}_\delta(V^{(\varepsilon)}, \lambda)^\flat$, such that the quotient $\overline{G}$-module*

$$\overline{I}_\delta(V^{(\varepsilon)}, \lambda)^\sharp := \overline{I}_\delta(V^{(\varepsilon)}, \lambda)/\overline{I}_\delta(V^{(\varepsilon)}, \lambda)^\flat$$

is irreducible. Moreover we have

$$\overline{I}_\delta(V^{(+)}, \lambda)^\flat \not\simeq \overline{I}_\delta(V^{(-)}, \lambda)^\flat,$$

$$\overline{I}_\delta(V^{(+)}, \lambda)^\sharp \not\simeq \overline{I}_\delta(V^{(-)}, \lambda)^\sharp$$

as $\overline{G}$-modules.

Proof. Since $\overline{G} = SO(2m+1, 1)$ is generated by the identity component $G_0 = SO_0(2m+1, 1)$ and a central element $-I_{2m+2}$, any irreducible $\overline{G}$-module remains irreducible when restricted to the connected subgroup G_0. Then the equivalence (i) $\Leftrightarrow$ (iii) (also (ii) $\Leftrightarrow$ (iii)) and the last assertion in Proposition 15.4 follows from Hirai [16]. See also Collingwood [11, Lem. 4.4.1 and Thm. 5.2.1] for the computation of τ-invariants of irreducible representations and a graphic description of the socle filtrations of principal series representations. Finally the equivalence (iii) $\Leftrightarrow$ (iv) is immediate from the definitions (14.13) and (14.14) of $S(V)$ and $S_Y(V)$, respectively.

The last assertion about the $\overline{G}$-inequivalence follows from the Langlands theory [50] because $\mathrm{Re}\,\lambda \neq m$ and $V^{(+)} \not\simeq V^{(-)}$ as $SO(2m)$-modules.   □

In the following proposition, we treat the set of the parameters λ complementary to the one in Proposition 15.4.

Proposition 15.5. *Suppose $G = O(n+1, 1)$ with $n = 2m$ and $V \in \widehat{O(n)}$ is of type Y. Let $\delta \in \{\pm\}$. Assume that $\overline{I}_\delta(V^{(\pm)}, \lambda)$ are irreducible representations of $\overline{G} = SO(2m+1, 1)$, or equivalently, assume that*

$$\lambda \in (\mathbb{C} - \mathbb{Z}) \cup S(V) \cup S_Y(V) \cup \{m\}.$$

(1) *The following two conditions on $\lambda \in \mathbb{C}$ are equivalent:*

   (i) *The two $\overline{G}$-modules $\overline{I}_\delta(V^{(+)}, \lambda)$ and $\overline{I}_\delta(V^{(-)}, \lambda)$ are isomorphic to each other;*
   (ii) *$\lambda = m$.*

(2) *If $\lambda = m$ then the principal series representation $I_\delta(V, \lambda)$ of G is decomposed into the direct sum of two irreducible representations of G.*
(3) *If $\lambda \neq m$, then $I_\delta(V, \lambda)$ is irreducible as a representation of G.*

Proof.

(1) As in the proof of Proposition 15.4 (2), if $\mathrm{Re}\,\lambda \neq m$, then the Langlands theory [50] implies $\overline{I}_\delta(V^{(+)}, \lambda) \not\simeq \overline{I}_\delta(V^{(-)}, \lambda)$ because $V^{(+)} \not\simeq V^{(-)}$ as $SO(2m)$-modules.

   If $\mathrm{Re}\,\lambda = m$, then $\overline{I}_\delta(V^{(\pm)}, \lambda)$ are (smooth) irreducible tempered representations, and the equivalence (i) $\Leftrightarrow$ (ii) follows from Hirai [16]. This would follow also from the general theory of the "R-group" (Knapp–Zuckerman [25]).

(2) Since $\mathrm{Re}\,\lambda = m$ is the unitary axis of the principal series representation $I_\delta(V, \lambda)$ in our normalization (Section 2.3.1), the G-module $I_\delta(V, \lambda)$ decomposes into the direct sum of irreducible G-modules, say, $\Pi^{(1)}, \ldots, \Pi^{(k)}$, and then decomposes further into irreducible $\overline{G}$-modules when restricted to the subgroup $\overline{G} = SO(2m+1, 1)$. Therefore the cardinality k of irreducible G-summands satisfies either $k = 1$ (*i.e.*, $I_\delta(V, \lambda)$ is G-irreducible) or $k = 2$ because the summands $\overline{I}_\delta(V^{(\pm)}, \lambda)$ in (15.2) are irreducible as $\overline{G}$-modules by assumption. Since $\overline{I}_\delta(V^{(+)}, m) \simeq \overline{I}_\delta(V^{(-)}, m)$ by the first statement, we conclude $k \neq 1$ by Lemma 15.3 (2). Thus the second statement is proved.

(3) We prove that $I_\delta(V, \lambda)$ is irreducible by *reductio ad absurdum*. Suppose there were an irreducible proper submodule Π of $I_\delta(V, \lambda)$. Then Π would remain irreducible when restricted to the subgroup $\overline{G} = SO(2m+1, 1)$ because the restriction $\Pi|_{\overline{G}}$ must be isomorphic to one of the $\overline{G}$-irreducible summands $\overline{I}_\delta(V^{(\pm)}, \lambda)$ in (15.2). Then $\Pi \not\simeq \Pi \otimes \det$ as G-modules by Lemma 15.1. Therefore the direct sum $\Pi \oplus (\Pi \otimes \det)$ would be a G-submodule of $I_\delta(V, \lambda)$ because $I_\delta(V, \lambda) \simeq I_\delta(V, \lambda) \otimes \det$ when V is of type Y. In turn, its restriction to the subgroup $\overline{G}$ would yield an isomorphism $\overline{I}_\delta(V^{(+)}, \lambda) \simeq \overline{I}_\delta(V^{(-)}, \lambda)$ of $\overline{G}$-modules, contradicting the statement (1) of the proposition. Hence $I_\delta(V, \lambda)$ must be irreducible.   □

Applying Propositions 15.4 and 15.5 to the middle exterior representation $\bigwedge^m(\mathbb{C}^n)$ of $O(n)$ when $n = 2m$, we obtain the following.

Example 15.6. Let $G = O(n+1, 1)$ with $n = 2m$, and $\delta \in \{\pm\}$. As in (15.5), we write $\overline{I}_\delta^{(\pm)}(m, \lambda)$ for the $\overline{G}$-modules $\overline{I}_\delta(V^{(\pm)}, \lambda)$ when $V = \bigwedge^m(\mathbb{C}^{2m})$.

(1) The $\overline{G}$-modules $\overline{I}_\delta^{(\pm)}(m, \lambda)$ are reducible if and only if $\lambda \in (-\mathbb{N}_+) \cup (n + \mathbb{N}_+)$.
(2) The G-module $I_\delta(m, m)$ decomposes into a direct sum of two irreducible G-modules (see also Theorem 2.20 (1)).
(3) $I_\delta(m, \lambda)$ is irreducible if $\lambda \in \mathbb{Z}$ satisfies $0 \leq \lambda \leq n (= 2m)$ and $\lambda \neq m$.

We refer to Theorem 2.20 (also to Example 14.16) for the irreducibility condition of $I_\delta(i, \lambda)$ for general i $(0 \leq i \leq n)$; to Theorem 14.15 for that of $I_\delta(V, \lambda)$, which will be proved in the next section.

15.3   Proof of Theorem 14.15: Irreducibility Criterion of $I_\delta(V, \lambda)$

As an application of the results in the previous sections, we give a proof of Theorem 14.15 on the necessary and sufficient condition for the principal series representation $I_\delta(V, \lambda)$ of $G = O(n+1, 1)$ to be irreducible.

Proof of Theorem 14.15. Suppose first that V is of type X (Definition 2.6). Then the restriction $V|_{SO(n)}$ is irreducible as an $SO(n)$-module, and $I_\delta(V, \lambda)$ is G-irreducible if and only if the restriction $I_\delta(V,\lambda)|_{G_0}$ is G_0-irreducible by Lemma 15.3 (1). The latter condition was classified in Hirai [16], which amounts to the condition that $\lambda \notin \mathbb{Z}$ or $\lambda \in S(V)$. Thus Theorem 14.15 for V of type X is proved.

Next suppose V is of type Y. As in (15.1), we write $V|_{SO(n)} \simeq V^{(+)} \oplus V^{(-)}$ for the irreducible decomposition as $SO(n)$-modules. If $I_\delta(V,\lambda)$ is G-irreducible, then $\overline{I}_\delta(V^{(\pm)},\lambda)$ are $\overline{G}$-irreducible by Lemma 15.3 (2). Then the condition (iv) in Proposition 15.4 (1) implies that

$$\lambda \notin \mathbb{Z} \text{ or } \lambda \in S(V) \cup S_Y(V) \cup \{m\}. \tag{15.3}$$

Conversely, under the condition (15.3), Proposition 15.5 tells that $I_\delta(V,\lambda)$ is irreducible if and only if $\lambda \notin \mathbb{Z}$ or $\lambda \in S(V) \cup S_Y(V)$. Thus Theorem 14.15 is proved also for V of type Y. $\square$

15.4   Socle Filtration of $I_\delta(V,\lambda)$: Proof of Proposition 14.19

In this section, we complete the proof of Proposition 14.19 about the socle filtration of the principal series representation $I_\delta(V,\lambda)$ of $G = O(n+1,1)$ when it is reducible and $\lambda \neq \frac{n}{2}$ by using the restriction to the subgroups $\overline{G} = SO(n+1,1)$ or $G_0 = SO(n+1,1)$.

We begin with the case that $V \in \widehat{O(n)}$ is of type X (Definition 2.6).

Proof of Proposition 14.19 When V Is of Type X. In this case, for any nonzero subquotient Π of the principal series representation $I_\delta(V,\lambda)$ of $G = O(n+1,1)$, we have

$$\Pi \not\simeq \Pi \otimes \det$$

as G-modules because their K-types are different by Proposition 14.29. In turn, Lemma 15.2 implies that Π is irreducible as a G-module if and only if the restriction $\Pi|_{\overline{G}}$ is irreducible.

For n even, the restriction $\Pi|_{G_0}$ further to the identity component group $G_0 = SO_0(n+1,1)$ is still irreducible because $\overline{G} = SO(n+1,1)$ is generated by G_0 and a central element $-I_{n+2}$. Thus the assertion follows from the socle filtration of the principal series representation of G_0 in Hirai [16].

For n odd, since the restriction $V|_{SO(n)}$ stays irreducible, $I_\delta(V,\lambda)|_{G_0}$ is a principal series representation of $G_0 = SO_0(n+1,1)$. Therefore the restriction $\Pi|_{G_0}$ is a G_0-subquotient of a principal series representation of G_0, of which the length of composition series is either 2 or 3 by Hirai [16]. Inspecting the K-structure of $I_\delta(V,\lambda)$ from Proposition 14.29 again and the K_0-structure of subquotients of the principal series representation of $G_0 = SO_0(n+1,1)$ in [16], we see that the restriction $\Pi|_{G_0}$ is irreducible as a G_0-module if Π is not (the smooth representation of) a discrete series representation, whereas it is a sum of two (holomorphic and anti-holomorphic) discrete series representations of G_0 if Π is a discrete series representation.   □

Alternatively, one may reduce the proof for type X to the case $(V, \lambda) = (\bigwedge^i(\mathbb{C}^n), i)$ by using the translation functor, see Theorems 16.6 and 16.8 (1) in Appendix III.

As the above proof shows, we obtain the restriction formula of irreducible subquotients $I_\delta(V, \lambda)^\flat$ and $I_\delta(V, \lambda)^\sharp$ of the G-module $I_\delta(V, \lambda)$ (Proposition 14.19) to the normal subgroup $\overline{G} = SO(n + 1, 1)$ as follows.

Proposition 15.7. *Suppose $V \in \widehat{O(n)}$ is of type X and $\lambda \in \mathbb{Z} - S(V)$. Let $\delta \in \{\pm\}$.*

(1) *The principal series representation $\overline{I}_\delta(V, \lambda)$ of $\overline{G}$ has a unique proper sub-module, to be denoted by $\overline{I}_\delta(V, \lambda)^\flat$. In particular, the quotient $\overline{G}$-module $\overline{I}_\delta(V, \lambda)^\sharp := \overline{I}_\delta(V, \lambda) / \overline{I}_\delta(V, \lambda)^\flat$ is irreducible.*

(2) *The restriction of the irreducible G-modules $I_\delta(V, \lambda)^\flat$ and $I_\delta(V, \lambda)^\sharp$ to the normal subgroup $\overline{G}$ is given by*

$$I_\delta(V, \lambda)^\flat|_{\overline{G}} \simeq \overline{I}_\delta(V, \lambda)^\flat,$$

$$I_\delta(V, \lambda)^\sharp|_{\overline{G}} \simeq \overline{I}_\delta(V, \lambda)^\sharp.$$

We end this section with the restriction of $I_\delta(V, \lambda)^\flat$ and $I_\delta(V, \lambda)^\sharp$ to the subgroup $\overline{G}$ when V is of type Y:

Proposition 15.8. *Suppose $G = O(n + 1, 1)$ with $n = 2m$. Assume that $V \in \widehat{O(n)}$ is of type Y, $\delta \in \{\pm\}$, and $\lambda \in \mathbb{Z} - (S(V) \cup S_Y(V) \cup \{m\})$. Then the restriction of $I_\delta(V, \lambda)^\flat$ and $I_\delta(V, \lambda)^\sharp$ to the normal subgroup $\overline{G} = SO(n + 1, 1)$ decomposes into the direct sum of two irreducible $\overline{G}$-modules:*

$$I_\delta(V, \lambda)^\flat|_{\overline{G}} \simeq \overline{I}_\delta(V^{(+)}, \lambda)^\flat \oplus \overline{I}_\delta(V^{(-)}, \lambda)^\flat,$$

$$I_\delta(V, \lambda)^\sharp|_{\overline{G}} \simeq \overline{I}_\delta(V^{(+)}, \lambda)^\sharp \oplus \overline{I}_\delta(V^{(-)}, \lambda)^\sharp,$$

where we recall from Proposition 15.4 for the definition of the irreducible $\overline{G}$-modules $\overline{I}_\delta(V^{(\pm)}, \lambda)^\flat$ and $\overline{I}_\delta(V^{(\pm)}, \lambda)^\sharp$.

Proof of Proposition 14.19 for V of Type Y. By Lemma 15.2 and by the structural results on $\overline{G}$-modules $\overline{I}_\delta(V^{(\pm)}, \lambda)$ in Proposition 15.4, the proof is reduced to the following lemma.   □

Lemma 15.9. *Under the assumption of Proposition 15.8, any G-submodule Π of $I_\delta(V, \lambda)$ satisfies*

$$\Pi \simeq \Pi \otimes \det \qquad (15.4)$$

as G-modules.

Proof. Since V is of type Y, $V \simeq V \otimes \det$ as $O(n)$-modules, hence we have natural G-isomorphisms

$$I_\delta(V,\lambda) \simeq I_\delta(V,\lambda) \otimes \det$$

by Lemma 2.14. We prove (15.4) by *reductio ad absurdum*. Suppose that the G-module Π were not isomorphic to $\Pi \otimes \det$. Then the direct sum representation $\Pi \oplus (\Pi \otimes \det)$ would be a G-submodule of $I_\delta(V,\lambda)$. In turn, the $\overline{G}$-module $\Pi|_{\overline{G}}$ would occur in $I_\delta(V,\lambda)|_{\overline{G}} \simeq \overline{I}_\delta(V^{(+)},\lambda) \oplus \overline{I}_\delta(V^{(-)},\lambda)$ at least twice. But this is impossible by Proposition 15.4. Thus Lemma 15.9 is proved.   □

15.5   Restriction of $\Pi_{\ell,\delta}$ to $SO(n+1,1)$

In this section we treat the case where $I_\delta(V,\lambda)$ is not irreducible as a G-module. We discuss the restriction of G-irreducible subquotients of $I_\delta(V,\lambda)$ to the subgroup $\overline{G} = SO(n+1,1)$.

We focus on the case when (σ, V) is the exterior representation on $V = \bigwedge^i(\mathbb{C}^n)$. In particular, irreducible representations that have the $\mathfrak{Z}_G(\mathfrak{g})$-infinitesimal character ρ, namely, the irreducible G-modules $\Pi_{\ell,\delta}$ ($0 \le \ell \le n+1$, $\delta \in \{\pm\}$) arise as G-irreducible subquotients of $I_\delta(V,\lambda)$. To be more precise, we recall from (2.35) that $\Pi_{\ell,\delta}$ are the irreducible subrepresentations of $I_\delta(\ell,\ell)$ for $0 \le \ell \le n$ and coincidently those of $I_{-\delta}(\ell-1,\ell-1)$ for $1 \le \ell \le n+1$.

Lemma 15.10. *For all $0 \le \ell \le n+1$ and $\delta = \pm$, the restriction of $\Pi_{\ell,\delta}$ to the subgroup $\overline{G} = SO(n+1,1)$ stays irreducible.*

Proof. The restriction $\Pi_{\ell,\delta}|_{\overline{G}}$ is irreducible by the criterion in Lemma 15.2 (1) because $\Pi_{\ell,\delta} \otimes \det \simeq \Pi_{n+1-\ell,-\delta} \not\simeq \Pi_{\ell,\delta}$ by Theorem 2.20 (5).   □

We denote by $\overline{\Pi}_{\ell,\delta}$ the restriction of the irreducible G-module $\Pi_{\ell,\delta}$ ($0 \le \ell \le n+1$, $\delta = \pm$) to the subgroup $\overline{G} = SO(n+1,1)$. By a little abuse of notation, we write $\overline{I}_\delta(i,\lambda)$ for the restriction of $I_\delta(i,\lambda)$, to the subgroup $\overline{G}$. Then the $SO(n)$-isomorphism $\bigwedge^i(\mathbb{C}^n) \simeq \bigwedge^{n-i}(\mathbb{C}^n)$ induces a $\overline{G}$-isomorphism

$$\overline{I}_\delta(i,\lambda) \simeq \overline{I}_\delta(n-i,\lambda).$$

Special attention is needed in the case when n is even and $n = 2i$. In this case, the $O(n)$-module $\bigwedge^i(\mathbb{C}^n)$ is of type Y (see Example 2.8), and it splits into the direct sum of two irreducible $SO(n)$-modules:

$$\bigwedge^{\frac{n}{2}}(\mathbb{C}^n) \simeq \bigwedge^{\frac{n}{2}}(\mathbb{C}^n)^{(+)} \oplus \bigwedge^{\frac{n}{2}}(\mathbb{C}^n)^{(-)}.$$

We set

$$\overline{I}_\delta^{(\pm)}(\frac{n}{2},\lambda) := \mathrm{Ind}_P^{\overline{G}}(\bigwedge^{\frac{n}{2}}(\mathbb{C}^n)^{(\pm)} \otimes \delta \otimes \mathbb{C}_\lambda).$$

As in (15.2), the restriction $I_\delta(\frac{n}{2}, \lambda)|_{\overline{G}}$ is the direct sum of two $\overline{G}$-modules:

$$\overline{I}_\delta(\frac{n}{2}, \lambda) \simeq \overline{I}_\delta^{(+)}(\frac{n}{2}, \lambda) \oplus \overline{I}_\delta^{(-)}(\frac{n}{2}, \lambda) \quad \text{for all } \lambda \in \mathbb{C}. \tag{15.5}$$

If $I_\delta(\frac{n}{2}, \lambda)$ is G-irreducible, then Lemma 15.3 (2) tells that the representations $\overline{I}_\delta^{(\pm)}(\frac{n}{2}, \lambda)$ of the subgroup $\overline{G}$ are irreducible, and that they are not isomorphic to each other.

On the other hand, if $\lambda = i\ (=\frac{n}{2})$, then the principal series representation $I_\delta(i, \lambda)$ is not irreducible as a G-module but splits into the direct sum of two irreducible G-modules (see Theorem 2.20 (1)):

$$I_\delta(\frac{n}{2}, \frac{n}{2}) \simeq \Pi_{\frac{n}{2}+1,-\delta} \oplus \Pi_{\frac{n}{2},\delta},$$

which are not isomorphic to each other. Moreover, the tensor product with χ_{--} switches $\Pi_{\frac{n}{2}+1,-\delta}$ and $\Pi_{\frac{n}{2},\delta}$ (Theorem 2.20 (5)). Hence we have a $\overline{G}$-isomorphism $\overline{\Pi}_{\frac{n}{2}+1,-\delta} \simeq \overline{\Pi}_{\frac{n}{2},\delta}$, which are $\overline{G}$-irreducible by Lemma 15.2 (1). Therefore, for n even, we have the following isomorphisms as $\overline{G}$-modules:

$$\overline{\Pi}_{\frac{n}{2},\delta} \simeq \overline{\Pi}_{\frac{n}{2}+1,-\delta} \simeq \overline{I}_\delta^{(+)}(\frac{n}{2}, \frac{n}{2}) \simeq \overline{I}_\delta^{(-)}(\frac{n}{2}, \frac{n}{2}) \quad \text{for } \delta = \pm. \tag{15.6}$$

Similarly to Theorem 2.20 about the $O(n+1,1)$-modules $\Pi_{\ell,\delta}$, we summarize the properties of the restriction $\overline{\Pi}_{\ell,\delta} = \Pi_{\ell,\delta}|_{\overline{G}}$ as follows.

Proposition 15.11. *Let $\overline{G} = SO(n+1,1)$ with $n \geq 1$.*

(1) $\overline{\Pi}_{\ell,\delta}$ *is irreducible as a $\overline{G}$-module for all $0 \leq \ell \leq n+1$ and $\delta = \pm$.*
(2) $\overline{\Pi}_{\ell,\delta} \simeq \overline{\Pi}_{n+1-\ell,-\delta}$ *as $\overline{G}$-modules for all $0 \leq \ell \leq n+1$ and $\delta = \pm$.*
(3) *Irreducible representations of $\overline{G}$ with $\mathfrak{Z}(\mathfrak{g})$-infinitesimal character $\rho_{\overline{G}}$ can be classified as*

$$\{\overline{\Pi}_{\ell,\delta} : 0 \leq \ell \leq \frac{n-1}{2}, \delta = \pm\} \cup \{\overline{\Pi}_{\frac{n+1}{2},+}\} \qquad \text{if } n \text{ is odd,}$$

$$\{\overline{\Pi}_{\ell,\delta} : 0 \leq \ell \leq \frac{n}{2}, \delta = \pm\} \qquad \text{if } n \text{ is even.}$$

(4) *Every $\overline{\Pi}_{\ell,\delta}\ (0 \leq \ell \leq n+1, \delta = \pm)$ is unitarizable.*

In the next statement, we use the same symbol $\overline{\Pi}_{\ell,\delta}$ to denote the irreducible unitary representation obtained by the Hilbert completion of $\overline{\Pi}_{\ell,\delta}$ with respect to a $\overline{G}$-invariant inner product.

(5) *For n odd, $\overline{\Pi}_{\frac{n+1}{2},+}$ is a discrete series representation of $\overline{G} = SO(n+1,1)$. For n even, $\overline{\Pi}_{\frac{n}{2},\delta}\ (\delta = \pm)$ are tempered representations. All the other representations in the list (3) are nontempered representations of $\overline{G}$.*

(6) *For n even, the center of* $\overline{G} = SO(n+1,1)$ *acts nontrivially on* $\overline{\Pi}_{\ell,\delta}$ *if and only if* $\delta = (-1)^{\ell+1}$. *For n odd, the center of* $\overline{G}$ *is trivial, and thus acts trivially on* $\overline{\Pi}_{\ell,\delta}$ *for any* ℓ *and* δ.

In Proposition 14.44, we gave a description of the underlying $(\mathfrak{g}, K)$-module of the irreducible G-module $\Pi_{\ell,\delta}$ in terms of cohomological parabolic induction. We end this section with analogous results for the irreducible $\overline{G}$-module $\overline{\Pi}_{\ell,\delta} = \Pi_{\ell,\delta}|_{\overline{G}}$ (see Proposition 15.11 (1)).

Proposition 15.12. *For* $0 \le i \le [\frac{n+1}{2}]$, *let* $\mathfrak{q}_i$ *be the* θ-*stable parabolic subalgebras with the Levi subgroup* $\overline{L}_i \simeq SO(2)^i \times SO(n-2i+1,1)$ *as in Definition 14.37 and write* $S_i = i(n-i)$, *see (14.35). Then the underlying* $(\mathfrak{g}, \overline{K})$-*modules of the irreducible* $\overline{G}$-*modules* $\overline{\Pi}_{\ell,\delta}$ ($0 \le \ell \le n+1$, $\delta \in \{\pm\}$) *are given by the cohomological parabolic induction as follows:*

$$(\overline{\Pi}_{i,+})_{\overline{K}} \simeq (\overline{\Pi}_{n+1-i,-})_{\overline{K}} \simeq \mathcal{R}_{\mathfrak{q}_i}^{S_i}(\mathbb{C}_{\rho(\mathfrak{u})})$$

$$\simeq (A_{\mathfrak{q}_i})_{+,+}|_{(\mathfrak{g},\overline{K})} \simeq (A_{\mathfrak{q}_i})_{-,-}|_{(\mathfrak{g},\overline{K})},$$

$$(\overline{\Pi}_{i,-})_{\overline{K}} \simeq (\overline{\Pi}_{n+1-i,+})_{\overline{K}} \simeq \mathcal{R}_{\mathfrak{q}_i}^{S_i}(\mathbb{C}_{\rho(\mathfrak{u})} \otimes \chi_{+-})$$

$$\simeq (A_{\mathfrak{q}_i})_{+,-}|_{(\mathfrak{g},\overline{K})} \simeq (A_{\mathfrak{q}_i})_{-,+}|_{(\mathfrak{g},\overline{K})}.$$

We notice that the four characters $\chi_{\pm\pm}$ of $O(n+1,1)$ induce the following isomorphisms $\chi_{--} \simeq \mathbf{1}$ and $\chi_{+-} \simeq \chi_{-+}$ when they are restricted to the last factor $SO(n-2i+1,1)$ of the Levi subgroup $\overline{L}_i$, whence Proposition 15.12 gives an alternative proof for the isomorphism

$$\overline{\Pi}_{i,\delta} \simeq \overline{\Pi}_{n+1-i,-\delta}$$

as $\overline{G}$-modules for $0 \le i \le n+1$ and $\delta = \pm$.

15.6   Symmetry Breaking for Tempered Principal Series Representations

In this section, we deduce a multiplicity-one theorem for tempered principal series representations $\overline{I}_\delta(\overline{V}, \lambda)$ and $\overline{J}_\varepsilon(\overline{W}, \nu)$ of $\overline{G} = SO(n+1,1)$ and $\overline{G}' = SO(n,1)$, respectively, from the corresponding result (see Theorem 3.30) for the pair $(G, G') = (O(n+1,1), O(n,1))$.

In [35, Chap. 2, Sect. 5], a trick analogous to Lemma 15.1 was used to deduce symmetry breaking for the pair $(G_0, G_0') = (SO_0(n+1,1), SO_0(n,1))$ from that for the pair (G, G') by using an observation that G_0 and G_0' are normal subgroups of G and G', respectively (cf. [35, page 26]). This is formulated in our setting as follows:

Proposition 15.13. *Let Π and π be continuous representations of $G = O(n+1,1)$ and $G' = O(n,1)$, respectively. Let $(\overline{G}, \overline{G'}) = (SO(n+1,1), SO(n,1))$. Then we have natural isomorphisms:*

$$\operatorname{Hom}_{\overline{G'}}(\Pi|_{\overline{G'}}, \pi|_{\overline{G'}}) \simeq \operatorname{Hom}_{G'}(\Pi|_{G'}, \pi) \oplus \operatorname{Hom}_{G'}(\Pi|_{G'}, \chi_{--} \otimes \pi)$$

$$\simeq \operatorname{Hom}_{G'}(\Pi|_{G'}, \pi) \oplus \operatorname{Hom}_{G'}((\Pi \otimes \chi_{--})|_{G'}, \pi).$$

For $\overline{V} \in \widehat{SO(n)}$ and $\overline{W} \in \widehat{SO(n-1)}$, we set

$$[\overline{V}|_{SO(n-1)} : \overline{W}] := \dim_{\mathbb{C}} \operatorname{Hom}_{SO(n-1)}(\overline{V}|_{SO(n-1)}, \overline{W}).$$

The main result of this section is the following.

Theorem 15.14 (tempered principal series representation). *Let $\overline{V} \in \widehat{SO(n)}$, $\overline{W} \in \widehat{SO(n-1)}$, $\delta, \varepsilon \in \{\pm\}$, and $(\lambda, \nu) \in (\sqrt{-1}\mathbb{R} + \frac{n}{2}, \sqrt{-1}\mathbb{R} + \frac{1}{2}(n-1))$ so that $\overline{I}_\delta(\overline{V}, \lambda)$ and $\overline{J}_\varepsilon(\overline{W}, \nu)$ are irreducible tempered principal series representations of $\overline{G} = SO(n+1,1)$ and $\overline{G'} = SO(n,1)$, respectively. Then the following conditions are equivalent:*

(i) $[\overline{V}|_{SO(n-1)} : \overline{W}] \neq 0.$
(ii) $\operatorname{Hom}_{SO(n,1)}(\overline{I}_\delta(\overline{V}, \lambda)|_{SO(n,1)}, \overline{J}_\varepsilon(\overline{W}, \nu)) \neq \{0\}.$
(iii) $\dim_{\mathbb{C}} \operatorname{Hom}_{SO(n,1)}(\overline{I}_\delta(\overline{V}, \lambda)|_{SO(n,1)}, \overline{J}_\varepsilon(\overline{W}, \nu)) = 1.$

For the proof, we use the following elementary lemma on branching rules of finite-dimensional representations of $O(n)$.

Lemma 15.15. *Suppose $\sigma \in \widehat{O(n)}$ and $\tau \in \widehat{O(n-1)}$ are of both type X (Definition 2.6). If $[\sigma|_{O(n-1)} : \tau] \neq 0$, then $[\sigma|_{O(n-1)} : \tau \otimes \det] = 0.$*

Proof of Lemma 15.15. Easy from Fact 2.12 and from the characterization in Lemma 2.7 of representations of type X by means of the Cartan–Weyl bijection (2.20). □

Proof of Theorem 15.14. There exist unique $V \in \widehat{O(n)}$ and $W \in \widehat{O(n-1)}$ such that $[V|_{SO(n)} : \overline{V}] \neq 0$ and $[W|_{SO(n-1)} : \overline{W}] \neq 0$. We divide the argument into the following three cases:

Case XX: Both V and W are of type X.
Case XY: V is of type X and W is of type Y.
Case YX: V is of type Y and W is of type X.

Then we have from (15.2)

$$I_\delta(V, \lambda)|_{\overline{G}} \simeq \begin{cases} \overline{I}_\delta(\overline{V}, \lambda) & \text{if } V \text{ is of type X,} \\ \overline{I}_\delta(\overline{V}, \lambda) \oplus \overline{I}_\delta(\overline{V}^\vee, \lambda) & \text{if } V \text{ is of type Y,} \end{cases}$$

and similarly for the restriction $J_\varepsilon(W, \nu)|_{\overline{G'}}$.

By Proposition 15.13, we have

$$\mathrm{Hom}_{\overline{G'}}(I_\delta(V,\lambda)|_{\overline{G'}}, J_\varepsilon(W,\nu)|_{\overline{G'}}) \simeq \bigoplus_{\chi \in \{1,\det\}} \mathrm{Hom}_{G'}(I_\delta(V,\lambda)|_{G'}, J_\varepsilon(W,\nu) \otimes \chi).$$

Applying the multiplicity-one theorem (Theorem 3.30) for tempered representations of the pair $(G,G') = (O(n+1,1), O(n,1))$ to the right-hand side, we get the following multiplicity formula:

$$\dim_{\mathbb{C}} \mathrm{Hom}_{\overline{G'}}(I_\delta(V,\lambda)|_{\overline{G'}}, J_\varepsilon(W,\nu)|_{\overline{G'}})$$
$$= [V|_{O(n-1)} : W] + [V|_{O(n-1)} : W \otimes \det]. \tag{15.7}$$

The right-hand side of (15.7) does not vanish if and only if $[\overline{V}|_{SO(n-1)} : \overline{W}] \neq 0$. In this case, we have

$$(15.7) = \begin{cases} 1 & \text{Case XX,} \\ 2 & \text{Case XY or Case YX,} \end{cases}$$

by Lemmas 2.13 and 15.15. Thus the conclusion holds in Case XX.

If V is of type Y, then the two $\overline{G}$-irreducible summands $\overline{I}_\delta(\overline{V},\lambda)$ and $\overline{I}_\delta(\overline{V}^\vee,\lambda)$ in the restriction $I_\delta(V,\lambda)|_{G'}$ are switched if we apply the outer automorphism of $\overline{G}$ by an element $g_0 := \mathrm{diag}(1,\cdots,1,-1,1) \in G = O(n+1,1)$. Since g_0 commutes with $\overline{G'}$, we obtain an isomorphism

$$\mathrm{Hom}_{\overline{G'}}(\overline{I}_\delta(\overline{V},\lambda)|_{\overline{G'}}, \overline{J}_\varepsilon(\overline{W},\nu)) \simeq \mathrm{Hom}_{\overline{G'}}(\overline{I}_\delta(\overline{V}^\vee,\lambda)|_{\overline{G'}}, \overline{J}_\varepsilon(\overline{W},\nu)).$$

Hence the conclusion holds for Case YX.

Similar argument holds for Case XY where W is of type Y. Therefore Theorem 15.14 is proved.   □

15.7 Symmetry Breaking from $\overline{I}_\delta(i,\lambda)$ to $\overline{J}_\varepsilon(j,\nu)$

In this section, we give a closed formula of the multiplicity for the restriction $\overline{G} \downarrow \overline{G'}$ when (σ, V) is the exterior tensor $\bigwedge^i(\mathbb{C}^n)$. For the admissible smooth representations $\overline{I}_\delta(i,\lambda)$ of $\overline{G} = SO(n+1,1)$ and $\overline{J}_\varepsilon(j,\nu)$ of $\overline{G'} = SO(n,1)$, we set

$$m(i,j) \equiv m(\overline{I}_\delta(i,\lambda), \overline{J}_\varepsilon(j,\nu)) := \dim_{\mathbb{C}} \mathrm{Hom}_{\overline{G'}}(\overline{I}_\delta(i,\lambda)|_{\overline{G'}}, \overline{J}_\varepsilon(j,\nu)).$$

In order to state a closed formula for the multiplicity $m(i, j)$ as a function of $(\lambda, \nu, \delta, \varepsilon)$, we introduce the following subsets of $\mathbb{Z}^2 \times \{\pm 1\}$:

$$L := \{(-i, -j, (-1)^{i+j}) : (i, j) \in \mathbb{Z}^2, 0 \le j \le i\} = L_{\text{even}} \cup L_{\text{odd}},$$

$$L' := \{(\lambda, \nu, \gamma) \in L : \nu \ne 0\}.$$

In the theorem below, we shall see

$$m(i, j) \in \{1, 2, 4\} \quad \text{if } j = i - 1 \text{ or } i,$$

$$m(i, j) \in \{0, 1, 2\} \quad \text{if } j = i - 2 \text{ or } i + 1,$$

$$m(i, j) = 0 \quad \text{otherwise.}$$

By Proposition 15.13 and Lemma 3.36, the multiplicity formula for $(\overline{G}, \overline{G}')$ is derived from the one for (G, G') by using Proposition 15.13, which amounts to

$$\text{Hom}_{\overline{G}'}(\overline{I}_\delta(i, \lambda)|_{\overline{G}'}, \overline{J}_\varepsilon(j, \nu))$$

$$\simeq \text{Hom}_{G'}(I_\delta(i, \lambda)|_{G'}, J_\varepsilon(j, \nu)) \oplus \text{Hom}_{G'}(I_\delta(n - i, \lambda)|_{G'}, J_\varepsilon(j, \nu)).$$

The right-hand side was computed in Theorem 3.25. Hence we get an explicit formula of the multiplicity for the restriction of nonunitary principal series representations in this setting:

Theorem 15.16 (multiplicity formula). *Suppose $n \ge 3$, $0 \le i \le [\frac{n}{2}]$, $0 \le j \le [\frac{n-1}{2}]$, δ, $\varepsilon \in \{\pm\} \equiv \{\pm 1\}$, and $\lambda, \nu \in \mathbb{C}$.*

Then the multiplicity $m(i, j) = \dim_{\mathbb{C}} \text{Hom}_{\overline{G}'}(\overline{I}_\delta(i, \lambda)|_{\overline{G}'}, \overline{J}_\varepsilon(j, \nu))$ is given as follows.

(1) *Suppose $j = i$.*

    (a) *Case $i = 0$.*

$$m(0, 0) = \begin{cases} 2 & \text{if } (\lambda, \nu, \delta\varepsilon) \in L, \\ 1 & \text{otherwise.} \end{cases}$$

    (b) *Case $1 \le i < \frac{n}{2} - 1$.*

$$m(i, i) = \begin{cases} 2 & \text{if } (\lambda, \nu, \delta\varepsilon) \in L' \cup \{(i, i, +)\}, \\ 1 & \text{otherwise.} \end{cases}$$

(c) *Case* $i = \frac{n}{2} - 1$ *(n: even).*

$$m(\frac{n}{2} - 1, \frac{n}{2} - 1) = \begin{cases} 2 & \textit{if } (\lambda, \nu, \delta\varepsilon) \in L' \cup \{(i, i, +)\} \cup \{(i, i+1, -)\}, \\ 1 & \textit{otherwise.} \end{cases}$$

(d) *Case* $i = \frac{n-1}{2}$ *(n: odd).*

$$m(\frac{n-1}{2}, \frac{n-1}{2}) = \begin{cases} 4 & \textit{if } (\lambda, \nu, \delta\varepsilon) \in L' \cup \{(i, i, +)\}, \\ 2 & \textit{otherwise.} \end{cases}$$

(2) *Suppose* $j = i - 1$.

(a) *Case* $1 \le i < \frac{n-1}{2}$.

$$m(i, i-1) = \begin{cases} 2 & \textit{if } (\lambda, \nu, \delta\varepsilon) \in L' \cup \{(n-i, n-i, +)\}, \\ 1 & \textit{otherwise.} \end{cases}$$

(b) *Case* $i = \frac{n-1}{2}$ *(n: odd).*

$$m(\frac{n-1}{2}, \frac{n-3}{2}) = \begin{cases} 2 & \textit{if } (\lambda, \nu, \delta\varepsilon) \in L', \\ 2 & \textit{if } (\lambda, \nu, \delta\varepsilon) \in \{(n-i, n-i, +)\} \cup \{(i, i+1, -)\}, \\ 1 & \textit{otherwise.} \end{cases}$$

(c) *Case* $i = \frac{n}{2}$ *(n: even).*

$$m(\frac{n}{2}, \frac{n}{2} - 1) = \begin{cases} 4 & \textit{if } (\lambda, \nu, \delta\varepsilon) \in L' \cup \{(n-i, n-i, +)\}, \\ 2 & \textit{otherwise.} \end{cases}$$

(3) *Suppose* $j = i - 2$.

(a) *Case* $2 \le i < \frac{n}{2}$.

$$m(i, i-2) = \begin{cases} 1 & \textit{if } (\lambda, \nu, \delta\varepsilon) = (n-i, n-i+1, -), \\ 0 & \textit{otherwise.} \end{cases}$$

(b) *Case* $i = \frac{n}{2}$ *(n: even).*

$$m(\frac{n}{2}, \frac{n}{2} - 2) = \begin{cases} 2 & \textit{if } (\lambda, \nu, \delta\varepsilon) = (\frac{n}{2}, \frac{n}{2} + 1, -), \\ 0 & \textit{otherwise.} \end{cases}$$

(4) *Suppose $j = i + 1$.*

    (a) *Case $i = 0$ and $n > 3$.*

$$m(0,1) = \begin{cases} 1 & \text{if } \lambda \in -\mathbb{N}, \nu = 1, \text{ and } \delta\varepsilon = (-1)^{\lambda+1}, \\ 0 & \text{otherwise.} \end{cases}$$

    (b) *Case $1 \le i < \frac{n-3}{2}$.*

$$m(i,i+1) = \begin{cases} 1 & \text{if } (\lambda, \nu, \delta\varepsilon) = (i, i+1, -), \\ 0 & \text{otherwise.} \end{cases}$$

    (c) *Case $i = \frac{n-3}{2}$ and $n > 3$, odd.*

$$m\left(\frac{n-3}{2}, \frac{n-1}{2}\right) = \begin{cases} 2 & \text{if } (\lambda, \nu, \delta\varepsilon) = (\frac{n-3}{2}, \frac{n-1}{2}, -), \\ 0 & \text{otherwise.} \end{cases}$$

    (d) *Case $i = 0$ and $n = 3$.*

$$m(0,1) = \begin{cases} 2 & \text{if } \lambda \in -\mathbb{N}, \nu = 1, \text{ and } \delta\varepsilon = (-1)^{\lambda+1}, \\ 0 & \text{otherwise.} \end{cases}$$

(5) *Suppose $j \notin \{i-2, i-1, i, i+1\}$. Then $m(i,j) = 0$ for all $\lambda, \nu, \delta, \varepsilon$.*

Remark 15.17 (multiplicity-one property). In [55] it is proved that

$$\dim_{\mathbb{C}} \operatorname{Hom}_{\overline{G'}}(\Pi|_{\overline{G'}}, \pi) \le 1$$

for any irreducible admissible smooth representations Π and π of $\overline{G} = SO(n+1,1)$ and $\overline{G'} = SO(n,1)$, respectively. Thus Theorem 3.25 fits well with their multiplicity-free results for $\lambda, \nu \in \mathbb{C} - \mathbb{Z}$, where $\overline{I}_\delta(i,\lambda)$ and $\overline{J}_\varepsilon(j,\nu)$ are irreducible admissible representations of $\overline{G}$ and $\overline{G'}$, respectively, except for the cases $n = 2i$ or $n = 2j+1$. In the case $n = 2i$ or $n = 2j+1$, the multiplicity is counted twice as we saw in (15.5) and (15.6), and thus the statements (1-d), (2-c), (3-b), and (4-c) in Theorem 3.25 fit again with [55].

Remark 15.18 (generic multiplicity-two phenomenon). In addition to the subgroup $\overline{G'} = SO(n,1)$, the Lorentz group $O(n,1)$ contains two other subgroups of index two, that is, $O^+(n,1)$ (containing orthochronous reflections) and $O^-(n,1)$ (containing anti-orthochronous reflections) with terminology in relativistic space-time for $n = 3$. Our results yield also the multiplicity formula for such pairs by using an analogous result to Proposition 15.13, and it turns out that a generic multiplicity-one statement fails if we replace $(\overline{G}, \overline{G'}) = (SO(n+1,1), SO(n,1))$ by $(O^-(n+1,1), O^-(n,1))$. In fact, the multiplicity $m(\Pi, \pi)$ is generically equal to 2 for irreducible representations Π and π of $O^-(n+1,1)$ and $O^-(n,1)$, respectively, as is expected by the general theory [39, 42] because there are two open orbits in $P'^- \backslash G^- / P^-$ in this case.

15.8   Symmetry Breaking Between Irreducible Representations of $\overline{G}$ and $\overline{G}'$ with Trivial Infinitesimal Character ρ

Similar to the notation $\overline{\Pi}_{i,\delta}$ for the restriction of the irreducible representation $\Pi_{i,\delta}$ of $G = O(n+1,1)$ to the special orthogonal group $\overline{G} = SO(n+1,1)$, we denote by $\overline{\pi}_{j,\varepsilon}$ the restriction of the irreducible representation $\pi_{j,\varepsilon}$ $(0 \le j \le n,$ $\varepsilon = \pm)$ of $G' = O(n,1)$ to the special orthogonal group $\overline{G}' = SO(n,1)$. Then $\overline{\Pi}_{i,\delta}$ $(0 \le i \le n+1, \delta = \pm)$ and $\overline{\pi}_{j,\varepsilon}$ $(0 \le j \le n, \varepsilon = \pm)$ exhaust irreducible admissible smooth representations of $\overline{G}$ and $\overline{G}'$ having $\mathfrak{Z}(\mathfrak{g})$-infinitesimal character $\rho_{\overline{G}}$ and $\mathfrak{Z}(\mathfrak{g}')$-infinitesimal character $\rho_{\overline{G}'}$ respectively, by Lemma 15.10.

In this section, we deduce the formula of the multiplicity

$$\dim_{\mathbb{C}} \mathrm{Hom}_{\overline{G}'}(\overline{\Pi}_{i,\delta}|_{\overline{G}'}, \overline{\pi}_{j,\varepsilon})$$

for the symmetry breaking for $(\overline{G}, \overline{G}') = (SO(n+1,1), SO(n,1))$ from the one for $(G, G') = (O(n+1,1), O(n,1))$.

In view of the $\overline{G}$-isomorphism $\overline{\Pi}_{\frac{n+1}{2},+} \simeq \overline{\Pi}_{\frac{n+1}{2},-}$ for n even and the $\overline{G}'$-isomorphism $\overline{\pi}_{\frac{n}{2},+} \simeq \overline{\pi}_{\frac{n}{2},-}$ for n odd, we shall use the following convention

$$+ \equiv - \text{ for } \delta \text{ if } n+1 = 2i; \quad + \equiv - \text{ for } \varepsilon \text{ if } n = 2j \tag{15.8}$$

when we deal with the representations $\overline{\Pi}_{i,\delta}$ $(0 \le i \le [\frac{n+1}{2}])$ and $\overline{\pi}_{j,\varepsilon}$ $(0 \le j \le [\frac{n}{2}])$.

Owing to Proposition 15.13, Theorem 2.20 tells that

$$\mathrm{Hom}_{\overline{G}'}(\overline{\Pi}_{i,\delta}|_{\overline{G}'}, \overline{\pi}_{j,\varepsilon}) \simeq \mathrm{Hom}_{G'}(\Pi_{i,\delta}|_{G'}, \pi_{j,\varepsilon}) \oplus \mathrm{Hom}_{G'}(\Pi_{n+1-i,-\delta}|_{G'}, \pi_{j,\varepsilon}).$$

Applying Theorems 4.1 and 4.2 about symmetry breaking for the pair $(G, G') = (O(n+1,1), O(n,1))$ to the right-hand side, we determine the multiplicity

$$m(\overline{\Pi}, \overline{\pi}) \quad \text{for all } \overline{\Pi} \in \mathrm{Irr}(\overline{G})_\rho \text{ and } \overline{\pi} \in \mathrm{Irr}(\overline{G}')_\rho$$

for the pair $(\overline{G}, \overline{G}') = (SO(n+1,1), SO(n,1))$ of special orthogonal groups as follows.

Theorem 15.19. *Suppose* $0 \le i \le [\frac{n+1}{2}]$, $0 \le j \le [\frac{n}{2}]$, *and* $\delta, \varepsilon = \pm$ *with the convention* (15.8), *then*

$$\dim_{\mathbb{C}} \mathrm{Hom}_{\overline{G}'}(\overline{\Pi}_{i,\delta}|_{\overline{G}'}, \overline{\pi}_{j,\varepsilon}) = \begin{cases} 1 & \text{if } \delta \equiv \varepsilon \text{ and } j \in \{i-1, i\}, \\ 0 & \text{otherwise.} \end{cases}$$

Chapter 16
Appendix III: A Translation Functor
for $G = O(n+1, 1)$

In this chapter, we discuss a translation functor for the group $G = O(n+1,1)$, which is not in the Harish-Chandra class if n is even, in the sense that $\mathrm{Ad}(G)$ is not contained in the group $\mathrm{Int}(\mathfrak{g}_{\mathbb{C}})$ of inner automorphisms. Then the "Weyl group" W_G is larger than the group generated by the reflections of simple roots. This causes some technical difficulties when we extend the idea of translation functor which is usually formulated for reductive groups in the Harish-Chandra class or reductive Lie algebras, see [20, 53, 59, 65] for instance.

16.1 Some Features of Translation Functors for Reductive Groups that Are Not of Harish-Chandra Class

For n even, say $n = 2m$, we write $\mathfrak{h}_{\mathbb{C}}$ ($\simeq \mathbb{C}^{m+1}$) for a Cartan subalgebra of $\mathfrak{g}_{\mathbb{C}}$. Then we recall from Section 2.1.4:

- the Weyl group $W_{\mathfrak{g}} \simeq \mathfrak{S}_{m+1} \ltimes (\mathbb{Z}/2\mathbb{Z})^m$ for the root system $\Delta(\mathfrak{g}_{\mathbb{C}}, \mathfrak{h}_{\mathbb{C}})$ is of index two in the Weyl group $W_G \simeq \mathfrak{S}_{m+1} \ltimes (\mathbb{Z}/2\mathbb{Z})^{m+1}$ for the disconnected group G;
- the $\mathfrak{Z}_G(\mathfrak{g})$-infinitesimal character for the irreducible admissible representation of G is parametrized by $\mathfrak{h}_{\mathbb{C}}^*/W_G$, but not by $\mathfrak{h}_{\mathbb{C}}^*/W_{\mathfrak{g}}$;
- $\rho_G = (m, \cdots, 1, 0)$ is not "W_G-regular", although it is "$W_{\mathfrak{g}}$-regular" (Definition 2.1).

We can still use the idea of a translation functor, but we need a careful treatment for disconnected groups G which are not in the Harish-Chandra class. In fact, differently from the usual setting for reductive Lie groups in the Harish-Chandra class, we are faced with the following feature:

© Springer Nature Singapore Pte Ltd. 2018
T. Kobayashi, B. Speh, *Symmetry Breaking for Representations
of Rank One Orthogonal Groups II*, Lecture Notes in Mathematics 2234,
https://doi.org/10.1007/978-981-13-2901-2_16

- translation from a W_G-regular (in particular, $W_{\mathfrak{g}}$-regular) dominant parameter to the trivial infinitesimal character ρ_G does not necessarily preserve irreducibility, see Theorem 16.8.

This means that translation inside the same "$W_{\mathfrak{g}}$-regular Weyl chamber" may involve a phenomenon as if it were "translation from the wall to regular parameter", *cf.*, [53].

In what follows, we retain the terminology "regular" for $W_{\mathfrak{g}}$ but not for W_G as in Definition 2.1 (in particular, ρ_G is regular in our sense), whereas we need to use W_G (not $W_{\mathfrak{g}}$) in describing $\mathfrak{Z}_G(\mathfrak{g})$-infinitesimal characters of G-modules.

16.2 Translation Functor for $G = O(n+1, 1)$

In this section we fix some notation for a *translation functor* for the group $G = O(n+1, 1)$. Usually, a translation functor is defined in the category of $(\mathfrak{g}, K)$-modules of finite length. However, we also consider a translation functor in the category of **admissible representations of finite length of moderate growth**.

16.2.1 Primary Decomposition of Admissible Smooth Representations

Let Π be an admissible smooth representation of G of finite length. For $\mu \in \mathfrak{h}_{\mathbb{C}}^* / W_G$, we define the μ-primary component $P_\mu(\Pi)$ of Π by

$$P_\mu(\Pi) := \bigcup_{N>0} \bigcap_{z \in \mathfrak{Z}_G(\mathfrak{g})} \mathrm{Ker}(z - \chi_\mu(z))^N,$$

where we recall the Harish-Chandra isomorphism (2.15)

$$\mathrm{Hom}_{\mathbb{C}\text{-alg}}(\mathfrak{Z}_G(\mathfrak{g}), \mathbb{C}) \simeq \mathfrak{h}_{\mathbb{C}}^* / W_G, \quad \chi_\mu \leftrightarrow \mu.$$

Then $P_\mu(\Pi)$ is a G-module with generalized $\mathfrak{Z}_G(\mathfrak{g})$-infinitesimal character μ, and Π is decomposed into a direct sum of finitely many primary components:

$$\Pi = \bigoplus_\mu P_\mu(\Pi) \quad \text{(finite direct sum)}.$$

By abuse of notation, we use the letter P_μ to denote the G-equivariant projection $\Pi \to P_\mu(\Pi)$ with respect to the direct sum decomposition.

16.2.2 Translation Functor $\psi_\mu^{\mu+\tau}$ for $G = O(n+1, 1)$

Let $G = O(n+1, 1)$ and $m = [\frac{n}{2}]$. We recall that $W_G \simeq \mathfrak{S}_{m+1} \ltimes (\mathbb{Z}/2\mathbb{Z})^{m+1}$ acts on $\mathfrak{h}_\mathbb{C}^* \simeq \mathbb{C}^{m+1}$ as a permutation group and by switching the signatures of the standard coordinates. For $\tau \in \mathbb{Z}^{m+1}$, we define τ_{dom} to be the unique element in $\Lambda^+(m+1)$ (see (2.17)) in the W_G-orbit through τ, i.e.,

$$\tau_{\text{dom}} = w\,\tau \qquad \text{for some } w \in W_G. \tag{16.1}$$

Let $F^{O(n+1,1)}(\tau_{\text{dom}})_{+,+}$ be the irreducible finite-dimensional representation of $G = O(n+1, 1)$ of type I (Definition 14.2) defined as in (14.3).

Definition 16.1 (translation functor $\psi_\mu^{\mu+\tau}$). For $\mu \in \mathbb{C}^{m+1}$ and $\tau \in \mathbb{Z}^{m+1}$, we define the translation functor $\psi_\mu^{\mu+\tau}$ by

$$\psi_\mu^{\mu+\tau}(\Pi) := P_{\mu+\tau}(P_\mu(\Pi) \otimes F^{O(n+1,1)}(\tau_{\text{dom}})_{+,+}). \tag{16.2}$$

Then $\psi_\mu^{\mu+\tau}$ is a covariant functor in the category of admissible smooth representations of G of finite length, and also in the category of $(\mathfrak{g}, K)$-modules of finite length. Clearly, we have

$$\psi_\mu^{\mu+\tau} = \psi_{w\mu}^{w\mu+w\tau} \qquad \text{for all } w \in W_G. \tag{16.3}$$

In defining the translation functor $\psi_\mu^{\mu+\tau}$ in (16.2), we have used only finite-dimensional representations of type I (Definition 14.2) of the disconnected group $G = O(n+1, 1)$. We do not lose any generality because taking the tensor product with the one-dimensional characters χ_{ab} ($a, b \in \{\pm\}$) yields the following isomorphism as G-modules:

$$\psi_\mu^{\mu+\tau}(\Pi) \otimes \chi_{ab} \simeq P_{\mu+\tau}(P_\mu(\Pi) \otimes F^{O(n+1,1)}(\tau_{\text{dom}})_{a,b}). \tag{16.4}$$

We shall use a finite-dimensional representation $F(V, \lambda)$ (Definition 16.17) which is not necessarily of type I in Theorems 16.22 and 16.23, which are a reformulation of the properties (Theorems 16.6 and 16.8, respectively) of the translation functor (16.2) via (16.4).

The translation functor $\psi_{\mu+\tau}^\mu$ is the adjoint functor of $\psi_\mu^{\mu+\tau}$. In our setting, since $(-\tau)_{\text{dom}} = \tau_{\text{dom}}$, the functor $\psi_{\mu+\tau}^\mu$ takes the following form:

$$\psi_{\mu+\tau}^\mu(\Pi) = P_\mu(P_{\mu+\tau}(\Pi) \otimes F^{O(n+1,1)}(\tau_{\text{dom}})_{+,+}).$$

16.2.3   The Translation Functor and the Restriction $G \downarrow \overline{G}$

We retain the notation of Appendix II, and denote by $\overline{G}$ the subgroup $SO(n+1, 1)$ in $G = O(n+1, 1)$. Then $\overline{G} = SO(n+1, 1)$ is in the Harish-Chandra class for all n. For the group $\overline{G}$, we shall use the notation $\overline{P}_\mu$ and $\overline{\psi}_\mu^{\mu+\tau}$ instead of P_μ and $\psi_\mu^{\mu+\tau}$, respectively. To be precise, for $\tau \in \mathbb{Z}^{m+1}$ where $m = [\frac{n}{2}]$, we write $\overline{\tau}_{\text{dom}}$ for the unique element in the orbit $W_{\mathfrak{g}}\tau$ which is dominant with respect to the positive system $\Delta^+(\mathfrak{g}_{\mathbb{C}}, \mathfrak{h}_{\mathbb{C}})$. We denote by $F^{SO(n+1,1)}(\overline{\tau}_{\text{dom}})_+$ the irreducible representation of $\overline{G} = SO(n+1, 1)$ obtained by the restriction of the irreducible holomorphic representation of $SO(n+2, \mathbb{C})$ having $\overline{\tau}_{\text{dom}}$ as its highest weight. For an admissible smooth representation $\overline{\Pi}$ of $\overline{G}$ of finite length, the translation functor $\overline{\psi}_\mu^{\mu+\tau}$ is defined by

$$\overline{\psi}_\mu^{\mu+\tau}(\overline{\Pi}) := \overline{P}_{\mu+\tau}(\overline{P}_\mu(\overline{\Pi}) \otimes F^{SO(n+1,1)}(\overline{\tau}_{\text{dom}})_+). \tag{16.5}$$

We collect some basic facts concerning the primary components for G-modules and $\overline{G}$-modules. The following lemma is readily shown by comparing (2.15) of the Harish-Chandra isomorphisms for G and $\overline{G}$.

Lemma 16.2. *Let Π be an admissible smooth representation of finite length of $G = O(n+1, 1)$. We set $m := [\frac{n}{2}]$ as before. Suppose $\mu \in \mathfrak{h}_{\mathbb{C}}^* \simeq \mathbb{C}^{m+1}$.*

(1) *If n is odd or if n is even and at least one of the entries $\mu_1, \cdots, \mu_{m+1}$ is zero, then there is a natural isomorphism of $\overline{G}$-modules:*

$$P_\mu(\Pi)|_{\overline{G}} \simeq \overline{P}_\mu(\Pi|_{\overline{G}}).$$

(2) *If n is even and all of μ_j are nonzero, then we have a direct sum decomposition of a $\overline{G}$-module:*

$$P_\mu(\Pi)|_{\overline{G}} = \overline{P}_\mu(\Pi|_{\overline{G}}) \oplus \overline{P}_{\mu'}(\Pi|_{\overline{G}}),$$

where we set $\mu' := (\mu_1, \cdots, \mu_m, -\mu_{m+1})$.

Now the following lemma is an immediate consequence of Lemma 16.2 and of the definition of the translation functors $\psi_\mu^{\mu+\tau}$ and $\overline{\psi}_\mu^{\mu+\tau}$, see (16.2) and (16.5).

Lemma 16.3. *Let $G = O(n+1, 1)$ and $\overline{G} = SO(n+1, 1)$. Let Π be an admissible smooth representation of G of finite length.*

(1) *Suppose n is odd. Then we have a canonical $\overline{G}$-isomorphism:*

$$\psi_\mu^{\mu+\tau}(\Pi)|_{\overline{G}} \simeq \overline{\psi}_\mu^{\mu+\tau}(\Pi|_{\overline{G}}). \tag{16.6}$$

(2) *Suppose that n is even. If all of μ, τ and $\mu + \tau$ contain 0 in their entries, then we have a canonical $\overline{G}$-isomorphism (16.6).*

16.2.4 Some Elementary Properties of the Translation Functor $\psi_\mu^{\mu+\tau}$

Some of the properties of the translation functors remain true for the disconnected group $G = O(n+1, 1)$.

Proposition 16.4. *Suppose $\mu \in \mathfrak{h}_\mathbb{C}^* (\simeq \mathbb{C}^{m+1})$ and $\tau \in \mathbb{Z}^{m+1}$.*

(1) $\psi_\mu^{\mu+\tau}$ *is a covariant exact functor.*
(2) *Suppose μ and $\mu + \tau$ belong to the same Weyl chamber with respect to $W_\mathfrak{g}$. If $\mu + \tau$ is regular (Definition 2.1), then $\psi_\mu^{\mu+\tau}(\Pi)$ is nonzero if Π is nonzero.*

Proof.

(1) The first statement follows directly from the definition, see Zuckerman [65].
(2) By Lemma 16.2 and the branching law from $G = O(n+1, 1)$ to the subgroup $\overline{G} = SO(n+1, 1)$, we have

$$\psi_\mu^{\mu+\tau}(\Pi)|_{\overline{G}} \supset \overline{\psi}_\mu^{\mu+\tau}(\Pi|_{\overline{G}}).$$

Since $\overline{G}$ is in the Harish-Chandra class, $\overline{\psi}_\mu^{\mu+\tau}(\Pi|_{\overline{G}})$ is nonzero under the assumption on μ and τ. Hence $\psi_\mu^{\mu+\tau}(\Pi)$ is a nonzero G-module. $\qquad\square$

Remark 16.5. The regularity assumption for $\mu + \tau$ in Proposition 16.4 is in the weaker sense (*i.e.*, $W_\mathfrak{g}$-regular), and not in the stronger sense (*i.e.*, W_G-regular).

16.3 Translation of Principal Series Representation $I_\delta(V, \lambda)$

We discuss how the translation functors affect induced representations of $G = O(n+1, 1)$. We recall that G is not in the Harish-Chandra class when n is even.

16.3.1 Main Results: Translation of Principal Series Representations

Theorem 16.6. *Suppose $G = O(n+1, 1)$ and $(V, \lambda) \in \mathcal{R}ed$, see (14.9), or equivalently, $V \in \widehat{O(n)}$ and $\lambda \in \mathbb{Z} - (S(V) \cup S_Y(V))$, see Theorem 14.15. Let $i := i(V, \lambda) \in \{0, 1, \ldots, n\}$ be the height of (V, λ) as in (14.17), and $r(V, \lambda) \in \mathbb{Z}^{m+1}$ as* in (14.10). We write $V = F^{O(n)}(\sigma)_\varepsilon$ with $\sigma \in \Lambda^+(m)$ and $\varepsilon \in \{\pm\}$, where $m := [\frac{n}{2}]$. We define a character χ of G by

$$\chi \equiv \chi(V, \lambda) := \begin{cases} 1 & \text{if } \varepsilon(\frac{n}{2} - i) \geq 0, \\ \det & \text{if } \varepsilon(\frac{n}{2} - i) < 0. \end{cases} \tag{16.7}$$

Then there is a natural G-isomorphism:

$$\psi_{\rho^{(i)}}^{r(V, \lambda)}(I_\delta(i, i)) \otimes \chi \simeq I_{(-1)^{\lambda - i}\delta}(V, \lambda).$$

Remark 16.7. The conclusion of Theorem 16.6 does not change if we replace the definition (16.7) with $\chi = \det$ when $i = \frac{n}{2}$. In fact, V is of type Y if the height $i(V, \lambda)$ equals $\frac{n}{2}$, and thus $V \otimes \det \simeq V$ as $O(n)$-modules (Lemma 2.9). Then there is an isomorphism of G-modules

$$I_\delta(V, \lambda) \otimes \det \simeq I_\delta(V, \lambda)$$

for any $\delta \in \{\pm\}$ by Lemmas 2.14 and 14.28.

The translation functor $\psi_{r(V, \lambda)}^{\rho^{(i)}}$ is the adjoint functor of $\psi_{\rho^{(i)}}^{r(V, \lambda)}$. Even when the infinitesimal character of $I_\delta(V, \lambda)$ is W_G-regular (in particular, $W_{\mathfrak{g}}$-regular) (Definition 2.1), the translation functor $\psi_{r(V, \lambda)}^{\rho^{(i)}}$ does not always preserve *irreducibility* if G is not of Harish-Chandra class as we see in the following theorem.

Theorem 16.8. *Retain the setting and notation of Theorem 16.6. In particular, we recall that $(V, \lambda) \in \mathcal{Red}$, $i = i(V, \lambda)$ is the height of (V, λ), and $\chi \equiv \chi(V, \lambda)$, see (16.7).*

(1) *If $(V, \lambda) \in \mathcal{Red}_I$ (Definition 14.17), i.e., if V is of type X (in particular, if n is odd) or if $\lambda = \frac{n}{2}$, then there is a natural G-isomorphism:*

$$\psi_{r(V, \lambda)}^{\rho^{(i)}}(I_{(-1)^{\lambda - i}\delta}(V, \lambda)) \otimes \chi \simeq I_\delta(i, i).$$

(2) *If $(V, \lambda) \in \mathcal{Red}_{II}$, i.e., if V is of type Y and $\lambda \neq \frac{n}{2}$, then n is even, $i \neq \frac{n}{2}$ and there is a natural G-isomorphism:*

$$\psi_{r(V, \lambda)}^{\rho^{(i)}}(I_{(-1)^{\lambda - i}\delta}(V, \lambda)) \otimes \chi \simeq I_\delta(i, i) \oplus I_\delta(n - i, i).$$

In Section 16.4, we will introduce an irreducible finite-dimensional representation $F(V, \lambda)$ by taking the tensor product of $F^{O(n+1,1)}(\tau_{\text{dom}})_{+,+}$ with an appropriate one-dimensional character of G, see Definition 16.17. Then, the signatures in Theorems 16.6 and 16.8 can be reformulated in a simpler form by using $F(V, \lambda)$, see Theorems 16.22 and 16.23.

16.3.2 Strategy of the Proof for Theorems 16.6 and 16.8

If n is odd, then $G = \langle SO(n+1,1), -I_{2n+2}\rangle$ is in the Harish-Chandra class, and therefore Theorems 16.6 and 16.8 are a special case of the general theory, see [59, Chap. 7] for instance. Moreover, the translation functor behaves as we expect from the general theory for reductive groups in the Harish-Chandra class when it is applied to the induced representation $I_\delta(V,\lambda)$ if $(V,\lambda) \in \mathcal{R}ed_{\mathrm{I}}$, see Theorem 16.8 (1). We note that $\mathcal{R}ed = \mathcal{R}ed_{\mathrm{I}}$ and $\mathcal{R}ed_{\mathrm{II}} = \emptyset$ if n is odd (Remark 14.18).

On the other hand, its behavior is somewhat different if $(V,\lambda) \in \mathcal{R}ed_{\mathrm{II}}$, see Theorem 16.8 (2) and Proposition 16.31 for instance. Main technical complications arise from the fact that we need the primary decomposition for the generalized $\mathfrak{Z}_G(\mathfrak{g})$-infinitesimal characters parametrized by $\mathfrak{h}^*_{\mathbb{C}}/W_G$ where W_G is larger than the group generated by the reflections of simple roots if n is even, for which $G = O(n+1,1)$ is not in the Harish-Chandra class.

Our strategy is to use partly the relation of translation functors for $G = O(n+1,1)$ and the subgroup $\overline{G} = SO(n+1,1)$ which is in the Harish-Chandra class.

Theorem 16.6 is proved in Section 16.6 as a consequence of the following two propositions.

Proposition 16.9. *Suppose that $(V,\lambda) \in \mathcal{R}ed$. Retain the notation as in Theorem 16.6. Then the G-module $\psi^{r(V,\lambda)}_{\rho^{(i)}}(I_\delta(i,i)) \otimes \chi$ contains $I_{(-1)^{\lambda-i}\delta}(V,\lambda)$ as a subquotient. Equivalently, the G-module $P_{r(V,\lambda)}(I_\delta(i,i) \otimes F(V,\lambda))$, see Definition 16.17 below, contains $I_\delta(V,\lambda)$ as a subquotient.*

We recall from (16.7) that the character $\chi \equiv \chi(V,\lambda)$ is trivial when restricted to the subgroup $\overline{G} = SO(n+1,1)$.

Proposition 16.10. *Suppose that $(V,\lambda) \in \mathcal{R}ed$. Retain the notation as in Theorem 16.6. Then $\psi^{r(V,\lambda)}_{\rho^{(i)}}(I_\delta(i,i))|_{\overline{G}}$ is isomorphic to $I_{(-1)^{\lambda-i}\delta}(V,\lambda)|_{\overline{G}}$ as a $\overline{G}$-module.*

Similarly, Theorem 16.8 is proved in Section 16.7 by using analogous results, namely, Propositions 16.33 and 16.34 in Section 16.7.

16.3.3 Basic Lemmas for the Translation Functor

We use the following well-known lemma, which holds without the assumption that G is of Harish-Chandra class.

Lemma 16.11. *Let F be a finite-dimensional representation of G, $V \in \widehat{O(n)}$, $\delta \in \{\pm\}$, and $\lambda \in \mathbb{C}$. Then there is a G-stable filtration*

$$\{0\} = I_0 \subset I_1 \subset \cdots \subset I_k = I_\delta(V,\lambda) \otimes F$$

such that

$$I_j/I_{j-1} \simeq \mathrm{Ind}_P^G(V_{\lambda,\delta} \otimes F^{(j)}) \quad (1 \leq j \leq k)$$

where $F^{(j)}$ is a P-module such that the unipotent radical N_+ acts trivially and that $F^{(j)}|_{MA}$ is isomorphic to a subrepresentation of the restriction $F|_{MA}$ to the Levi subgroup MA.

For the sake of completeness, we give a proof.

Proof. Take a P-stable filtration

$$\{0\} = F_0 \subset F_1 \subset \cdots \subset F_k = F$$

such that the unipotent radical N_+ of P acts trivially on

$$F^{(j)} := F_j/F_{j-1} \quad (1 \leq j \leq k).$$

As in (2.25), we denote by $V_{\lambda,\delta}$ the irreducible P-module which is an extension of the MA-module $V \boxtimes \delta \boxtimes \mathbb{C}_\lambda$ with trivial N_+ action. We define G-modules I_j $(0 \leq j \leq k)$ by

$$I_j := \mathrm{Ind}_P^G(V_{\lambda,\delta} \otimes F_j|_P).$$

Then there is a natural filtration of G-modules

$$0 = I_0 \subset I_1 \subset \cdots \subset I_k = \mathrm{Ind}_P^G(V_{\lambda,\delta} \otimes F|_P)$$

such that

$$I_j/I_{j-1} \simeq \mathrm{Ind}_P^G(V_{\lambda,\delta} \otimes (F_j/F_{j-1}))$$

as G-modules. Since the finite-dimensional representation F of G is completely reducible when viewed as a representation of the Levi subgroup MA, the MA-module $F^{(j)} = F_j/F_{j-1}$ is isomorphic to a subrepresentation of the restriction $F|_{MA}$. Now Lemma 16.11 follows from the following G-isomorphism:

$$\mathrm{Ind}_P^G(V_{\lambda,\delta} \otimes F|_P) \simeq \mathrm{Ind}_P^G(V_{\lambda,\delta}) \otimes F. \qquad \square$$

Similarly to Lemma 16.11, we have the following lemma for cohomological parabolic induction. Retain the notation as in Section 14.9.1.

Lemma 16.12. *Suppose that $\mathfrak{q} = \mathfrak{l}_{\mathbb{C}} + \mathfrak{u}$ is a θ-stable parabolic subalgebra of $\mathfrak{g}_{\mathbb{C}}$ with Levi subgroup L, see (14.31), and that W a finite-dimensional $(\mathfrak{l}, L \cap K)$-module. Let F be a finite-dimensional representation of G, and*

$$\{0\} = F_0 \subset F_1 \subset \cdots \subset F_k = F$$

a $(\mathfrak{q}, L)$-*stable filtration such that the nilpotent radical* $\mathfrak{u}$ *acts trivially on* $F^{(j)} :=$ F_j/F_{j-1}. *Then there is a natural spectral sequence*

$$\mathcal{R}_{\mathfrak{q}}^p(W \otimes F^{(j)} \otimes \mathbb{C}_{\rho(\mathfrak{u})}) \Rightarrow \mathcal{R}_{\mathfrak{q}}^p(W \otimes F \otimes \mathbb{C}_{\rho(\mathfrak{u})}) \simeq \mathcal{R}_{\mathfrak{q}}^p(W \otimes \mathbb{C}_{\rho(\mathfrak{u})}) \otimes F$$

as $(\mathfrak{g}, K)$-*modules.*

The proof is similar to the case where G is in the Harish-Chandra class, see [59, Lem. 7.23].

By the definition (16.2) of the translation functor $\psi_\mu^{\mu+\tau}$, we need to estimate possible $\mathfrak{Z}_G(\mathfrak{g})$-infinitesimal characters of $\mathrm{Ind}_P^G(V_{\lambda,\delta} \otimes F^{(j)})$ in Lemma 16.11 or that of $\mathcal{R}_{\mathfrak{q}}^p(W \otimes F^{(j)} \otimes \mathbb{C}_{\rho(\mathfrak{u})})$ in Lemma 16.12.

In order to deal with reductive groups that are not in the Harish-Chandra class, we use the following lemma which is formulated in a slightly stronger form than [59, Lem. 7.2.18], but has the same proof.

Lemma 16.13. *Let* $\mathfrak{h}_\mathbb{C}$ *be a Cartan subalgebra of a complex semisimple Lie algebra* $\mathfrak{g}_\mathbb{C}$, $W_\mathfrak{g}$ *the Weyl group of the root system* $\Delta(\mathfrak{g}_\mathbb{C}, \mathfrak{h}_\mathbb{C})$, $\Delta^+(\mathfrak{g}_\mathbb{C}, \mathfrak{h}_\mathbb{C})$ *a positive system,* $\langle, \rangle$ *a* $W_\mathfrak{g}$-*invariant inner product on* $\mathfrak{h}_\mathbb{R}^* := \mathrm{Span}_\mathbb{R} \Delta(\mathfrak{g}_\mathbb{C}, \mathfrak{h}_\mathbb{C})$, *and* $|| \cdot ||$ *its norm.*
Suppose that ν *and* $\tau \in \mathfrak{h}_\mathbb{R}^*$ *satisfy*

$$\langle \nu, \alpha^\vee \rangle \in \mathbb{N}_+ \qquad (^\forall \alpha \in \Delta^+(\mathfrak{g}_\mathbb{C}, \mathfrak{h}_\mathbb{C})),$$

$$\langle \nu + \tau, \alpha^\vee \rangle \in \mathbb{N} \qquad (^\forall \alpha \in \Delta^+(\mathfrak{g}_\mathbb{C}, \mathfrak{h}_\mathbb{C})).$$

If $\gamma \in \mathfrak{h}_\mathbb{C}^*$ *satisfies the following two conditions:*

$$\nu + \gamma = w(\nu + \tau) \qquad \text{for some } w \in W_\mathfrak{g}, \tag{16.8}$$

$$||\gamma|| \leq ||\tau||, \tag{16.9}$$

then $\gamma = \tau$.

Remark 16.14. In [59, Lem. 7.2.18], γ is assumed to be a weight occurring in the irreducible finite-dimensional representation of G (in the Harish-Chandra class) with extremal weight τ instead of our assumption (16.9).

16.4   Definition of an Irreducible Finite-dimensional Representation $F(V, \lambda)$ of $G = O(n+1, 1)$

For $(V, \lambda) \in \mathcal{RInt}$, i.e., for $V \in \widehat{O(n)}$ and $\lambda \in \mathbb{Z} - S(V)$, we defined in Chapter 14

$i(V, \lambda) \in \{0, 1, \ldots, n\}$,   height of (V, λ) (Definition 14.26),

$r(V, \lambda) \in \mathbb{C}^{[\frac{n}{2}]+1}$,   giving the $\mathfrak{Z}_G(\mathfrak{g})$-infinitesimal character of $I_\delta(V, \lambda)$,

see (14.10).

In this section we introduce an irreducible finite-dimensional representation $F(V, \lambda)$ of $G = O(n+1,1)$ which contains important information on signatures.

16.4.1   Definition of $\sigma^{(i)}(\lambda)$ and $\widehat{\sigma^{(i)}}$

We begin with some combinatorial notation.

Definition 16.15. Let $m := [\frac{n}{2}]$. For $1 \le i \le n$, $\sigma = (\sigma_1, \cdots, \sigma_m) \in \Lambda^+(m)$, and $\lambda \in \mathbb{Z}$, we define $\sigma^{(i)}(\lambda) \in \mathbb{Z}^{m+1}$ as follows.

Case 1. $n = 2m$

$$\sigma^{(i)}(\lambda) := \begin{cases} (\sigma_1 - 1, \cdots, \sigma_i - 1, i - \lambda, \sigma_{i+1}, \cdots, \sigma_m) & \text{for } 0 \le i \le m-1, \\ (\sigma_1 - 1, \cdots, \sigma_m - 1, |\lambda - m|) & \text{for } i = m, \\ (\sigma_1 - 1, \cdots, \sigma_{n-i} - 1, \lambda - i, \sigma_{n-i+1}, \cdots, \sigma_m) & \text{for } m+1 \le i \le n. \end{cases}$$

Case 2. $n = 2m + 1$

$$\sigma^{(i)}(\lambda) := \begin{cases} (\sigma_1 - 1, \cdots, \sigma_i - 1, i - \lambda, \sigma_{i+1}, \cdots, \sigma_m) & \text{for } 0 \le i \le m, \\ (\sigma_1 - 1, \cdots, \sigma_{n-i} - 1, \lambda - i, \sigma_{n-i+1}, \cdots, \sigma_m) & \text{for } m+1 \le i \le n. \end{cases}$$

Moreover we define $\widehat{\sigma^{(i)}} \in \mathbb{Z}^m$ to be the vector obtained by removing the $\min(i + 1, n - i + 1)$-th component from $\sigma^{(i)}(\lambda) \in \mathbb{Z}^{m+1}$.

Case 1. $n = 2m$

$$\widehat{\sigma^{(i)}} := \begin{cases} (\sigma_1 - 1, \cdots, \sigma_i - 1, \sigma_{i+1}, \cdots, \sigma_m) & \text{for } 0 \le i \le m-1, \\ (\sigma_1 - 1, \cdots, \sigma_m - 1) & \text{for } i = m, \\ (\sigma_1 - 1, \cdots, \sigma_{n-i} - 1, \sigma_{n-i+1}, \cdots, \sigma_m) & \text{for } m+1 \le i \le n. \end{cases}$$

Case 2. $n = 2m + 1$

$$\widehat{\sigma^{(i)}} := \begin{cases} (\sigma_1 - 1, \cdots, \sigma_i - 1, \sigma_{i+1}, \cdots, \sigma_m) & \text{for } 0 \le i \le m, \\ (\sigma_1 - 1, \cdots, \sigma_{n-i} - 1, \sigma_{n-i+1}, \cdots, \sigma_m) & \text{for } m+1 \le i \le n. \end{cases}$$

Definition-Lemma 16.16. Let $m := [\frac{n}{2}]$. For $(V, \lambda) \in \mathcal{RInt}$, i.e., for $V \in \widehat{O(n)}$ and $\lambda \in \mathbb{Z} - S(V)$, we write $V = F^{O(n)}(\sigma)_\varepsilon$ with $\sigma \in \Lambda^+(m)$ and $\varepsilon \in \{\pm\}$. We set

$$\sigma(\lambda) := \sigma^{(i)}(\lambda), \tag{16.10}$$

where $i := i(V, \lambda) \in \{0, 1, \ldots, n\}$ is the height of (V, λ) as in (14.17). Then we have

$$\sigma(\lambda) \in \Lambda^+(m + 1).$$

Proof. Suppose $n = 2m$ (even). Let $\lambda \in \mathbb{Z}$. By the definition of $R(V; i)$ (Definition 14.23), we have the following equivalences:

- for $0 \le i \le m - 1$,

$$\lambda \in R(V; i) \Leftrightarrow \sigma_i - i > -\lambda > \sigma_{i+1} - i - 1$$

$$\Leftrightarrow \sigma_i - 1 \ge i - \lambda \ge \sigma_{i+1};$$

- for $i = m$,

$$\lambda \in R(V; m) \Leftrightarrow -\sigma_m < \lambda - m < \sigma_m$$

$$\Leftrightarrow \sigma_m - 1 \ge |\lambda - m|;$$

- for $m + 1 \le i \le n$,

$$\lambda \in R(V; i) \Leftrightarrow \sigma_{n-i+1} - 1 < \lambda - i < \sigma_{n-i}$$

$$\Leftrightarrow \sigma_{n-i} - 1 \ge \lambda - i \ge \sigma_{n-i+1}.$$

Thus in all cases $\sigma^{(i)}(\lambda) \in \Lambda^+(m + 1)$.

The proof for n odd is similar.   □

16.4.2 Definition of a Finite-dimensional Representation $F(V, \lambda)$ of G

We are ready to define a finite-dimensional representation, to be denoted by $F(V, \lambda)$, for $(V, \lambda) \in \mathcal{RInt}$.

Definition 16.17 (a finite-dimensional representation $F(V, \lambda)$)**.** Suppose that $(V, \lambda) \in \mathcal{RInt}$, i.e., $V \in \widehat{O(n)}$ and $\lambda \in \mathbb{Z} - S(V)$. We write $V = F^{O(n)}(\sigma)_\varepsilon$ with $\sigma \in \Lambda^+(m)$ and $\varepsilon \in \{\pm\}$ where $m := [\frac{n}{2}]$. We set $i := i(V, \lambda)$, the height of (V, λ) as in (14.17), and $\sigma(\lambda) \in \Lambda^+(m + 1)$ as in Definition-Lemma 16.16.

We define an irreducible finite-dimensional representation $F(V, \lambda)$ of $G = O(n+1,1)$ as follows:

- for V of type Y and $\lambda = \frac{n}{2} (= m)$,

$$F(V, \lambda) := F^{O(n+1,1)}(\sigma(\lambda))_{+,+}$$
$$= F^{O(n+1,1)}(\sigma_1 - 1, \cdots, \sigma_m - 1, 0)_{+,+};$$

- for V of type X or $\lambda \neq \frac{n}{2}$,

$$F(V, \lambda) := \begin{cases} F^{O(n+1,1)}(\sigma(\lambda))_{\varepsilon,(-1)^{\lambda-i}\varepsilon} & \text{if } i \leq \frac{n}{2}, \\ F^{O(n+1,1)}(\sigma(\lambda))_{-\varepsilon,(-1)^{\lambda-i-1}\varepsilon} & \text{if } i > \frac{n}{2}, \end{cases} \qquad (16.11)$$

see (14.5) for notation.

By using the character $\chi \equiv \chi(V, \lambda)$ of G as defined in (16.7), we obtain a unified expression

$$F(V, \lambda) \simeq F(\sigma(\lambda))_{+,(-1)^{\lambda-i}} \otimes \chi. \qquad (16.12)$$

Remark 16.18. We note that (16.11) is well-defined. In fact, if V is of type Y (Definition 2.6), then ε is not uniquely determined because there are two expressions for V:

$$V \simeq F^{O(n)}(\sigma)_+ \simeq F^{O(n)}(\sigma)_-,$$

see Lemma 14.4 (1). On the other hand, the $(m+1)$-th component of $\sigma(\lambda)$ does not vanish except for the case $i = \lambda = m$ by Definition 16.15. Hence we obtain an isomorphism of $O(n+1,1)$-modules:

$$F^{O(n+1,1)}(\sigma(\lambda))_{a,b} \simeq F^{O(n+1,1)}(\sigma(\lambda))_{-a,-b}$$

for any $a, b \in \{\pm\}$ by Lemma 14.4 (2).

By Definition 14.2, the following lemma is clear.

Lemma 16.19. *Suppose that V is of type X or $\lambda \neq \frac{n}{2}$. Then there is a natural isomorphism of $O(n+1,1)$-modules:*

$$F(V \otimes \det, \lambda) \simeq F(V, \lambda) \otimes \det.$$

Lemma 16.20. *The following two conditions on $(V, \lambda) \in \mathcal{RInt}$ (i.e., $V \in \widehat{O(n)}$ and $\lambda \in \mathbb{Z} - S(V)$) are equivalent:*

(i) $F(V, \lambda) \otimes \det \simeq F(V, \lambda)$ *as G-modules;*
(ii) V *is of type Y (Definition 2.6) and $\lambda \neq \frac{n}{2}$.*

In particular, for $(V, \lambda) \in \mathcal{R}ed$, *(i) holds if and only if* $(V, \lambda) \in \mathcal{R}ed_{\mathrm{II}}$ *(Definition 14.17)*.

Proof. Any of the conditions (i) or (ii) implies that n is even, say, $n = 2m$. Let us verify (ii) $\Rightarrow$ (i). If we write $V = F^{O(n)}(\sigma)_{\varepsilon}$ for some $\sigma = (\sigma_1, \cdots, \sigma_m) \in \Lambda^+(m)$ and $\varepsilon \in \{\pm\}$, then $\sigma_m \neq 0$ because V is of type Y. On the other hand, the height $i := i(V, \lambda)$ is not equal to m because $\lambda \neq m$, hence the $(m+1)$-th component of $\sigma^{(i)}(\lambda)$ equals $\sigma_m (\neq 0)$ by Definition 16.15. Thus there is a natural G-isomorphism $F(V, \lambda) \otimes \det \simeq F(V, \lambda)$. The converse implication is similarly verified. $\qquad \square$

Example 16.21. Let $(V, \lambda) = (\bigwedge^{\ell}(\mathbb{C}^n), \ell)$ for $\ell = 0, 1, \cdots, n$. We set $m = [\frac{n}{2}]$ as usual. Then

$$i(V, \lambda) = \ell, \ \sigma(\lambda) = 0 (\in \mathbb{Z}^{m+1}), \text{ and } \widehat{\sigma^{(i)}} = 0 \in \mathbb{Z}^m.$$

Moreover, we have an isomorphism of G-modules:

$$F(V, \lambda) \simeq \mathbf{1} \quad \text{for } 0 \leq \ell \leq n.$$

16.4.3 Reformulation of Theorems 16.6 and 16.8

By using the finite-dimensional representation $F(V, \lambda)$ of $G = O(n+1, 1)$ (Definition 16.17), Theorems 16.6 and 16.8 may be reformulated in simpler forms, respectively, as follows.

Theorem 16.22. *For* $(V, \lambda) \in \mathcal{R}ed$ *(Definition 14.8), we set* $i := i(V, \lambda)$, *the height of* (V, λ) *as in* (14.17). *Then there is a natural* G-*isomorphism:*

$$P_{r(V, \lambda)}(I_{\delta}(i, i) \otimes F(V, \lambda)) \simeq I_{\delta}(V, \lambda).$$

Theorem 16.23. *Suppose* $(V, \lambda) \in \mathcal{R}ed$. *Retain the notation as in Theorem 16.22.*

(1) *If* $(V, \lambda) \in \mathcal{R}ed_{\mathrm{I}}$ *(Definition 14.17), then there is a natural* G-*isomorphism:*

$$P_{r(V, \lambda)}(I_{\delta}(V, \lambda) \otimes F(V, \lambda)) \simeq I_{\delta}(i, i).$$

(2) *If* $(V, \lambda) \in \mathcal{R}ed_{\mathrm{II}}$, *then there is a natural* G-*isomorphism:*

$$P_{r(V, \lambda)}(I_{\delta}(V, \lambda) \otimes F(V, \lambda)) \simeq I_{\delta}(i, i) \oplus I_{\delta}(n - i, i).$$

16.4.4   Translation of Irreducible Representations $\Pi_{\ell,\delta}$

We recall from (2.35) that $\Pi_{\ell,\delta}$ $(0 \leq \ell \leq n+1,\ \delta \in \{\pm\})$ are irreducible admissible smooth representations of G with trivial infinitesimal character ρ_G, and from (14.28) that $\Pi_\delta(V, \lambda)$ is an irreducible admissible smooth representation of G with $\mathfrak{Z}_G(\mathfrak{g})$-infinitesimal character $r(V, \lambda) \mod W_G$. We also recall that $\rho^{(i)} \equiv \rho_G \mod W_G$ for all $0 \leq i \leq n$. In this section, we determine the action of translation functor $\psi_{\rho^{(i)}}^{r(V,\lambda)}$ on irreducible representations.

Theorem 16.24. *Suppose that $(V, \lambda) \in \mathcal{R}ed$. Let $i := i(V, \lambda)$ be the height of (V, λ), and $F(V, \lambda)$ be the irreducible finite-dimensional representation of G (Definition 16.17). Then there is a natural G-isomorphism:*

$$P_{r(V,\lambda)}(\Pi_{i,\delta} \otimes F(V, \lambda)) \simeq \Pi_\delta(V, \lambda).$$

Proof. Since the translation functor is a covariant exact functor (Proposition 16.4 (1)), the exact sequence of G-modules

$$0 \to \Pi_{i,\delta} \to I_\delta(i, i) \to \Pi_{i+1,-\delta} \to 0$$

(Theorem 2.20 (1)) yields an exact sequence of G-modules

$$0 \to P_{r(V,\lambda)}(\Pi_{i,\delta} \otimes F(V, \lambda)) \to I_\delta(V, \lambda) \to P_{r(V,\lambda)}(\Pi_{i+1,-\delta} \otimes F(V, \lambda)) \to 0,$$

where we have used Theorem 16.22 for the middle term. Since the first and third terms do not vanish by Proposition 16.4 (2), we conclude the following isomorphisms of G-modules:

$$I_\delta(V, \lambda)^\flat \simeq P_{r(V,\lambda)}(\Pi_{i,\delta} \otimes F(V, \lambda)),$$

$$I_\delta(V, \lambda)^\sharp \simeq P_{r(V,\lambda)}(\Pi_{i+1,-\delta} \otimes F(V, \lambda))$$

because $I_\delta(V, \lambda)$ has composition series of length two (Corollary 14.22). Hence Theorem 16.24 follows from the definition (14.28) of $\Pi_\delta(V, \lambda)$. $\square$

16.4.5   Proof of Theorems 16.22 and 16.23

In this subsection, we explain that Theorem 16.6 is equivalent to Theorem 16.22; Theorem 16.8 is equivalent to Theorem 16.23.

For this we begin with the following lemma which clarifies some combinatorial meaning of the height $i(V, \lambda) \in \{0, 1, \ldots, n\}$ and the dominant integral weight $\sigma(\lambda) \in \Lambda^+(m+1)$ in Definition 16.17. Here we recall $m = [\frac{n}{2}]$.

Lemma 16.25. *Suppose* $V = F^{O(n)}(\sigma)_\varepsilon$ *with* $\sigma \in \Lambda^+(m)$ *and* $\varepsilon \in \{\pm\}$. *For* $0 \leq i \leq n$ *and* $\lambda \in \mathbb{Z}$, *we set*

$$\tau^{(i)}(V, \lambda) := r(V, \lambda) - \rho^{(i)} \in (\frac{1}{2}\mathbb{Z})^{m+1}, \tag{16.13}$$

see (14.10) *and Example 14.9 for the notation.*

(1) *Then* $\tau^{(i)}(V, \lambda) \in \mathbb{Z}^{m+1}$ *is given by*

$$\begin{cases} (\sigma_1 - 1, \cdots, \sigma_i - 1, \sigma_{i+1}, \cdots, \sigma_m, \lambda - i) & \text{for } 0 \leq i \leq m, \\ (\sigma_1 - 1, \cdots, \sigma_{n-i} - 1, \sigma_{n-i+1}, \cdots, \sigma_m, \lambda - i) & \text{for } m+1 \leq i \leq n. \end{cases}$$

(2) *Assume that* $\lambda \in \mathbb{Z} - S(V)$, *and we take* i *to be the height* $i(V, \lambda)$ *of* (V, λ) *as in* (14.17). *Then,*

$$r(V, \lambda) \text{ and } \rho^{(i)} \text{ belong to the same Weyl chamber for } W_{\mathfrak{g}}.$$

(3) *Let* $\sigma(\lambda)$ *be as defined in Definition 16.17. Then we have*

$$\tau^{(i)}(V, \lambda)_{\text{dom}} = \sigma(\lambda). \tag{16.14}$$

Proof. (1) Clear from the definition (14.10) of $r(V, \lambda)$ and $\rho^{(i)}$.
(2) The assertion is verified by inspecting the definition (14.17) of the height $i(V, \lambda)$.
(3) The statement follows from Definition-Lemma 16.16. □

Now we determine the action of the translation functor $\psi_{\rho^{(i)}}^{r(V,\lambda)}$. We recall that the principal series representation $I_\delta(i, i)$ $(0 \leq i \leq n)$ has the trivial $\mathfrak{Z}_G(\mathfrak{g})$-infinitesimal character, which is $W_{\mathfrak{g}}$-regular but not always W_G-regular. We apply the translation functor (16.2) to $I_\delta(i, i)$ for an appropriate choice of i.

Proposition 16.26. *Let* $m = [\frac{n}{2}]$. *Suppose* $G = O(n+1, 1)$, $\delta \in \{\pm\}$, $V = F^{O(n)}(\sigma)_\varepsilon$ *with* $\sigma \in \Lambda^+(m)$ *and* $\varepsilon \in \{\pm\}$, *and* $\lambda \in \mathbb{Z} - (S(V) \cup S_Y(V))$. *Let* $i := i(V, \lambda) \in \{0, 1, \ldots, n\}$ *be as in* (14.17). *We define* $r(V, \lambda) \in \mathbb{C}^{m+1}$ *as in* (14.10) *and* $\sigma(\lambda) \in \Lambda^+(m+1)$. *Then we have*

$$\psi_{\rho^{(i)}}^{r(V,\lambda)}(I_\delta(i, i)) = P_{r(V,\lambda)}(I_\delta(i, i) \otimes F^{O(n+1,1)}(\sigma(\lambda))_{+,+}).$$

Proof. Since $I_\delta(i, i)$ has the trivial $\mathfrak{Z}_G(\mathfrak{g})$-infinitesimal character, $P_{\rho^{(i)}}(I_\delta(i, i)) = I_\delta(i, i)$ by (14.12). Since $r(V, \lambda) = \rho^{(i)} + \tau^{(i)}(V, \lambda)$ by (16.13), and since $\sigma(\lambda) = \tau^{(i)}(V, \lambda)_{\text{dom}}$ by (16.14), the definition of the translation functor shows

$$\psi_{\rho^{(i)}}^{\rho^{(i)} + \tau^{(i)}(V,\lambda)}(I_\delta(i, i)) = P_{r(V,\lambda)}(I_\delta(i, i) \otimes F^{O(n+1,1)}(\sigma(\lambda))_{+,+}).$$

Thus Proposition 16.26 is proved. □

It follows from Proposition 16.26 and from the definition of $F(V, \lambda)$ (Definition 16.17) that Theorem 16.6 is equivalent to Theorems 16.22 and 16.8 is equivalent to Theorem 16.23.

16.5   Proof of Proposition 16.9

In this section we complete the proof of Proposition 16.9. By Lemma 16.11, the proof reduces to some branching laws for the restriction of finite-dimensional representations of $G = O(n+1, 1)$ with respect to $MA \simeq O(n) \times SO(1, 1)$ and to the study of their tensor product representations, see Proposition 16.29.

16.5.1   Irreducible Summands for $O(n+2) \downarrow O(n) \times O(2)$ and for Tensor Product Representations

Before working with Proposition 16.29 in the noncompact setting, we first discuss analogous branching rules for the restriction with respect to a pair of compact groups $O(n+2) \supset O(n) \times O(2)$:

Lemma 16.27 ($O(n+2) \downarrow O(n) \times O(2)$). *Let* $\mu = (\mu_1, \cdots, \mu_{m+1}) \in \Lambda^+(m+1)$, *where* $m := [\frac{n}{2}]$ *as before. For* $1 \le k \le m+1$, *we set*

$$\mu'_{(k)} := (\mu_1, \cdots, \mu_{k-1}, \widehat{\mu_k}, \mu_{k+1}, \cdots, \mu_{m+1}) \in \Lambda^+(m).$$

Then the $O(n+2)$-module $F^{O(n+2)}(\mu)_+$ (see (14.3)) contains the $(O(n) \times O(2))$-module

$$\bigoplus_{k=1}^{m+1} F^{O(n)}(\mu'_{(k)})_+ \boxtimes F^{O(2)}(\mu_k)_+$$

when restricted to the subgroup $O(n) \times O(2)$.

Proof. Take a Cartan subalgebra $\mathfrak{h}_{\mathbb{C}}$ of $\mathfrak{gl}(n+2, \mathbb{C})$ such that $\mathfrak{h}_{\mathbb{C}} \cap \mathfrak{o}(n+2, \mathbb{C})$ is a Cartan subalgebra of $\mathfrak{o}(n+2, \mathbb{C})$. We identify $\mathfrak{h}_{\mathbb{C}}^*$ with $\mathbb{C}^{n+1}$ via the standard basis $\{f_j\}$ as before, and choose a positive system $\Delta^+(\mathfrak{gl}(n+2, \mathbb{C}), \mathfrak{h}_{\mathbb{C}}) = \{f_i - f_j : 1 \le i < j \le n+2\}$. Then

$$\widetilde{\mu} := (\mu_1, \cdots, \mu_{m+1}, 0^{n+1-m}) \in \Lambda^+(n+2)$$

is a dominant integral with respect to the positive system. Let $v_{\widetilde{\mu}}$ be a (nonzero) highest weight vector of the irreducible representation $(\tau, F^{U(n+2)}(\widetilde{\mu}))$ of the unitary group $U(n+2)$. By definition, the $O(n+2)$-module $F^{O(n+2)}(\mu)_+$, see (14.3), is the unique irreducible $O(n+2)$-summand of $F^{U(n+2)}(\widetilde{\mu})$ containing the highest weight vector $v_{\widetilde{\mu}}$. We now take a closer look at the $U(n+2)$-module $F^{U(n+2)}(\widetilde{\mu})$. Fix $1 \leq k \leq m+1$. Iterating the classical branching rule for $U(N) \supset U(N-1) \times U(1)$ for $N = n+2$, $n+1$, we see that the restriction $F^{U(n+2)}(\widetilde{\mu})|_{U(n) \times U(2)}$ contains

$$F^{U(n)}(\widetilde{\mu'_{(k)}}) \boxtimes W$$

as an irreducible summand, where

$$\widetilde{\mu'_{(k)}} := (\mu_1, \cdots, \widehat{\mu_k}, \cdots, \mu_{m+1}, 0^{n-m}) \in \Lambda^+(n)$$

and W is an irreducible representation of $U(2)$ which has a weight $(\mu_k, 0)$. Since all the weights of an irreducible finite-dimensional representation are contained in the convex hull of the Weyl group orbit through the highest weight, we conclude that $(\mu_k, 0)$ is actually the highest weight of the $U(2)$-module W. Hence the $(U(n) \times U(2))$-module

$$F^{U(n)}(\widetilde{\mu'_{(k)}}) \boxtimes F^{U(2)}(\mu_k, 0)$$

occurs as an irreducible summand of the $U(n+2)$-module $F^{U(n+2)}(\widetilde{\mu})$. We now consider the following diagram of subgroups of $U(n+2)$, and investigate the restriction of the $U(n+2)$-module $F^{U(n+2)}(\widetilde{\mu})$.

$$U(n+2) \supset U(n) \times U(2)$$
$$\cup \qquad\qquad \cup$$
$$O(n+2) \supset O(n) \times O(2)$$

By our choice of the Cartan subalgebra $\mathfrak{h}_{\mathbb{C}}$, we observe that there exists $w_k \in O(n+2)$ such that $\mathrm{Ad}(w_k)\mathfrak{h}_{\mathbb{C}} = \mathfrak{h}_{\mathbb{C}}$ and

$$w_k\widetilde{\mu} = (\mu_1, \cdots, \widehat{\mu_k}, \cdots, \mu_{m+1}, 0^{n-m}, \mu_k, 0) \in \mathbb{Z}^{n+2},$$

where we write $w_k\widetilde{\mu}$ simply for the contragredient action of $\mathrm{Ad}(w_k)$ on $\widetilde{\mu} \in \mathfrak{h}_{\mathbb{C}}^*$ ($\simeq \mathbb{C}^{n+2}$). In particular, the $O(n+2)$-submodule $F^{O(n+2)}(\mu)_+$ of the restriction $F^{U(n+2)}(\widetilde{\mu})|_{O(n+2)}$ contains the weight vector $v_{w_k\widetilde{\mu}} := \tau(w_k)v_{\widetilde{\mu}}$ for the weight $w_k\widetilde{\mu}$. Since $w_k\widetilde{\mu}$ is an extremal weight, the weight vector in $F^{U(n+2)}(\widetilde{\mu})|_{O(n+2)}$ is unique up to scalar multiplication. Hence $v_{w_k\widetilde{\mu}}$ is contained also in the submodule $F^{U(n)}(\widetilde{\mu'_{(k)}}) \boxtimes F^{U(2)}(\mu_k, 0)$. Thus we conclude that the irreducible $O(n+2)$-module $F^{O(n+2)}(\mu)_+$ contains

$$F^{O(n)}(\mu'_{(k)}) \boxtimes F^{O(2)}(\mu_k)$$

as an $(O(n) \times O(2))$-summand when restricted to the subgroup $O(n) \times O(2)$ of $O(n+2)$.   □

Let $m = [\frac{n}{2}]$ as before. Let $V = F^{O(n)}(\sigma)_\varepsilon$ with $\sigma \in \Lambda^+(m)$ and $\varepsilon \in \{\pm\}$. Suppose $\lambda \in \mathbb{Z} - S(V)$. We recall from Definition-Lemma 16.16 that $\sigma(\lambda) \in \Lambda^+(m+1)$.

Lemma 16.28. *For $0 \le i \le n$, the following $(O(n) \times O(2))$-module*

$$(\textstyle\bigwedge^i(\mathbb{C}^n) \boxtimes \mathbf{1}) \otimes F^{O(n+2)}(\sigma(\lambda))_\varepsilon |_{O(n) \times O(2)}$$

contains

$$V \boxtimes F^{O(2)}(|\lambda - i|)_+ \qquad\qquad if\ i \le \frac{n}{2}$$

$$(V \otimes \det) \boxtimes F^{O(2)}(|\lambda - i|)_+ \qquad if\ i \ge \frac{n}{2}$$

as an irreducible summand.

We note that $V \simeq V \otimes \det$ as $O(n)$-modules if $i = \frac{n}{2}$ by Lemmas 2.9 and 14.28.

Proof. It suffices to prove Lemma 16.28 for $\varepsilon = +$ by using a similar argument to (3.22) for the pair $(O(n+2), O(n) \times O(2))$ and for $\chi = \det$. Then Lemma 16.28 is derived from the following two branching laws of compact Lie groups.

- $O(n+2) \downarrow O(n) \times O(2)$:
  By Lemma 16.27, the $O(n+2)$-module $F^{O(n+2)}(\sigma(\lambda))_+$ contains

  $$F^{O(n)}(\widehat{\sigma^{(i)}})_+ \boxtimes F^{O(2)}(|\lambda - i|)$$

  as an irreducible summand when restricted to the subgroup $O(n) \times O(2)$, see Definition 16.15 for the notation $\widehat{\sigma^{(i)}}$.
- Tensor product for $O(n)$:
  The tensor product representation

  $$\textstyle\bigwedge^i(\mathbb{C}^n) \otimes F^{O(n)}(\widehat{\sigma^{(i)}})_+$$

  contains

  $$V \simeq F^{O(n)}(\sigma)_+ \qquad if\ i \le \frac{n}{2},$$

  $$V \otimes \det \simeq F^{O(n)}(\sigma)_- \qquad if\ i \ge \frac{n}{2}$$

  as an irreducible component.   □

16.5.2 Irreducible Summand for the Restriction $G \downarrow MA$ and for Tensor Product Representations

We recall that the Levi subgroup MA of the parabolic subgroup P in $G = O(n + 1, 1)$ is expressed as

$$MA \simeq O(n) \times SO(1, 1) \simeq O(n) \times \mathbb{Z}/2\mathbb{Z} \times \mathbb{R}.$$

The goal of the subsection is to prove the following proposition.

Proposition 16.29 (tensor product and the restriction $O(n + 1, 1) \downarrow MA$). *Suppose that $(V, \lambda) \in \mathcal{R}ed$, i.e., $V \in \widehat{O(n)}$ and $\lambda \in \mathbb{Z} - (S(V) \cup S_Y(V))$. Let $i = i(V, \lambda)$ be the height of (V, λ) (see (14.17)), and $F(V, \lambda)$ be the irreducible $O(n+1, 1)$-module as in Definition 16.17. Then the MA-module*

$$(\textstyle\bigwedge^i(\mathbb{C}^n) \boxtimes \delta \boxtimes \mathbb{C}_i) \otimes F(V, \lambda)|_{MA} \tag{16.15}$$

contains

$$V \boxtimes \delta \boxtimes \mathbb{C}_\lambda$$

as an irreducible component.

In what follows, we use a mixture of notations in describing irreducible finite-dimensional representations (see Sections 2.2 and 14.1). To be precise, we shall use:

- $\Lambda^+(O(n + 2))$ ($\subset \mathbb{Z}^{n+2}$), see (2.20), to denote irreducible holomorphic finite-dimensional representations of the *complex* Lie group $O(n + 1, \mathbb{C})$ as in Section 2.2;
- $\Lambda^+(m + 1)$ ($\subset \mathbb{Z}^{m+1}$) and signatures to denote irreducible finite-dimensional representations of the *real* groups $O(n + 2)$ and $O(n + 1, 1)$ where $m := [\frac{n}{2}]$, as in Section 14.1.

 See (14.3) for the relationship among these representations.

Proof of Proposition 16.29. We write $V = F^{O(n)}(\sigma)_\varepsilon$ as before where $\sigma \in \Lambda^+(m)$, $\varepsilon \in \{\pm\}$, and $m = [\frac{n}{2}]$. By Weyl's unitary trick for the disconnected group $O(n + 1, 1)$, see (14.3), the restrictions of the holomorphic representation $F^{O(n+2, \mathbb{C})}(\sigma(\lambda), 0^{n+1-m})$ to the subgroups $O(n + 2)$ and $O(n + 1, 1)$ are given respectively by

$$F^{O(n+2)}(\sigma(\lambda))_+,$$

$$F^{O(n+1, 1)}(\sigma(\lambda))_{+, +}.$$

Then Lemma 16.28 implies that the holomorphic $(O(n,\mathbb{C}) \times O(2,\mathbb{C}))$-representation

$$(\textstyle\bigwedge^i(\mathbb{C}^n) \boxtimes \mathbf{1}) \otimes F^{O(n+2,\mathbb{C})}(\sigma(\lambda), 0^{n+1-m})|_{O(n,\mathbb{C}) \times O(2,\mathbb{C})}$$

contains

$$F^{O(n,\mathbb{C})}(\sigma, 0^{n-m}) \boxtimes F^{O(2,\mathbb{C})}(|\lambda - i|, 0) \qquad \text{if } i \leq \frac{n}{2},$$

$$(F^{O(n,\mathbb{C})}(\sigma, 0^{n-m}) \otimes \det) \boxtimes F^{O(2,\mathbb{C})}(|\lambda - i|, 0) \quad \text{if } i \geq \frac{n}{2}$$

as an irreducible summand. Because the restriction of the first factor to compact real form $O(n)$ is isomorphic to $F^{O(n)}(\sigma)_+$ or $F^{O(n)}(\sigma)_-$ according to whether $i \leq \frac{n}{2}$ or $i \geq \frac{n}{2}$. Taking the restriction to another real form $O(n) \times O(1,1)$ of $O(n,\mathbb{C}) \times O(2,\mathbb{C})$, we see that the $(O(n) \times O(1,1))$-module

$$(\textstyle\bigwedge^i(\mathbb{C}^n) \boxtimes \mathbf{1}) \otimes F^{O(n+1,1)}(\sigma(\lambda))_{+,+}|_{O(n) \times O(1,1)}$$

contains

$$F^{O(n)}(\sigma)_+ \boxtimes F^{O(2,\mathbb{C})}(|\lambda - i|, 0)|_{O(1,1)} \quad \text{if } i \leq \frac{n}{2},$$

$$F^{O(n)}(\sigma)_- \boxtimes F^{O(2,\mathbb{C})}(|\lambda - i|, 0)|_{O(1,1)} \quad \text{if } i \geq \frac{n}{2}$$

as an irreducible summand.

Since $V = F^{O(n)}(\sigma)_\varepsilon$, the definition of $F(V,\lambda)$ (Definition 16.17) implies that the MA-module

$$(\textstyle\bigwedge^i(\mathbb{C}^n) \boxtimes \mathbf{1}) \otimes F(V,\lambda)|_{MA}$$

contains

$$V \boxtimes (F^{O(2,\mathbb{C})}(|\lambda - i|, 0)|_{SO(1,1)} \otimes \chi_{\varepsilon,(-1)^{\lambda-i}\varepsilon}|_{SO(1,1)})$$

as an MA-module. Here we have used that $MA \simeq O(n) \times SO(1,1)$ and that $\chi_{a,b}|_{SO(1,1)} \simeq \chi_{-a,-b}|_{SO(1,1)}$. Hence Proposition 16.29 is derived from the following lemma on the restriction $O(2,\mathbb{C}) \downarrow SO(1,1)$.   □

Let $\mathbb{C}_k$ denote the holomorphic character of $SO(2,\mathbb{C})$ on $\mathbb{C}e^{ik\theta}$.

Lemma 16.30 $(O(2,\mathbb{C}) \downarrow SO(1,1))$.

$$F^{O(2,\mathbb{C})}(k,0)|_{\mathbb{Z}/2\mathbb{Z} \times \mathbb{R}} \simeq \begin{cases} (-1)^k \boxtimes (\mathbb{C}_k \oplus \mathbb{C}_{-k}) & \text{for } k \in \mathbb{N}_+, \\ \mathbf{1} \boxtimes \mathbf{1} & \text{for } k = 0, \end{cases}$$

where we identify $SO(1,1) \simeq \{\pm I_2\} \times SO_0(1,1)$ *with* $\mathbb{Z}/2\mathbb{Z} \times \mathbb{R}$. *In particular, the* $SO(1,1)$*-module*

$$F^{O(2,\mathbb{C})}(|\lambda - i|, 0)|_{SO(1,1)} \otimes \chi_{\varepsilon,(-1)^{\lambda-i}\varepsilon}|_{SO(1,1)}$$

$$\simeq F^{O(2,\mathbb{C})}(|\lambda - i|, 0)|_{SO(1,1)} \otimes \chi_{-\varepsilon,(-1)^{\lambda-i-1}\varepsilon}|_{SO(1,1)}$$

contains

$$\mathbf{1} \boxtimes \mathbb{C}_{\lambda-i}$$

as an irreducible summand.

Proof. For $k \in \mathbb{N}_+$, the holomorphic representation $F^{O(2,\mathbb{C})}(k,0)$ is a two-dimensional representation of $O(2, \mathbb{C})$, which is isomorphic to $\mathrm{Ind}_{SO(2,\mathbb{C})}^{O(2,\mathbb{C})}(\mathbb{C}e^{ik\theta})$. Its restriction to the connected subgroup $SO(2, \mathbb{C})$ decomposes into a sum of two characters of $SO(2, \mathbb{C})$:

$$F^{O(2,\mathbb{C})}(k,0)|_{SO(2,\mathbb{C})} \simeq \mathbb{C}e^{ik\theta} \oplus \mathbb{C}e^{-ik\theta},$$

on which the central element $-I_2$ acts as the scalar multiplication of $(-1)^k = (-1)^{-k}$. Since $SO(1, 1)$ is generated by the central element $-I_2$ and the identity component $SO_0(1, 1)$, Lemma 16.30 follows. $\qquad\square$

16.5.3 Proof of Proposition 16.9

Proof of Proposition 16.9. Let $F(V, \lambda)$ be the finite-dimensional representation of G as in Definition 16.17. Filter $F(V, \lambda)$ as in Lemma 16.11. We may assume in addition that each $F^{(j)} = F(V, \lambda)_j / F(V, \lambda)_{j-1}$ is irreducible as an MA-module. Then by Proposition 16.29, $I_\delta(V, \lambda)$ occurs as a subquotient of the G-module $Pr_{(V,\lambda)}(I_\delta(i,i) \otimes F(V, \lambda))$. Hence the second assertion of Proposition 16.9 is shown. By Proposition 16.26, the first assertion follows. $\qquad\square$

16.6 Proof of Theorem 16.6

In this section we complete the proof of Theorem 16.6 and also its reformulation Theorem 16.22. By Proposition 16.9, it suffices to show Proposition 16.10 is an isomorphism in the level of $\overline{G}$-modules instead of the isomorphism in Theorem 16.6 as G-modules.

We divide the argument according to the decomposition

$$\mathcal{R}ed = \mathcal{R}ed_{\mathrm{I}} \sqcup \mathcal{R}ed_{\mathrm{II}},$$

where we recall from Definition 14.17:

- $(V, \lambda) \in \mathcal{R}ed_{\mathrm{I}}$, if V is of type X or $\lambda = \frac{n}{2}$;
- $(V, \lambda) \in \mathcal{R}ed_{\mathrm{II}}$, if V is of type Y and $\lambda \neq \frac{n}{2}$.

As we shall show in the proof of Proposition 16.10 below, the following assertion holds with the notation therein.

Proposition 16.31. *There is a natural isomorphism, as $\overline{G}$-modules*

$$
\psi_{\rho^{(i)}}^{r(V,\lambda)} (I_\delta(i,i))|_{\overline{G}} \simeq
\begin{cases}
\overline{\psi}_{\rho^{(i)}}^{r(V,\lambda)} (I_\delta(i,i)|_{\overline{G}}) & \text{if } (V,\lambda) \in \mathcal{R}ed_{\mathrm{I}}, \\
\bigoplus_{\xi=\pm} \overline{\psi}_{\rho^{(i)}}^{r(V^{(\xi)},\lambda)} (I_\delta(i,i)|_{\overline{G}}) & \text{if } (V,\lambda) \in \mathcal{R}ed_{\mathrm{II}}.
\end{cases}
$$

16.6.1   Case: $(V, \lambda) \in \mathcal{R}ed_{\mathrm{I}}$

In this subsection, we discuss the case where V is of type X or $\lambda = \frac{n}{2}$.

Proof of Proposition 16.10 for $(V, \lambda) \in \mathcal{R}ed_{\mathrm{I}}$. If n is odd, then Proposition 16.10 follows from Lemma 16.3 (1).

Hereafter we assume n is even, say $n = 2m$. We claim that Proposition 16.10 follows from Lemma 16.3 (2) if V is of type X (Definition 2.6) or $\lambda = m$. To see this, it is enough to verify that all of $\rho^{(i)}$, $r(V, \lambda)$, and $\tau^{(i)}(V, \lambda) = r(V, \lambda) - \rho^{(i)}$, see (16.13), contain 0 in their entries. This is automatically true for $\rho^{(i)}$ as $\rho^{(i)} \in W_G \rho^G$ (Example 14.9 (3)) and n is even. For $r(V, \lambda)$, one sees from (14.10) that the m-th component vanishes if V is of type X and the $(m+1)$-th component vanishes if $\lambda = m$. For $\tau^{(i)}(V, \lambda)$, one see from the formula of $\tau^{(i)}(V, \lambda)$ in Lemma 16.25 that an analogous assertion holds because $\lambda = m \ (= \frac{n}{2})$ implies that the height $i(V, \lambda)$ equals m by Definition 14.26. Hence Proposition 16.10 for $(V, \lambda) \in \mathcal{R}ed_{\mathrm{I}}$ is shown. □

16.6.2   Case: $(V, \lambda) \in \mathcal{R}ed_{\mathrm{II}}$

In this subsection, we discuss the case where V is of type Y and $\lambda \neq \frac{n}{2}$. In this case, n is even $(= 2m)$, $i := i(V, \lambda) = m$, and the restriction of V to $SO(n)$ is a sum of two irreducible representations of $SO(n)$:

$$
V = V^{(+)} \oplus V^{(-)},
$$

as in (15.1). We extend the definition (14.10) of $r(V, \lambda)$ to irreducible representations $V^{(\pm)}$ of $SO(n)$ with $n = 2m$ by

$$
r(V^{(\pm)}, \lambda) := (\sigma_1 + m - 1, \cdots, \sigma_{m-1} + 1, \pm\sigma_m, \lambda - m) \in \mathbb{Z}^{m+1}.
$$

Then $r(V^{(\pm)}, \lambda)$ viewed as an element of $\mathfrak{h}_{\mathbb{C}}^* / W_{\mathfrak{g}}$ is the $\mathfrak{Z}(\mathfrak{g})$-infinitesimal character of the principal series representation $\overline{I}_\delta(V^{(\pm)}, \lambda)$ of $\overline{G} = SO(n+1, 1)$.

As in (16.13), we set

$$\tau^{(i)}(V^{(\pm)}, \lambda) := r(V^{(\pm)}, \lambda) - \rho^{(i)}.$$

Inspecting the definition (14.17) of the height $i := i(V, \lambda)$, we see that both $r(V^{(\pm)}, \lambda)$ and $\rho^{(i)}$ belong to the same Weyl chamber with respect to the Weyl group $W_{\mathfrak{g}}$ (*not* W_G) as in Lemma 16.25.

By Lemma 16.20, the irreducible finite-dimensional G-module $F(V, \lambda)$ decomposes into a direct sum of two irreducible $\overline{G}$-modules, which we may write as

$$F(V, \lambda)|_{\overline{G}} = \overline{F}(V^{(+)}, \lambda) \oplus \overline{F}(V^{(-)}, \lambda).$$

To be precise, we set $\sigma^{(+)}(\lambda) := \sigma(\lambda)$ (Definition 16.17), and define $\sigma^{(-)}(\lambda)$ by replacing the $(m+1)$-th component σ_m with $-\sigma_m$. For instance, if $\lambda < m$, then the height $i = i(V, \lambda)$ is smaller than m and

$$\sigma^{(+)}(\lambda) = (\sigma_1 - 1, \cdots, \sigma_i - 1, i - \lambda, \sigma_{i+1}, \cdots, \sigma_{m-1}, \sigma_m),$$
$$\sigma^{(-)}(\lambda) = (\sigma_1 - 1, \cdots, \sigma_i - 1, i - \lambda, \sigma_{i+1}, \cdots, \sigma_{m-1}, -\sigma_m).$$

Then $\overline{F}(V^{(\pm)}, \lambda)$ are the irreducible $\overline{G}$-modules such that

$$\overline{F}(V^{(\pm)}, \lambda) \otimes \chi_{+, (-1)^{\lambda-i}}|_{SO(n+1,1)}$$

extends to irreducible holomorphic finite-dimensional representations of the connected complex Lie group $SO(n+2, \mathbb{C})$ with highest weights $\sigma^{(\pm)}(\lambda)$.

Proof of Proposition 16.10 for $(V, \lambda) \in \mathcal{Red}_{\mathrm{II}}$. By the definition (16.2) of the translation functor and by Lemma 16.2, there is a natural $\overline{G}$-isomorphism:

$$\psi_{\rho^{(i)}}^{r(V, \lambda)}(I_\delta(i, i))|_{\overline{G}}$$

$$\simeq (\overline{P}_{r(V^{(+)}, \lambda)} + \overline{P}_{r(V^{(-)}, \lambda)})(I_\delta(i, i)|_{\overline{G}} \otimes (\overline{F}(V^{(+)}, \lambda) \oplus \overline{F}(V^{(-)}, \lambda))).$$
$$\tag{16.16}$$

We claim for $\xi, \eta \in \{\pm\}$:

$$\overline{P}_{r(V^{(\xi)}, \lambda)}(I_\delta(i, i)|_{\overline{G}} \otimes \overline{F}(V^{(\eta)}, \lambda)) = \overline{\psi}_{\rho^{(i)}}^{r(V^{(\xi)}, \lambda)}(I_\delta(i, i)|_{\overline{G}}) \quad \text{if } \xi\eta = +, \tag{16.17}$$

$$\overline{P}_{r(V^{(\xi)}, \lambda)}(I_\delta(i, i)|_{\overline{G}} \otimes \overline{F}(V^{(\eta)}, \lambda)) = 0 \quad \text{if } \xi\eta = -. \tag{16.18}$$

The first claim (16.17) holds by definition (16.5). To see the vanishing (16.18) of the cross terms in (16.18), suppose that

$$\rho^{(i)} + \gamma = w(\rho^{(i)} + \tau^{(i)}(V^{(\xi)}, \lambda))$$

for some weight γ in $F(V^{(\eta)}, \lambda)$ and for some $w \in W_{\mathfrak{g}}$. Then we have

$$||\gamma|| \le ||\tau^{(i)}(V^{(\eta)}, \lambda)|| = ||\tau^{(i)}(V^{(\xi)}, \lambda)||.$$

Hence we can apply Lemma 16.13 and conclude

$$\gamma = \tau^{(i)}(V^{(\xi)}, \lambda).$$

By the vanishing (16.18) of the cross terms in (16.16), we obtain the following $\overline{G}$-isomorphisms:

$$\psi^{r(V,\lambda)}_{\rho^{(i)}}(I_\delta(i,i))|_{\overline{G}} \simeq \bigoplus_{\xi \in \{\pm\}} \overline{\psi}^{r(V^{(\xi)}, \lambda)}_{\rho^{(i)}}(I_\delta(i,i)|_{\overline{G}})$$

$$\simeq \bigoplus_{\xi \in \{\pm\}} \overline{I}_{(-1)^{\lambda - i}\delta}(V^{(\xi)}, \lambda),$$

which is isomorphic to the restriction of the principal series representation $I_{(-1)^{\lambda - i}\delta}(V, \lambda)$ of G to the subgroup $\overline{G}$ by (15.2).   □

16.7   Proof of Theorem 16.8

In this section, we show Theorem 16.8, or its reformulation, Theorem 16.23. The proof is similar to that of Theorem 16.6, hence we give only a sketch of the proof with focus on necessary changes. A part of the proof is carried out separately according to the decomposition

$$\mathcal{Red} = \mathcal{Red}_{\mathrm{I}} \sqcup \mathcal{Red}_{\mathrm{II}} \quad \text{(Definition 14.17)}.$$

The following lemma is a counterpart of Proposition 16.29.

Lemma 16.32 (tensor product and $G \downarrow MA$). *Suppose $(V, \lambda) \in \mathcal{Red}$. Let $i := i(V, \lambda)$ be the height of (V, λ), see (14.17), and $F(V, \lambda)$ be the irreducible finite-dimensional representation of $G = O(n+1,1)$ as in Definition 16.17. Then the MA-module*

$$(V \boxtimes \delta \boxtimes \mathbb{C}_\lambda) \otimes F(V, \lambda)|_{MA}$$

contains

$$\bigwedge{}^i(\mathbb{C}^n) \boxtimes \delta \boxtimes \mathbb{C}_i \qquad\qquad\qquad if\ (V,\lambda) \in \mathcal{R}ed_{\mathrm{I}},$$

$$(\bigwedge{}^i(\mathbb{C}^n) \boxtimes \delta \boxtimes \mathbb{C}_i) \oplus (\bigwedge{}^{n-i}(\mathbb{C}^n) \boxtimes \delta \boxtimes \mathbb{C}_i) \qquad if\ (V,\lambda) \in \mathcal{R}ed_{\mathrm{II}},$$

as an irreducible component.

Proof. The proof is similar to that of Proposition 16.29 except that there is a G-isomorphism $F(V,\lambda) \otimes \det \simeq F(V,\lambda)$ by Lemma 16.20 if $(V,\lambda) \in \mathcal{R}ed_{\mathrm{II}}$. In this case, the height $i = i(V,\lambda)$ is not equal to $\frac{n}{2}$ by Lemma 14.28 (3). Thus both the $O(n)$-modules $\bigwedge^i(\mathbb{C}^n)$ and $\bigwedge^{n-i}(\mathbb{C}^n) \simeq \bigwedge^i(\mathbb{C}^n) \otimes \det$ occur simultaneously in $V \otimes F(V,\lambda)|_{O(n)}$. □

Theorem 16.23, or equivalently, Theorem 16.8 is deduced from the following two propositions.

Proposition 16.33. *Suppose* $(V,\lambda) \in \mathcal{R}ed$. *(Definition 14.8), equivalently,* $V \in \widehat{O(n)}$ *and* $\lambda \in \mathbb{Z} - (S(V) \cup S_Y(V))$. *Then the G-module* $P_{r(V,\lambda)}(I_\delta(V,\lambda) \otimes F(V,\lambda))$ *contains*

$$I_\delta(i,i) \qquad\qquad for\ (V,\lambda) \in \mathcal{R}ed_{\mathrm{I}},$$

$$I_\delta(i,i)\ and\ I_\delta(n-i,i) \quad for\ (V,\lambda) \in \mathcal{R}ed_{\mathrm{II}},$$

as subquotients.

Proof. As in the proof of Proposition 16.9 in Section 16.5.3, Proposition 16.33 follows readily from Lemma 16.11 by using Lemma 16.32. □

Proposition 16.34. *Suppose* $(V,\lambda) \in \mathcal{R}ed$, *namely, suppose* $V \in \widehat{O(n)}$ *and* $\lambda \in \mathbb{Z} - (S(V) \cup S_Y(V))$. *Then there is a natural isomorphism of $\overline{G}$-modules:*

$$P_{r(V,\lambda)}(I_\delta(V,\lambda) \otimes F(V,\lambda))|_{\overline{G}} \simeq \begin{cases} I_\delta(i,i)|_{\overline{G}} & for\ (V,\lambda) \in \mathcal{R}ed_{\mathrm{I}}, \\ I_\delta(i,i)|_{\overline{G}} \oplus I_\delta(n-i,i)|_{\overline{G}} & for\ (V,\lambda) \in \mathcal{R}ed_{\mathrm{II}}. \end{cases}$$

Proof. The proof is similar to that of Proposition 16.10, again by showing the vanishing of the cross terms as in (16.18). □

References

1. J. Adams, D. Barbasch, D.A. Vogan, Jr., *The Langlands Classification and Irreducible Characters for Real Reductive Groups*. Progr. Math., vol. 104 (Birkhäuser Boston, Boston, MA, 1992)
2. A. Aizenbud, D. Gourevitch, Multiplicity one theorem for $(GL_{n+1}(\mathbb{R}), GL_n(\mathbb{R}))$. Selecta Math. (N.S.) **15**, 271–294 (2009)
3. J. Arthur, Classifying automorphic representations, in *Current Developments in Mathematics 2012* (Int. Press, Somerville, MA, 2013), pp. 1–58
4. M.W. Baldoni Silva, A.W. Knapp, Unitary representations induced from maximal parabolic subgroups. J. Funct. Anal. **69**, 21–120 (1986)
5. N. Bergeron, L. Clozel, Spectre automorphe des variétés hyperboliques et applications topologiques, in *Astérisque*, vol. 303 (Soc. Math. France, 2005), xx+218 pp.
6. N. Bergeron, J. Millson, C. Mœglin, Hodge type theorems for arithmetic manifolds associated to orthogonal groups. Int. Math. Res. Not. IMRN **2017**(15), 4495–4624
7. N. Conze-Berline, M. Duflo, Sur les représentations induites des groupes semi-simples complexes. Compos. Math. **34**, 307–336 (1977)
8. R. Beuzart-Plessis, A local trace formula for the Gan–Gross–Prasad conjecture for unitary groups: the archimedean case. ArXiv:1506.01452
9. A. Borel, N.R. Wallach, *Continuous Cohomology, Discrete Subgroups, and Representations of Reductive Groups: Second edition*. Math. Surveys Monogr., vol. 67 (Amer. Math. Soc., Providence, RI, 2000), xviii+260 pp.; First ed., Ann. of Math. Stud., vol. 94 (Princeton Univ. Press, Princeton, NJ, 1980), xvii+388 pp.
10. M. Burger, J.-S. Li, P. Sarnak, Ramanujan duals and automorphic spectrum. Bull. Am. Math. Soc. (N.S.) **26**, 253–257 (1992)
11. D.H. Collingwood, *Representations of Rank One Lie Groups*. Res. Notes in Math., vol. 137 (Pitman (Advanced Publishing Program), Boston, MA, 1985), vii+244 pp.
12. W.T. Gan, B.H. Gross, D. Prasad, J.-L. Waldspurger, Sur les conjectures de Gross et Prasad. I, in *Astérique*, vol. 346 (Soc. Math. France, 2012), xi+318 pp.
13. I.M. Gel'fand, M.A. Naĭmark, *Unitary Representations of Classical Groups*. Trudy Mat. Inst. Steklov., vol. 36 (Izdat. Nauk SSSR, 1950), 288 pp.
14. B.H. Gross, D. Prasad, On the decomposition of a representations of SO_n when restricted to SO_{n-1}. Can. J. Math. **44**, 974–1002 (1992)
15. B.H. Gross, N. Wallach, Restriction of small discrete series representations to symmetric subgroups, in *The Mathematical Legacy of Harish-Chandra, Baltimore, MD, 1998*. Proc. Sympos. Pure Math., vol. 68 (Amer. Math. Soc., Providence, RI, 2000), pp. 255–272

© Springer Nature Singapore Pte Ltd. 2018
T. Kobayashi, B. Speh, *Symmetry Breaking for Representations of Rank One Orthogonal Groups II*, Lecture Notes in Mathematics 2234,
https://doi.org/10.1007/978-981-13-2901-2

16. T. Hirai, On irreducible representations of the Lorentz group of n-th order. Proc. Jpn. Acad. **38**, 258–262 (1962). https://projecteuclid.org/euclid.pja/1195523378

17. R.E. Howe, E.-C. Tan, Homogeneous functions on light cones: the infinitesimal structure of some degenerate principal series representations. Bull. Am. Math. Soc. (N.S.) **28**, 1–74 (1993)

18. A. Ichino, T. Ikeda, On the periods of automorphic forms on special orthogonal groups and the Gross–Prasad conjecture. Geom. Funct. Anal. **19**, 1378–1425 (2010)

19. A. Ichino, S. Yamana, Periods of automorphic forms: the case of $(GL_{n+1} \times GL_n, GL_n)$. Compos. Math. **151**, 665–712 (2015)

20. J.C. Jantzen, Moduln mit einem höchsten Gewicht, in *Lecture Notes in Math.*, vol. 750 (Springer, Berlin/Heidelberg/New York, 1979)

21. A. Juhl, *Families of Conformally Covariant Differential Operators, Q-Curvature and Holography*. Progr. Math., vol. 275 (Birkhäuser Verlag, Basel, 2009)

22. A.W. Knapp, E.M. Stein, Intertwining operators for semisimple groups. Ann. Math. (2) **93**, 489–578 (1971)

23. A.W. Knapp, E.M. Stein, Intertwining operators for semisimple groups. II. Invent. Math. **60**, 9–84 (1980)

24. A.W. Knapp, D.A. Vogan, Jr., *Cohomological Induction and Unitary Representations*. Princeton Math. Ser., vol. 45 (Princeton Univ. Press, Princeton, NJ, 1995), xx+948 pp. ISBN: 978-0-691-03756-6

25. A.W. Knapp, G.J. Zuckerman, Classification of irreducible tempered representations of semisimple groups. Ann. Math. (2) **116**, 389–455 (1982); II, ibid, 457–501

26. T. Kobayashi, Proper action on a homogeneous space of reductive type. Math. Ann. **285**, 249–263 (1989). https://doi.org/10.1007/BF01443517

27. T. Kobayashi, *Singular Unitary Representations and Discrete Series for Indefinite Stiefel Manifolds* $U(p, q; \mathbb{F})/U(p - m, q; \mathbb{F})$. Mem. Amer. Math. Soc., vol. 462 (Amer. Math. Soc., Providence, RI, 1992), 106 pp. ISBN: 978-0-8218-2524-2. http://www.ams.org/books/memo/0462/

28. T. Kobayashi, Discrete decomposability of the restriction of $A_q(\lambda)$ with respect to reductive subgroups. III. Restriction of Harish-Chandra modules and associated varieties. Invent. Math. **131**, 229–256 (1998). http://dx.doi.org/10.1007/s002220050203

29. T. Kobayashi, Restrictions of generalized Verma modules to symmetric pairs. Transform. Groups **17**, 523–546 (2012)

30. T. Kobayashi, F-method for constructing equivariant differential operators, in *Geometric Analysis and Integral Geometry, Contemp. Math.*, vol. 598 (Amer. Math. Soc., Providence, RI, 2013), pp. 139–146. http://dx.doi.org/10.1090/conm/598/11998

31. T. Kobayashi, F-method for symmetry breaking operators. Differ. Geom. Appl. **33**, 272–289 (2014). Special issue "The Interaction of Geometry and Representation Theory. Exploring New Frontiers" (in honor of Michael Eastwood's 60th birthday)

32. T. Kobayashi, Shintani functions, real spherical manifolds, and symmetry breaking operators, in *Developments and Retrospectives in Lie Theory*. Dev. Math., vol. 37 (Springer, 2014), pp. 127–159. http://dx.doi.org/10.1007/978-3-319-09934-7_5

33. T. Kobayashi, A program for branching problems in the representation theory of real reductive groups, in *Representations of Lie Groups. In Honor of David A. Vogan, Jr. on his 60th Birthday*. Progr. Math., vol. 312 (Birkhäuser, 2015), pp. 277–322. http://dx.doi.org/10.1007/978-3-319-23443-4_10

34. T. Kobayashi, Residue formula for regular symmetry breaking operators, in *Contemp. Math.*, vol. 714 (Amer. Math. Soc., Province, RI, 2018), pp. 175–193 http://doi.org/10.1090/com/714/14380. Available also at arXiv:1709.05035. http://arxiv.org/abs/1709.05035

35. T. Kobayashi, T. Kubo, M. Pevzner, *Conformal Symmetry Breaking Operators for Differential Forms on Spheres*. Lecture Notes in Math., vol. 2170 (Springer, 2016), iv+192 pp. ISBN: 978-981-10-2657-7. http://dx.doi.org/10.1007/978-981-10-2657-7

36. T. Kobayashi, T. Matsuki, Classification of finite-multiplicity symmetric pairs. Transform. Groups **19**, 457–493 (2014). http://dx.doi.org/10.1007/s00031-014-9265-x. Special issue in honor of Dynkin for his 90th birthday

37. T. Kobayashi, B. Ørsted, Analysis on the minimal representation of $O(p, q)$, III. Ultrahyperbolic equations on $\mathbb{R}^{p-1,q-1}$. Adv. Math. **180**, 551–595 (2003). http://dx.doi.org/10.1016/S0001-8708(03)00014-8

38. T. Kobayashi, B. Ørsted, P. Somberg, V. Souček, Branching laws for Verma modules and applications in parabolic geometry. I. Adv. Math. **285**, 1796–1852 (2015). http://dx.doi.org/10.1016/j.aim.2015.08.020

39. T. Kobayashi, T. Oshima, Finite multiplicity theorems for induction and restriction. Adv. Math. **248**, 921–944 (2013). http://dx.doi.org/10.1016/j.aim.2013.07.015

40. T. Kobayashi, M. Pevzner, Differential symmetry breaking operators. I. General theory and F-method. Selecta Math. (N.S.) **22**, 801–845 (2016). http://dx.doi.org/10.1007/s00029-015-0207-9

41. T. Kobayashi, M. Pevzner, Differential symmetry breaking operators. II. Rankin–Cohen operators for symmetric pairs. Selecta Math. (N.S.) **22**, 847–911 (2016). http://dx.doi.org/10.1007/s00029-015-0208-8

42. T. Kobayashi, B. Speh, *Symmetry Breaking for Representations of Rank One Orthogonal Groups*. Mem. Amer. Math. Soc., vol. 238 (Amer. Math. Soc., Providence, RI, 2015), v+112 pp. ISBN: 978-1-4704-1922-6. http://dx.doi.org/10.1090/memo/1126

43. T. Kobayashi, B. Speh, Symmetry breaking for orthogonal groups and a conjecture by B. Gross and D. Prasad, in *Geometric Aspects of the Trace Formula, Simons Symposia*, ed. by W. Müller et al. (Springer, 2018), pp. 245–266. https://doi.org/10.1007/978-3-319-94833-1_8. Available also at arXiv:1702.00263. http://arxiv.org/abs/1702.00263

44. B. Kostant, Verma modules and the existence of quasi-invariant differential operators, in *Noncommutative Harmonic Analysis, Marseille–Luminy, 1974*. Lecture Notes in Math., vol. 466 (Springer, Berlin, 1975), pp. 101–128

45. M. Krämer, Multiplicity free subgroups of compact connected Lie groups. Arch. Math. (Basel) **27**, 28–36 (1976)

46. S.S. Kudla, J.J. Millson, Geodesic cyclics and the Weil representation. I. Quotients of hyperbolic space and Siegel modular forms. Compos. Math. **45**, 207–271 (1982)

47. S.S. Kudla, J.J. Milllson, The theta correspondence and harmonic forms. I. Math. Ann. **274**, 353–378 (1986)

48. S.S. Kudla, J.J. Millson, The theta correspondence and harmonic forms. II. Math. Ann. **277**, 267–314 (1987)

49. R.P. Langlands, *On the Functional Equation Satisfied by Eisenstein Series*. Lecture Notes in Math., vol. 544 (Springer, New York, 1976)

50. R.P. Langlands, On the classification of irreducible representations of real algebraic groups, in *Representation Theory and Harmonic Analysis on Semisimple Lie Groups*. Math. Surveys Monogr., vol. 31 (Amer. Math. Soc., Providence, RI, 1989), pp. 101–170

51. C. Mœglin, J.-L. Waldspurger, Sur les conjectures de Gross et Prasad. II, in *Astérique*, vol. 347 (Soc. Math. France, 2012)

52. J. Peetre, Une caractérisation abstraite des opérateurs différentiels. Math. Scand. **7**, 211–218 (1959)

53. B. Speh, D.A. Vogan, Jr., Reducibility of generalized principal series representations. Acta Math. **145**, 227–299 (1980)

54. B. Sun, The nonvanishing hypothesis at infinity for Rankin–Selberg convolutions. J. Am. Math. Soc. **30**, 1–25 (2017)

55. B. Sun, C.-B. Zhu, Multiplicity one theorems: the Archimedean case. Ann. Math. (2) **175**, 23–44 (2012). http://dx.doi.org/10.4007/annals.2012.175.1.2

56. Y.L. Tong, S.P. Wang, Geometric realization of discrete series for semisimple symmetric spaces. Invent. Math. **96**, 425–458 (1989)

57. F. Trèves, *Topological Vector Spaces, Distributions and Kernels* (Academic Press, New York–London, 1967), xvi+624 pp.

58. J.A. Vargas, Restriction of some discrete series representations. Algebras Groups Geom. **18**, 85–99 (2001)

59. D.A. Vogan, Jr., *Representations of Real Reductive Lie Groups*. Progr. Math., vol. 15 (Birkhäuser, Boston, MA, 1981), xvii+754 pp.
60. D.A. Vogan, Jr., The local Langlands conjecture, in *Representation Theory of Groups and Algebras*. Contemp. Math., vol. 145 (Amer. Math. Soc., Providence, RI, 1993), pp. 305–379
61. D.A. Vogan, Jr., G.J. Zuckerman, Unitary representations with nonzero cohomology. Compos. Math. **53**, 51–90 (1984)
62. N.R. Wallach, *Real Reductive Groups. I*. Pure Appl. Math., vol. 132 (Academic Press, Boston, MA, 1988), xx+412 pp. ISBN: 0-12-732960-9; II, ibid, vol. 132-II (Academic Press, Boston, MA, 1992), xiv+454 pp. ISBN: 978-0127329611
63. S.P. Wang, Correspondence of modular forms to cycles associated to $O(p, q)$. J. Differ. Geom. **22**, 151–213 (1985)
64. H. Weyl, *The Classical Groups. Their Invariants and Representations*. Princeton Landmarks Math. (Princeton Univ. Press, Princeton, NJ, 1997)
65. G. Zuckerman, Tensor products of finite and infinite dimensional representations of semisimple Lie groups. Ann. Math. (2) **106**, 295–308 (1977)

List of Symbols

Symbols

1, trivial one-dimensional representation, 31

$\mathbf{1}^{\mathcal{I}}$, **120**

$\mathbf{1}^{\mathcal{I}}_{\lambda}$, **124**, 127, 146

$\mathfrak{a}$, maximally split abelian subspace, **14**

$\gamma(\mu, a)$, **46**

$(\delta, V, \lambda)^{\uparrow} = (\delta^{\uparrow}, V^{\uparrow}, \lambda^{\uparrow})$, **271**

$(\delta, V, \lambda)^{\downarrow} = (\delta^{\downarrow}, V^{\downarrow}, \lambda^{\downarrow})$, **271**

ε_I, **112**, 114

ι_K, **121**

ι^*_{λ}, 122, 144

$\iota_{j \to i}$, **145**

$\mu^{\flat}(i, \delta)$, **28**, 127, 142

$\mu^{\sharp}(i, \delta)$, **28**, 127, 142

Ξ, isotropic cone, **16**

$\Lambda^+(O(N))$, **20**, 86

$\bigwedge^{\ell}(\mathbb{C}^N)$, exterior tensor, **22**

$\Lambda^+(N)$, dominant weight, **20**, 226, 252

$\Pi_i = \Pi_{i,+}$, **32**

$\Pi_{i,\delta}$, irreducible representations of G, 30, **225**, 251, 281, 282, 298

$\Pi_i(F)$, standard sequence starting with F, **225**, 232, 274

$\Pi_{\delta}(V, \lambda)$, **273**, 275

χ_{+-}, 56

$\chi_{--} = \det$, 17, 56, 227, 289, 292

χ_{-+}, 229

$\chi_{\pm\pm}$, one-dimensional representation of $O(n+1, 1)$, 8, **17**, 31, 57, 213, 253, 290

$\chi(V, \lambda)$, **312**, 313

ρ_G, **19**, 30, 59, 225, 258

$\rho^{(i)}$, **258**

$\sigma(\lambda) (= \sigma^{(i)}(\lambda))$, **317**

$\tau \prec \sigma$, **23**

τ_{dom}, **309**

$\tau^{(i)}(V, \lambda)$, **321**

Ψ_{sp}, special parameter in $\mathbb{C}^2 \times \{\pm\}^2$, **3**, 40, 97, 107, 139, 186, 231, 248

ψ_n, **38**, 38, 72, 85, 114, 137

$\psi_n(\cdot; \lambda)$, 113

$\psi^{\mu+\tau}_{\mu}$, **309**

A

$a^{i,j}_+(\lambda, \nu)$, 145

$\widehat{\mathcal{A}}_{\lambda,\nu,+}$, 82, 147, 150

$\widehat{\mathcal{A}}_{\lambda,\nu,-}$, 82, 153

$\widetilde{\mathcal{A}}^{V,W}_{\lambda,\nu,+}$, **38**, **88**

$\widetilde{\mathcal{A}}^{V,W}_{\lambda,\nu,-}$, **38**, **88**

$\widetilde{\mathcal{A}}^{i,j}_{\lambda,\nu,\pm}$, 137

$\widetilde{\widetilde{\mathcal{A}}}^{V,W}_{\lambda,\nu,\pm}$, **77**

$(\widetilde{\widetilde{\mathcal{A}}}^{V,W}_{\lambda,\nu,\pm})_{\infty}$, **77**

$\widetilde{\mathbb{A}}^{V,W}_{\lambda,\nu,\pm}$, **39**, 63

$\frac{\partial^{k+l}}{\partial\lambda^k \partial\nu^l}|_{\lambda=\lambda_0 \atop \nu=\nu_0} \widetilde{\mathbb{A}}^{V,W}_{\lambda,\nu,\pm}$, **93**

$\widetilde{\mathbb{A}}^{i,j}_{\lambda,\nu,\pm}$, **137**, 143

$\widetilde{\mathbb{A}}^{i,j}_{\lambda,\nu,+}$, 144

$\widetilde{\mathbb{A}}^{i,j}_{\lambda,\nu,-}$, 152

$\widetilde{\widetilde{\mathbb{A}}}^{V,W}_{\lambda,\nu,\pm}$, 40, **94**

$\widetilde{\widetilde{\mathbb{A}}}^{i,j}_{\lambda,\nu,+}$, **49**, 160, 162

$\widetilde{\widetilde{\mathbb{A}}}^{i,j}_{\lambda,\nu,-}$, **49**

$A_{\mathfrak{q}}(\lambda)$, 32, 52, 205, **280**, 281, 300

$A_{\mathfrak{q}_i}$, **280**

© Springer Nature Singapore Pte Ltd. 2018

T. Kobayashi, B. Speh, *Symmetry Breaking for Representations
of Rank One Orthogonal Groups II*, Lecture Notes in Mathematics 2234,
https://doi.org/10.1007/978-981-13-2901-2

Index

© Springer Nature Singapore Pte Ltd. 2018
T. Kobayashi, B. Speh, *Symmetry Breaking for Representations
of Rank One Orthogonal Groups II*, Lecture Notes in Mathematics 2234,
https://doi.org/10.1007/978-981-13-2901-2

LECTURE NOTES IN MATHEMATICS

Editors in Chief: J.-M. Morel, B. Teissier;

Editorial Policy

1. Lecture Notes aim to report new developments in all areas of mathematics and their applications – quickly, informally and at a high level. Mathematical texts analysing new developments in modelling and numerical simulation are welcome.

   Manuscripts should be reasonably self-contained and rounded off. Thus they may, and often will, present not only results of the author but also related work by other people. They may be based on specialised lecture courses. Furthermore, the manuscripts should provide sufficient motivation, examples and applications. This clearly distinguishes Lecture Notes from journal articles or technical reports which normally are very concise. Articles intended for a journal but too long to be accepted by most journals, usually do not have this "lecture notes" character. For similar reasons it is unusual for doctoral theses to be accepted for the Lecture Notes series, though habilitation theses may be appropriate.

2. Besides monographs, multi-author manuscripts resulting from SUMMER SCHOOLS or similar INTENSIVE COURSES are welcome, provided their objective was held to present an active mathematical topic to an audience at the beginning or intermediate graduate level (a list of participants should be provided).

   The resulting manuscript should not be just a collection of course notes, but should require advance planning and coordination among the main lecturers. The subject matter should dictate the structure of the book. This structure should be motivated and explained in a scientific introduction, and the notation, references, index and formulation of results should be, if possible, unified by the editors. Each contribution should have an abstract and an introduction referring to the other contributions. In other words, more preparatory work must go into a multi-authored volume than simply assembling a disparate collection of papers, communicated at the event.

3. Manuscripts should be submitted either online at www.editorialmanager.com/lnm to Springer's mathematics editorial in Heidelberg, or electronically to one of the series editors. Authors should be aware that incomplete or insufficiently close-to-final manuscripts almost always result in longer refereeing times and nevertheless unclear referees' recommendations, making further refereeing of a final draft necessary. The strict minimum amount of material that will be considered should include a detailed outline describing the planned contents of each chapter, a bibliography and several sample chapters. Parallel submission of a manuscript to another publisher while under consideration for LNM is not acceptable and can lead to rejection.

4. In general, **monographs** will be sent out to at least 2 external referees for evaluation.

   A final decision to publish can be made only on the basis of the complete manuscript, however a refereeing process leading to a preliminary decision can be based on a pre-final or incomplete manuscript.

   Volume Editors of **multi-author works** are expected to arrange for the refereeing, to the usual scientific standards, of the individual contributions. If the resulting reports can be forwarded to the LNM Editorial Board, this is very helpful. If no reports are forwarded or if other questions remain unclear in respect of homogeneity etc, the series editors may wish to consult external referees for an overall evaluation of the volume.

5. Manuscripts should in general be submitted in English. Final manuscripts should contain at least 100 pages of mathematical text and should always include

   – a table of contents;
   – an informative introduction, with adequate motivation and perhaps some historical remarks: it should be accessible to a reader not intimately familiar with the topic treated;
   – a subject index: as a rule this is genuinely helpful for the reader.
   – For evaluation purposes, manuscripts should be submitted as pdf files.

6. Careful preparation of the manuscripts will help keep production time short besides ensuring satisfactory appearance of the finished book in print and online. After acceptance of the manuscript authors will be asked to prepare the final LaTeX source files (see LaTeX templates online: https://www.springer.com/gb/authors-editors/book-authors-editors/manuscriptpreparation/5636) plus the corresponding pdf- or zipped ps-file. The LaTeX source files are essential for producing the full-text online version of the book, see http://link.springer.com/bookseries/304 for the existing online volumes of LNM). The technical production of a Lecture Notes volume takes approximately 12 weeks. Additional instructions, if necessary, are available on request from lnm@springer.com.

7. Authors receive a total of 30 free copies of their volume and free access to their book on SpringerLink, but no royalties. They are entitled to a discount of 33.3 % on the price of Springer books purchased for their personal use, if ordering directly from Springer.

8. Commitment to publish is made by a *Publishing Agreement*; contributing authors of multiauthor books are requested to sign a *Consent to Publish form*. Springer-Verlag registers the copyright for each volume. Authors are free to reuse material contained in their LNM volumes in later publications: a brief written (or e-mail) request for formal permission is sufficient.

Addresses:
Professor Jean-Michel Morel, CMLA, École Normale Supérieure de Cachan, France
E-mail: moreljeanmichel@gmail.com

Professor Bernard Teissier, Equipe Géométrie et Dynamique,
Institut de Mathématiques de Jussieu – Paris Rive Gauche, Paris, France
E-mail: bernard.teissier@imj-prg.fr

Springer: Ute McCrory, Mathematics, Heidelberg, Germany,
E-mail: lnm@springer.com

Printed in the United States
By Bookmasters